WIRELESS NETWORKING

Charles N. Thurwachter, Jr.
Purdue University

Upper Saddle River, New Jersey
Columbus, Ohio

Library of Congress Cataloging-in-Publication Data
Thurwachter, Charles N.
  Wireless networking / Charles N. Thurwachter, Jr.
    p. cm.
  ISBN 0-13-088366-2
  1. Wireless communication systems. 2. Computer networks. I. Title.
TK5103.2 .T49 2002

621.382'1—dc21

2001021759

Editor in Chief: Stephen Helba
Assistant Vice President and Publisher: Charles E. Stewart, Jr.
Product Manager: Scott Sambucci
Production Editor: Tricia L. Rawnsley
Design Coordinator: Diane Ernsberger
Text Designer: Carlisle Communications, Ltd.
Cover Designer: Jeff Vanik
Production Manager: Matthew Ottenweller
Electronic Text Management: Karen L. Bretz

This book was set in Palatino by Carlisle Communications, Ltd., and was printed and bound by R.R. Donnelley & Sons Company. The cover was printed by The Lehigh Press, Inc.

Pearson Education Ltd., *London*
Pearson Education Australia Pty. Limited, *Sydney*
Pearson Education Singapore Pte. Ltd.
Pearson Education North Asia Ltd., *Hong Kong*
Pearson Education Canada, Ltd., *Toronto*
Pearson Educación de Mexico, S.A. de C.V.
Pearson Education—Japan, *Tokyo*
Pearson Education Malaysia Pte. Ltd.
Pearson Education, *Upper Saddle River, New Jersey*

10 9 8 7 6 5 4 3 2 1
ISBN 0-13-088366-2

Preface

The subject of wireless communications is very broad and can be approached from many different perspectives. In most cases, an entire textbook could be devoted to each section of this text. The information provided in this textbook is useful for an overview course on the subject of wireless networking. This coverage allows the instructor to choose specific areas on which to focus the course. It also makes this textbook a good reference for the nonspecialist practitioner who does not have shelf space for several more specialized references.

Several unique chapters bring together information that is often hard to locate. Examples include chapters dedicated to wireless layer 2 protocols; data protocols used in wireless networks; special considerations when using TCP/IP with an underlying wireless network; and WAP, 3G cellular networks, and Bluetooth—three emerging standards.

Wireless communications systems have an air interface rather than a wireline interface. Wireless characteristics have a profound impact on layers 1, 2, 3, and 4 protocols and services. Both LAN and WAN wireless systems are explored in detail. Wireless WAN and LAN techniques are widely available to consumers and businesses for the first time, thus wireless networking is emerging as its own specialized area of study.

Traditional wireless WAN technologies, terrestrial broadband, and satellites are introduced. Whenever possible, actual systems are used to illustrate how the technology works. Cellular technology (the emerging wireless WAN technology) is introduced, summarized, and further explained in several chapters focusing on each group of cellular systems. A chapter on the rapidly expanding LAN technology for wireless networks, 802.11, is provided. Thus, both traditional and modern wireless systems and networks are examined.

Several chapters provide a fundamental description of the wireless WAN and LAN infrastructure. Once these topics are covered, specialized data communications protocols used primarily in the cellular environment, are discussed in detail. The implications of the mobility of wireless devices on IP and the specific effects on TCP flow control are defined and explored. Existing solutions to these problems are summarized and compared.

Chapters on wireless networking focus on each of the lower four layers of the protocol architecture and each discuss their relative strengths and weaknesses. This coverage is unique and allows the reader to set any one particular area of interest in context with the other layer protocols with which it interacts. Students must understand that, to achieve peak performance in a wireless network, each layer has special characteristics that need to be addressed.

Part Four consists of three chapters that provide an introduction to the emerging standards and specifications important to wireless networking. The content of these chapters is not intended to be comprehensive. Instead, they offer an overview of operation and a perspective on how these emerging standards and specifications might fit into an evolving communications architecture.

■ PART 1: FUNDAMENTALS OF WIRELESS NETWORKING

Part 1 is not intended to be read in its entirety by every reader. Depending on each reader's focus, some chapters may be skipped entirely while others will need to be carefully considered. The intent of Part 1 is to group together several disparate but central concepts and allow the reader to choose the chapters of most interest and applicability.

Chapter 1, *Technology Fundamentals,* discusses several basic wireless concepts. It introduces the concepts of wavelength, frequency set in an electromagnetic spectrum, frequency bands, gain and attenuation, and power radiation. This chapter also summarizes the use of decibel and logarithmic techniques. Bandwidth, information capacity, Shannon's limit, licensed and unlicensed bands of operation, the importance of the ISM bands, FCC rule part 15, and propagation modes are covered.

Chapter 2, *Multiple-Access Methods,* summarizes multiple access methods, which are the air interfaces of wireless systems. After a general introduction to wireless topologies and duplexing, the rest of the chapter is devoted to air interfaces. All major air interface methods are discussed, including FDMA, TDMA, FDM/TDMA, and CDMA. Additionally, the spread spectrum process is examined and compared to CDMA, and both FHSS and DSSS techniques are illustrated.

Chapter 3, *Protocol Architectures,* examines the three protocol models referred to throughout the text. The chapter begins with a summary of basic protocol concepts and terminology. After this introduction, the OSI, TCP/IP, and SS7 protocol models are described. Where appropriate, layer protocol specifics are used to illustrate the concepts needed in later chapters.

Chapter 4, *Wireless Layer Protocols,* is a survey and examination of protocols that are important for wireless network communications. Knowledge of this material is assumed in later parts of the text, especially Part 3. Because many wireless systems will be transitioning to IP as a layer 3 protocol, the chapter closes with a closer examination of the protocol than was provided in Chapter 3.

■ PART 2: THE WIRELESS PHYSICAL LAYER

Part 2 is primarily of interest to readers who want to learn more about engineering and implementation techniques. Part 2 is written from an engineering or engineering technology viewpoint, and goes into some detail on the actual implementation practices used in wireless devices. Thus, it addresses issues such as modulation, coding, and antenna design. In other words, wireless systems are discussed from a physical layer perspective. These concepts are fundamental to any wireless system.

Part 2 is divided into two groups of chapters. The first group, Chapters 5, 6, and 7, explores the digital and pulse modulation techniques used by wireless communications systems. This discussion is followed by a chapter that introduces the error-correction codes used in wireless systems. This group of chapters is about technology, not systems. The second group of chapters in this part, Chapters 8, 9, and 10, discusses traditional wireless systems from a technological perspective. The second group is a three-chapter examination of wireless systems with a common theme of link analysis. The first chapter in this group explores antenna technology and gives a brief introduction to link analysis. The next chapter focuses on broadband, terrestrial, line-of-sight, microwave wireless links. This chapter includes a somewhat more complex link analysis. Part 2 closes with a chapter on satellite communications systems. It explores not only the communication applications of satellites but also the basics of satellite technology and orbital mechanics. Finally, a link analysis, which ties together common concepts in these related fields, is performed.

Chapter 5, *Digital Modulation,* covers digital modulation. The chapter begins with an introduction to the central concepts of bandwidth efficiency, symbol rate, and M-ary encoding. To form a basis for comparision as well as to illustrate basic relationships between information bandwidth and transmission bandwidth, traditional DSB-AM is examined. An emphasis on transmission bandwidth is used throughout the chapter as a comparison of different techniques. The chapter then shifts to digital techniques, beginning with ASK, to build a bridge from the previous AM discussion. FSK is then discussed. Both coherent and noncoherent FSK are examined; the close relationship between coherent FSK and MSK is shown; the problem of wide transmission bandwidth of MSK is identified; and the solution used by many wireless systems, GMSK, is introduced. PSK at representative levels of modulation density is explored next, and the equivalence of QPSK and QAM is shown. Constellation diagrams are introduced, as are enhancements to traditional PSK methods important to wireless systems such as DQPSK and $\pi/4$ DQPSK. The chapter then moves to a discussion of error distance and higher modulation density techniques, and closes with an overview of BER graphs and how to use them.

Chapter 6, *Pulse Modulation,* explores pulse modulation, focusing on PAM and PCM. Specific wireless examples are given. The chapter begins with an introduction to sampling theory, aliasing, and the effects of variation on sample interval and sample resolution. PAM is then presented as the basis of most pulse modulation techniques. The way in which it naturally flows into PCM is then demonstrated. Many important aspects of PCM are discussed, including quantization error, nonuniform quantization techniques, the use of code rather than binary to represent the sample value, quantization noise, dynamic range, and power utilization. The chapter closes with a brief overview of related pulse modulation techniques, including DPCM, ADPCM, and the various forms of delta modulation.

Chapter 7, *Error-Correcting Codes,* contains material not often found in many wireless textbooks. If and when this material *is* discussed, the treatment of coding is often buried in complex mathematics. This important subject, which has links to modulation and bandwidth efficiency, is critical for understanding how wireless systems work and is discussed in a simplified way. Chapter 7 sets error-correcting coding methods in a general channel coding context. After a brief introduction to modulo 2 arithmetic, the chapter defines block codes and immediately introduces parity and Hamming codes. Specific examples are worked out and a step-by-step method is provided. Parity techniques, first through Hamming enhancements and then later in application to the Viterbi algorithm, link the concepts together. The interleaving technique and cyclic codes are then introduced, along with a brief introduction to the polynomial notation used. CRC codes are also explored. Again, specific examples are worked out and a step-by-step method is provided. The chapter then briefly makes appropriate connections to cyclic codes and concludes with an examination of the classic convolutional coding technique, trellis coding. An example and a step-by-step method are again provided to ensure that the reader understands how these techniques operate. To conclude this chapter, the Viterbi algorithm is shown as one method of an optimal decoding technique for trellis-coded information.

Chapter 8, *Antennas,* introduces the subject of antennas for the nonspecialist. The treatment begins with an examination of the terms and definitions used in this field. The basic operation of an antenna is presented as a way to give the reader a feel for an antenna's physical function. The important subjects of ERP, radiation resistance, beamwidth, front-to-back ratio, and channel attenuation are each addressed in their own sections. The chapter then turns to an overview of many classic antenna types, including various versions of the dipole, helical, parabolic, and horn reflector antenna designs. Simplified algebraic relationships give the reader a way of gauging the impact on and appropriateness of a particular antenna design for a system. The chapter closes with a brief introduction to link analysis, a common theme across this and the next two chapters.

Chapter 9, *Terrestrial Broadband Wireless,* delves deeper into the engineering of a typical terrestrial, broadband, line-of-sight communications link. This treatment of link engineering is traditional and, when coupled with the discussion in Chapter 8, provides most of the information and techniques that would be used to generate a first approximation for a design. Not all details are presented here, but the dominant effects are discussed, along with a way to lump together many small effects not individually examined. Noise effects are developed and are treated in some detail, as are techniques in amplifier cascading and curvature of the earth. Obstructions in an otherwise clear line-of-sight environment are addressed using the Fresnel zone technique.

Chapter 10, *Satellite Communications Technology,* explores many aspects of satellite technology and communications using satellites. A brief history is provided, and orbital particulars of GEO satellites are discussed. GEO and LEO satellites are compared from a communications system perspective, and the strengths and weaknesses of satellites are examined in relation to other communications systems. The chapter then turns to a discussion of satellite technology and construction. The primary satellite subsystems are described and communications bands are identified. Polarization issues are touched on briefly. The multiple-access procedures used in satellites are, in some cases, somewhat different from those used terrestrially. The next sections discuss the dominant techniques in use, from the traditional FDMA and SCPC techniques to more modern methods using DAMA and spot beam technology. Of course, to use any GEO satellite, you must be able to point the antenna at the satellite, so the next sections discuss how to find a satellite. Terminology and techniques specific to this practice are introduced. The concept of look angle is presented and broken down into its two components, azimuth and elevation. These concepts will be unfamiliar to many readers, so the more familiar concepts of longitude and latitude are discussed first. A block diagram of an earth station is provided. Several satellite systems are then described as a way to illustrate the techniques examined earlier. These systems include TVRO systems, Internet access, SPADE (to illustrate DAMA), VSAT systems, and the current GPS system. Future evolution of GPS, based on currently available information, is also presented. Finally, the MSAT and Inmarsat systems are presented, and the chapter concludes with the third and final installment of link budget calculations.

■ PART 3: CELLULAR SYSTEMS AND PROTOCOL PERSPECTIVES

Part 3 explores cellular technology from a system and protocol perspective. Cellular communications systems are examined at this point, and the true focus of the textbook emerges. The section begins with an overview of cellular basics. The next two chapters focus on the three classes of cellular deployments: FDM, FDM/TDMA, and CDMA based air interfaces. These chapters focus on the voice telephony mechanisms. Chapter 14 extends the discussion into cellular provision of data services. The major techniques and protocols used for the provision of data services are examined. Chapter 15 describes the operation of the IEEE 802.11 wireless LAN.

Except for the final chapter, the chapters in this part focus on layers 1 and 2 from a protocol perspective in a cellular environment. Layers 3 and 4 must be addressed when using any type of wireless subnet, and they are the focus of Chapter 16. While the TCP/IP protocol stack receives the most attention, the SS7 protocol stack is also briefly addressed.

Chapter 11, *Cellular Basics,* serves as an introduction to cellular systems and provides the foundation for the rest of the chapters in Part 3. A model of a cellular system featuring six components is presented and used to tie together the more detailed discussions in the following chapters. One of the key advantages to this approach is that each cellular system is described in the same terms and functional subsystems. This presentation allows easy comparison of the essential similarities and differences of the various approaches to the same problem. The basics of engineering a cellular network are presented and worked through. The concepts of frequency planning, cell sizing and cell splitting, frequency reuse, and others are illustrated. Basic operational concepts, such as antenna choices, system evolution with technology, and QOS from transmission and revenue perspectives, are discussed. Cellular system concepts such as blocking, overlay, location, handoff, roaming, and the near-far effect are defined and clearly delineated. After briefly comparing cellular systems to the Ricochet approach, the chapter turns to various business, marketing, and IT issues that present themselves when an organization moves to wireless solutions. Issues such as business requirements, the concepts of push and pull, wireless data security, virus susceptibility, and wireless PBX technology are discussed. When you read about these topics, remember that the vast majority of wireless networked communications will be supported and enabled by cellular based systems. The chapter closes with a section on cellular system economics. Representative costs are provided, as is a framework for a running example or project, which allows the reader to work through some of the decisions facing cellular service providers when new technology emerges. For example, should they upgrade? Should they continue to extract revenue from an existing deployment? If nothing else, this section gives a sense of the economics involved in a modern cellular system.

Chapter 12, *Frequency and Time Division Multiple-Access Cellular Systems,* is a key chapter in the text. It describes two of the three currently deployed cellular systems: those based on pure FDMA and those based on the hybrid FDM/TDMA air interface. The first approach is illustrated using the AMPS and EAMPS analog systems, while the second is illustrated using the widely deployed GSM system. The approach first outlined in Chapter 11 is followed for both air interface techniques. After a single system is described in detail, other deployed systems that follow the same approach are identified. This presentation approach allows an examination of almost every cellular system deployed by using only three specific examples, each discussed in depth.

Because GSM is most representative of a modern cellular system, the bulk of the chapter focuses on this system, with discussion of topics such as framing and control channels and their interface, as well as how logical and physical channels are partitioned. The security techniques used in GSM, which many believe will be very similar to those used in 3G systems, are discussed in detail. Because GSM is so important, a separate section on the protocol components of layer 2 and layer 3 is provided at the end of the chapter. It helps tie together the control channel structure and how layer 3 protocols are used to address mobility, physical channel effects, and call management in a GSM-compatible mobile station.

Chapter 13, *Code Division Multiple Access Based Cellular,* examines the third type of air interface: CDMA. Again, the approach followed allows easy comparisons among the various cellular systems. The deployed system chosen for this chapter is the IS-95A system. Because IS-95B is a straightforward extension of IS-95A, it is also addressed by pointing out where the two systems diverge. Note that the technical descriptions of CDMA, and spread spectrum techniques in general, are not discussed in this chapter as it was previously presented in Chapter 2. The critical differences between systems based on a CDMA approach and those that are not are centralized in the areas of frequency planning and capacity planning. These elements are illustrated through example and discussion. The central limiting factor to capacity in a CDMA-based system, interference, is compared to the central limiting factor found in systems using the more traditional air interfaces of FDMA and FDM/TDMA, namely, the limited resource of frequency allocation. Other, more subtle differences in a CDMA approach are also discussed. Topics such as the difference between hard and soft handoffs, power limits, and power control on both forward and reverse channels are explored. The techniques of security unique to IS-95 are also shown.

Chapter 14, *Cellular Data Services,* makes this textbook unique compared with other wireless networking textbooks. This chapter is dedicated solely to cellular data services and their evolution. While data services are touched on in previous chapters, their primary focus is on supporting voice services. Chapter 14 is concerned only with data services in a wireless networking context. Many believe that much of the interest in wireless devices is based on their ability to support data services. After a discussion of circuit-switched data services and the layer protocol RLP, which is critical to this service offering, the rest of the chapter is dedicated to packet data services. CDPD, SMS, and GPRS are explored in detail. The approach taken is a functional layer protocol one. The service is described, the protocol stack is shown, and then a layer-by-layer examination of the functions and limitations of each protocol is provided. This discussion should allow an interested reader to become knowledgeable about implementing such an environment in a corporate network.

Chapter 15, *IEEE 802.11 Wireless LAN,* is an in-depth examination of the wireless LAN protocol IEEE 802.11. While not normally considered a cellular-based system, many aspects of it correspond directly to those identified in the previous chapters in Part 3. Almost all aspects of this system necessary to understand and deploy it are provided. Note that important elements of the operation of this LAN are described in Chapter 4, where important layer protocols are described. The protocol stack, topology, and components are detailed. The definitions of handoff and roaming, as applicable to this technology, are described. While each physical layer option in the standard is discussed, the 802.11a: 11 Mbps DSSS option receives the most attention. Station and distribution services are detailed, as is the MAC frame structure. Both network and power management are touched on. In an effort to capture future trends in this environment, IAPP (an emerging standard) is identified and briefly discussed, as is 802.11a.

Chapter 16, *Layers 3 and 4 in a Wireless Environment,* is also unique for a textbook on the subject of wireless networking and communications. This chapter focuses exclusively on the implications at layer 3 and layer 4 of the protocol stack when using wireless as a subnet in an enterprise network. While the bulk of the chapter assumes that the protocols at these layers will be IP and TCP, respectively, some attention is also given to using an SS7 approach, which is common in the existing circuit-switched cellular architectures. Critical issues must be addressed when moving to a packet based network using any wireless communications environment. This chapter identifies these issues and proposes solutions. As you might expect, no single solution is applicable in every case. Some fundamental concepts inherent in the TCP layer protocol make implementing wireless subnets difficult and challenging. It is important to have some basic understanding of TCP (a topic covered in Part 1) prior to reading this chapter.

■ PART 4: EMERGING SYSTEMS

Part 4 attempts to provide some guidance about the future development of wireless systems. Writing about this topic is always a difficult task, and the time lag in a textbook from writing to publication makes the task particularly challenging. It is almost certain that at least some of what is presented in this part will be outdated by the time you read it. Nevertheless, in this rapidly changing field, some attempt to address emerging technologies and systems must be made. The three topics covered in this final part explore emerging technology (WAP, 3G cellular systems, and Bluetooth) from a protocol perspective.

Chapter 17, *Emerging Standards: WAP*, describes the wireless application protocol (WAP) that has received so much attention. WAP is a protocol, based on HTML, that standardizes the way a WAP client interacts with the Internet.

Chapter 18, *Emerging Standards: 3G Cellular Systems*, attempts to clarify the terminology associated with 3G cellular evolution by describing what 3G really is and what its goals are, sorted in priority order. The standards bodies that define 3G systems are described and the relationships among them are identified. The ways in which the air interface and data standards currently used are expected to transition to those that will be used in a 3G environment is summarized.

Chapter 19, *Emerging Specification: Bluetooth*, covers the emerging specification called Bluetooth. This specification is still developing, but the information currently available is presented. A protocol architecure is shown, and this context is used to describe the Bluetooth operation and interfaces to other systems.

■ INSTRUCTIONAL APPROACH

The overall organization of *Wireless Networking* lends itself to two instructional approaches to this general subject. Portions of Part 1 and all of Part 2 are well suited to a first course in wireless technology where the focus is on implementation-related channel characteristic effects. This approach is essentially a physical layer perspective of wireless communications systems. Systems that demonstrate these effects, terrestrial and satellite line-of-sight communications, close the discussion.

Portions of Part 1 and all of Part 3 lend themselves well to a protocol or systems-based course on wireless networking. Such a course would focus on the air interface and protocol elements of a wireless channel. Systems that demonstrate these choices, cellular and wireless LANs, are covered.

Clearly, the line is not sharp in this artificial structure I have created. Several good examples of air interface schemes are used in satellites, and these examples are discussed in Part 2. Similarly, there are good examples of protocol effects in network segments that use satellites, and these examples are discussed in Part 3.

A few comments about the organization and topic selection in this presentation are appropriate. First, I have attempted to provide a rough guide to subject matter groupings. Individual professors and motivated readers can and should "cherry pick" from different chapters, depending on particular interests.

Second, this field is rapidly evolving. When I teach classes in this subject area, I find myself bringing to class topical articles as I come across them. Students find these articles interesting, and I think it gives them a practical idea about what is most applicable in the networks they will design and use in the near future. Readers of *Wireless Networking* in a noninstructional setting will also find a broad range of sources helpful.

Third, in an attempt to provide a textbook with broad coverage of most basic wireless systems, in-depth detail and mathematical rigor have occasionally been omitted. Entire textbooks focus on specific subject matter or discuss system implementation in much greater detail. In my view, this availability of information is another reason to "cherry pick" and to use supplementary material from both the current media and other textbooks where appropriate.

◼ ACKNOWLEDGMENTS

I want to thank the following professionals, who were generous enough to review the manuscript of *Wireless Networking* prior to publication: Sami Al-Salman, DeVry Institute of Technology; Lawrence Bernstein, Stevens Institute of Technology; Robert Borns, Purdue University; Sue Garrod, Purdue University; Byron Todd, Tallahassee Community College; and Frank Whetten, Embry Riddle University.

I also want to thank Jeong Yeop Yi, a graduate student who made my drawings comprehensible and who eliminated my use of American slang. All of this work was critical to the success of this book.

I want to thank the staff at Prentice Hall, whose efforts in turning a manuscript into a professional publication should not go unremarked. Specifically, I want to thank Charles Stewart, Editor; and Tricia Rawnsley, Production Editor.

Special thanks also go to Marianne L'Abbate, freelance copy editor and proofreader. Among many other things, she made sure all my pronouns were clearly defined for the reader.

Jeff Honchell at Purdue University was kind enough to review the pages of this book after Chuck Thurwachter's untimely passing. We, at Prentice Hall, are quite grateful for his help.

When we feel sad, may we remember his trumpeting laughter;
When we feel weary, may we remember his boundless enthusiasm;
And, when we feel like we don't have time for each other,
    may we remember he's no longer with us

Charles N. Thurwachter, Jr.   1953–2001

Contents

Fundamentals of Wireless Networking

1

Technology Fundamentals

•

■ INTRODUCTION

This chapter will introduce basic ideas central to understanding wireless communications. Many topics, some covered in more detail in other chapters, are introduced. To begin, we will review briefly some abbreviations in scientific notation. Then, the electromagnetic spectrum is introduced along with the related concepts of frequency and wavelength. A decibel review, with applications to gain attenuation and signal-to-noise calculations, is then provided. Next is a discussion on bandwidth and information capacity, Hartley's law, and the Shannon limit, followed by an examination of the frequency bands used in wireless systems, both licensed and unlicensed. Frequency standardization and the results of WARC-2000 are briefly summarized. FCC part 15 is discussed briefly, along with propagation modes and isotropic power radiation.

■ SCIENTIFIC NOTATION ABBREVIATIONS

Engineers and technicians commonly work with numbers that vary widely in their magnitude. Scientific notation is often utilized to express the powers of 10 so that these values can be written in a concise form. The textual abbreviations and symbols for common notations are summarized in Table 1-1.

Table 1-1
Scientific Notation Abbreviations

Power of 10	Abbreviation	Symbol
−9	nano	n
−6	micro	μ
−3	milli	m
+3	kilo	k
+6	Mega	M
+9	Giga	G

■ ELECTROMAGNETIC SPECTRUM

The *electromagnetic spectrum* is defined as the total range of frequencies of electromagnetic radiation. The word *electromagnetic* illustrates the fact that radiation at various frequencies is emitted by electrical charges while in motion.

The word *spectrum* literally means a range of values or a set of related quantities. The electromagnetic spectrum is usually represented as a chart, similar to the one shown in Figure 1-1. This chart shows the first meaning clearly: there is a range of values of frequencies extending from 30 Hz to 300 GHz. The electromagnetic spectrum extends both below and above the chart, but the range in the figure shows the range of frequencies that virtually all communications systems use today. Near each frequency span is a two- or three-letter abbreviation for the band; all are defined below.

It is also useful to have a chart that shows where in these bands certain common communications systems lie. Figure 1-2 gives this information. The frequency ranges are approximate, designed to give a general idea where they lie with respect to each other. Each of the frequency ranges listed below will be discussed.

Audiophile band: 30–20,000 Hz

Voice band: 300–3,400 Hz

Broadcast AM radio: 540–1710 kHz

LF cordless telephone: 43–50 MHz

Broadcast VHF TV: 54–216 MHz (Channels 2–13)

Broadcast FM radio: 88–108 MHz

Broadcast UHF TV: 470–800 MHz (Channels 14–69)

Analog cellular telephone: 824–894 MHz (old UHF channels 73–83)

ISM bands: 902–928 MHz and 2.4–2.48 GHz and 5.7–5.85 GHz

Digital cellular telephone: 1,710–1,880 MHz

Figure 1-1

Electromagnetic Spectrum (*Reprinted by permission of Pearson Education, Inc., Upper Saddle River, NJ*)

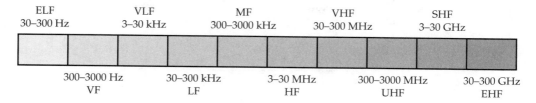

Figure 1-2

Communications System Operating Frequencies (*Reprinted by permission of Pearson Education, Inc., Upper Saddle River, NJ*)

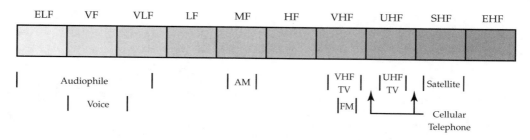

Satellite links: 4–8 GHz (C-band: 1-meter-diameter dishes); 12–18 GHz (Ku-band: half-meter-diameter dishes)

■ FREQUENCY AND WAVELENGTH

The second meaning of electromagnetic spectrum, a range of related values, also applies. It is easy to see that a higher frequency must be related in some way to a lower frequency. For example, 20 Hz is exactly twice 10 Hz, but exactly how is this relationship worked out? Is there a relationship to anything else that depends on the frequency or will change in some way as the frequency is changed? The answer is yes. Frequencies are related to each other and inversely related to their wavelengths. Equation 1-1 defines this relationship:

$$c = \lambda f \qquad\qquad \textbf{(1-1)}$$

c = speed of light = 3×10^8 m/s
λ = wavelength
f = frequency

Before exploring this equation in more detail, it is important to recognize the relationships among frequency, wavelength, and period. Figure 1-3 details these relationships.

Figure 1-3
Frequency, Wavelength, and Period (*Reprinted by permission of Pearson Education, Inc., Upper Saddle River, NJ*)

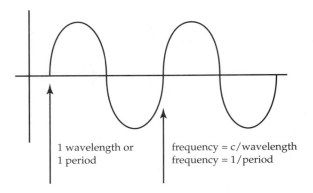

1 wavelength or
1 period

frequency = c/wavelength
frequency = 1/period

Table 1-2
Frequency, Wavelength, and Period of Some Common Wireless Systems

System	Approximate Center Frequency (MHz)	Wavelength (m)	Period (ns)
Broadcast AM	1	300	1,000
Broadcast FM	100	3	10
802.11 wireless LAN	2,000	0.15	0.5
K band satellite	15,000	0.02	0.07

The term *wavelength* is appropriate if the length of one cycle of the waveform is measured in units of length, such as meters. The term *period* is appropriate if the time it takes for one cycle of the waveform is measured in units of time, such as nanoseconds (ns). Some representative wireless systems are listed in Table 1-2 along with a summary of these attributes. If one of the attributes, the wavelength, period, or frequency, of any electromagnetic signal is known, the other two can be determined. Equation 1-1 showed how the speed of light, wavelength, and frequency are related. The period of any waveform in seconds is the inverse of its frequency expressed in hertz (Hz).

■ **EXAMPLE 1-1**   What is the shortest wavelength used in broadcast AM radio? What is the period of this waveform? The shortest wavelength would be associated with the highest frequency, which is given as 1,710 kHz.

$$\lambda = \frac{c}{f} = \frac{3.00 \times 10^8}{1,710 \times 10^3} = 175 \text{ meters}$$

$$\text{Period} = \frac{1}{f} = \frac{1}{1,710 \times 10^3} = 584 \text{ ns}$$

You should also realize that the wavelength of a signal makes a difference in how the antenna designed to receive it is constructed. It should make sense to you that an antenna designed to receive wavelengths thousands of meters long would be different than an antenna designed to receive wavelengths just a few meters long. You have probably seen many different kinds of antennas; the frequency that they are designed to receive determines their shape. For example, all television antennas are alike, all cellular telephone antennas are alike, and so on. See Chapter 8 for further discussion on this topic.

The next subjects to be explored are bandwidth and information capacity. However, prior to this discussion, a review of the concepts of decibels, gain, attenuation, and signal-to-noise ratio must be provided.

■ DECIBELS

The decibel is a base 10 logarithmic measure; the decibel is just the power of 10. For example, if one were to ask how many decibels 1,000 represents, the answer would be 30 db, or 10 times 3, the power of ten that represents 1,000. The factor of 10 comes from the definition of a decibel, which is shown in Equation 1-2.

$$dB = 10 \log(x) \tag{1-2}$$

■ **EXAMPLE 1-2**   Find the number of decibels represented by 10,000.

$$? \, dB = 10 \log(10{,}000)$$

$$? \, dB = 10 \log(10^4) = 10(4) = 40 \, dB$$

While the above example illustrates the concept of a decibel, in communications, this concept is used to relate one value to another. A more precise definition would be the one shown in Equation 1-3.

$$? \, dB = 10 \log\left(\frac{P_1}{P_2}\right) \tag{1-3}$$

Usually decibels are applied when relating one power to another. Inside the braces in Equation 1-3 are two powers, P_1 and P_2. The decibel represents the *value of the ratio* of the two powers. This representation is an easy way to talk about two powers without regard to the actual values, but to the value of their ratio. This concept is important and is used throughout communications.

■ **EXAMPLE 1-3**   System 1 has an input power of 1 W and an output power of 100 W. System 2 has an input power of 5 mW and an output power of 0.5 W. Compare the two systems' power ratios using the decibel.

$$1{:}10 \log\left(\frac{100 \, W}{1 \, W}\right) = 20 \, dB \qquad 2{:}10 \log\left(\frac{0.5 \, W}{0.005 \, W}\right) = 20 \, dB$$

Example 1-3 illustrated the use of the decibel to relate the gain of two systems, independent of the actual input and output values. Gain and a related concept, attenuation, are introduced in a later section.

There are some common numerical values of decibels that everyone should be familiar with. They are used frequently as a kind of shorthand in discussing power ratios generally. You may want to check these values with your calculator. Note that the decibel value is not linear with the power ratio, as shown in Table 1-3. It is also important to realize that the decibel is a dimensionless number, like values for radian and degree. Values of wavelength, frequency, period, power, and voltage have dimensions and units because they are all physical quantities, not just numbers describing a ratio of two physical quantities.

Because it is just a form of a number, a decibel can be converted to a pure number using the antilogarithm. One can always use the antilogarithm to convert back and forth between the ratios shown in Table 1-2 and the decibel values. Since the decibel is an expression of the power of 10, you convert back to a pure number from a decibel by just taking the decibel value and evaluating it as a power of 10. Example 1-4 illustrates this process.

■ **EXAMPLE 1-4**   Confirm that the power ratio of 2 is given by 3 dB.

$$3 \text{ dB} = 10 \log\left(\frac{2}{1}\right) \Rightarrow \frac{3}{10} = \log(2) \Rightarrow 10^{\left(\frac{3}{10}\right)} = 10^{\log(2)} \Rightarrow 1.995 = 2$$

The result in Example 1-4 is very close to the value in Table 1-3. This illustrates that many of the values in Table 1-3 are approximate. For the vast majority of calculations you will do, near accuracy is acceptable.

There are many alternate definitions for power using the decibel concept. We will explore one, the dBm. It is used where the reference power is defined as 1 mW and most commonly used to rate the output power of instruments and systems. It is also a common rating on power meters. The dBm is defined as:

$$\text{dBm} = 10 \log\left(\frac{\text{power level (watts)}}{10^{-3}}\right) \tag{1-4}$$

Table 1-3
Common dB Values

Power Ratio	dB Value	Power Ratio	dB Value
1	0 dB	8	9 dB
2	3 dB	10	10 dB
3	5 dB	100	20 dB
4	6 dB	1,000	30 dB

Table 1-4
Decibel–Volts–Watts Conversion Table, 50 Ω Termination

dBm	Vrms (Volts)	Vpp (Volts)	Power (mW)
+30	7.10	20.0	1,000
+20	2.25	6.36	100
+15	1.15	3.25	25
+10	0.71	2.01	10
+5	0.40	1.13	3.2
0	0.225	0.64	1.0
−5	0.125	0.35	0.32
−10	0.071	0.20	0.10

It is important to emphasize that dBm is just another way to talk about a power level, while dB is another way to talk about gain or attenuation, never a power value. Remember, dB is a pure, unitless number, and cannot be used to give a power reading. Since it is a value that expresses the value of the power relative to 1 mW, dBm is a perfectly acceptable way of discussing powers.

■ **EXAMPLE 1-5**   The power measured at the output of a certain communication system is 15 watts. What is the output power expressed in dBm?

$$? \, dBm = 10 \log\left(\frac{15}{0.001}\right) = 41.8 \, dBm$$

When using equipment to measure or drive communication systems, make sure you know what units the powers are calibrated in. One of the frustrating things when measuring waveforms on an oscilloscope and relating them to power and dBm is that you always have to have the conversions written down somewhere. Then you need your calculator to compute them. Table 1-4 does this conversion for you for one particular case: where the measurement impedance is 50 Ω. Most high-frequency test equipment is terminated in this impedance.

■ GAIN AND ATTENUATION

Discussing the performance of a system will depend on having a way to talk about how much power was used to transmit the information and how much power was received. The gain (rise) or attenuation (loss) of a system is defined by the ratio of the output power over the input power. This power gain or loss is measured in units of the decibel (dB). Applying this to the definitions of gain and attenuation, we have:

$$\text{Gain or attenuation in dB} = 10 \log \frac{\text{average power output}}{\text{average power input}} = 10 \log \frac{P_{out}}{P_{in}} \quad \textbf{(1-5)}$$

Note that gain and attenuation are found by the same relationship. Gain is always a positive value; attenuation is always a negative value. An example or two will illustrate how these formulas are applied.

■ **EXAMPLE 1-6**   A communication system has a specified transmit power of 10 watts. At the receiver, you have measured the received power as 1 mW. What is the attenuation of the channel?

$$\text{Attenuation} = 10 \log\left(\frac{P_{out}}{P_{in}}\right) = 10 \log\left(\frac{0.001}{10}\right) = 10 \log(10^{-4}) = 10(-4) = -40 \text{ dB}$$

■ **EXAMPLE 1-7**   A communication system has a specified gain of 30 dB. You have measured the output power of the system with a power meter and have determined its value to be 100 watts. What is the input power being applied?

$$30 \text{ db} = 10 \log\frac{100 \text{ W}}{P_{in}} \Rightarrow \frac{30}{10} = \log\frac{100}{P_{in}} \Rightarrow 10^3 = \frac{100}{P_{in}} \Rightarrow P_{in} = \frac{100}{10^3} = 0.1 \text{ W}$$

It is important to emphasize again that a gain is always positive and an attenuation is always negative. That concept should make sense to you: gain is getting bigger, so the ratio of powers, as expressed by the decibel, is positive; attenuation is getting smaller, so the ratio of powers, as expressed by the decibel, is negative. For any two powers, there is either gain or attenuation, never both.

Before the subject of decibels is complete, it is important to realize that the dB and dBm can be combined in ways to describe the gain and attenuation of any system. Sometimes one can combine the two quantities directly and obtain a measure of the gain or attenuation of a system. To understand when this is possible, it is necessary to explore a little deeper how these quantities are used.

Voltage, current, and power can be amplified. Power amplification means that the power applied to the input of an amplifier is increased by some amount when it appears on the output. Power amplification is usually described in dBm when talking about the input and output powers. When describing the power amplification of an amplifier, dB is the preferred term. Therefore, it is permissible to add dBm and dB to describe gain and subtract them to describe attenuation. These relationships are shown in Equation 1-6 and are illustrated in Example 1-8.

$$\text{Gain (dBm)} = \text{dBm} + \text{dB} \qquad \text{Attenuation (dBm)} = \text{dBm} - \text{dB} \qquad \textbf{(1-6)}$$

■ **EXAMPLE 1-8**   A certain communication system has an input power of 2 dBm. The gain of the system is 5 dB. What is the output power?

$$\text{Gain (dBm)} = 2 \text{ dBm} + 5 \text{ dB} = 7 \text{ dBm}$$

Figure 1-4

Power Amplification (*Reprinted by permission of Pearson Education, Inc., Upper Saddle River, NJ*)

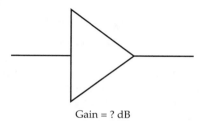

Gain = ? dB

Example 1-8 illustrates a handy way of quickly finding the output power once the input power is known. The symbol shown in Figure 1-4 is used to represent power gain. It is a shorthand way to describe voltage or current gain. When calculating system performance, a particular decibel value is assigned to it.

Finally, you cannot always add and subtract dB and dBm freely. It is always permissible to add or subtract two quantities expressed in dB. The resulting unit is the dB. You can always add and subtract two quantities expressed in dB and dBm. The resulting unit is dBm. While it is permissible to subtract two quantities in dBm from each other, it is never permissible to add two quantities expressed in dBm! In the first case, the resulting units are dB and express a comparison between two powers. In the second case, the operation has no meaning. If at least one of the quantities is expressed in dB, then the operation is always permissible.

■ SIGNAL TO NOISE

Signal to noise (S/N) is the ratio of signal strength, or power, to noise power. Both quantities must always be measured in the same bandwidth for the ratio to have meaning. Whatever the bandwidth of the signal, you must measure the noise in that same bandwidth to get an accurate result.

In communication systems, we are concerned with powers. Even though you may not understand how a transmitter or receiver works, it is possible to measure a key performance figure of the receiver using S/N measurements. Signal to noise is a key performance parameter of any communication system.

Signal to noise is only strictly defined after reception; thus, it is a good way of talking about the performance of the receiver. One can ask, What is the signal-to-noise performance of the receiver? Signal to noise measures the relative power of the signal and the noise. It is defined in Equation 1-7:

$$\frac{S}{N}(dB) = 10 \log\left(\frac{S \text{ (watts)}}{N \text{ (watts)}}\right) \tag{1-7}$$

■ BANDWIDTH

One of the most fundamental concepts of the electromagnetic spectrum that is important to communications is bandwidth. Bandwidth is defined as the range of frequencies used to send the information from the source to the destination. Sometimes it is also defined as that portion or band of the electromagnetic spectrum occupied or required by a signal. The bandwidth of a signal is found by subtracting the lower frequency value from the upper frequency value, as shown in Equation 1-8:

$$BW = f_{upper} - f_{lower} \tag{1-8}$$

As you can see, the "width" of frequencies is captured by the bandwidth calculation. The wider the bandwidth, the more frequency space that is used to send the information and hence the more information that can be sent.

■ **EXAMPLE 1-9**   What is the bandwidth of the voice band?

$$BW = 3400 \text{ Hz} - 300 \text{ Hz} = 3100 \text{ Hz}$$

Bandwidth and Information Capacity

Sometimes bandwidth can be confused with the information capacity. Information capacity is defined as the amount of information that can be sent from a source to a destination. As a general principle, the more bandwidth used by a communications system, the more information content can be sent and the larger the information capacity. Just how much more information can be sent for a given amount of increased bandwidth depends on the modulation technique. In normal use, the information capacity of a communications system or device is referred to as the speed of that system or device. To see an example, think for a moment about the telephone modem used with a personal computer. When the speed of a modem is discussed, what one really means is the amount of information capacity it has. Since all electromagnetic waves travel at the same speed, independent of their wavelength or frequency, why is the speed of the modem expressed in bits per second?

What is being expressed is the information capacity of the modem. The larger the information capacity, the larger the speed specification in bits per second. It is also true that all of the modems designed to work with the wireline telephone system use the same bandwidth. The voice bandwidth of the telephone system, which is 300 Hz to 3400 Hz, is fixed. Many different speed modems are available; these are examples of many different information capacities operating in the same bandwidth.

Bandwidth and Carrier Frequency

If you accept as a general principle that the wider the bandwidth, the more information can be sent, you may want to ask, Why not just keep increasing the bandwidth again and again? Higher information capacity modems (faster modems) seem easy when looked at from this perspective. Since more information is being sent, paying more money for the modems would be justified if higher information capacity was provided. But this scenario is limited primarily by one of two factors: either the frequency bandwidth is fixed, or there is limited frequency bandwidth that is economically profitable to utilize.

A fixed frequency bandwidth means that the bandwidth cannot be expanded by any means. Usually this situation results from the limitations of the medium that is used for the communications channel. The best example is the traditional telephone system. Without an entire refit, the traditional telephone system has a limited bandwidth, determined by the components used to construct it. The physical nature of the components limits the bandwidth.

The other possibility is that the communications channel is fixed by law. Television broadcasts and AM and FM radio all have bandwidths determined today by law. Law, unlike physical reality, can be changed. A good example is the plight of UHF-TV channels 73 to 83. Prior to about 1975, these stations believed that the frequencies they were assigned by the FCC would always be theirs, by law. Around that time, cellular telephone systems were a gleam in the eye of several telecommunications companies, but no appropriate frequency space was available. They lobbied for changes in the law and won. Today this frequency band is, by law, theirs.

If the frequency bandwidth is fixed, then one must look to modulation or coding techniques to increase the information capacity. Since the frequency bandwidth of the telephone system is fixed, both of these techniques have been exploited to achieve the high information capacity modems currently offer for voice-grade telephone lines.

Although there are many reasons why some additional frequency bandwidth might be unprofitable to exploit, the primary one is the fact that the higher the carrier frequency selected, the higher the cost of the system. High-frequency components are generally more costly than low-frequency components. The pressure to make the best economic use of the limited usable spectrum available tends to keep the channel bandwidth allocations narrow. Clever engineers must figure out how to pack in more information using less bandwidth.

This approach keeps areas open for future innovation and also helps with international standardization. The less spectrum needed to dedicate to some new service, the less chance that some system is already using that spectrum. Different countries allocate the electromagnetic spectrum differently within their borders. Because an international service must cross these boundaries, careful selection of carrier frequencies and bandwidth will give the service a chance to succeed.

Although using a higher carrier frequency results in higher system cost, there are some advantages. The first is that, as in most industries, new technology frees up new resources. Only within the last thirty years or so could cost-effective systems be built operating at carrier frequencies above a few hundred megahertz. Earlier, the technology simply could not produce the components necessary to construct devices that operated at these frequencies. This situation illustrates a general rule: the higher the carrier frequency, the fewer systems that can compete for the available spectrum. Fewer competitors makes less competition and fewer previously licensed spectrum "off limits" signs. It is also an advantage that there are more available channels of a given bandwidth at a higher carrier frequency than at a lower frequency in a given percentage bandwidth of the carrier. For example, FM broadcast radio uses a carrier frequency 100 times higher than AM broadcast radio (100 MHz versus 1,000 kHz). This situation allows 100 times more 10 kHz channels in the FM band than is possible in the AM band.

The telephone system voice bandwidth is 3100 Hz; this bandwidth cannot be changed without very great investment and a long time. Thus, engineers must find ways to stretch the information bandwidth (information capacity) beyond the frequency bandwidth. The frequency bandwidth is the range of frequencies used; the information bandwidth (information capacity) is the information throughput that can be sent in a specific frequency bandwidth. Much of modern communications design is about increasing the information bandwidth (information capacity) without increasing the frequency bandwidth.

Again, *bandwidth* has two meanings but usually the context makes the meaning clear. When one talks about a communications channel, frequency bandwidth is usually meant; when one talks about data rate or information transfer, information capacity or information bandwidth is usually meant. Now we are ready to describe the fundamental limit on all communications systems' information capacity. This limit was first defined by Hartley, and was expressed as Hartley's law; then it was redefined and enhanced by Shannon in the form of Shannon's limit.

Information Capacity

Although we have discussed in a general way how bandwidth and information capacity are related, there was no explicit discussion of the development of a relationship between the two. Hartley's law was developed to express this relationship. Hartley's law, expressed in Equation 1-9, says that as the bandwidth or duration of a transmission increases, the amount of information communicated also increases:

$$I \propto BT \qquad \qquad \textbf{(1-9)}$$

I = information capacity
B = bandwidth
T = time or duration of transmission

This statement of Hartley's law illustrates that just as distance equals rate times time, information capacity equals bandwidth times time. It would be convenient to use Hartley's law to estimate the data rate one can expect to achieve for a given channel bandwidth and duration of transmission. This estimation can be done, but first it must be made clear how Hz and bps are related. Both have an important impact on how the bandwidth of a digital signal is determined.

Bandwidth of Binary Signals

The purpose of this section is to examine how the bandwidth of a digital signal is determined. To take the next step, think about how digital data is expressed and compare it to how an analog signal's frequency is determined. Figure 1-5 should make this comparison clear.

As we illustrated earlier, when the frequency of a waveform is calculated, the period is just inverted. Note that with digital signals, however, an alternating 1 and 0 combine to create one "period." Consider what the fastest digital signal change would be. Alternating 1s and 0s, like 10101010, is as fast as it can get. Of course, this is just half of the bit rate! In other words, there are two bits in every square wave period. This result means that the baseband bandwidth of a digital signal is half the bit rate for any binary modulation scheme. This relationship is expressed by Equation 1-10.

$$f_m = \frac{DR}{2} \qquad \qquad \textbf{(1-10)}$$

f_m = equivalent modulation frequency
DR = data rate

Example 1-10 will make this relationship clear.

■ **EXAMPLE 1-10**   A binary signal is to be amplitude modulated. The data rate of the binary modulation signal is 10 kbps. What is the baseband bandwidth and the double-sided transmit bandwidth of the signal?

$$\text{Baseband BW:} \quad f_m = \frac{DR}{2} = \frac{10 \text{ kbps}}{2} = 5 \text{ kHz}$$

$$\text{Transmit BW:} \quad BW = 2f_m = (2)(5 \text{ kHz}) = 10 \text{ kHz}$$

Pay close attention to the subscripts shown in Example 1-10. To avoid confusion, when the double-sided transmit bandwidth is being referred to in this chapter, there will be no subscript on the BW symbol. When bandwidth is referred to, it will mean the double-sided transmit bandwidth unless specifically noted otherwise.

The modulating data rate will be assumed to be converted to a frequency if it is stated in Hz, and it will be assumed to be in binary if it is stated in bps. Be careful when converting between them or your bandwidth calculations will always be incorrect.

This maximum rate of change in the binary modulating waveform is referred to as the Nyquist rate. In this text, except where explicitly noted otherwise, the binary waveform is at this rate. In all bandwidth calculations, the maximum bandwidth will result. The modulating signal, the binary waveform, will usually have an arbitrary data pattern; if the Nyquist rate is used for the bandwidth calculations, sufficient bandwidth will

Figure 1-5
Frequency Calculations for Analog and Digital Waves (*Reprinted by permission of Pearson Education, Inc., Upper Saddle River, NJ*)

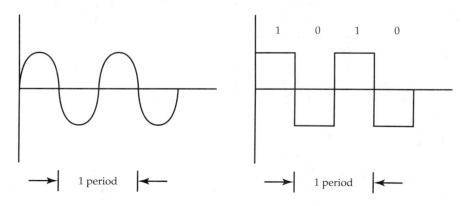

Figure 1-6
Example 1-11 (*Reprinted by permission of Pearson Education, Inc., Upper Saddle River, NJ*)

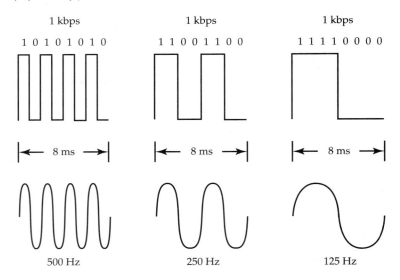

always be enough to pass all information. Example 1-11 illustrates the relationship between binary waveforms and frequency.

■ **EXAMPLE 1-11**   A binary waveform is formed of alternating 8-bit words shown below:

```
10101010        11001100        11110000
```

Determine the frequency components and verify that, by using the Nyquist rate for bandwidth calculations, all frequency components of the original data pattern are of a lower frequency than the Nyquist rate. Assume that the data rate of the binary modulating waveform is 1 kbps.

Figure 1-6 should make the relationship clear. If the first pattern is used for finding the baseband bandwidth by applying Equation 1-4, then all other patterns will be of a lower frequency. Note that data rate is expressed in bps and frequency is expressed in Hz.

Now Hartley's law can be reevaluated with this fact in mind. Just multiply the bandwidth by an additional factor of two. This rewriting of the relationship for digital signals is shown in Equation 1-11.

$$I \propto 2BT \tag{1-11}$$

■ **SHANNON'S LIMIT**

Twenty years after Hartley's law, in 1948, Claude Shannon wrote one of the most famous papers in communications. It related information capacity to bandwidth and signal-to-noise ratio rather than time of transmission. This was a critical insight and opened the door to many modern modulation techniques. Equation 1-12 expresses Shannon's limit.

$$I = 3.32B \log_2\left(1 + \frac{S}{N}\right) \tag{1-12}$$

S/N is the signal-to-noise ratio and the logarithm is the common logarithm, or base 2 logarithm. The really interesting thing about Shannon's limit is that if each symbol transmitted represents more than one bit, information capacity is limited only by the signal-to-noise ratio of the channel. Shannon's limit allows the possibility of representing more than one input change with one output change. This is a powerful concept and is at the basis of all modern digital communications. Once it is rejected that one input change can result in only one output change, the linear relationship of information capacity and bandwidth shown by Hartley's law is also rejected.

■ **EXAMPLE 1-12**   What is the information capacity predicted by Hartley's law and Shannon's limit for a bandwidth of 3,100 Hz and a signal-to-noise ratio of 30 dB? First, find the signal level that is implied by the logarithmic value of signal to noise.

$$S/N \text{ (dB)} = 10 \log\left(\frac{S}{N}\right) = 30 = 10 \log\left(\frac{S}{1}\right) \Rightarrow S = 10^{30/10} = 1,000$$

As expected, 30 dB corresponds to a power ratio of 1,000. Hartley's law says that the S/N does not matter and if there is 3,100 Hz of bandwidth, 6,200 bps of information can be transmitted. This situation does not leave too much room for innovation. The factor of 2 comes from the observation that 2 bps corresponds to 1 Hz, which is represented in the equation below:

$$I = 2BT = 2(3,100 \text{ Hz})(1 \text{ s}) = 6,200 \text{ bps}$$

Shannon's limit says that much more information capacity is available if something other than binary modulation is applied.

$$I = (3.32)(3,100 \text{ Hz})\log(1 + 1,000) = 30.88 \text{ kbps}$$

By not restricting consideration to binary modulation techniques, a large increase in the maximum information capacity is achieved for this signal-to-noise ratio. For larger signal-to-noise ratios, the difference would grow larger. Digital modulation techniques that exploit the idea of representing more than one input change with one output change are explored in Chapter 5.

■   **FREQUENCY BANDS**

Table 1-5 summarizes the ten frequency bands or ranges that are defined and discussed here. Note that each is designated to be an entire decade of frequencies. They all use

Table 1-5
Frequency Bands and Propagation Mode

Band	Frequency Range	Wavelength Range (Meters)	Propagation Mode	Nickname
ELF	30–300 Hz	10,000,000–1,000,000	GW	ELF
VF	300–3,000 Hz	1,000,000–100,000	GW	VF
VLF	3–30 kHz	100,000–10,000	GW	
LF	30–300 kHz	10,000–1,000	GW	
MF	300–3,000 kHz	1,000–100	GW	
HF	3–30 MHz	100–10	SW	Short-wave
VHF	30–300 MHz	10–1	LOS	VHF
UHF	300–3,000 MHz	1–0.1	LOS	UHF
SHF	3–30 GHz	0.1–0.01	LOS	Microwave
EHF	30–300 GHz	0.01–0.001	LOS	Millimeter

numbers divisible by three. Thus, wavelength calculations end up as integer powers of 10 meters because the speed of light = 30,000,000 meters/s. Table 1-4 is also used to introduce the concept of propagation mode, which will be discussed below. Be forewarned, the band names are uninspired but should be easy to remember because of their simplicity. Each of the bands' primary applications is reviewed below. Virtually all the wireless systems that will be examined in this text use the UHF and SHF bands.

ELF, extremely low frequencies (30–300 Hz): These are the power-line frequencies and the lower end of human hearing. Certain home-control systems use the power-line frequencies of 50 Hz (European) and 60 Hz (United States) to communicate with "smart" lamps, toasters, home security systems, etc.

VF, voice frequencies (300–3,000 Hz): This is essentially the same frequency band used by the telephone system (300–3,400 Hz) and contains 95 percent of the frequency content of human speech. The so-called audiophile band (30–20,000 Hz) is the extreme limit of human hearing. Human voice transmissions are contained in a much narrower band, primarily that band discussed here.

VLF, very low frequencies (3–30 kHz): These are the upper limits of human hearing, although most people by the time they are twenty-five have lost all hearing above 15 kHz. The only technological use for this band of frequencies today is for communicating with submarines covertly over very long ranges.

LF, low frequencies (30–300 kHz): These frequencies are not used widely as they are less effective for long-range communication than are the frequencies in the VLF band.

MF, medium frequencies (300–3000 kHz): The most common application for this band is broadcast AM radio, which uses the frequency band 540–1660 kHz.

HF, high frequencies (3–30 MHz): This band is widely used for long-range communication making use of the sky wave propagation effect. Military (long-distance aircraft), political (Voice of America, BBC), commercial (CB radio), and amateur radio (ham radio) users all make use of this frequency band, made unique by the composition of the earth's atmosphere. A different atmosphere, or none, and we would not have sky wave propagation effects in this frequency band.

VHF, very high frequencies (30–300 MHz): Broadcast FM radio uses the band 88–108 MHz. The traditional VHF-TV channels (channels 2–13) broadcast in this band. Traditional two-way FM radios use this band, and it also has applications in the military sector.

UHF, ultra high frequencies (0.3–3 GHz): Many new services make use of this band. Traditional users were the UHF-TV channels (14–83). Today many new uses are emerging, including cellular telephones and all kinds of paging systems. The L-band (1.5 GHz), used for low-cost satellite communication links with a wide variety of mobile applications, is located here. The two lower industrial, scientific, and medical (ISM) bands, 900 MHz and 2.4 GHz, are located here. Cellular telephone systems, wireless LANs, and the vast majority of new wireless technologies are located in this band. (Because of the critical nature of the ISM bands to wireless communications systems, these bands will be discussed in their own section below.) The low end of terrestrial microwave links is also located in this band.

SHF, super high frequencies (3–30 GHz): While the lower end of this range is used by microwave ovens to heat food, the uses of this band for communications is broad. Satellites use several bands, including the C-band (4–6 GHz) and the Ku band (14–16 GHz). The 5 GHz ISM band is located in this range and will be exploited in the future for a wide variety of applications. The majority of terrestrial microwave links, used in various private and public communication systems, also use this band for LOS transmission over a range of about forty miles.

EHF, extremely high frequencies (30–300 GHz): This is the next frontier for communications. Active experimentation is ongoing in this band, but component costs are extremely high so any widespread commercial use is some time away. The nickname for this band comes directly from the wavelength range; it is measured in millimeters. At the time this nickname was first applied, the use of wavelengths of these lengths was considered remarkable.

Above 300 GHz, one enters the so-called optical spectrum, infrared, visible spectrum, and ultraviolet. Although communication systems use these frequency bands, they are chiefly optical systems that make use of fiber-optic cable. These frequencies are not good choices for soft channel transmissions because of the interference caused by the environment, specifically the sun.

The infrared band is used inside buildings for communications; some optical wireless LAN systems use reflected infrared signals to connect PCs together. Virtually all handheld remote control units use infrared frequencies to control various appliances in the home. Examples include the remote controls for audiovisual systems (television, home theater, CD player, etc.). Additionally, infrared wavelengths produce heat. A heat lamp uses an infrared lamp, and it is often used for relaxing muscle aches or in a bathroom-ceiling lamp. The lamps that keep burgers hot until customers arrive are also infrared lamps. Infrared wavelengths are not the same as the ultraviolet radiation that tans you on a summer day at the beach.

The visible spectrum is used to communicate by human beings naturally. We have special receptors, our eyes, that can see these high frequencies, and we use them in many different ways to communicate. Visible light wavelengths are used in fiber-optic hard channels and rarely in unguided channels by lasers. Even setting aside the issue of interference from the sun, the use of these frequencies in unguided channels is not considered a good choice because of the danger of blinding system users if they look into the laser.

The ultraviolet spectrum is not generally used to communicate. For many, ultraviolet wavelengths mean radiation. A tanning studio uses ultraviolet lamps. You get a sunburn, exhibited by the reddening and/or browning of your skin, from the ultraviolet waves from the sun. Fluorescent lights like those in the ceiling of the classroom you attend also emit a fair amount of ultraviolet light, as do computer monitor screens. Sometimes special glasses are used by those especially sensitive to light in this frequency range. It is not visible, but specific lens materials can be manufactured to block that portion of the spectrum above visible light and thus protect sensitive eyes.

Frequencies above the ultraviolet region are interesting physical phenomena but are not usable by components manufactured today, so they are not used in communication systems. As technology improves, however, it seems likely that the usable frequency range will increase and these bands will come into use in the future. The UHF and SHF bands constitute the frequency span where the vast majority of wireless data systems operate. Any wireless device that does not require a license to operate makes use of one of the ISM bands. These devices include wireless telephones and wireless LANs. Licensed wireless systems in these bands include all satellite and terrestrial microwave links.

ISM Bands

Many of the products that use the technologies for business computing needs and that are discussed in this book would not be possible were it not for two actions taken by the United States federal government. The first is an action taken by the Federal Communications Commission (FCC) in 1985. In that year, they authorized unlicensed operation in three frequency bands: the industrial, scientific, and medical (ISM) bands. The other action taken was by the U.S. armed forces. Prior to the mid 1980s, use of the modulation technique spread spectrum was not permitted by commercial enterprises. This technology was reserved for use by the military exclusively. When this restriction was removed, it opened the doors to the vast majority of wireless LAN technologies, the newer satellite telephones, and many other mobile and fixed wireless applications.

Operation in these frequency bands is permitted without license, a major step forward to enabling commercial products. While there are certain restrictions, such as transmit power limitations to minimize interference to other unlicensed services, they are minor. More important, they likely have been addressed by the vendor of the product, freeing the user from these concerns.

Table 1-6
ISM Bands: Frequency Allocation for the United States and Canada

900 MHz Band	2.4 GHz Band	5.7 GHz Band
902–928 MHz	2.400–2.4835 GHz	5.725–5.850 GHz

Table 1-7
ISM Bandwidths

902 MHz Band	2.4 GHz Band	5.7 GHz Band
26 MHz	83.5 MHz	125 MHz

Without the ability to operate in an unlicensed manner, every producer of a deployed service would have to apply and receive a federal license to use the atmosphere. This process is cumbersome and time consuming and surely would have stifled much innovation. Imagine reapplying every time you moved your radio equipment! However, there are advantages to operating in a licensed band, and we will discuss these issues in more detail later.

Table 1-6 defines the exact frequency allocations available in each of the three ISM bands for the United States and Canada. Note that the bandwidths available in each of these bands is different, which follows this general principle: as you move higher in frequency, more bandwidth is available as an absolute measure but less as a percentage of the carrier frequency. The more bandwidth that is available, the higher the data rates that can be supported. Table 1-7 shows the bandwidth implied by Table 1-6.

Licensed and Unlicensed Bands

Many important wireless technologies need a licensed band to operate in. Two that do—radio frequency (RF) terrestrial links and satellite links—are discussed in Chapters 9 and 10. Both technologies rely on well-understood interference patterns from other sources to achieve their performance level. In a licensed environment, engineering studies allow the system operating in a licensed spectrum to achieve any conceivable level of performance.

In an unlicensed environment, the interference patterns cannot be determined. Without a record of where, and at what power, potentially interfering transmissions may exist, there is no way to determine what level of performance will be achieved. Clearly, when a wireless service can document and provide guaranteed levels of performance, that service can be sold for a higher fee. Most licensed wireless services provide a service level agreement (SLA). It offers the user a known level of performance that, if not achieved, results in a refund of some kind from the service provider. Unlicensed wireless systems have no way of determining the level of performance at any time and so cannot provide this kind of agreement.

The primary advantage to unlicensed operation is that no license application is filed with governmental agencies. Not filing paperwork is a major advantage in several applications, and indeed many emerging wireless technologies, such as the IEEE 802.11 wireless LAN standard and the Bluetooth specification, would not be feasible if they operated in a licensed band.

■ FREQUENCY STANDARDIZATION

Because communication systems operate in many different wavelengths, it should come as no surprise that government standardization of names and frequency bands is required. Each frequency band or wavelength range has certain types of transmissions licensed to operate in that band or range. Typically, there is also a maximum power limit set for each type of transmission, but that subject is outside the scope of this discussion. Several government agencies meet and discuss what services are allowed in each frequency band.

In the United States, the agency that controls this activity is called the Federal Communications Commission (FCC). Most other countries have similar agencies with similar names. These agencies have authority to set standards only within their own boundaries. As you can easily imagine, however, when one is dealing with wavelengths longer than some countries are big, coordination is important. Economic considerations, regional-based content, and interference issues resulting from electromagnetic energy leaking beyond national boundaries are big issues.

Worldwide standardization makes sense if the goal is to have a portable telephone or radio that works anywhere in the world. If country A is concerned that country B's television transmissions are interfering with its own, this topic would also be of great interest to both governing agencies and individual consumers. Consequently, these agencies send representatives to conventions occasionally to work these issues out. These global meetings are fairly rare and might occur only every ten years or so. Individual working groups meet often, and working papers and draft standards are issued quite regularly, but the official documents and decisions are done at these meetings.

Companies that manufacture communication equipment are very interested in the outcomes of these meetings. Therefore, they contribute a lot of technical and legal talent to try to make the outcomes favorable for the particular line of products or services they offer. This area can be interesting to work in only if you are patient, like to travel, and can both read and write long, detailed technical/legal documents. Writing standards can be a thankless task.

Worldwide coordination for spectrum allocation is performed by the International Telecommunications Union—Radio (ITU-R). The ITU-R is defined by an internationally signed treaty that lays out its responsibilities and rights and obligations. Approximately 150 member states have signed the treaty. Each member state belongs to one of three regions:

Region 1: Europe, Russia and surrounding territories, Asia Minor, and Africa

Region 2: The entire Western Hemisphere, including Hawaii

Region 3: Australia, New Zealand, and the rest of Asia (excluding Asia Minor)

The ITU-R is one of four main groups coordinated under an organization called the Plenipotentiary Conference:

1. The General Management (the former General Secretariat), is responsible for management and oversight. It translates and publishes all documents.
2. The International Telecommunications Union—Radio (ITU-R) has the task of studying the technical and operating issues related to radio transmission. This key technical committee concerns itself with frequency allocation for wireless devices.
3. The International Telecommunications Union—Telecommunications (ITU-T) has the task of studying the technical, operating, and tariff issues related to wireline devices. This key technical committee concerns itself with the operating characteristics of wireline devices.
4. The International Telecommunications Union—Development (ITU-D) has the task of studying technical issues related to the development of new technologies.

These four groups are divided into many subcommittees and study groups that do the actual work of the committee. Often these smaller groups are made up of representatives of commercial organizations that have a stake in the outcome of the standards defined by the various committees. More detail on the subcommittees can be found in Chapter 18.

There are actually two levels of coordination in both the wireline and wireless areas. The first is regional. Conferences may be for a single region or all regions at once. Our concern is with the wireless conferences. If all regions participate, then the conference is truly global in nature and is called a World Administrative Radio Conference (WARC). In 2000 the tenth such conference was held. Its decisions will affect wireless system operations for years to come.

It is important to note that different regions have different rules for the same spectrum. Only in a relatively few frequency bands is there global harmonization. As wireless systems grow in application, pressure will grow to harmonize these rules. In many cases, however, critical elements of a region's communications infrastructure may require specific regional allocations that conflict with the principle of global harmonization. This situation is, and will be, an ongoing struggle for the internationalization of wireless systems.

WARC 2000

The tenth WARC meeting was held in Istanbul during May 2000. To give an example of the size and importance of the meeting, the U.S. delegation included 150 people, primarily diplomats, engineers, and lawyers. The two key U.S. goals were to globalize the spectrum for the global positioning system (GPS) and spectrum use issues with third generation (3G) cellular systems.

The result on the GPS front was that two bands in the so-called L5 band are dedicated for GPS systems. The L5 band is to be used for the upgraded GPS systems now just being deployed. The Region 1 allocation was for 1,188–1,215 MHz; the Region 2 allocation was for 1,164–1,188 MHz. Both the U.S. military GPS system and the new commercially owned system envisioned by the Europeans, called Galileo, will coexist. Region 3, which used to have a GPS system deployed by the USSR, was not addressed by the conference. Presumably, at some future conference, whichever system becomes dominant from an application perspective in Region 3 will become standardized.

To complement the GPS capability, the L2 already allocated for satellite navigation was also expanded worldwide, to 1,215 MHz–1,300 MHz. The L1 band, which was also being eyed for GPS use and is currently harmonized for aircraft control and navigation, was left untouched and is dedicated to its current application. The L1 region is from 1,159–1,610 MHz.

The 3G cellular controversy was bigger. Both Europe and the United States had directly competing interests to advance here. The Europeans, who came out on top in the end, wanted specific rules that would encourage a single 3G cellular standard for the entire world. The U.S. position was that a generic rule that allowed the marketplace to decide would better serve the industry and the consumer. In the end, the United States stood virtually alone on this issue, and the European position was embraced by the vast majority of voting delegates. In WARC, each signatory to the treaty gets one vote only, regardless of how many delegates participate in the conference. The result was that the 3G bandwidth was allocated to a bandwidth of 160 MHz, with each region's spectrum slightly offset from the others. Additionally, it was agreed that the air interface would not be part of the definition of 3G. Instead, each country is free to standardize a specific air interface of its own choosing. (For much more on what 3G really means and how the technology is likely to develop, see Chapter 18.)

FCC Part 15

The management of the electromagnetic spectrum in the United States is the responsibility of the FCC. Part 15 regulates the emissions of products intended to operate in the license-free ISM bands. The operation of products in these bands will have been tested to ensure adherence to the rules of part 15. If the device causes interference, it is the responsibility of the installer or operator to address any interference issues that may arise from device configuration or use.

Licensed users have priority and use the same frequency band, but they are not subject to the part 15 rules because they are licensed. If such a user identifies an interference source, he or she can issue a complaint that will be addressed by the FCC. Remember, the FCC issues federal regulations so violations are addressed by federal authorities. It is possible for a user to configure an ISM band device with an antenna configuration that violates the part 15 rules that apply to these devices. In that case, it is the responsibility of the operator, or installer if the configuration is professionally installed, to address the problem. It is easy to imagine this situation leading to a significant service outage.

It is also easy to imagine that no one could detect the interference. However, ham radio operators have license privileges in all the 900 and 2.4 GHz bands. These operators are technically inclined and often have as part of their station apparatus easy ways to measure the interference in the band they are operating in. Additionally, a wireless service provider, which has SLAs with its customers, may observe that an interference from an unlicensed source is preventing their SLA provisions from being met. With the financial incentives provided by the SLA, the provider would be quick to address the cause of the interference.

It is often tempting to increase the antenna gain to address complaints about device performance. In most wireless data applications, as the power that the receiver delivers moves below certain thresholds, the throughput performance degrades. An easy fix is just to add a bit more gain. As the antenna gain is increased, FCC part 15 rules require that the output power where the antenna is connected to the device be reduced, thereby limiting overall power density in the spectrum used. If this is not done correctly, interference may result to a licensed operator and could result in changes being required to the unlicensed system deployed. As a rule of thumb, if only omnidirectional antennas are used and those antennas have no more than 6 dBi of gain, there is no reason to be concerned. A summary of the part 15 rules that must be followed are shown below. You can obtain the complete document from the FCC.

Omnidirectional antenna applications in the 900 MHz and 2.4 GHz bands:
1. Maximum transmitter output is 1 W.
2. Maximum EIRP is 4 W. For every additional dB of antenna gain over 6 dBi, the transmitter output must be decreased by 1 dBm.

Directional antenna applications in the 2.4 GHz band:
3. For point-to-point operation, for every 3 additional dB of antenna gain over 6 dBi, the transmitter output must be decreased by 1 dBm.

All antenna applications in the 5.4 GHz band are acceptable; there are no current rules governing this band. This situation is likely to change when 5.4 GHz devices appear in the marketplace in the coming years.

■ PROPAGATION MODES

The propagation characteristics listed in Table 1-4 show how the electromagnetic waves propagate through the atmosphere in different ways in unguided channels. There are three modes of propagation that waves can take: ground wave (GW), sky wave (SW), and line of sight (LOS). These modes of propagation are depicted in Figure 1-7. Each will be examined, and some simple calculations will be performed.

Ground Wave (GW) Propagation

In the GW mode of propagation, the electromagnetic waveform essentially follows the contour of the earth. All transmissions below about 3 MHz use this method to travel from source to destination. The most common example would be broadcast AM radio, which uses the frequency band 540 kHz to 1,710 kHz. This propagation mode can be used to communicate over long distances if both the source and destination are on the earth. Since electromagnetic waves in this frequency band are scattered by the atmosphere, they do not penetrate the upper atmosphere. Therefore, these frequencies are not used for earth-to-space or space-to-earth communications.

GW propagation uses the diffraction properties of the lower atmosphere to propagate from source to destination. The diffraction can be thought of as scattering the electromagnetic wave. This frequency band also has the dubious distinction of being the one

Figure 1-7
GW, SW, and LOS Modes of Propagation (*Reprinted by permission of Pearson Education, Inc., Upper Saddle River, NJ*)

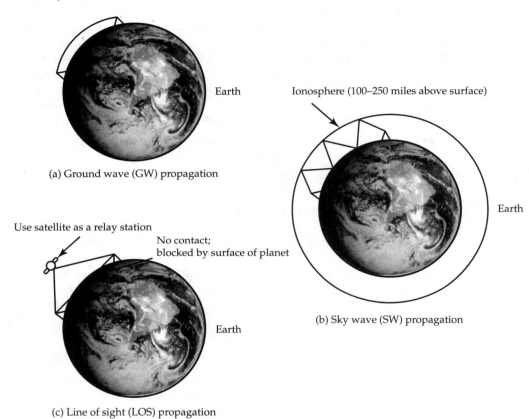

(a) Ground wave (GW) propagation

Earth

Ionosphere (100–250 miles above surface)

Earth

(b) Sky wave (SW) propagation

Use satellite as a relay station

No contact;
blocked by surface of planet

Earth

(c) Line of sight (LOS) propagation

most susceptible to variation due to the time of day or presence of the sun. Since it is most dependent on the characteristics of the atmosphere, the frequency band diffracts through it. If the atmosphere changes, it will affect how the waves travel. The sun has a profound effect on the upper atmosphere, so a different behavior of transmission in the frequency band during the day and evening is observed.

Sky Wave (SW) Propagation

SW makes use of the reflecting properties of the earth's atmosphere at altitudes from 100 to 250 miles for frequencies in the band 3 to 30 MHz. When an electromagnetic wave reflects, or bends, part of its energy passes through the atmosphere and is lost in space. However, most of it is bent back toward the surface of the earth. The wave essentially skips along, jumping over large areas, only to appear again several hundred miles away. The areas that are skipped have a special name: each is called a skip zone. This propagation mode allows waves transmitted in this frequency range to be heard on the other side of the globe, opposite the point from which they were transmitted, which explains why this band is so popular for long-distance communications.

Line of Sight (LOS) Propagation

LOS is used by frequencies above 30 MHz. Here the atmosphere of our planet does not refract or bend the electromagnetic wave hardly at all. Typically less than 1 percent of the power is reflected back toward the surface of the earth. Therefore, these waves travel in a straight line and pass through any portion of the atmosphere easily. The two primary bands of interest in wireless communications are the UHF and SHF bands. Both use LOS propagation.

This characteristic of waveforms with a frequency exhibiting LOS propagation is critical to communications from the earth's surface to beyond its atmosphere. Similarly, the straight-line propagation is helpful when communicating between stations located in outer space. When communicating between two terrestrial sites, however, long-distance communications is made difficult by the curvature of the earth. Therefore, when using frequencies that propagate using the LOS mode on earth, towers are often required to raise the antennas to a sufficient height so that a physical line of sight exists between them.

Because of this phenomenon, LOS radio horizon has been developed. This equation calculates the needed height of an antenna. Either the transmit antenna alone can be raised, or both the transmit and receive antennas can be raised. For most consumer applications, it is impractical to require the consumer to raise a tower. The receivers must be able to see the transmitters, which explains why the transmit antennas for broadcast FM, broadcast television, cellular telephones, and others who transmit using these frequencies for their carrier waves need to use tall towers. To calculate how tall they must be, use Equation 1-13.

$$d = \sqrt{2h}$$

(1-13)

$$d = \text{LOS radio horizon (miles)}$$
$$h = \text{antenna height (feet)}$$

This equation does not use the same units for both sides. It is a handy rule, often used by engineers who are practiced at ignoring the fact that the units are not the same once it has been demonstrated that the equation produces practical results. Please be careful when applying it.

■ **EXAMPLE 1-13**   You must calculate how tall an antenna must be for a new television station. The only data you can get is from the marketing department, and they say that it must reach customers up to 50 miles away. You must apply Equation 1-13:

$$d = \sqrt{2h} \text{ or } h = \frac{d^2}{2}$$

$$50 = \sqrt{2h} \Rightarrow h = \frac{50^2}{2} = 1250 \text{ feet}$$

■ **EXAMPLE 1-14** Another question has appeared in your in-basket. Due to local ordinances, the highest tower possible on this site is 1,000 feet. What is the impact of this requirement? Are there any other solutions that will let the original viewing area be maintained? You make the following calculation:

$$d = \sqrt{2(1,000)} = 44.7 \text{ miles}$$

You told them that 1,250 feet were required. You ask yourself, "What if all viewers used antennas on the tops of their roofs?" Assume a roof height of 3 meters and that the antenna mounted on a pole will add another 1 meter. What difference will this make? Remembering that 3.2 feet equals 1 meter, produce the following:

$$d_1 = \sqrt{2(1,000)} = 44.7 \text{ miles}$$
$$d_2 = \sqrt{2(4)(3.2)} = 5.1 \text{ miles}$$
$$d_{\text{total}} = d_1 + d_2 = 49.8 \text{ miles}$$

It's not quite up to specification, but it's close. Maybe some people on the fringe will build their antennas just a little higher, or maybe they will add a preamplifier since they are at the extreme limit of reception anyway. You make it clear in your presentation to your manager what the engineering tradeoffs were.

■ **EXAMPLE 1-15** Another good question to ask is, "How much lower can one make a transmit antenna if one uses a receive antenna?" Assume the transmit antenna has an initial height of 100 meters. Then compute the new height if a receive antenna 10 meters high is added to the communication system.

$$d = \sqrt{2h} = \sqrt{2(100)(3.2)} = 25.3 \text{ miles}$$
$$25.3 \text{ miles} = \sqrt{2h_{tx}} + \sqrt{2h_{rx}} \Rightarrow \sqrt{2h_{tx}} = 25.3 - \sqrt{2(10)(3.2)} = 25.3 - 8 = 17.3$$
$$h_{tx} = \frac{(17.3)^2}{2} = 149.6 \text{ ft} = 46.8 \text{ meters}$$

As you can see in Example 1-15, the addition of a 10 meter receive antenna saved 100 − 46.8 = 53.2 meters in height for the transmit antenna. This example illustrates that it is always more efficient to construct two smaller towers, one for transmitting and one for receiving, than it is to construct just one big transmit tower where possible.

■ **ISOTROPIC POWER RADIATION**

Equation 1-14 is taken directly from the inverse square law of physics for radiation in an isotropic medium. *Isotropic* means that everything is the same everywhere you look. This

Figure 1-8
Isotropic Radiation from an Antenna

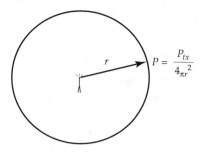

situation is true only for free space, but it is a good assumption to make for determining the maximum distance that a transmitter and receiver can be separated and still expect to communicate. It is always true that this calculation gives a maximum distance.

$$P_{rx} = \frac{P_{tx}}{4\pi r^2}$$ (1-14)

P_{rx} = power arriving at receiver
P_{tx} = power transmitted
r = distance from transmitter

To understand Equation 1-14 better, recall that the surface area of a sphere is given by the equation:

$$SA_{\text{sphere}} = 4\pi r^2$$

If the power is radiating isotropically, that is, in the same way in every direction, it should expand in a sphere. Then the power should fade away as the volume of the sphere grows because the same power must now cover a larger surface area. If this equation still does not seem clear to you, imagine a day at the beach and your child wants a beach ball inflated. There are two balls that you can choose to inflate, the first is a small ball, the second is a larger one. Which is easier, or requires less power, to inflate? The small ball is easier to inflate because when it is completely inflated, its surface area is smaller. Make the analogy between the power radiating and your breath, and the surface area of the beach ball is the surface of the energy radiating out from an antenna. The smaller the radius of the beach ball, the less power is required to inflate it. Figure 1-8 illustrates the concept.

■ **EXAMPLE 1-16**   How far away from a transmitter can the receiver be when the transmitted power is 1,000 watts and the minimum receive power necessary is 1 mW?

$$P_{rx} = \frac{P_{tx}}{4\pi r^2} = 10^{-3}\,\text{W} = \frac{1,000\,\text{W}}{4\pi r^2} \Rightarrow r = \sqrt{\frac{1,000}{4\pi 10^{-3}}} = 282 \text{ meters}$$

Remember that this distance is a maximum. For most applications, it will be significantly less because of the relatively bad assumption that your locality is as free from obstruction as free space, or the possibility that you are not using an isotropically radiating antenna at the transmitter. Some types of antennas direct more of their power in one direction than another. Consequently, the receiver can be farther away in some directions than it is in others. For more detail on this subject, refer to Chapter 8.

It is often interesting to calculate just how far specific transmit powers will carry. The following example will illustrate how far an isotropically radiating antenna will carry for a typical low-power radio station. While different receivers need different amounts of power to operate well, the estimate in Example 1-17 is what one would need for a typical situation.

■ **EXAMPLE 1-17** Assume a receiver sensitivity of 100 μ V, an antenna resistance of 100 Ω, and a transmit power of 1 kW. With this information, calculate the necessary receive power and radius.

$$P_{rx} = \frac{E^2}{R} = \frac{(100 \times 10^{-6})^2}{100} = 1 \times 10^{-10}\,\text{W} \Rightarrow r = \sqrt{\frac{1,000}{4\pi 10^{-10}}} = 9 \times 10^5\,\text{meters} = 540\,\text{miles}$$

Remember that this radius is a maximum. In most cases, the signal will not reach nearly this far.

■ **SUMMARY**

This chapter has introduced a wide range of issues important to the study of wireless communications systems. The purpose of this chapter is to bring together several basic concepts. The reader can choose which need to be studied more closely and which are already familiar.

REVIEW QUESTIONS

1. Discuss why the HF band is unique.

2. In what band do most cellular and wireless LAN systems operate?

3. How are frequency and wavelength related?

4. Discuss the relationships among bandwidth, information capacity, and the signal-to-noise ratio.

5. Discuss the relationship between the signal-to-noise ratio and information capacity.

6. What are the key advantages and disadvantages to operating in unlicensed bands?

7. What are the key advantages and disadvantages to operating in a licensed band?

8. Discuss the importance of FCC rule 15.

9. Compare all three propagation modes. Which is preferred for cellular systems? Why?

10. What antenna configuration will maximize the range for systems using LOS propagation mode?

11. Find the wavelength of each of the following frequencies.
    a) 10 Hz   c) 10 MHz
    b) 1 kHz   d) 1 GHz

12. Find the frequency of each of the following wavelengths.
    a) 1 cm   c) 1 km
    b) 1 m   d) 1 mile

13. Find the bandwidth for each of the following pairs (lower, upper) of signal frequencies.
    a) 1 kHz, 10 kHz   c) 0 Hz, 3 kHz
    b) 10 MHz, 20 MHz   d) 1 GHz, 1.1 GHz

14. For each of the following bandwidths, find the upper frequency given the lower frequency (BW, lower).
    a) 1 kHz, 1 kHz   c) 10 kHz, 1 GHz
    b) 1 kHz, 1 MHz   d) 60 Hz, 0 Hz

15. Find the number of decibels represented by each of the following numbers.
    a) 10   c) 10,000
    b) 1,000   d) 1,000,000

16. Find the gain or attenuation exhibited by a communications system that has the following input and output powers (in, out).
    a) 10 mW, 10 W   d) 5 V at 50 Ω, 10 V at 50 Ω
    b) 10 μ W, 10 W e) 10 V at 100 Ω, 5 V at 50 Ω
    c) 1 W, 1 mW   f) 10 V at 1 mA, 1 W

17. A communications system has a 30 dB gain. Find the output power for each of the following input powers.
    a) 1 W   d) 5 V at 50 Ω
    b) 1 mW   e) 50 V at 600 Ω
    c) 1 μW   f) 10 V at 1 mA

18. Find the S/N ratios for the following pairs of signal and noise levels (S, N).
    a) 1 W, 1 mW   c) 1 mW, 10 W
    b) 100 W, 1 mW   d) 1 V at 50 Ω, 25 V at 50 Ω

19. For a 1 W input power, find the power ratios expressed in decibels for the following output powers.
    a) 2 W   c) 10 W
    b) 5 W   d) 100 W

20. Translate the following values into 50 Ω dBm.
    a) 100 dBm   d) 3.25 V (p-p)
    b) 20 dBm   e) 2 dB
    c) 0.4 V (RMS)   f) −10 dB

21. Find the output power in dBm for each input power. Assume that the gain of the circuit is 3 dB.
    a) 10 dBm   c) 3 dBm
    b) −10 dBm   d) −30 dBm

22. What happens to the information capacity for each of the following situations?
    a) Channel bandwidth doubles.
    b) Information capacity doubles.
    c) Channel bandwidth halves.
    d) Information capacity halves.

23. What frequency band and propagation mode does each of the following service use?
    a) Cellular telephone
    b) Broadcast AM radio
    c) Broadcast VHF television
    d) Broadcast FM radio
    e) Long-range terrestrial communications
    f) Satellite communications

24. How high would a transmitting tower have to be for each of the following distances from the tower? Assume that there is no receive antenna.
    a) 1 mile   c) 40 miles
    b) 25 miles   d) 200 miles

25. For each of the distances below, calculate the new height of the transmit antenna required with the addition of a receive antenna 50 feet high.
    a) 1 mile   c) 40 miles
    b) 25 miles   d) 200 miles

26. Discuss the relationship between frequency and wavelength.

27. Discuss the concept of information capacity and relate it to bandwidth.

28. Discuss how the signal-to-noise ratio affects information capacity.

29. Why is FCC rule part 15 important to unlicensed transmissions?

Multiple-Access Methods

■ INTRODUCTION

This chapter summarizes the techniques of wireless multiple access, commonly referred to as the air interface. Discussion begins with a summary of the two major wireless topologies and how some minor enhancements to the traditional concepts of simplex and duplex best describe many wireless systems. The following sections of the chapter discuss in some detail the forms that a multiple access wireless system may use, including FDMA, TDMA, FDM/TDMA, and CDMA, and the two forms of spread spectrum, DSSS and FHSS. This important chapter lays the foundation for much of the material in Parts Two and Three of the text.

■ WIRELESS MULTIPLE ACCESS

One of the most fundamental concepts of wireless networks is that they are essentially broadcast by nature. Any radio frequency transmission from an omnidirectional antenna radiates out in all directions. Through the use of directional antennas, this energy can be directed in one preferred direction, but the transmission format is still a broadcast. (See Chapter 8 for more information on antennas and their characteristics.)

For some wireless network elements, the broadcast feature is more difficult to see or understand. Satellite links have a multicast capability, but it is achieved through spot beams that broadcast over a small geographical region using specialized antennas. Some terrestrial microwave systems are called point to point, but these systems are just broadcasts that are narrowly focused, again using specialized antennas.

This fundamental nature produces a couple of interesting results. First, the number and location of the receiving stations is unknown. Second, without knowing the location of the receiving stations, the transmitter cannot know the power level at the receivers. Of course, most LAN networks, using a shared medium, wireline or wireless, are fundamentally broadcast in nature, which explains why network topology and access methodologies are so fundamental for understanding them. The next sections focus on these two topics as they are applied to wireless networks.

■ TOPOLOGIES

Two dominant topologies are used in wireless networks, star and distributed. Both are logical topologies, illustrating logical relationships. But there is only one physical topology, or signal flow topology, in wireless networks: the broadcast, or star, physical topology.

Star Topology

The star topology is the most widely used topology in both wireline LANs and wireless networks. The star topology features a central hub. In wireline networks, this topology is often referred to as the client-server logical topology. To understand this concept, think of every node as acting like both a hub and a station. Many wireless systems explored in this book, including terrestrial microwave links, cellular telephones, the IEEE 802.11 wireless LANs, as well as all VSAT satellite networks, use this topology. Several important reasons underly its use.

First, a centralized hub allows easy connection to other networks. It can function as a gateway, as in most cellular telephone systems, or as a bridge, as in the IEEE 802.11 wireless LAN. Second, power control in wireless transmissions is often a critical issue. Chapter 1 showed how a radio signal fades with distance from the transmitting antenna. The central hub allows a single point of control for all transmissions from it as well as a convenient and logical location to coordinate network management. This silver lining of the star topology is, of course, also the sow's ear: the single point of control is also a single point of failure.

The last major characteristic of a star architecture in a wireless environment is that a round trip transmission from the star hub to a destination station and back again is a maximum of two hops. A hop in wireless terminology is any single transmission link, very similar to the way it is used to refer to inter-router links when discussing WAN architectures.

Distributed Topology

This topology is often referred to as the peer-to-peer logical topology. It is used primarily by advanced satellite networks where satellite-to-satellite communication is possible and by ad-hoc networking applications with no centralized hub. Bluetooth is a good example of a system that uses this topology.

Figure 2-1
Logical Topologies: (a) Star Topology; (b) Distributed Topology

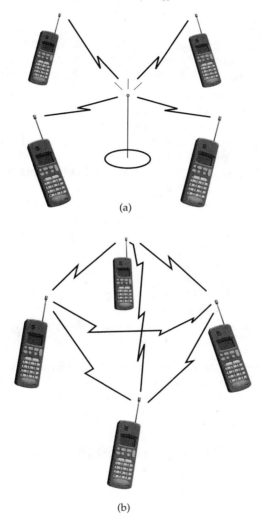

(a)

(b)

One primary advantage to this approach is the very fast connectivity between any two stations. There is no centralized hub to pass the signal through. Another way of stating this is that the minimum delay from any node to any other is one hop. The other main advantage is that there is no single failure point, as is the case in the star topology.

The two main disadvantages of this approach both stem from the fact that there is no central hub. There is no single easy location for control and management. Because there is no logical location for the gateway or server, one of the nodes must be designated to provide this function, and many applications require this designation. Figure 2-1 illustrates the two logical topologies.

■ SIMPLEX AND DUPLEX

Communication systems are often discussed with the implicit assumption that the information transfer is in one direction only. This assumption is not generally true. Many times it is an advantage, or a requirement, to have two-way communication. Try to imagine telephone conversations if the telephone communication system was one way. This observation leads us to divide all wireline communication systems into three types.

1. Simplex (SX) communication systems transfer information in one direction, all the time. Broadcast radio is an example of simplex communication. The radio station transmitter sends a signal to your radio. Your radio does not communicate back to the transmitter. Broadcast television is also a familiar example of a simplex communication system.
2. Half-duplex (HDX) communication systems transfer information in two directions, but only in one direction at one time. These systems usually alternate communication flow, first in one direction and then in another. A good example of a half-duplex communication system is a walkie-talkie. One person presses the transmit button, speaks, and says over; then the other person presses his or her transmit button, speaks, and says over to transfer the communication channel back to the first speaker. Each person can speak, but that transfer can take place in only one direction at a time.
3. Full-duplex (FDX) communications systems are those systems where information is transferred in both directions at the same time. The traditional telephone system is a good example of a full-duplex system. Both users can speak at the same time, even if this reduces the intelligibility of the conversation.

For an illustration of the three modes of communication, see Figure 2-2.

In a wireless environment, these three simple modes take on a new level of complexity. In Figure 2-2 and the text discussion, the channel was assumed to be common for communication in either direction. Many wireless systems use two different frequency bands for com-

Figure 2-2
Simplex and Duplex (Half-Duplex and Full-Duplex) Modes of Communication

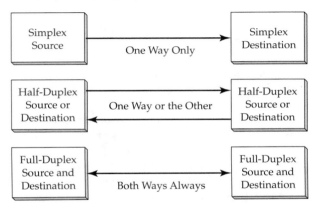

munication in each direction. For example, in most first- and second-generation cellular telephone systems, the frequency used to communicate from the cellular telephone to the base station is offset from the frequency used to communicate from the base station to the cellular telephone by 45 MHz. As a result of this offset, the half-duplex and full-duplex situations need to be refined. The following four techniques are used for this refinement. The definitions have been modified to accommodate a remote mobile unit and a fixed base station.

Simplex (SX): One frequency is used and only the base station can transmit to the mobile. The mobile is only a receiver. This situation is exactly similar to the previous simplex definition; listening to the radio in a car is a good example.

Single half-duplex (SHDX): One frequency is used; however, the base and the mobile can both transmit. This concept is exactly similiar to the previous half-duplex definition; a walkie-talkie captures the idea.

Double half-duplex (DHDX): Two frequencies are used, and both the base and the mobile can transmit. The fundamental refinement in this technique over the previous SHDX is that two frequencies are used. The base, when transmitting to the mobile, uses one frequency; the mobile, when transmitting to the base, uses another. In this approach, however, the communication is still half duplex. The equipment is such that each unit may only transmit or receive, not both simultaneously. Thus, the equipment is still half duplex in nature.

Full duplex (FDX): Either two frequencies or two physical lines are required, and both the base station and mobile station can transmit and receive simultaneously. For true simultaneity in a wireless system, two antennas are required at each device because a single antenna cannot be used both to transmit and to receive at the same time.

A good question to ask is, "Which of these environments describes the cellular telephone system?" You might say that it must be the FDX system because you can simultaneously speak and receive, just like you can with a wireline telephone. However, there is only one antenna on a cellular telephone. In all digital cellular telephones, the transmit and receive functions do not operate exactly simultaneously, and hence the DHDX model is the correct answer. The two functions are offset slightly in time so that a single antenna can be used for the mobile device. At the base station, two antennas are occasionally used to enhance overall performance, but the use of them on any particular channel is not quite simultaneous.

This slight offset between transmit and receive functions is a major cost savings in the manufacturing of cellular telephones. It also conserves battery power by lowering the peak power requirements that the battery must be designed to achieve. Separating the two functions allows power to be diverted to either the transmit or receive function and not both at the same time. While the transmit function by far consumes the most power, not having it occur simultaneously with the receive function enhances battery life.

■ AIR INTERFACE

In this section, we discuss the three basic ways that the air interface of multiple access wireless systems is performed. Specific system attributes are covered in Chapters 11 to 13.

Figure 2-3
FDMA

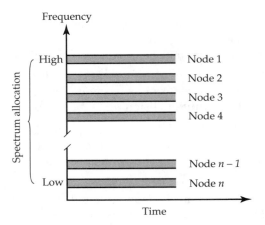

FDMA

The basic principle of frequency division multiple access (FDMA) is that each station is assigned a unique frequency band used by that station to communicate. In the FDMA approach, *each signal utilizes a small portion of the bandwidth of the common channel continuously*. FDM multiplexes in frequency by taking several data sources and shifting them so that they are adjacent in frequency. By combining the bandwidths of each channel and adding them together, you can obtain a rough idea of the total bandwidth of the shared common channel. This channel is used to send the entire multiplexed group of data sources. Figure 2-3 illustrates the idea.

FDMA makes use of the fact that if you have a wide bandwidth channel, you can FDMA many narrow bandwidth channels into it. FDMA is the simplest and oldest multiplexing technology. Its main disadvantage is the complexity required to implement because it is inherently an analog approach.

TDMA

The basic principle of the time division multiple access (TDMA) approach is that each station is assigned a unique time slot in which that station communicates. In the TDMA approach, *each signal utilizes the entire bandwidth of the common channel for short periods of time*. TDM multiplexes in time by taking several data sources and shifting them so that they are adjacent in time. By combining the duration of each channel and adding them together, you can obtain a rough idea of the total frame time of the shared common channel. This channel is used to send the entire multiplexed group of data sources. Figure 2-4 illustrates the idea.

The greatest advantage of TDMA is that it is the access methodology used in the PSTN. Interface into these systems is straightforward and uses similar technology. It requires strict time synchronization, but this technology is well developed. The availability of GPS signals renders this potential concern secondary.

Figure 2-4
TDMA

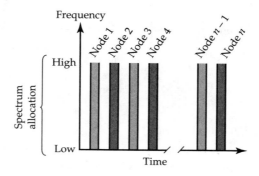

Figure 2-5
Hybrid FDM/TDMA

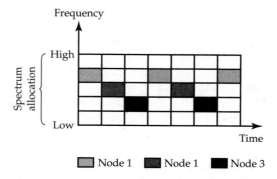

Hybrid FDM/TDMA

Hybrid FDM/TDMA systems combine the two above techniques to allow the time frequency space to be shared by several stations simultaneously. Note that in each of the above classic techniques, one dimension of the available frequency and time is completely dedicated to a single station. In the FDMA approach, each signal uses a portion of the bandwidth for a continuous period of time. In the TDMA approach, each signal uses the total bandwidth for a short period of time. In hybrid FDM/TDMA systems, neither of these situations exist.

Instead of a signal being static in one dimension, either frequency or time, in hybrid approaches it is dynamic in both. It occupies a portion of the available frequency for a short time. Essentially, the available bandwidth is divided up into several narrow channels, then each of these channels is dedicated to a TDMA frame. This process results in a graph, shown in Figure 2-5, that is similar to those in Figures 2-3 and 2-4.

CDMA

The basic principle of code division multiple access (CDMA) is that each station is assigned a unique code word used to distinguish between messages. In CDMA, neither

Figure 2-6
CDMA Modulator Block Diagram

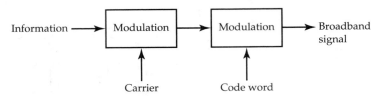

Figure 2-7
CDMA Demodulator Block Diagram

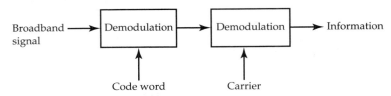

the frequency band nor time slot is used to distinguish a station. Rather, stations are distinguished by the code word assigned to them. CDMA allows simultaneous use of the frequency and time space by all signals and is the preferred choice today for an air interface to a wireless station.

In CDMA, each station transmits across the entire allocated spectrum whenever it wants to. Thus, many different signals can occupy the same frequency space at the same time. As you will see in the following sections on spread spectrum, CDMA and spread spectrum are intimately related. In fact, CDMA can be seen as just another way of looking at spread spectrum. The CDMA code word used is the chip word discussed in the next sections. It is discussed as a code word if you think of these systems as CDMA systems; it is discussed as a chip if you think of these systems as spread spectrum systems. Either perspective is equally valid.

Fundamentally, the information spectrum is modulated again by the code word. One way to visulize the process is as a second modulation step. This process is illustrated in Figure 2-6 and is known as spreading. Immediately, the links to spread spectrum are apparent. The number of bits in the code word determine the amount of spreading or bandwidth that the signal uses. As mentioned above, in CDMA, each station occupies the same frequency band at the same time. The stations are differentiated from each other by the use of the code word. If the code words are significantly different from one another, the receivers will be able to use that difference to pick out the signal intended for them and ignore the others. Picking the intended signal is done at the receiver by passing all received signals through an inverse of the second modulation step. If the received signal has not been spread by the same code word that the receiver expects, this first step in demodulation results in very low energy passing through to the second demodulation step. The signal is regarded as noise and ignored. This process is shown in Figure 2-7.

As you might imagine, the choice of the code word is critical. Longer code words mean more bandwidth and a better ability to distinguish one transmission from an-

Table 2-1
Example 2-1 (Autocorrelation)

Time Shift	Bit 1	Bit 2	Bit 3	Bit 4	Autocorrelation
0	1	1	−1	−1	
0	1	1	−1	−1	4
0	1	1	−1	−1	
+1	−1	1	1	−1	2
0	1	1	−1	−1	
+2	−1	−1	1	1	0
0	1	1	−1	−1	
+3	1	−1	−1	1	2
0	1	1	−1	−1	
−1	1	−1	−1	1	2
0	1	1	−1	−1	
−2	−1	−1	1	1	0
0	1	1	−1	−1	
−3	−1	1	1	−1	4

other. Usually the code word length is limited by the bandwidth allocated to the system. All units in any one system use the same length code word. As a result, code word choices usually mean selecting words of a certain length so that optimum performance results.

There are two key items to choosing code words in a set. The first applies to each individual code word. The second applies to the code word set as a whole. The first is to what degree each individual code word autocorrelates to itself. Autocorrelation is the multiplication of one code word by the same code word shifted in time. If a code word does not autocorrelate with itself for any time shift other than zero, it is a better code word. In other words, a code word works best if a large autocorrelation result is obtained only at a zero time shift with the same code word.

Autocorrelation is an important characteristic for wireless communications because one of the most common problems is receiving delayed versions, or echos, of a signal at the receiver. This delay can lead to significant performance limitations and has been the primary limiting factor on performance when wireless systems are applied in reflective environments such as buildings, factories, warehouses, etc. Specific code words perform better than others. Gold codes are well known for their excellent autocorrelation properties.

■ **EXAMPLE 2-1** This example will illustrate the autocorrelation process for a short 4-bit code word. Table 2-1 shows the autocorrelation result for each time shift. The number in the far right column represents the autocorrelation of the two time shifted, but identical, codes being correlated. The best codes result when the value in that column is large only for the zero time shift case. The code word is 1, 1, −1, −1. Each pair of rows in the table shows the code word unshifted paired with the code word shifted, according to the far left-hand column. The far right-hand column shows the autocorrelation result, which is just the number of bit positions where the two rows are identical. In the first pair of rows, a time shift of zero is shown; thus, every bit is identical, and the maximum autocorrelation of 4 results. An ideal result would be when all other time shifted autocorrelation results have the value of zero.

As you can see, 1, 1, −1, −1 is a fairly poor choice for a code word. There are two time shifts where the autocorrelation is zero, a highly desirable result, but there are four time shifts where the autocorrelation result is half the peak. As a rule of thumb, a good code word will have the largest autocorrelation peak of at least three times any other, and that peak should occur only at a time shift of zero.

As indicated earlier, the code words that make up a set must also have some properties relative to each other. If the receiver is receiving delayed replicas of transmissions intended for it, it is likely it is also receiving delayed replicas of transmissions intended for other stations. For optimum performance, the code word set as a whole should also exhibit a property known as orthogonality. A code word set is said to be orthogonal when any two codes are multiplied by each other and the result is zero. Another way of looking at orthogonality is to view it as a kind of autocorrelation across the entire code set. The best known code set that exhibits this property is the Walsh code family, which is used in several cellular systems that employ CDMA.

As mentioned above, however, the length of the code word directly affects the bandwidth of the system. One interesting property of orthogonality is that, in any family of codes or code word set, there can be only as many orthogonal codes as there are bits in the code word. Therefore, since the number of bits in the code word is defined by the system spectrum allocation, and the number of orthogonal code words in any set is limited by the number of bits in the code word, this situation indirectly limits the number of code words in any code word set because regulatory agencies set limits on the bandwidth of any system.

As a result, most wireless systems cannot use true orthogonal codes because they would set too low a limit on the number of stations. A compromise is usually reached by using nearly orthogonal codes called pseudorandom codes, or PN codes. The number of code words is no longer limited, but interference from other transmissions becomes a greater problem.

If the codes are not orthogonal, some transmissions received, but not intended for a station, will correlate at the receiver at the same time as an intended transmission. Remember, in CDMA, stations transmit whenever they choose, so many simultaneous signals exist in the system frequency allocation. A transmission intended for another station could be mistaken for the intended transmission, especially if the received power of the unintended signal is greater than the intended signal.

This situation requires precise power control so that the power coming to the receiver from any transmitter is very nearly the same. The receiver uses the power of the signal to discriminate between transmissions intended for itself, and the power of the signal is multiplied by the autocorrelation peak. Thus, if both transmissions arrive at the same power, slight differentiation helps the receiver to distinguish between the two signals. The usual tolerance in most cellular systems is about ± 1 dB. Above this threshold, error rates increase dramatically for the reason outlined above.

Commercial CDMA   While the above explanation is the classic description of CDMA, both examples that we will explore in this text, IS-95 cellular and IEEE 802.11 WLAN, use CDMA somewhat differently. In the classic form of CDMA described above, the number of code words is equal to the number of logical channels. In effect, each receiver must be unique: it must have a unique code word implemented in it.

This requirement is fine for military applications where the transmission is intended for only one receiver and cost is not the primary concern. For many commercial systems, however, cost must be considered. For example, imagine if every cell phone that used CDMA as an air interface worked only in a small area, the cell where its code word is being used. In that case, as soon as the cell phone moved out of that particular cell, it would not operate because its code word is not being used in any other cell. A cellular user would have to purchase cell phones for each cell that he or she moved into. This cost, not to mention the inconvenience, would render the cost of cellular telephone service prohibitive.

The other possibility would be to make each cell phone operate with any code word possible. An equivalence exists between the number of code words in a set and the length of the code word. For this option to be feasible in a widely deployed system like a cellular telephone system with millions of users, the code word would have to be extremely long. Each cellular telephone would also have to contain the entire set of code words, each millions of bits long. This kind of memory requirement would make cellular telephones very expensive.

The two major applications of CDMA techniques in the current market are the IS-95 cellular systems and the IEEE 802.11 WLAN systems. Both address this problem of unique receivers in different ways. The IS-95 system uses FDM techniques to complement the CDMA approach. Instead of one wide band transmission, IS-95 limits the bandwidth of the transmission to 1.23 MHz. Essentially, it allocates another nonoverlapping bandwidth for every new set of subscribers. This technique gets around the limitation of code words. The Walsh code family is used with a code word length of 64 bits. Therefore, in every cell where there are more than sixty-four subscribers, the system shifts the entire spectrum up or down by the transmission bandwidth (here, 1.23 MHz) of the system. This shift effectively yields several narrowband CDMA systems, each with sixty-four users per cell. Hence, only sixty-four possible codes exist, and sixty-four is a small enough number for every receiver, or cell phone, to interpret any code that happens to be available in any cell in which the phone is present.

In the IEEE 802.11 WLAN, this idea is carried to its logical extreme. Here, the same code word is used by all stations; hence, every receiver is identical and very cost effective. In a LAN, the intent is a shared medium, with all transmissions heard by all stations on the segment. This feature can be accomplished only if all stations use the same code word. As in any LAN environment, individual stations are differentiated by the MAC address. Hence, in an 802.11 system, the only consideration for the code word is its autocorrelation properties, and the length is defined by dividing the spectrum by the bit rate. This process will be demonstrated below when spread spectrum is discussed in more detail.

Spread Spectrum   Spread spectrum is not actually a unique air interface but refers to one of several implementation approaches of CDMA. Spread spectrum systems get their name from the nature of the broadband transmission resulting from the second step in the modulation process. This process is referred to as spreading the signal, hence, the name spread spectrum. This text will describe the two basic spread spectrum approaches, direct sequence and frequency hopping.

Spread spectrum technology preceded CDMA but is implemented very similarly. It was originally developed for military applications. The basic concept was to find a way to implement secure communications that resisted jamming. This feature was critical for at least two reasons. First, there is an obvious advantage to communicating to units in the field covertly. Second, it was very desirable to have a way for those units in the field to transmit back without their radio signals acting like a beacon for enemy fire.

Spread spectrum technology offered a way to address these two issues, and in just the way you might imagine from the name. It spreads the transmitted spectrum over a much wider bandwidth than would otherwise be required. The bandwidth expansions range from a ratio of about ten at the low end to over a thousand times what would be required at the high end. Most commercial systems spread in the range of ten to fifty times.

This spreading is performed in a similar way to the CDMA approach described above. The signal to be transmitted is modulated in the normal way and then passed through a second modulator, where the original modulated signal is modulated by a second signal. The second modulation spreads the signal to a degree that depends on the character of the second signal.

It is important to note that the second modulation step adds no information content to the signal. Its only purpose is to spread the signal over a very wide bandwidth. Additionally, this technique of spreading the spectrum can be applied to any first-stage modulation technique. Again, it is an entirely separate step with no impact on the information content and it is not represented in the modulated signal, other than spreading out that modulated signal over a much wider bandwidth.

Spread spectrum systems are characterized by a concept called a chip rate, which is just a way to describe how often the sequence of bits repeats. The entire sequence of bits, or chip, must repeat for every symbol of the first modulation step. If the modulation technique is binary, every bit of the input data word is spread by the entire chip. Longer sequences spread the spectrum more than do shorter sequences. In any single system, such as cellular telephony or wireless LAN, all stations will use bit sequences with the same chip rate.

The other main characteristic of merit for all spread spectrum systems is the processing gain. The processing gain gives a direct way of determining the "gain" provided by the spreading process. It is also often referred to as the spreading gain or ratio. This gain contributes directly to the overall system signal-to-noise ratio. The better the processing gain, the better any given system will perform in a given interference environment. Again, while processing gain is a significant characteristic, all stations in any system will feature the same processing gain because it is directly related to chip rate. For any system, it can be estimated conveniently with Equation 2-1, and it is demonstrated in Example 2-2.

$$\text{Processing gain} = G_p = \frac{\text{channel BW}}{\text{symbol rate}} \qquad \textbf{(2-1)}$$

■ **EXAMPLE 2-2**   Find the processing gain of IEEE 802.11 wireless LAN at data rates of 1 Mbps (DBPSK) and 2 Mbps (DQPSK) spread spectrum system. Due to the increased bandwidth efficiency of the 2 Mbps system, the symbol rate stays the same in both cases and is 1 Mbps. IEEE 802.11 systems use a channel bandwidth of 11 MHz. Processing gain is always best expressed as a logarithmic quantity.

$$G_p = \frac{11 \text{ MHz}}{1 \text{ Mbps}} = 11 \rightarrow G_p = 10 \log(11) = 10.4 \text{ dB}$$

Direct Sequence Spread Spectrum (DSSS)   Direct sequence systems get their name from the form of the second modulation step. In these systems, it is always a sequence of 1s and 0s formed in such a way that they appear to be a random set of values. This set of values, called a chip and analogus to a code word, may be orthogonal or pseudorandom. Both *appear* random. This sequence of bits is the signal that is applied in the second modulation step and is shown in Figure 2-8.

There are two steps in a DSSS modulation. The first is the normal modulation that would occur in any broadband system. The second is where the spreading occurs. The sequence of bits referred to above is applied directly to a second modulator. Since the pseudorandom bit stream is applied directly to the modulator, the origin of the name *direct sequence* becomes clear. This process is shown in Figure 2-9.

Figure 2-8
DSSS Bit Representation

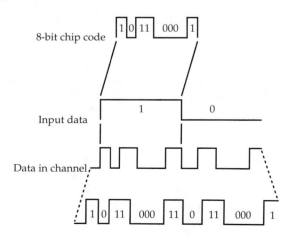

Figure 2-9
DSSS Modulator Block Diagram

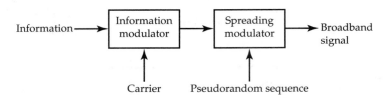

Figure 2-10
DSSS Prior to and After Spreading

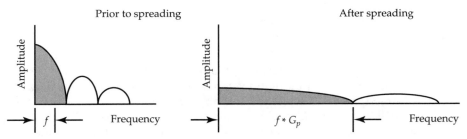

The shaded area is the same, but the signal after spreading is at very low peak powers.

In summary, a DSSS system spreads the original modulated signal out over a bandwidth defined by the processing gain. The resulting spectrum is spread out over the entire bandwidth of the channel. Since the total power in the signal must stay the same whether it is spread or not, the power level at any one frequency location in the spectrum is very low. When in the channel, direct sequence spread spectrum signals appear as low-level noise signals. Only when they are correlated in the receiver does the processing gain come into play and the signal is recovered for conventional demodulation. The processing gain then reduces the power in any one frequency location during transmission and then recovers that same gain in any one frequency location during demodulation. To make this clear, Figure 2-10 illustrates the spectral signal after spreading relative to the signal prior to spreading.

DSSS systems have a built-in error-recovery mechanism. During transmission, if one or more chips is lost due to channel effects, the receiver can employ statistical techniques to recover the data without the requirement of retransmission. The processing gain of DSSS systems depends on the chip length: the longer the chip, the greater the probability that the original data can be recovered.

Finally, to another receiver not designed to receive a DSSS signal, the spectrum appears as nothing more than low-power noise and is ignored or rejected by the front end of the receiver. To another receiver designed to receive a DSSS signal, but not the particular chip that was used to modulate the signal received, the spectrum fails to correlate during the first-stage demodulation. The signal again appears as low-power noise and is rejected. In the case where the receiver has the correct chip, the received signal is en-

Figure 2-11
FHSS Modulator Block Diagram

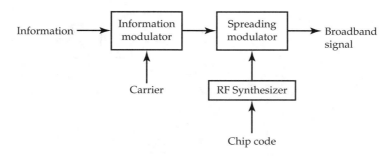

Figure 2-12
FHSS: (a) Prior to Spreading; (b) After Spreading

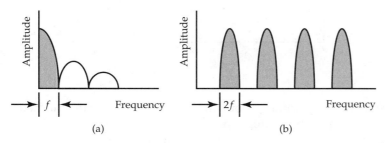

hanced by the processing gain and is passed, reduced in bandwidth and with sufficient power, to the second demodulation step.

Frequency Hopping Spread Spectrum (FHSS) Frequency hopping systems are very similar to DSSS systems in their implementation, and both are very similar to CDMA systems. FHSS systems also use a chip word consisting of a series of bits to spread the spectrum, but in this case, they do not spread it directly. Instead, a sequence of 1s and 0s drives radio frequency (RF) frequency synthesizer that "hops" about in frequency output, depending on the pattern of bits driving it. The frequency location of the signal is determined by the pattern of bits driving the frequency generator. A block diagram of an FHSS modulator is shown in Figure 2-11. Note the similarity to Figure 2-9.

The FHSS spectrum, both prior to and after the second modulation step, is shown in Figure 2-12. Note that each of the individual spectra shown exists in its own moment of time; no two are present simultaneously. Ideally, and over some long period of time, the relatively narrow, original modulated spectrum will occupy every possible frequency location in the entire spread spectrum that has a correspondance to a chip code.

There are two basic types of FHSS systems. Classification into one or the other depends on whether the hopping rate, analogous to chip rate for DSSS systems, is higher or lower than the input data rate. If the hopping rate is higher, then the system is known as a fast frequency hopping (FFH) system. If the chip rate is lower, the system is known as a slow frequency hopping (SFH) system.

Figure 2-13
FHSS Frequency Slot Representation

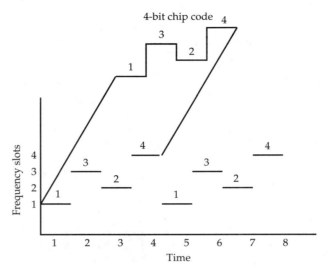

An alternate way of thinking about an FHSS system is for the modulated signal to take on a specific frequency slot for some duration of time and then hop to a new one. Figure 2-13 illustrates this for an FHSS system with a 4-bit chip code defining four discrete frequency locations. In Figure 2-13, the modulated signal cycles through the chip code twice. Note how each bit in the chip code defines a new frequency location.

Just like any spread spectrum system, an FHSS system is characterized by its processing gain. The processing gain is defined in the same way as shown earlier. To a receiver not designed to receive an FHSS signal, the FHSS spectrum appears to be a short duration impulse noise and is usually rejected by the front end of the receiver. To a receiver designed to receive an FHSS signal, but without knowledge of the chip, the front end of the receiver is never in the right frequency location at the right time. To a receiver designed to receive an FHSS signal and with knowledge of the chip, the frequency synthesizer places the front end of the receiver in the right frequency location at the right time and thus gathers the individual pulses that, taken together, comprise the signal. In fact, during World War II this method was used to communicate covertly between the United States and Great Britain.

■ SUMMARY

All wireless systems rely on some form of air interface. The engineering details of each approach may or may not be suitable for every course of study of wireless, but their basic concepts are critical. Hybrid FDM/TDMA and CDMA environments will dominate the future product releases of wireless devices.

REVIEW QUESTIONS

1. What two logical topologies dominate wireless networks?

2. Compare FDMA and TDMA.

3. Discuss the advantages of FDM/TDMA over pure FDMA or pure TDMA.

4. Discuss why precise transmit power control is critical to CDMA cellular systems.

5. Compare and contrast classic CDMA, IS-96 CDMA, and 802.11 CDMA.

6. How do the number of code words and the number of channels change in each form of CDMA?

7. Compare and contrast DSSS and FHSS.

8. Discuss the importance of processing gain.

9. What are the two major topologies used in wireless networks?

10. Discuss the significance of the two half-duplex techniques described in wireline and wireless environments.

11. Compare and contrast FDMA and TDMA as air interfaces.

12. What advantages does a hybrid FDM/TDMA interface offer?

13. Discuss the basic principle of CDMA.

14. How do CDMA cellular networks address the problem of unique receivers?

15. How do IEEE 802.11 WLAN networks address the problem of unique receivers?

16. Explain the significance of processing gain.

17. Describe the spread spectrum technique called DSSS.

18. Describe the spread spectrum technique called FHSS.

19. Compare and contrast DSSS and FHSS.

20. Discuss the similarities of and differences between CDMA and spread spectrum.

Protocol Architectures

■ INTRODUCTION

This chapter explores the protocols and reference models used in wireless and wireline networking. The chapter begins with a review of basic protocol architectural concepts. Next, the OSI, TCP/IP, and SS7 reference models are described using these concepts. A detailed treatment of layer 2 protocols is found in Chapter 4, along with more detail on IP. Chapter 16 examines in detail the functionality and limitations of the TCP/IP protocol stack in a wireless context. The material in this chapter is fundamental to those discussions.

■ PROTOCOL CONCEPTS AND TERMINOLOGY

Layering

Today's computer and communications networks are designed in a highly structured way. To reduce the design complexity, most networks are organized in a series of layers, each built on top of the former. The number, operation, and content of each layer is different for each network. *In all networks, the function of each layer is to provide services to the next highest layer, insulating the upper layer from the implementation concerns of the lower layers.*

This idea is at the heart of any protocol model. By separating the functionality into several specific tasks or layers, it becomes possible to define each layer as a set of operations that stands alone. If these layers are designed correctly, all interaction between the layers is performed at the layer boundaries. This setup greatly simplifies the interaction between layers that is necessary to accomplish the larger job of defining the overall protocol. The set of layers is called the protocol stack.

Layering is exactly like dividing a job into a series of subtasks. For example, imagine that you have just bought a race car and want to enter it in a race. While it is possible to do all the jobs yourself, it would be much better to divide the jobs among various specialists. You might hire a driver to drive it, a mechanic to repair it, a marketer to obtain sponsors, etc. This same idea is applied to the task of building a communications protocol. For a communications protocol, what is divided up is the job of getting messages from the source to the destination. Each layer has its own specialization and expertise, just like each individual of the race team. The driver would be in the position to tell if something might be not quite right with the engine. He would not be the best person on the team to fix the engine, so he would inform the mechanic of the problem and request that the mechanic fix it.

Service Access Point (SAP)

In any protocol model, each layer has its own specialized functionality. When it needs the functionality of another layer, it requests it. It makes this request at the layer interface that exists between each pair of layers. It is important to realize that this interface defines which operations and services the lower layer offers to the upper one.

In protocols (which are really software programs), the responsibility for different tasks is divided into separate subroutines. While there are many subroutines in each layer, the idea of a layer as a subroutine of the entire communications task is a good one. When you write a program to add two numbers and print the result, it is usually a good idea to separate the tasks. You might write a subroutine to add two numbers and another subroutine to print the results. It would not make much sense to have both subroutines be able to print the results.

This idea is central to a service access point. At each layer boundary, there is at least one service access point, or SAP. At the SAPs, all communications between layers occurs. Further, it is always the upper layer that requests services from the lower one, never the other way around. Each layer has one or more services that it can offer to the layer above.

One can think of the service access points as a kind of counter where the customer, the layer above, comes to order lunch. The counter has many items on its menu, and the customer, after reviewing this menu, requests a specific item. Figure 3-1 illustrates the layer interface, or boundary, and the location of the SAP.

Service

As you may have realized from the previous discussion, protocol terminology has a special use for the word *service*. A service is the function provided to the layer above that

Figure 3-1
SAP Interface

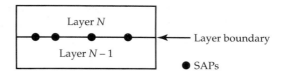

uses the services of the layer below. A layer protocol functions by having each layer provide service to the layer above through the services available from the layer below. Each layer provides its service by passing special units of exchange, called interface data units, or IDUs, across the layer boundary at SAP locations.

The layer above is called the service user and the layer below is called the service provider. In Figure 3-1, the layer N would be the service user and the layer $N - 1$ would be the service provider. It is important to make four key points about services and how layer protocols communicate using them:

1. While layers request services from the layer below, each layer communicates with its companion layer. For example, the source location network layer communicates with the destination location network layer through the services provided by the layers below each of them. This communication is called virtual communication. In virtual communication, the two layers communicating are known as peers; that is, they communicate using the same protocol as their peer layer. Two peer layer protocols maintain a virtual communication. This communication uses the protocol that the layers have in common.
2. Services are made up of operation primitives, of which there are four classes: requests, responses, indications, and confirmations.
3. Services and protocols are *not* the same. Services are made up of operations; protocols are the set of rules for exchanging information between the source and destination. The source and destination systems use protocols to implement their services.
4. There are two broad types of service a layer can offer to the layer above it: connection-oriented service (COS) and connectionless service (CLS).

Virtual Communications

Suppose there are two computers at two different locations. Each one has a layered protocol stack that it is using to exchange information with the other. Here, the only concern is the communication protocol stack, not the ultimate program running on each system. The protocol stack is broken up into layers, with the functionality of each layer distinct from any other layer. It stands to reason, then, that only that similar layer at the other end of the communication would understand the message from its counterpart. This virtual communication between similar layers is called a layer protocol.

Each layer has its own protocol. For example, the network layers of two systems conduct a conversation using the services of the data link layer. The physical path that the information follows is made possible by the communicating layer requesting services from the layer below it. This setup holds for each successive layer as one proceeds down the network model. At each layer, the peer entities communicate with each other and use the services provided by the layer below to implement that communication.

A protocol communicates with a virtual communication between each similar layer at both source and destination. In Figure 3-2, the service boundaries and virtual communications are shown for two systems using a layered protocol to communicate. The physical path is shown as the bold line; the virtual communication is the dotted line. Three layer protocols are illustrated in this figure: the layer 1 protocol, physical; the layer

Figure 3-2
Virtual Communications

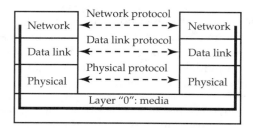

Table 3-1
Service Primitives

Service Primitive Class	Action
Request	Request service
Response	Respond to a request
Indication	Indication of some action in response to a request
Confirm	Confirmation of a response

2 protocol, data link; and the layer 3 protocol, network. The SAP locations are at the boundaries of the boxes delineating the layers of the two network architectures.

Operation Primitives

As was stated earlier, services are made up of four classes of primitive operations, referred to as primitives. These primitive classes are summarized in Table 3-1. Requests and indications are paired, along with responses and confirmations. When one system makes a request, the other system first indicates that it has heard the request. When one system generates a response, the other system must confirm that response.

There are various levels of reliability in these responses and confirmations. Sometimes, silence is taken as the response or confirmation, compared to some active response or confirmation. When the two classes of services that all layers provide are examined, this distinction will become clear.

Additionally, these service primitives must pass across the layer boundaries at the SAP locations. These transfers are always done by units of data known as protocol data units, or PDUs. The convention for naming the units of data exchanged between each layer and the one immediately above it is the layer name coupled with the term *protocol data unit* (PDU). Essentially, each layer appends the notation to its name. For example, the network unit of exchange is the network protocol data unit (NPDU).

Services Versus Protocols

As was stated earlier, services and protocols are *not* the same. Services are made up of operations; protocols are the set of rules for exchanging information between the source and the destination. The source and destination systems use protocols to implement their services. The service can transcend changes in the protocol. Take, for example, the telephone system. The *protocol* for connecting to someone for whom you do not have the telephone number has changed with the evolution of the system. The *service* associated with the need for connecting to someone for whom you do not have the telephone number remains the same.

In the early days of the telephone, if you wanted the telephone number of a friend, you called the operator and asked to be connected to the party by name. The operator, a human being, would physically connect your telephone with the telephone circuit that was wired to, let's say, Mr. Garland. (Some of you may have seen this process acted out in old movies, where the operator sat in front of a patch panel and physically connected the two circuits.)

As technology improved, the operator sat in front of a computer screen and when you called for a number, a database was searched and the operator read the number listed for the party you wanted to reach. Later still, the human being still searched for the number, but the response was replaced by a computer voice providing you with the correct telephone number.

As you can see from this example, the service aspect remained the same but the protocol changed. This example illustrates one of the great advantages of separating the two concepts. Technology is changing faster than ever, and the way of doing something, the protocol, may change quickly. However, the need for the service remains. As long as the service is separated from the protocol, this evolution can proceed naturally. If it is not, systems become outdated and disappear. The reasons for this phenomenon are complex but are captured by the idea that when the service and protocol are tightly coupled, a change in one causes a change in the other. Many times this close relationship and subsequent change alters the very nature of the service, and it becomes unattractive to many of its users.

Protocol Data Units

Each layer communicates to the layers immediately adjacent to it, its peer layers, by passing a unit of data called a protocol data unit, or PDU. These PDUs contain the data and other information necessary for the destination peer layer to interpret the actions of the source peer layer to perform the virtual communication. However, the physical path of the data is first down a network stack, across a physical medium, and up a network stack.

At the source, each layer, on receiving a PDU from the layer above, adds a header to the PDU and passes it down one layer. Similarly, at the destination, each layer strips the header applied by its peer counterpart at the source and passes the remaining PDU up one layer. At each layer, the PDUs have specific names derived from the layer where the header is applied. As stated earlier, the PDU exchanged between network layers in their virtual communication is called the network protocol data unit, or NPDU. The PDU names at each layer are shown in Figure 3-3. For this figure, the OSI protocol model is used.

As you can see, the PDU exchanged between each layer protocol is composed of two logical parts. The first is the data component. At the source, the data component originates from the application running on the station and is destined for the application running on the destination station. From a protocol perspective, the data at each layer is passed down from the layer above.

Figure 3-3
PDU Construction

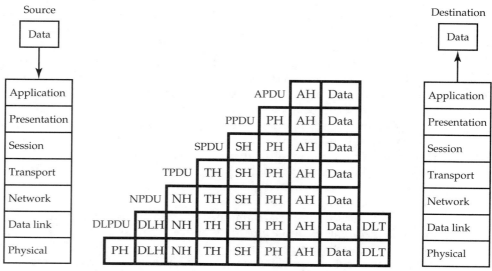

Actual composition of bits transferred through the physical media

The second part is the header, which is added by each layer. Each layer adds a header only, except for the data link layer, which also appends a trailer. In Figure 3-3, the terms DLH and DLT represent the data link layer header and trailer, respectively. Other layers use the first letter of the OSI layer name to represent headers in a similar fashion.

The data link layer is unique in at least two ways. First, as noted above, only the data link layer appends both a header and trailer. This layer is also the only one to append two separate headers, one for the MAC sublayer and the other for the LLC sublayer. These features will be explored in more detail when the data link layer is also discussed in more detail.

■ SERVICE CLASSES

In the following sections, pay careful attention to the use of the words *communication* and *connection*. They have different meanings and it is important to keep the distinction clear. A communication can occur using either of the two broad types of service classifications. A connection is a specific type of service classification. There are two broad types of service that a layer can offer to the layer above it:

1. Connection-oriented service (COS)
2. Connectionless service (CLS)

COS

The most sophisticated service any layer can provide to the layer above it is connection-oriented service (COS). COS is modeled after the telephone system. To talk, you pick up the telephone handset, dial the number, talk, and hang up. In data communications, the service user (the layer above) first establishes the connection, uses it, and then terminates the connection.

Conceptually, the connection acts like a tube. The source places messages in at one end, and the destination takes them out at the other end. Therefore, with COS, the source and destination machines establish a connection before any data is transferred. With COS, first-in-first-out (FIFO) is guaranteed. COS provides the upper layer with a reliable data stream.

Since COS is the most sophisticated service that a layer can provide, its implementation is also the most complex. The more complex the implementation, the more service primitives are required to implement it.

CLS

The other main class of service a layer can offer is the connectionless service (CLS). CLS is modeled after the postal system. To send a message, you address each letter and drop it into a mailbox. Suppose you dropped two letters in the mailbox, one after the other. CLS does not guarantee which letter arrives first. Each letter is addressed individually and each follows its own path to the destination. In data communications, the service user (the layer above) never establishes a connection. Instead, the message is sent and either arrives at its destination or not. Only the destination can tell if the message arrives.

Conceptually, the communication has none of the attributes of a connection. It does not guarantee FIFO. Each message must be addressed individually, and each may follow a different path from the original source to the ultimate destination.

There are two types of connectionless service. The first is one step down in reliability from COS and is called acknowledged CLS. Think of sending a registered letter via the postal service and requesting a return receipt. When the receipt comes back, the sender is absolutely sure the letter was delivered to the intended party. When this service is used, there are no connections established, but each message sent is individually, acknowledged. In this way, the sender knows whether or not a message has arrived safely. If the return receipt has not arrived within a specified time interval, the message can be sent again. In an acknowledged connectionless environment, both the source and the destination know if a message has been received.

The least reliable service is unacknowledged CLS. This service is often called datagram service because it is analogous with the telegram service, which does not provide an acknowledgment sent to the sender. This service consists of having the source send messages to the destination without having the destination acknowledge them. No connection is established beforehand or released afterward. In an unacknowledged connectionless environment, only the destination knows that a message has been received.

If part of a message is lost due to noise on the line, no attempt is made to recover it at the layer. This class of service is only appropriate when the error rate is very low and recovery is left to the higher layers. However, sometimes the delay associated with reliable messaging is unacceptable. Unacknowledged CLS may be appropriate for real-time traffic applications where late data is worse than bad data. One example would be digitized voice service; here, bad data is preferable to late data.

To review, there are actually two "levels of effort" for CLS:

1. Acknowledged connectionless service
2. Unacknowledged connectionless service

Both forms of CLS require that each message sent by the source to the destination carries the full destination address, and each is routed through the system independent of any other message. Normally, the first to be sent arrives first; however, this sequence is not guaranteed. The second message sent may arrive prior to the first one sent.

This situation creates a unique problem for both forms of CLS. A mechanism must be in place to make sure that the destination can determine what message was sent first so it can order them correctly. This task is accomplished by numbering each frame sent over the connection, and the layer providing the service guarantees that each frame is received exactly once and is received in the correct order. The numbers assigned to each frame are called sequence numbers.

You may wonder why a message might be received more than once when using CLS. If an acknowledged CLS is used, then the delay in sending the acknowledgment must be considered. Each message travels through the communications system independent of any other. If the delay is too long, then the source might assume that the message never arrived because it did not receive an acknowledgment and it might send another copy. The destination must have a mechanism to address this situation. The mechanism is the same as the one used to ensure that the messages are in the correct order: the sequence number. The sequence number is the same in any duplicated message, so the destination system knows to discard any messages received with duplicate sequence numbers.

For example, suppose you are having a conversation with someone and suddenly, in the middle of the conversation, the person stops acknowledging your statements. Once the acknowledgments stop, most people will repeat the last few sentences to make sure they were heard before continuing the conversation. Essentially, the acknowledgment did not arrive and so part of the message is sent again.

If an unacknowledged connectionless service is used, this problem with delayed acknowledgments is not a factor. The problem of what to do when the destination does not receive the entire message or receives parts of the message out of order remains. Therefore, the destination must still have some way to make sure it received all the parts of the message sent by the source. Since only the destination can know if all the parts of the message arrived, it must specifically request retransmission of any parts that it missed. Remember, since all this activity occurs as a protocol between peer layers, the higher layers have no knowledge of these operations. It is an example of a virtual communication between peer layer protocols. All the layer above knows is that a message was requested from the layer below and now it has arrived, identified as complete by the layer below.

Different services have different reliability factors, called quality of service (QOS) parameters. These levels of service are summarized in Table 3-2. Each service can be characterized by a quality of service (QOS) parameter. Usually, a reliable service has the

Table 3-2
Service Levels

Service Type	Common Analog	Reliability
Connection-oriented	Telephone	High
Acknowledged connectionless	Return receipt	Medium
Unacknowledged connectionless	Telegram	Low

destination acknowledge the receipt of each message so the source is sure it arrived. One example would be a file transfer. Clearly, such a transfer would be unacceptable if a random byte or two is misplaced. It is generally true that there is a tradeoff between message reliability and end-to-end transmission delay. Acknowledged service always takes longer than unacknowledged service, all things being equal.

■ OPEN SYSTEM INTERCONNECTION (OSI)

A network is really made up of two parts, the underlying media network topology and hardware, along with the protocol stack that runs on the network. The application running on the device attached to the network uses the protocol stack to communicate over the media topology. Protocols can be quite complicated when viewed as a whole, so a standardized approach to analyze and explain them will be used. This approach breaks the protocol into sets of specific responsibilities. Most protocols in use today can be discussed in terms of a model based on this approach. The model is called OSI.

OSI is a reference model that, if followed, offers a way for products manufactured by different vendors to communicate in a standard way. If vendor A and vendor B make a communications product and they need to communicate, they can follow the OSI model and avoid designing a custom gateway. OSI, or any reference model designed for communications, is only concerned with the exchange of information between end systems. If both end systems follow the OSI model, they are called open systems (they are open to communications).

OSI is an excellent way to classify communications products. OSI has all the terms for describing how service and destination systems communicate and the nomenclature for the architecture of that communications functionality. Thus, it is an excellent way of sorting them and knowing what can work and what cannot. OSI is not the only reference model. Others are SNA, ISDN, SS7, and TCP/IP.

OSI is an international standard recognized by the International Organization for Standardization (ISO). As stated earlier, the goal of the standard is to give end systems manufactured by different vendors a way to share information seamlessly. Like all protocol models, the OSI model accomplishes this task using a concept called layering. The layering concept is a way to talk about and design pieces of the functionality that will allow internetworking between different end systems. Essentially, the goal is to divide the

Figure 3-4
OSI Model

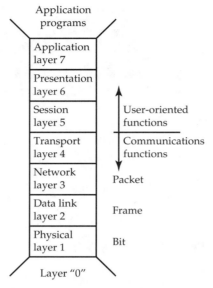

job into several (seven or fewer) subtasks, define what each subtask does, and describe how it communicates with the other subtasks.

Although there are seven layers to the OSI model, these seven layers are really classified into two groups. Layers 1 to 4 cover communications functions, and layers 5 to 7 cover user-oriented functions. All traditional communications services use layers 1 to 4 to accomplish their tasks. For example, the entire telephone system can be described using only the first four layers. Figure 3-4 illustrates the seven-layer OSI model. It also shows the traditional names of the PDUs exchanged at the three lowest layers in the OSI model. The terms *packet*, *frame*, and *bit* are applied, respectively, to the lower three layers: network, data link, and physical. These names are the traditional names of PDUs in many wide area networks (WANs) where the lower three layers are operated by the carrier. With regional bell holding companies (RBHCs), users have little or no control over the WAN, so the need to be able to define a set of primitives that will work independently of the WAN makes this conceptual boundary between user-oriented functions and communications-oriented functions important.

One more layer is associated with communications: layer 0. Layer 0 corresponds to the actual medium used: metallic, optical, or atmospheric. It defines the physical topology of the network. For many of the communications systems explored in this text, the medium will be the air around us.

It is also important to understand that the application layer does *not* house the application programs that run on a computer connected to the network. These programs reside above the communications structure. The OSI model is a way to talk about communications, not a way to define application software. The application layer just provides services to the application program, just like any layer provides services to the layer above it.

Figure 3-5
Physical Layer Virtual Communication

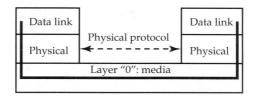

Physical Layer (Layer 1)

The physical layer provides the physical transfer of data bits between adjacent nodes. It focuses on electrical engineering and technology issues almost exclusively. This layer specifies items such as the bit rate, modulation technique, pin-out of interfaces, and other similar issues. The physical layer deals in bits. The important idea is that any modulation technique can be used as a physical layer. Additionally, as mentioned above, any medium can be used for the so-called layer 0. The only requirement is that the modulation technique and the media and topology be compatible. The medium is explicitly not part of the physical layer.

At layer 1, the virtual communication occurs between the modems that implement the physical layer. The physical layer uses the service provided by the layer below to physically transfer the communication between stations. The SAP between the physical layer and the medium is the modem's attachment to the medium. Only the two physical layers can understand and demodulate the information sent from the other. Modulation and demodulation are the service primitives of the physical layer protocol.

The information to be sent is transferred to the physical layer by the layer above, the data link layer. This layer uses the SAP between the data link layer and the physical layer to request that the physical layer send the information. This process is shown in Figure 3-5.

The information transferred between the two physical layers is referred to as bits. As described earlier, there is a convention for units of exchange between all peer layers called protocol data units, or PDUs.

Note that the physical layer is responsible for node-to-node transmission. In a point-to-point environment such as a broadband terrestrial transmission or satellite transmission, the physical layer is point-to-point. In a typical LAN environment, the data link layer, with its media access control (MAC) addressing, is responsible for node-to-node addressing. This situation also holds true in most wireless LANs, such as the IEEE 802.11 system.

For most purposes, and in all shared media environments, the data link layer is therefore responsible for node-to-node addressing. Because most of the systems (except those identified in the previous paragraph) follow this approach, in the next section the data link layer will be responsible for node-to-node addressing.

Data Link Layer (Layer 2)

The data link layer of the OSI model covers the transfer of bits between stations on the same LAN segment. A LAN segment is a unit of a network where all stations share a common address set, in this case, the MAC address. In an environment where stations have a MAC address associated with them, the data link layer is responsible for node-to-node addressing on the LAN segment. Other good examples include transmission between a

Figure 3-6
Message Path Showing Virtual Communication in the Data Link Layer

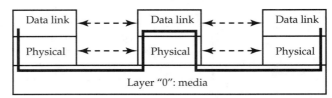

PC and its modem in a packet-switched environment such as X.25, or transmission between two modems in a circuit-switched environment.

Having made the point that layer 2 is responsible for node-to-node addressing on a LAN segment, we should emphasize that the data link layer, like any layer, is providing services to the layer above. In this case, the layer above is the network layer. The network layer also does node-to-node addressing, as we will see later, but on the LAN segment itself, the data link layer actually delivers the frame to the destination node. It provides that service to the network layer.

A good example from daily life might be all the houses on a street. Because all have the same street name, all that is really needed to distinguish one house from another is the house number. To illustrate this concept, Figure 3-6 shows how the frame is passed up to each data link layer because it is the only layer that can recognize the house addresses. The physical layer cannot tell whether or not the message is destined for its upper layers. Its functionality is confined to modulating and demodulating.

In general, the data link layer performs three main tasks:

1. Insulates the network layer from the physical layer
2. Defines how the bandwidth of the LAN segment is shared among the stations connected to it
3. Performs framing, flow control, and error detection and correction at the link layer

In the access methodologies that will be explored in Chapter 4, the data link layer is the origin of any acknowledgments that are sent back to the sender indicating that the message was received. In some other protocols, this situation is not the case. In these protocols, this functionality is placed at the network layer.

In the IEEE project 802, the data link layer was broken into two sublayers: the first specified the medium access control (MAC) sublayer, and the second specified the logical link control (LLC) sublayer. This layering was done to separate two distinct functionalities of the data link layer. The MAC functionality has to do with accessing the particular physical layer specified. This protocol module is particular to the physical layer specified. It talks almost exclusively about items such as frame formats and air interface. The second part is the section of the data link layer that is independent of the physical layer: the LLC. This portion specifies the services provided to the network layer, independent of the particular MAC utilized.

As mentioned, only the data link layer appends two separate headers, one for the MAC sublayer and the other for the LLC sublayer. The reason is that the MAC layer header specifies the particular physical address of the node, commonly called the MAC address. The second header, applied by the LLC layer, specifies the particular SAP address or memory location where the service interface to the network layer lies. Thus, the data link layer, in a multiple access environment such as a LAN, specifies a two-part address, identifying both the location of the station and the location of the memory address where the encapsulated network layer packet is to be placed.

The data link layer truly does insulate the network layer from any contact with the physical layer. The LLC is common to all IEEE 802 protocols and it is the source of all services provided to the layer above. Since the only interface between the layers occurs at SAP locations, and SAPs are where services are passed, the combined activities of the MAC and LLC effectively isolate the upper layers. Chapter 4 has much more to say about the particular layer 2 protocols used in wireless networks.

Network Layer (Layer 3)

The network layer (NL) is concerned with how to route packets from the source all the way to the destination. Thus, it must know about the topology of the network and choose a path or paths through it. The NL provides a link between networks. Therefore, the network layer is responsible for end-to-end addressing. As part of its function, it must count the packets to avoid overloading sections of the network through which it is trying to route. If the network is associated with a public utility such as the telephone system, this layer is where packets are counted and billing information originates. Protocols at this layer must be consistent or at least communicate for proper operation of a network.

The network layer is the first layer in the model where the packets must be routed all the way from the source to the destination (end to end). Getting packets all the way through the communication system involves a lot of detail about how the communication system works. In most scenarios, this task will involve many intermediate steps or hops. Each of these hops is traditionally implemented in a router. If the communication system is reliable, then it makes sense to take as little time as possible at this layer to check and make sure the packet was received at the destination. On the other hand, if the communication system is not reliable, then this layer must include sufficient complexity to take care of the various methods of ensuring reliable message transfer. These methods are exactly the same as the three levels of service identified above in the LLC sublayer discussion. Recall that there are three broad classes of service *any layer* can offer the layer above it.

The network layer is just another layer. Therefore, the network layer can be implemented in all three ways. However, many networks use the PSTN as a major part of the WAN. Since the modern telephone system is quite reliable, many networks use an unacknowledged connectionless protocol at layer 3. They leave error recovery for the few packets that do become lost to the next layer up in the OSI model, the transport layer. The OSI network and transport layers are not widely deployed, so additional details about the functions of network layer addressing and all discussion of the transport layer will be done in the following section, which discusses the TCP/IP model.

■ ADDRESS RESOLUTION

The distinction between addressing functions by layer protocols is sometimes not clear. In the last several sections, both layers 1 and 2 have been identified as being responsible for addressing in different environments. This section brings those ideas together and makes the distinctions clear. A detailed treatment of the mechanisms that support end-to-end addressing will be presented in the section on the most widely adapted layer 3 protocol, IP.

Node-to-node addressing is used to identify layer 1 addressing in networks with a dedicated point-to-point media environment. The same term is used for layer 2 addressing in a shared-media environment. Shared-media environments are always characterized by the existence of a MAC sublayer in the layer 2 protocol, for example, the family of protocols called Ethernet. Layer 3 protocols (here, IP) provide end-to-end addressing.

The physical layer is responsible for node-to-node addressing in most point-to-point networks. (For this reason, *point-to-point* and *node-to-node* are often used interchangeably.) Good examples include a typical terrestrial or satellite microwave link. Since the ultimate source and destination systems are defined by the physical connection itself, the physical layer is responsible for "addressing" to the two end systems.

At layer 2, there are really two separate situations. The specific case of an X.25 network has a layer 2 protocol, but it is limited to a frame definition, using link access procedure balanced (LAPB). Because there is no shared media, there is no MAC layer. Therefore, when there is no access method, the physical layer performs node-to-node addressing, even though the address is actually transmitted in the layer 2 frame. Only networks that share the medium among several stations, each contending for access, need a MAC sublayer 2 protocol. When there is no sharing, there is no need for multiple access.

Like many packet protocols, X.25 uses virtual circuits. These circuits define the endpoints that the information takes from source to destination. Because those endpoints are defined during the connection establishment phase, the physical connection at layer 1 defines the end-to-end address (although not the path itself, as in dedicated point-to-point networks).

When in a shared-medium environment, like the vast majority of LANs and all IEEE 802 LANs, the MAC layer takes over the responsibility for node-to-node addressing on the LAN segment. However, it takes this responsibility as a service to layer 3. In the specific case of IP, this layer 3 protocol uses the address resolution protocol (ARP) to map the host portion of the IP address to a MAC address. The layer 2 protocol then actually passes the frame to the ultimate destination. ARP is part of the widely used layer 2 protocol family called Ethernet and is used to resolve the host portion of the IP address into a MAC address that the layer 2 protocol can understand.

The layer 3 protocol (in this case, IP) must address the fact that in the typical environment, not every station attached to a LAN can hear every other station. They are separated by WAN segments, routers, etc. When this separation blocks transparency at layer 2, then an end-to-end address resolution and routing mechanism is usually required to ensure communication between the ultimate source and destination systems. For example, a router, because it operates on layer 3, blocks transparency at layer 2, so a layer 3 protocol is required. Transparency at a particular layer is defined as a layer protocol in which all communications addressed to a particular system pass using addresses understood at that layer. Examples are broadcast and multicast communications when the address of the destination system is included in the address range. This issue

Figure 3-7
Address Resolution

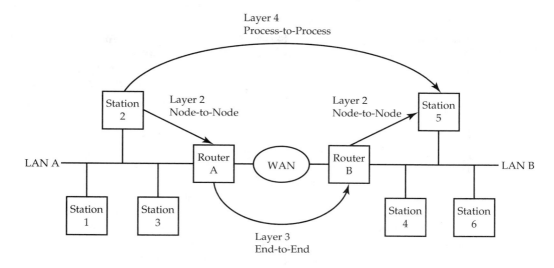

of transparency can be used to understand how layer 2 protocols work with layer 3 protocols to pass a communication from ultimate source to ultimate destination.

Figure 3-7 illustrates how this process works. In the figure, each station and router has a unique layer 2 (MAC) address and a unique layer 3 (IP) address. Since every frame sourced by any station on LAN A is seen by all other stations and router ports on LAN A, there is no role for the layer 3 address. It is extremely likely, however, that the application running on the source station will have used an IP address in its message to identify the source and destination systems. Therefore, even though the layer 3 protocol has no role in the actual delivery, ARP has a role because it maps the host portion of the IP address to a MAC address.

When a frame has a destination address that does not lie on LAN A, the situation is similar but requires additional steps. For example, a frame addressed to station 5 coming from station 2 would not be seen by station 5 because the router placed between the two LANs would block this transparency. In this case, the router would examine the IP address in the packet that arrives at its port connected to LAN A to determine if it has a path to a station on LAN B that uses the IP network prefix contained in the IP address of the packet. When it does, the router would then employ IP, the layer 3 protocol, which has knowledge of what addresses are connected to the various router ports, to perform the routing necessary to find the destination system with the correct IP network prefix address, in this case, LAN B. Of course, in this simple example, the routing is internal to the router itself because it redirects the packet to the correct port that attaches to LAN B. Once this connectivity is established, the router would transfer the packet to the port connected to LAN B, apply the ARP to map the host portion of the IP address to a MAC address, and use the services of the layer 2 protocol to deliver the frame.

As should be clear, the host portion of the IP address and the MAC address are both actual physical addresses that identify the unique ultimate source and destination, or end, systems. Because IP is used over several layer 2 protocols, where a MAC layer may or may not exist, the vast majority of applications use an IP host address to identify a unique system. This situation explains why IP is referred to as an end-to-end protocol, even though it uses a layer 2 protocol to make the address translation and to perform the actual frame transport in the vast majority of LAN environments. This situation is consistent with an understanding of how protocol stacks work, because it is one example of a service that a lower layer can provide to the layer above.

◼ TCP/IP

TCP/IP is the protocol architecture used on the Internet. It is the most popular, nonproprietary communications architecture in the marketplace. TCP/IP was developed as part of the ARPANET research project sponsored by the Department of Defense (DOD). Since this project was begun prior to the definition of international standards for developing communication protocols (the OSI model), it does not conform to the model. Although TCP/IP was not developed under the OSI model, a correspondence can be made. This correspondence is illustrated in the next section.

Relation to OSI Model

As you can see in Figure 3-8, a rough correspondence exists between the OSI concept of transport service users and the TCP/IP application or process layer. Similarly, a correspondence exists between the OSI transport layer and the TCP/IP host-to-host layer, commonly implemented as TCP and functioning as the transport provider. Finally, the OSI concept of the lower three layers providing the network service to the transport layer translates well to the TCP/IP concept of a network access layer. However, TCP/IP uses

Figure 3-8
TCP/IP Model

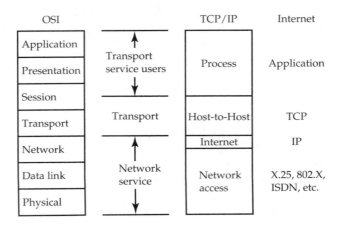

Figure 3-9
TCP Processes

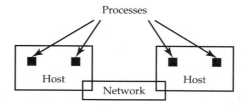

Table 3-3
TCP/IP Protocol Tasks

Layer Number	Layer Name	Function	Name
4	Process layer	Protocols used to support the application	
3	Host-to-host layer	Reliability mechanisms	TCP
2	Internet layer	Routing of data between networks	IP
1	Network access layer	Exchange of data between a host and a network	

an entirely different nomenclature to talk about communications. Since it was developed before the days of networked PCs, the standard is closely associated with the idea of a host-to-host communication between mainframe computers servicing terminals.

TCP/IP is a communications architecture composed of three basic components: processes, hosts, and networks. Processes are the fundamental elements that communicate, for example, a file transfer operation or a remote log-in. Processes execute on hosts, and hosts are connected by networks, which transfer data from one host to another. The transfer of data from one process to another involves first getting the data to the host in which the process resides, and then getting it to the process within the host. Therefore, a communications network needs to be concerned only with routing data between hosts, and the hosts are concerned with directing data to processes. See Figure 3-9.

Four separate tasks are involved in the TCP/IP communications architecture. These tasks are shown in Table 3-3. Therefore, TCP/IP protocol stack is really composed of four protocols combined and named for the middle two. (Actually, there are five, the fifth being an additional process layer protocol. Both the file transfer protocol (FTP) and simple mail transfer protocol (SMTP) are usually present in any implementation.) As in any layer architecture, many different network access layers can be utilized, and the resulting protocol stack continues to be called TCP/IP.

TCP/IP again illustrates the power of layered architectures. The network access layer uses the protocol appropriate to a specific network, for example, Ethernet, Token Ring, or X.25. The process layer is also home to many different types of protocols, such as FTP or SMTP. One term, TCP/IP, stands for many different communications options. As long as the process layer protocols can communicate, all implementation variations of TCP/IP will communicate.

Each host contains software at the network access, internet, and host-to-host layers and software at the process layer for one or more processes. Relays between networks need the network access layer to interface to the networks that they attach to, and they need the internet layer to perform the routing and relaying function. To understand the role of a relay in TCP/IP, recognize that the term is used more generically than in the OSI model. A relay in TCP/IP can include devices of the protocol converter, bridge, switch, router, or gateway classes.

Like any protocol, each entity must have an unique address for successful communication. Actually, two levels of addressing are needed. Each host must have a unique address on the network, and each process on a host must have an address unique for that host. This convention allows TCP, the host-to-host protocol, to deliver data to the proper address.

When any two processes need to communicate, the identification is done via an addressing scheme. First, each network must maintain a unique address for each host attached to that network, and that address allows the network to route packets through the network and deliver them to the intended host. This first element of the two-part address, which identifies the network and the location of the host on the network, is called the subnetwork attachment point address. The term *subnetwork* is used to identify a physically distinct network that may be part of a complex system of networks interconnected by relays. At the internet level, it is the responsibility of the internet protocol to deliver datagrams across multiple networks from source to destination. Hence, the internet protocol must be provided with a unique global network address for each host. Here the role of the internet protocol as an end-to-end protocol is clearly identified using the network address.

Once a data unit is delivered to a host, it is passed to the host-to-host layer for delivery to the ultimate user (process). This second element of the two-part address identifies the location of the process or port in the host. Because there may be multiple users, each is identified by a port number unique to that host. Thus, the combination of port and global network addresses identifies a unique process within an environment of multiple networks and hosts. This three-level address (network, host, port) is referred to as a SOCKET.

Internet Protocol (IP)

This section will focus on the internet protocol (IP) and addressing within it. First, the word *internet* as used here and in the above paragraphs has nothing to do with the familiar Internet and World Wide Web (WWW). The Internet you are probably quite familiar with is composed of hundreds of interconnected networks, most using TCP/IP, including the two unclassified segments of the defense data network (DDN), MILNET, and ARPANET. As the term is used here, an internet is nothing more than an interconnected set of networks connected by relays. Also confusing to some is that the term *internet* is used to describe the TCP/IP protocol stack.

IP provides a connectionless, or datagram, end-to-end service between hosts; that is, IP does not set up a logical connection between hosts and does not guarantee that all data units will be delivered or that those delivered will be in the proper order. Recall the earlier discussion on the differences between a connectionless and connection-oriented mechanism. A connectionless service is one that corresponds to the datagram mechanism of a packet-switching network. On the other hand, a connection-oriented service is one that corresponds to the physical circuit mechanism of a circuit-switching network.

The decision to have the internet layer provide an unreliable connectionless service evolved gradually from an earlier connection-oriented service: the ARPA internet evolves. This network contains many networks, not all of them reliable. By putting all the reliability mechanisms into the transport layer (TCP), it was possible to have reliable end-to-end connections even when some of the underlying networks were not very dependable.

A major segment of IP is concerned with addressing. This subject is complex, using labels where traditional distinctions are blurred. To illustrate, a distinction is generally made among names, addresses, and routes. A name specifies what an object is, an address specifies where it is, and a route indicates how to get to an address. On a single network, the name and address distinction is arbitrary; either one can be used to identify a unique object.

Each protocol stack uses a different method to accomplish addressing of individual stations. IP assigns a unique number to every station, in addition to any addressing that might exist in the layers below, such as the MAC address used in many networks. This IP address is 32 bits long and is composed of two parts, the network part and the host part. Each of these parts is converted into a decimal number and is referred to as an IP number.

On an internet, the distinction among names, addresses, and routes can be unclear. Applications continue to use names, and individual networks continue to use addresses and, if necessary, routes. To transfer data through a relay, two entities must be identified: the destination network and destination host. The relay requires the address of the network to perform its function. This address can be specified in three ways:

1. The application can refer to a network by a unique number; in effect, the name and address are the same.
2. The internet logic in the host can translate a network name into a network address.
3. A global addressing scheme can be used; that is, there is a unique identifier for each host in the internet. For routing purposes, each relay would need to derive network addresses from host addresses.

Ethernet uses the third approach, and so does the Internet. For Ethernet, the main advantage of this approach is that the network address can be hardwired into the device. The main disadvantages are that a central authority must manage the assignment of names and that unnecessarily long address fields must be carried across networks.

Therefore, a relay will typically receive an internet packet with a reference in the form NET.HOST, where NET is a network address. The identifier HOST is usually a name and an address. To the higher layer software that generated the packet, HOST is an address translated from an application-level name. When a relay must deliver a datagram to a host on an attached network, however, HOST must be translated into a subnetwork attachment point address because different networks will have different address field lengths, as well as many other problems in mapping addresses. Hence, HOST is treated as a name by the relay.

The example NET.HOST illustrates a two-level hierarchical addressing scheme. The IP address is broken up into two parts: the network part and the HOST part. Note that by appending a service access point (SAP) address on the end, the address becomes a three-level addressing scheme, NET.HOST.SAP. The SAP is an individual service access point in a host. Conceptually, SAP is a more general term for a port. It has the same function as the SAP does in OSI terminology. With this identifier, the internet protocol can be viewed as process-to-process rather than host-to-host.

As discussed in an earlier section, addressing and the location where it is performed corresponds to layer complexity and function. In the above example, with SAP in the internet layer, the internet protocol is responsible for multiplexing and demultiplexing datagrams for software modules that use the internet service. The advantage, of course, is that the next higher layer can now be simplified, a useful convention for vendors building products where cost and complexity are real issues.

How does the station software determine the NET.HOST identifier of a desired destination? This task is done by a directory service, generally located on a server. Each server contains part, or all, of the name/address directory for internet hosts. Note that a directory service is a database for the local address span only. It is not a routing table for reaching outside the locally connected hosts.

Routing is perhaps the most well-known responsibility of the internet protocol. Note that for successful operation of a router between two networks, both networks must use the same internet protocol. All routing strategies fall into two classes, but they are discussed a little differently in the TCP/IP world. The terms *fixed* and *adaptive* correspond to terms used in the OSI, *static* and *dynamic.*

IP also provides some minimal error control. Basically, it discards datagrams it deems inappropriate. This error control always takes place at the relay points in the network. The router may discard for several reasons, including lifetime expiration, congestion, and bit error. Note that notification of the source is not guaranteed. For example, if the address field is damaged due to a bit error, identification of the source is not possible. As with any layer protocol, IP is defined in two parts:

1. The interface to higher layers (TCP) specifies the services that IP provides.
2. The actual protocol format and mechanisms specify the host–router and router–router interactions.

The first part is sometimes discussed as the IP services and is generally less precisely defined than the protocol itself. Its definition is biased toward the functional rather than a precise operational definition. This bias is based on the fact that, as the interaction between layers occurs within a single system, generally made by a single manufacturer, room is left for innovation to perform this interaction in the most efficient manner possible.

The inner workings of the IP protocol can best be understood by defining the IP datagram; however, such a definition is not the focus of this section. The basic idea is that all services that a protocol can provide must fit into the datagram format. Since the network layer is a source-to-destination layer, it must have an addressing mechanism to distinguish the true source and destination from any intermediate systems that the packet flows through en route to its ultimate destination. IP has a standard addressing technique that is applied almost universally. It uses IP addresses, and they are end-to-end addresses.

Figure 3-10
IP Address Formats

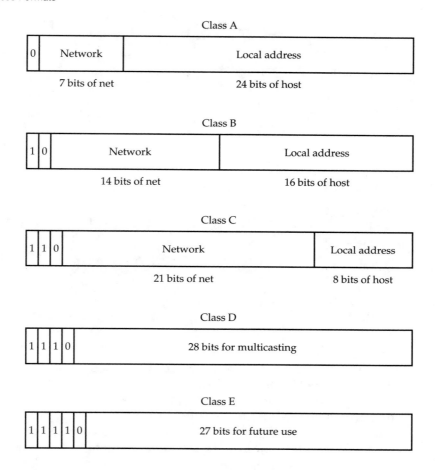

As you have probably guessed by now, IP uses the term *datagram* for "packets." The source and destination addresses are of one of four classes: A, B, C, or D. The idea is that the larger the local installation (local address space), the more bits are needed in any address to accommodate the number of local stations. As you can see in Figure 3-10, class A addresses are designed for very large installations; class B, for large organizations; and class C, for smaller ones. Class D addresses are reserved for multicasting applications. A fifth class, E, is reserved for future use.

This structure allows various network sizes to accommodate a uniform length for the address field. It simplifies the IP address resolution problem and, at the same time, recognizes that many more small organizations want an officially recognized address than do large ones. These IP routing addresses can be purchased. As you might imagine, it is much easier to obtain class C addresses than class A. The address formats are shown for each of the classes in Figure 3-10. These addresses are assigned by two different

Table 3-4
IP Address Ranges

Class	Starting Address	Ending Address	Number of Hosts
A	001.000.000.000	126.255.255.255	16,777,215
Loop back	127.000.000.000	127.255.255.255	—
B	128.000.000.000	191.255.255.255	65,535
C	192.000.000.000	223.255.255.255	255
D	224.000.000.000	239.255.255.255	—

Table 3-5
IP Address Ranges Valid for Assignments to Routers

Class	Starting Address	Ending Address	Number of Hosts
A	001.000.000.001	126.254.254.254	16,777,214
Loop back	127.000.000.000	127.255.255.255	—
B	128.000.000.001	191.255.254.254	65,534
C	192.000.000.001	223.255.255.254	254
D	224.000.000.001	239.255.255.254	—

groups. The network portion is assigned uniquely and globally. The host or local address portion is assigned by the local network administrator.

Most of you have encountered such addresses written in decimal form. This nomenclature is called dotted decimal notation. Each byte of the 4-byte address is written in decimal form so that it is more compact and easily recognized by human beings. Since the first byte is constrained by the status of the first few bits used to classify the addresses, a numeric range is defined for each class. See Table 3-4. Class A addresses are often reserved for network providers or large subnets, of which the Internet is composed. Class B addresses are often found in large corporations and some educational institutions. By far the largest class, class C addresses are most common and are used by many different organizations.

While technically correct, Table 3-6 implies that every address in the class can be used to designate an IP address, which is not the case. Some reserved addresses cannot be assigned to any station. Table 3-5 illustrates the ranges that can actually be assigned to a station.

IP addresses composed of all 1s are reserved for broadcasts, while addresses composed of all 0s are reserved for segment or subnet addresses (router addresses). Class D addresses are intended to specify multicast host groups. Finally, there is an address range called class E, which is not shown in Tables 3-4 and 3-5. This address class is reserved for future use. Subnets, as used in IP addressing, are described next.

Table 3-6
TCP Flags

Flag	Meaning
URG	Urgent pointer field; when set indicates no urgent data
ACK	Acknowledgment field; always set once a connection is established
PSH	Push field; indicates the source initiated a push operation
RST	Reset field; destination desires to reset connection and reestablish
SYN	Synchronize field; indicates synchronized sequence numbers
FIN	Finish field; indicates source has no more data, so terminate connection

■ SUBNETS

A subnet is a set of addresses used to identify a smaller portion of a network that makes up a large network. For example, when you are assigned an address from your Internet service provider (ISP), you are essentially made a subnet of the ISP's network, which is done by segmenting the host portion of the address. The approach used is to divide this range of addresses into two groups, subnet and host. For example, for a class C address, one might break the 8-bit address range into two 4-bit groups. The first 4 bits, or the 4 most significant bits (MSB), would identify the subnet, and the host address for the last 4 bits would be for the subnet. Breaking down the address range is usually done so that the network administrator can manage the address fields more easily.

A company might decide to set the host address ranges as departments within the organization. The administration, purchasing, engineering, production, and sales departments might all have specific 4-bit host addresses assigned to each. Then each department would take the individual stations in their departments and assign to each one of the 4-bit subnet addresses. Here, sixteen departments could be identified uniquely, and each could have up to sixteen stations attached to the Internet.

Routers would use these subnet masks to route messages automatically to and from individual stations. For example, assume that engineering was assigned the host address of 1011, and the manager of that department was assigned subnet address 0001. The router would know how to route a message to and from that station by routing all messages with 10111111 addresses to engineering and those with address 10110001 to the manager's station. From this example, you can see that if the suffix of the address is all 1s, the packet will be routed to all nodes on any specific segment. Example 3-1 provides more detail about a similar situation.

■ EXAMPLE 3-1   A company has a class C network and its IP address (router's address) is 195.212.148.0. The network administrator wants to divide this class C network into four

departments through the use of subnets. The administrator chooses to perform this task using the following two principles:

1. The three most significant bits of the host address are used as extended addresses for the four departments' network addresses.
2. The five remaining bits are used as the host address.

To document this task, the administrator defines the subnet mask, each department's router IP address, and the host address range of each department.

Subnet mask:	11111111 11111111 11111111 1110000 *or* 255.255.255.2
Router address of department 1:	11000011 11010100 10010100 00100001 *or* 195.212.148.33
Router address of department 2:	11000011 11010100 10010100 01000001 *or* 195.212.148.65
Router address of department 3:	11000011 11010100 10010100 01100001 *or* 195.212.148.97
Router address of department 4:	11000011 11010100 10010100 10000001 *or* 195.212.148.129
Host address range of department 1:	195.212.148.34 to 195.212.148.64
Host address range of department 2:	195.212.148.66 to 195.212.148.96
Host address range of department 3:	195.212.148.98 to 195.212.148.128
Host address range of department 4:	195.212.148.130 to 195.212.148.158

Note that the address range of the department does not include the address of the router. This address must be dedicated to the router.

■ TRANSPORT CONTROL PROTOCOL (TCP)

In any protocol stack, the transport protocol serves a key function in the concept of a communications architecture. Refer again to Figure 3-8, which compares the TCP/IP model with that of the OSI model. Note that much of what happens at the transport layer repeats what could happen in the network layer if all real networks were flawless and had the same service primitives. TCP is an end-to-end protocol.

In TCP/IP architectures, everyone uses the same IP or internet protocol, and the host-to-host layer is implemented in many different ways. Therefore, for TCP/IP, the boundary between communications functions and user functions is placed at the TCP/IP boundary. The idea of separating communications functions or transport service providers from user functions or transport service users is the same.

Note that TCP is a true source-to-destination or end-to-end layer. Here, a program from a source host communicates with a similar program, called a process, on the destination host. In the internet layer below, the protocol is between each host or router and its immediate neighbor. Again, just like the network layer in OSI, the internet layer or IP functions similarly.

Figure 3-11
TCP PDU

2	2	4	4		2		2	2	2	4	
SPORT	DPORT	SN	AN	DO	RSV	FLAGS	WS	CS	UP	Optional	Data

To facilitate the discussion below, the TCP header is shown in Figure 3-11. The TCP header is normally 20 bytes long. This length can be exceeded if optional elements are included. These elements are typically used to specify the maximum segment size that will be accepted. The number above the field indicates the length of that field in bytes. Each field is identified briefly in the list below:

SPORT: Source port; identifies the source system port number.
DPORT: Destination port; identifies the destination system port number.
SN: Sequence number; the sequence number of the first data byte in this segment. If the flag SYN is present, then this field represents the initial sequence number (ISN) and the first data byte is actually ISN + 1.
AN: Acknowledgment number; the sequence number that the source expects to receive. Therefore, it is the sequence number + 1 of the last successfully received byte of data.
DO: Data offset; a 4-bit field that represents the number of 32-bit words in the header, often called the header length. This field is required because of the optional fields, which would render the TCP header length variable.
RSV: Reserved; a 6-bit field held for future use.
FLAGS: Six 1-bit flags used to signal between the TCP entities. They are listed in Table 3-6. The SYN and FIN flags are essential in establishing and terminating, respectively, a TCP connection.
WS: Window size; contains the number of bytes the source is willing to accept from the sender.
CS: Check sum; a 1's complement of the TCP header and data.
UP: Urgent pointer; points to the bytes following the urgent data. Allows the destination to determine how much urgent data is contained in the packet.
Optional: Optional data and padding; must always be 32 bits in length.
Data: Note that this field is also optional; no data is required.

The four specific categories of service that TCP provides are:

1. Connection management
2. Data transport
3. Error reporting
4. Multiplexing

The first, connection management, is actually composed of three parts: the process of establishing a logical connection between two sockets, managing the connection during its lifetime, and eliminating the logical connection when done.

Managing the TCP Connection

The source and destination systems processes communicate using TCP sockets with sequence numbers. These elements ensure that TCP functions as a reliable COS. TCP actually implements reliability, independent of the reliability of any lower layers, through the use of retransmission. Each transmission that a source system sends must be acknowledged, or ACKed, by the destination system.

There is a limited time in which the source system will wait for such an acknowledgment. This time is called the retransmission time. If this time expires before the source system receives an acknowledgment of its transmission from the destination, TCP assumes that the IP packet has been lost and initiates a retransmission of that packet. Note that each packet is marked by sequence numbers, as is each acknowledgment. In this way, the source system knows which packet has been lost.

As you might imagine, the efficiency of communication is very sensitive to the length of time that TCP waits until it begins a retransmission. TCP calls this time the retransmission time out, or RTO value. RTO can be set to a static value or it can be adaptive. While almost all TCP implementations use an adaptive RTO, we will examine briefly how a static RTO would affect communications efficiency.

In a LAN, source and destination systems are usually quite close, the bandwidth is quite high, and the time it would take for a packet to travel between them is quite short. Clearly, when a packet is lost in this environment, efficiency would be obtained with a very small RTO. On the other hand, in a communication across the ocean using a standard telephone line, the source and destination systems are not close at all, the bandwidth is quite small, and the time it would take for a packet to travel between them is significant. In this case, efficiency would not be served by a small RTO. If RTO were too small, the source system might assume that the packet is lost when, in fact, the destination system received the packet and the acknowledgment has been sent; however, it has not had time to arrive back at the source.

It is not possible to set the value of RTO for each application. It would be much easier if the protocol itself could anticipate the response time and set the value of RTO in an adaptive manner. TCP performs this task by estimating the round trip delay of each socket pair established. Since the network that lies between the two end systems is not static, the time for a packet to travel between any two systems is not static either. For this reason, the round trip delay is constantly measured and adjusted to adapt to the changing conditions of the network. TCP starts a counter when a packet is sent and stops it when the acknowledgment is received. It then writes this value into a variable called the round trip time (RTT). As more communications are sent and received, the value is continually updated to adapt to the changing conditions of the internet that lies between the two communicating end systems.

Establishing and Terminating the TCP Connection

The flags in the TCP header are used to guarantee that a TCP connection is established and terminated reliably. This process is called the three-way handshake, indicating that three messages are exchanged between the source and destination systems to accom-

Figure 3-12
Three-Way Handshake for Establishing TCP Connection

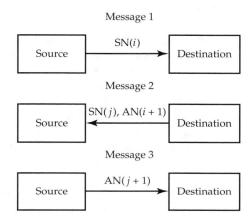

plish both the establishment and termination of the connection. These messages are sent through the use of the flag fields in the TCP header. Figure 3-12 illustrates the three messages and how the AN and SN counters work for the establishment phase.

Establishment of a TCP Connection (SYN Flag)

Message 1: To establish a connection, the source system sends a TCP message to the destination system. This message has the SYN flag set to indicate a desire to synchronize sequence numbers for a connection. The SN field contains this number.

Message 2: The destination system then responds with the ACK flag set, indicating that a connection has been established. Until the connection is terminated, the ACK flag will remain set. The SN field now contains a new sequence number, which will be used for the connection. An explicit confirmation is indicated by the incrementing of the original SN by 1 and placement of the new value in the AN field.

Message 3: In its first data transmission, the source system increments the sequence number by 1 and places the new number in the acknowledgment field. The connection is established.

Termination of a TCP Connection (FIN Flag)

Message 1: To terminate a connection, the source system sends a TCP message to the destination system. This message has the FIN flag set to indicate a desire to terminate the connection. The AN field contains the correct acknowledgment number, and the ACK flag is set.

Message 2: The destination system then responds with the FIN flag set, and the ACK flag is set. The SN field contains the correct sequence number.

Message 3: The source system sends its last message to the destination in that connection with the ACK flag not set, indicating the connection has terminated.

The second category (data transport) is the service category used to transport data from one host to another. Data transport takes care of issues like handling full- or half-duplex communications, flow control through the use of windows, and error control through the use of the 16-bit TCP checksum. Most TCP connections are full duplex.

Flow control of a TCP connection is a critical function that can have a great impact on the overall efficiency of wireless communications systems. Flow control in TCP is accomplished with two functional elements: buffers and windows. Every time a connection is established, each system designates a portion of its available memory as a buffer for that connection. Initially, the WS field in the TCP header contains that buffer size. This portion of memory is called the window in TCP terminology. However, as communications proceed, the content of the WS field changes. Instead of containing a static value of the initial buffer size, it adapts and instead contains the amount of buffer space available for the next message. This amount of buffer space is called the window advertisement because the WS field advertises to the other end system how much buffer is remaining. Each acknowledgment contains a window advertisement. The size of the window advertisement tells the source system how much data it can send to the destination in the next message. If the window advertisement is set to zero, then the source system will not send any data to the destination until it receives an acknowledgment with a nonzero value in the WS field.

This mechanism is what end systems use to adapt a connections data transfer rate to their own capability. Clearly, higher performance systems send data faster than lower performance systems can assimilate it. As will be discussed in detail in Chapter 16, this mechanism also has an effect on how rapidly the two end systems make use of the available bandwidth of the internet between them.

The third capability of TCP is its error reporting. TCP will report failure resulting from conditions for which TCP cannot compensate. If TCP can recover from the failure conditions, no error will be reported. TCP has robust capability to retransmit and thus can avoid error conditions. In a typical environment, very few TCP errors are reported due to packet failure.

Multiplexing in TCP is very similar to multiplexing anywhere. A TCP entity within a host can provide service to multiple processes simultaneously. TCP uses ports to provide multiplexing. First, a port is associated with each process within a host. This port is called a SAP, the same term used to describe a location where a protocol goes to obtain services from the layer below. The same activity occurs here. Each port has an address; this address is where a process goes to obtain communications services.

The address of this SAP is constructed according to the following rule. The host is identified by the address NET.HOST. Then, adding the port gives us what is called a socket. A socket is unique throughout the internet. Because it is an end-to-end service, TCP provides a logical connection between a pair of sockets. Clearly, there can be many sockets and, hence, many processes.

■ PROCESS LAYER

Earlier in the discussion about IP, two protocols, FTP and SMTP, were mentioned briefly. These protocol applications formed the basis of networking in the ARPANET environment. They shielded the user from the complexities of TCP and allowed him or her to transfer files and send and receive e-mail.

File Transfer Protocol (FTP)

File transfer protocol (FTP) is a piece of application software that uses the TCP services to exchange files between two hosts. FTP works quite differently than similar services in the OSI world. Since TCP/IP was designed before the OSI model was developed, it did not allocate parts of the job to other layers of the protocol stack. Everything associated with file transfer had to be performed in this location. Word size differences, bit order preferences, bit versus byte transfer procedures, etc., all had to be accommodated in this piece of software, FTP. As a result, the number of different types of files it could handle was quite limited.

Simple Mail Transfer Protocol (SMTP)

Simple mail transfer protocol (SMTP) is a piece of application software that forms the basis for all e-mail sent on the Internet today. It was originally defined in a now famous document, RFC 822. In its original form, this document was limited to sending text messages only. Just like FTP, SMTP must contain all the particulars associated with any e-mail transfer. As many of you no doubt realize, this limitation to sending text only no longer exists, and many updates to RFC 822 have occurred over the years. Like FTP, SMTP uses TCP services to accomplish its mission. The format for e-mail addresses is defined by the RFC 822 standard. All follow the same format, shown below:

NAME@DOMAIN

For example: Godzilla@Downtown.Tokyo.Jpn

You can see that the domain contains three parts. The first two parts are descriptive terms identifying the location. The third part is one of a few reserved abbreviations used throughout the Internet for routing mail. Although the third part is only one piece of the domain, often just this last part is called the domain. Several domains are now defined. It is very probable that more will be defined in the near future. The bulk of the domain registrations are currently performed by one organization and licensed by the U.S. government. There are six domain types used in the United States:

.COM—commercial organizations
.EDU—educational organizations
.GOV—government bodies and agencies
.MIL—military organizations
.NET—network organizations like your ISP
.ORG—nonprofit organizations

Many organizations based outside the United States use a two- or three-letter abbreviation for their country name as the final field in the domain name. For example, the United Kingdom uses UK, Canada uses CA, Japan uses JPN, etc.

■ SIGNALING SYSTEM SEVEN (SS7) MODEL

The final model discussed here is the signaling system seven (SS7) model. SS7 was developed by the international community as an alternate way to pass the signaling that the telephone system uses. SS7 is a protocol through which switching elements of the telephone system exchange information to do call setup, routing, and control. This signaling is part of every call, and every telephone system in every country needs it.

Signaling, as used in the telephone system, is the procedure used to perform the three tasks that every call must include: call initiation, voice or data transfer along with the route it takes, and call termination. These tasks are often referred to as call components. Signaling can be defined as the exchange of information between call components that is required to offer service. For example, dialing a phone number can be seen as signaling used to connect to that number address. The timer counting the minutes that the two parties are connected is the signaling (really monitoring) that is done during voice or data transfer. When you hang up, that can also be seen as signaling the data plane to terminate the call.

The information necessary to placing the call is carried on physical lines separate from those used for voice or data transfer. In essence, there are two operational planes: the first is the SS7 or control plane, the second is the voice data network or data plane. Traditionally, the network used the installed base of stored program control (SPC) switching systems for voice and data transfer and a separate network for service specific functions. In this environment, the service logic was hardwired, and the introduction of new services required the replacement of this same hardware. Additionally, if a subscriber was not serviced by an SPC, those service options that came with the SPC were not available to that subscriber.

Today, the network used for managing calls is entirely different from the one used for actually conducting telephone connections. This alternate physical network is referred to as the interoffice network. The switches are placed in a switching station, and the network that the switches use to communicate about how the traffic will be switched is the interoffice network. The signaling that this network uses is referred to as the common channel signaling network (CCSN). The protocol that the signaling follows is called SS7. Today, these concepts are merged, and this interoffice network is called the SS7 network.

By definition, an SS7 network uses out-of-band signaling. Out-of-band signaling does not take place over the same path as the voice traffic. In an SS7 network, the entire facility that manages the calls is outside the network that the traffic passes over. Another term used to refer to that portion of the telephone network that actually passes customer traffic is the switching fabric. It is composed of switches connected by packet data links. The concept of out-of-band signaling is illustrated in Figure 3-13. The abbreviation SSP stands for signal service point and represents a telephone switch.

Out-of-band signaling also has advantages in efficiency and utilization of the voice/data network. The three primary benefits are:

1. Support for the advanced features and services
2. Faster call setup times
3. The use of lines other than the voice/data network lines for nonrevenue tasks like signaling

Figure 3-13
Out-of-Band Signaling

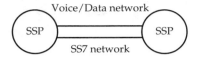

Figure 3-14
Relationship Between OSI and SS7 Protocol Models

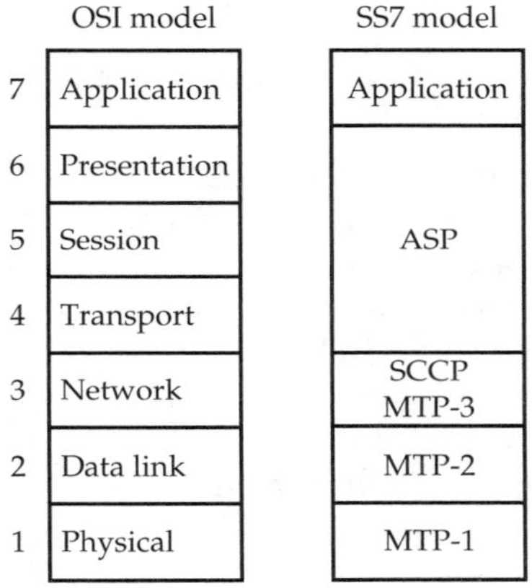

	OSI model	SS7 model
7	Application	Application
6	Presentation	ASP
5	Session	ASP
4	Transport	
3	Network	SCCP / MTP-3
2	Data link	MTP-2
1	Physical	MTP-1

The alternate physical network that the SS7 network uses to carry its information is a packet-switched network. The voice/data transfer uses the circuit-switched network. SS7 itself is a software product that was developed using the OSI model. The protocol has seven layers and is shown in Figure 3-14.

A layer labeled 0 defines the alternate physical network or data plane. Layer 0 in the interoffice network is composed of full-duplex 64 kbps digital data service (DDS) lines. These lines are usually multiplexed up into DS-1 lines for transfer between offices and switches.

Layer 1 is also known as the signaling data link. In the United States, it is usually implemented with a packet interface. Interfaces included in the standard are DS-1, V.35, DS-0, and DS-0A. The latter is defined below. Because of the modular nature of any layered protocol, however, this layer can be and is also implemented with satellite links. Layer 1 in the SS7 protocol is called message transfer part—level 1 (MTP-1).

Layer 2 corresponds to the data link layer. Like most layer 2 protocols it covers two tasks. The first is ensuring reliable communication between adjacent stations on a link. This task includes the functions of error detection, error correction, and flow control. Second is identifying and aligning frame types known as signal units. The structure of the signal units is patterned after the high-level data link control (HDLC) frame structure. The three signal units specified are the message signal unit (MSU), link status signal unit (LSSU), and fill-in signal unit (FISU). Layer 2 in the SS7 protocol is called message transfer part—level 2 (MTP-2).

The MSU is used to carry all call control, database query (more on that later), and network management communications in its data field. Each MSU is routed according to its destination address. Since only MSUs cross link boundaries, only MSUs have source and destination addresses. The LSU is used to carry a byte or two of link status information between signaling locations on a single link. The FISU is a frame used to monitor the link continuously. Essentially, whenever a station does not have an MSU or LSU to send, it sends an FISU. These frames are checked continuously for errors by examination of the cyclic redundancy check (CRC) field in the frame. This monitoring allows the stations to be constantly aware of the current quality of the line and to detect any break or degradation immediately. Fill-in frames are very appropriately named.

Layer 3 corresponds to the network layer, and its function here is similar to what would be expected of a network layer. Message transfer part—level 3 (MTP-3) is concerned with routing the message signal units and providing network management. As mentioned earlier, the routing is based on the destination address in the MSU. MTP-3 is also responsible for updating the routing tables when a link goes down or comes up. Using terminology introduced below, MTP-3 is responsible for determining the optimum data route among the SSP, STP, and SCP. It also facilitates load sharing between alternate routes. MTP-3 also provides sequence numbering, segmentation, and reassembly and flow control based on a sliding window algorithm.

The signaling connection control part (SCCP) corresponds to the upper portion of layer 3 and complements MTP-3. SCCP provides additional functions to MTP-3 for both connectionless and connection-oriented services. SCCP is the user part of the network layer. It provides functions and procedures particular to the specific user. For wireless networks, the SCCP provides additional functionality for mobile users.

The network layer is divided into two sublayers so that the SCCP functions that any particular user may need are executed only when required. For example, if the application requires only to address a node, then the additional user-oriented services of SCCP would not be called on. Most of the additional functions that are part of SCCP are additional layers of addressing. MTP-3 can address only the SS7 node itself. SCCP adds functionality that can identify the particular SCCP application at a particular node. In essence, SCCP provides the logical association between the MTP-3 node address and the various user application parts, which are located in the application layer and are discussed below.

SCCP modifies the basic network services offered by MTP-3. Five classes of service are provided by SCCP.

Class 0 service: basic connectionless service

Class 1 service: sequenced connectionless service (very similar to MTP-3)

Class 2 service: basic connection-oriented service

Class 3 service: flow control connection-oriented service

Class 4 service: error recovery and flow control connection-oriented service

Class 0 service is a basic connectionless service; the network layer packets are transported independently and may arrive out of order. Class 1 service adds sequence numbers to class 0. Class 2 service is the basic connection-oriented network layer service. A temporary connection or virtual circuit is set up. Segmentation and reassembly of packets larger than 255 bytes is handled automatically by this service class. Class 3 enhances class 2 by adding flow control, and class 4 enhances class 3 by adding error-recovery techniques. Most wireless networks use only SCCP class 0 and class 2 services. Class 0 messages are used on the A link for messages not directly addressed to a particular mobile station. Class 2 messages are often used to address individual users, relatively simultaneously, using multiple connections.

OSI layers 4, 5, and 6 correspond to the application service part (ASP) of the SS7 model. This layer of the SS7 protocol stack uses functionality equivalent to that found in the OSI model. The ASP builds on the services of the MTP and SCCP layer 3 in exactly the way that layers 4 to 6 build on the services of an OSI network layer.

Layer 7, the application layer, is where the transaction capabilities application part (TCAP) is located. It is also the location for the mobile applications part (MAP). The protocol that supports application services offered by the system are located here. For example, there is a TCAP part for each of the following telephone system services: enhanced 800 number service, alternate billing service (ABS), custom local area signaling service (CLASS), and others. Since TCAP messages must be delivered to a particular application inside a node, TCAP uses SCCP for transport.

Basically, TCAP is a set of protocol rules designed to ensure that consistent messages are passed between the switching nodes. The actual application programs lie above this layer. TCAP can directly use the services of SCCP when the ASP layer is null. TCAP provides a set of tools that can be used by an application at one node to exchange information with a similar tool present at another node. The primary use of TCAP is to invoke services like those listed below:

Enhanced 800 service: This 800 service allows the portability of 800 numbers. In the traditional network, once an 800 number was assigned, it was tied to a specific area code. This service addresses that shortcoming.

Alternate billing service (ABS): This service allows you to use a calling card to place a telephone call, allows you to bill to a third number, and allows you to place collect calls. Any time the calling party does not directly pay for the call, this service is used.

Custom local area signaling services (CLASS): This broad class of services includes call blocking, automatic redial, call forwarding, call waiting, call tracing, etc.

TCAP is divided into two sublayers, a component layer and a transaction layer. The component layer exchanges components between users. These components are either requests for action of a TCAP tool at the remote node or data responding to such a request. The transaction layer facilitates the exchange or transaction of messages that contain the components.

The mobile application part (MAP) is needed only for mobile user services. While not part of TCAP, it lies at the same protocol layer as TCAP and should be thought of as complementing TCAP for mobile users. MAP services support the exchange of data between components of a mobile network, thus supporting subscriber mobility and network control. Do not confuse MAP with systems management application process (SMAP). SMAP uses the services of TCAP and is the collection of all monitoring, control, and coordination functions above the application layer. SMAP uses the application entity of operations, maintenance, and administration part (OMAP), which is a peer component of TCAP.

MAP services and procedures include location and registration of the subscriber, access and retrieval of subscriber information during call setup, handoff, subscriber management, and supplementary services. MAP supports over forty different message types. The length of the message depends on the type of message sent. To give a sense of the relative complexity of MAP as compared to TCAP, contrast the number of different message types that each uses. Recall that TCAP supports traditional wireline telephone users and that it has only seven message types.

MAP complements TCAP. When a roaming cellular subscriber moves into a cell controlled by a different mobile switching center (MSC), the database responsible for maintaining service information on this subscriber uses MAP to request information from the home database of this particular subscriber. All MAP messages are encapsulated into TCAP messages for transport.

Interoffice Network

As discussed above, SS7 is an internationally developed and recognized standard for interoffice signaling. There are four main components of the SS7 or interoffice network: signal service point (SSP), signal control point (SCP), signal transfer point (STP), and signaling data link. The first three components are generically referred to as nodes. There is a special way of referring to any switch that has SS7 capability. These switches are called signaling points (SPs).

> SSP: This term is used for a switching station that is connected directly to the SS7 network and that has an SP. Simply put, SSPs are telephone switches. However, not all offices are connected directly to the SS7 network. When a central office has the special designation of being able to launch an SS7 query, it is called an SSP office. The signaling is called a query when a call is originated from a telephone attached to the office. When an office is not connected directly to the SS7 network, it is designated a non-SSP office. When a telephone call originates from a non-SSP office, that office must use the SS7 network to route this call over the interoffice network to an SSP office for connection to the network.

> SCP: An SCP is nothing more than a specialized computer that knows all about a certain database. A GSM home location register (HLR) is an example of an SCP. This database is associated with some service that the telephone system can provide using SS7. For example, there are databases for processing 800 calls, 900 calls, ANI, call forwarding, local number portability (LPN), etc. The parts of the SS7 protocol that perform these tasks are called parts. SCPs and SSPs communicate using the SS7 net-

work. For example, when you dialed an 800 number, the switching system was alerted to the need for input from a particular database contained in an SCP, or a particular "part" of the SS7 protocol.

Before going on to the next two components, we will say a few words about queries and databases, which form the heart of the SS7 network. The SS7 network is in place to pass queries and responses to and from SS7 capable switches. Databases respond to the queries, and these databases generate the responses back to the switches. In SS7 terminology, an SSP with an SP switch sends an MSP to an SCP, which generates an MSP back to the SP located in the SSP. The MSPs are passed through the SS7 network through STPs, defined next.

STP: These signal transfer points pass the signaling around the SS7 network. The signaling is transferred to the correct point through interpretation of the information contained in the SS7 address fields. The STPs are connected by 56-kbps DDS trunk lines and are really specialized packet switches. These 56-kbps DDS lines have a special name when they are used to connect to and from STPs and either SSPs or SCPs. They are called A links, and they are the primary traffic links used in the SS7 network. Each regional bell operating company (RBOC) must have two STPs for each local access transport area (LATA) and therefore two A links. One is used as an active backup for the other. Several types of links are used in the SS7 network, and these will be discussed next.

Signaling data link: There are actually six types of signaling data links used in the SS7 network. As mentioned earlier, all are bidirectional or full-duplex links. They transmit and receive traffic in both directions at the same time and data rate. Each type of link is used to connect between specific types of nodes in the SS7 network. Again, a node is just a term for a particular type of office or switch. Table 3-7 lists the six types of nodes and their use for connecting between specific types of nodes. The term in parentheses is the full name of each link. Some link types are used for multiple purposes. In Table 3-7, *hierarchy* refers to the class of office, and *different networks*

Table 3-7
SS7 Link Types

Link	Connection
A (access)	STP-SSP—same network
A (access)	STP-SCP—same network
B (bridges)	STP-STP—same hierarchy level—message traffic
C (crossovers)	STP-STP—same hierarchy level—administrative
D (diagonal)	STP-STP—different hierarchy level
E (extended)	STP-SSP—different networks
E (extended)	STP-SCP—different networks
F (fully associated)	SSP-SSP—same network—adjacent nodes

Figure 3-15
A, B, and C Links

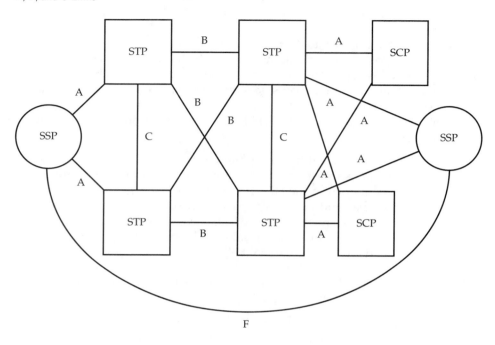

captures the distinction between communicating between nodes in a single country or telephone system and communicating between countries that have different telephone systems.

How A, B, and C links are connected is shown in Figure 3-13. These three types of links form the primary paths for all message traffic. D and E links function primarily as secondary or alternate routes. F links are used only in networks without STPs. In Figure 3-15, the F link shown would function if the rest of the network is disabled. Note that the A links are redundant and that the B links are multiply interconnected to bridge across SSPs. The A and B links carry the message traffic used by the switch fabric to control the voice/data traffic. Note that the C links, which primarily carry administrative traffic, do not connect to the SSP offices at all! The voice data network is not shown in Figure 3-15; only the SS7 network is shown. The voice data network would connect directly between SSPs shown. Coincidentally, this connectivity is the same as the F link, but the voice data link would carry no SS7 information.

There are actually three physical types of links used in the United States. They are all DDS services and we have discussed the first previously. All these links use the physical interfaces of either V.35 or RS-232. While the link data rates are at 56 or 64 kbps, they are multiplexed into DS-1 frames for actual transport through the network. It is probably best to think of these as logical rather than physical links. They are listed below:

Digital signal level 0 (DS0): A single 64-kbps channel operating at its clear channel data rate.

Digital signal level 0 A (DS0A): A single 56-kbps channel operating at its normal data rate. As in the voice network, the other 8 kbps are used for clocking and frame checking.

Digital signal level 0 B (DS0B): A DS0 channel composed of several multiplexed 9,600-bps administrative links.

Addressing in the SS7 Network

Just like any other network, SS7 has its own addressing scheme. SS7 addresses are strictly hierarchical and are composed of three levels:

1. Network
2. Cluster
3. Member (SSP)

Each of these address levels is an 8-bit number, from 0 to 255. Therefore, the total address length required to route to any SS7 signaling point is 24 bits. The network number is unique for each network. Examples of networks would be individual RBOCs, competitive local exchange carriers (CLECs), local exchange carriers (LECs), or international exchange carriers (IXCs). Network addresses 0 and 255 are reserved.

The next level of addressing is the cluster. A cluster is a group of signaling points. The lowest level of address is the individual signaling point or switch (SSP). It is called a member address but is often referred to as a point code.

A minimum network size is required to obtain a network level address. Smaller CLECs may not qualify for this address level. In this case, they are assigned to a cluster in a reserved group of network addresses (1 to 4). If they are really small, they are assigned a point code (member address) in network address 5. The cluster assignment is determined by the state in which it is located.

Intelligent Network (IN)

The last item to be addressed in this section is the difference between the SS7 or interoffice network and the intelligent network. While the transport of messages that form the intelligent network flows over the interoffice network using the SS7 protocol, the components that make up the interoffice network are distinct from the intelligent network. The intelligent network distinguishes the physical components of the switching network, such as STP, SCP, and SSP, from the service offerings themselves. The services offered to the subscriber are distinct from the switching mechanisms and protocols that facilitate and support them. It is a service-independent network. The intelligence is taken out of the switch and placed in computational nodes spread throughout the network.

The deployment of an IN in the telephone system has been gradual. The first step was to separate the service logic from the switching systems. This separation was done through the addition of the SCP component, which held the information necessary for the IN services. Now, of course, the SCP is considered part of the interoffice network, hence the confusion that often results in distinguishing the intelligent network from the interoffice or SS7 network.

■ SUMMARY

This chapter provided an overview of the OSI, TCP/IP, and SS7 protocol models and the ways they are used to describe how a communications architecture works. The idea of a service was introduced, and the distinction between services and protocols was made.

Although it was not discussed specifically, you should see that a layered reference model allows new layers to be added in response to changes in technology without altering the other layers. What is important is that the service interface stay the same. This advantage is key for any protocol developed under any reference model.

One important consequence of applying the OSI model to a communications protocol is high-layer compatibility with low-layer diversity. An upper layer of any layered protocol may remain constant when lower layers change. For example, a common application layer can be used throughout a product line. This feature has many benefits to the customer, user, and support staff because there is only one interface to the user. However, many of these products will have different connection requirements. In this situation, one common user interface to the customer of an organization's products can still accommodate the differences in product line requirements.

Peer layers in a communications architecture communicate by a virtual communication through the exchange of PDUs. These PDUs travel down the network architecture at the source and back up the network architecture at the destination. The bits transferred through the physical medium are appended at each layer. These bits surround and guide the application data from the source to the destination. This routing may include passage through many intermediate systems. Communication between and across these systems is coordinated by the appropriate layer.

The physical layer handles all electrical engineering questions and the interface to the physical medium. The data link layer handles the order in which stations talk and how communication between adjacent systems on the same network segment is conducted (node-to-node addressing). The network layer handles communication between the original source and destination systems (end-to-end addressing). The transport layer handles error recovery between the original source and ultimate destination systems and process-to-process addressing. The lower four layers of the OSI model are all that is necessary to describe any communication system.

REVIEW QUESTIONS

1. Define and discuss virtual communications and its importance.

2. Discuss the concept of service from a protocol perspective.

3. What are the two broad types of service that a layer can offer?

4. What do services consist of?

5. How are services used by source and destination systems?

6. How many classes of operation primitives are used to make up services?

7. Discuss the relationship between service and protocols as technology evolves.

8. Compare and contrast the roles of PDUs and IDUs.

9. What is another name for datagram service? Why is that name used?

10. Discuss why only the lower four layers of the OSI model are required by most communications systems.

11. Discuss the three-level address used in TCP/IP. What is each component used for?

12. Discuss, from a protocol point of view, how source and destination systems using the TCP/IP architecture communicate.

13. Discuss the importance of the RTO timer.

14. Describe why the interface between layer 4 and layer 5 of the OSI model is so important. What is the difference in services provided above and below this interface?

15. What are the two primary pieces of the data link layer?

16. What are the two types of services that any layer can offer to the layer above it?

17. Describe briefly how COS works.

18. Describe briefly how CLS works. What are its two variations?

19. What is the primary activity of the network layer?

20. List the seven layers of the OSI model.

21. Describe layer 0 and give an example.

22. Summarize the functions of each layer in the OSI model.

23. Define service access point (SAP) and interface data units (IDUs). Discuss and compare.

24. What is the meaning of virtual communication?

25. Define service and protocol. Discuss and compare.

26. State two service types and explain them.

27. List the four classes of operation primitives and describe their functions.

28. Describe the relationship among SAP, PDU, and service primitives.

29. What is a protocol data unit (PDU)?

30. Define MAC and LLC and explain their functions.

Wireless Layer Protocols

■ INTRODUCTION

This chapter will present, analyze, and discuss individual layer protocols that are referenced elsewhere in the text. Some will be the same as those used in various wireline protocol architectures; others will be unique to the wireless area. To understand fully the function of a newly developed protocol, it is instructive in many cases to begin with its wireline forebears. This chapter will focus exclusively on layer 2 to 3 protocols. Chapter 14 discusses, in detail, TCP limitations when used in a wireless environment. The discussion on CSMA/CA in this chapter describes the IEEE 802.11 wireless LAN approach and the hidden node problem, along with a solution. This material is required for the treatment of 802.11 in Chapter 15.

■ LAYER 2 PROTOCOLS

All members of the IEEE 802 family of layer 2 protocols are composed of two sublayer protocols; the medium access control (MAC) sublayer protocol and the logical link control (LLC) sublayer protocol. The MAC sublayer protocol forms the interface between the data link layer and the physical layer. The LLC sublayer forms the interface between the network layer and the data link layer. See Figure 4-1.

Figure 4-1
Data Link Sublayers

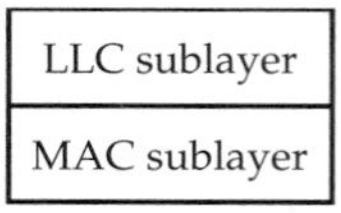

LLC sublayer
MAC sublayer

Because of this division of the data link layer into two distinct components, the functions of the data link layer are best described as flowing from the separate responsibilities of these two components. These sublayer protocols are discussed in the next two sections. Although these protocol components are properly referred to as sublayer protocols, for the rest of this discussion the *sub* prefix will be dropped and they will be referred to as the MAC layer and the LLC layer.

■ MEDIUM ACCESS CONTROL (MAC) LAYER

The name of the MAC layer reveals a lot about what it does. It controls the access of the network to the physical layer. Literally, it controls the access to the media in use by the network. This layer stands between the physical layer implementation, which is usually determined from the medium (or layer 0) that the network uses, and the media-independent software and hardware implementation that lies above it, the LLC layer. The MAC layer controls the way the bandwidth of the network is shared among the stations attached to it.

Not all protocols use a MAC layer. For example, if the protocol is designed to run on only one type of physical layer and is point-to-point, then there is no need for a MAC layer. A good example of a protocol that does not use a MAC layer is X.25. X.25 is designed for point-to-point communications in a twisted-pair physical layer. Therefore, there is no MAC layer. But the absence of a MAC layer doesn't mean that there is no data link layer. It just means that there is no shared medium that requires a specific access methodology to allocate the bandwidth. There is still a data link protocol and a frame definition, both of which are defined by the specific protocol used, the link access procedure (LAP).

The part of the data link layer that implements the access methodology is the MAC layer. It accomplishes this task by exchanging MAC frames. The MAC frame is the basic unit of exchange between devices on the network. Other frames are formed by building on the MAC frame, but the core of all data and control frames exchanged between devices is in this frame.

At the MAC sublayer of the data link layer, each protocol defines and uses a different frame format. Therefore, an understanding of the structure of this format can be very useful when diagnosing problems on a network. Many of the most powerful tools for diagnosing problems on a network are called protocol analyzers. They are designed to recognize the structure of the frame format.

Four key functions must be accomplished by the MAC frame structure defined by a protocol:

1. Synchronizing the destination station clock to the source station clock for accurate demodulation.
2. Identifying the start and end of each frame.

3. Identifying the source and destination address for the frame.
4. Checking for errors in the frame.

In principle, each frame format consists of two distinct portions: the first accomplishes the first function listed above, and a second addresses the rest. In the specific case of the IEEE 802.3, or Ethernet, family of standards, the frame format for this first portion is called a preamble. It is an alternating series of bits, several bytes long, that the station modem uses to synchronize its timing to that of the source station. In another example, the IEEE 802.5, or token ring, standards, no preamble is needed in the frame format because there is a constantly rotating frame, called a token, that maintains a common clock for each station.

The second function, identifying the start and end of each frame, can be broken into two separate tasks because the start flag and end flag can be treated differently. The start flag, called a start delimiter, is implemented in each of the IEEE family of MAC layers in the same way. A unique pattern of bits, 1 byte long, is used to signify the beginning of the data portion of the frame. The end flag, called the end delimiter, is also 1 byte long but is required in only some of the formats. For example, while the IEEE 802.5 format uses an end delimiter composed of a unique pattern of bits, the IEEE 802.3 format uses a length field placed in the second portion of the frame so that the end of the frame is determined by counting the number of bytes in the frame.

The third function, identifying the source station address and the destination station address, is accomplished in each format in the same way. In IEEE 802.3, 6 unique bytes identify the station address. These addresses are commonly referred to as MAC addresses and are implemented on the NIC card in the widely adopted IEEE 802.3 family of standards.

The last function, checking for errors in the frame, is done with a field that implements the cyclic redundancy code (CRC), which is explored in Chapter 8. In all IEEE 802 frame formats, the length of the CRC code is 4 bytes.

Every layer 2 MAC protocol has its own frame format or definition but, as has been pointed out above, all have to do the same kind of tasks in each protocol. Because of the operational similarity, they are all based on the high-level data link control (HDLC) format. This protocol is where the specific functional components of the frames that accomplish the tasks identified above are described.

High-Level Data Link Control (HDLC)

All modern data link protocols and their frame definitions are based on this protocol. HDLC is a bit level protocol because the bit positions themselves convey information. In HDLC, the information to accomplish the control tasks identified above are always placed at the same location. We will examine the evolution of this standard because it is so ubiquitous.

The first of these bit level protocols was designed by IBM for its synchronous network architecture (SNA) protocol and was known as the synchronous data link control (SDLC) protocol. This protocol was standardized, but at that time a proprietary name was deemed inappropriate for an American National Standards Institute (ANSI) national standard, so it was modified slightly and renamed advanced data communications control procedure (ADCCP). This national standard was then modified slightly again and renamed HDLC by the International Standards Organization (ISO). Because national standards bodies like to have unique items to standardize upon, the International Telecommunications Union (ITU) modified it again slightly, renamed it link access protocol (LAP), and used it in the widely adopted X.25 standard. Later it was modified again to link access protocol, balanced (LAPB), to track changes in the HDLC standard. When the integrated services digital network (ISDN) became available, LAPB was modified slightly and renamed again as link access protocol for the D-channel (LAPD). With the arrival of the wireless cellular standard, the global system for mobile communications (GSM), it was modified slightly again and renamed link access protocol for the D-channel, mobile (LAPDm).

It should be clear to you from this short history lesson that while protocols may have different names, they often have a history that allows you to capture the essence of how they work without knowing the precise details of the standard. In any bit level protocol, the position of the bits is descriptive, so these protocols are ideally suited for forming frame definitions. The HDLC frame will provide you with a good idea of how almost all data link layer frames are built. The modifications mentioned in the previous paragraph have to do with small changes in the field names and functions. For example, LAPDm modified LAPD by adding a few new, specific frame types, all based on the basic HDLC frame, required by the wireless channel and network layer protocols specific to wireless channels.

Each frame follows the HDLC protocol with small variations. As suggested above, some omit or change the use of one or more of the fields or differ in small details of interpretation. In each of the formats, the frame is transmitted in byte order, from left to right. The HDLC frame format used in IEEE 802 MAC frames is shown in Figure 4-2. Since the focus of this chapter is on wireless layer protocols at layers 2 and 3, the next section explores the important layer 2 protocol, LAPDm.

LAPD for Wireless—LAPDm As suggested above, LAPDm is very similar to LAPD, with the exception that new specific frame types are defined. These frame types are closely coupled with the network layer protocol service requests made to the data link layer. We will not go into the details of the coding of these protocol data units (PDUs), but we will briefly outline the frame types. Since LAPDm is a specific part of the GSM standard, some of the comments below apply only to that standard.

There are three broad classes of frame types defined by this protocol: data frames, command frames, and response frames. Data frames transport LLC data imported from the network layer. Command frames include all I-frames used to transport information (hence the name I-frame) from one station to another. A typical I-frame would be a frame

Figure 4-2
HDLC Frame Structure

Start delimiter	Source address	Destination address	Frame control	Data	CRC	End delimiter

Table 4-1
Sample GSM Control and Information Frames

Function	Name	Application
Unnumbered ACK	UA	Confirmation
Unnumbered Information	UI	Unconfirmed I frame
	I	Confirmed I frame
Receive Ready	RR	Okay to transmit
Reject	REJ	Negative confirmation

that supports the sequence numbers in an acknowledged communication. Another good example would be the frames that transport the window size negotiation between the two stations. I-frames are always command frames.

S-frames can occur as either command or response frames. A response to a window negotiation would be an example of an S-frame. In general, all signaling over the control channels and paging channels is accomplished with a combination of I-frames and S-frames. In GSM, a popular second-generation cellular standard, the broadcast control channel (BCCH), paging channel (PCH), and dedicated control channel (DCCH) channel signaling are all accomplished with the use of I-frames and S-frames. Table 4-1 defines a few of the more representative frames used in the GSM standard.

■ LOGICAL LINK CONTROL (LLC) LAYER

The other half of the data link layer protocol implements the service interface. In the IEEE 802 family of protocols, the LLC layer is always the same. Because the LLC layer is common to all the MAC protocols and the physical layer protocols below them, it looks the same to the next layer up the stack. Essentially, the LLC layer insulates higher layers from the differences between the various standards. This function is the first defining feature of an LLC layer.

While it is true that the capabilities of the lower layers differ in each protocol, the service interface remains constant. In other words, the way a layer 3 protocol requests services from a layer 2 protocol remains the same, independent of what that particular layer 2 protocol may offer as a service. For example, the data rate provided by 10BaseT or 16 Mbps token ring is an order of magnitude less than that provided by 100BaseT. Because they all use a common LLC layer, requesting data transport from layer 2 is the same to the network layer above.

The second feature is that the LLC layer offers a way to ensure a reliable data stream from source to destination station at the data link layer. This capability allows a layer 2 protocol to offer a reliable data stream service to the network layer above it. This capability is not widely implemented outside the IEEE 802 family of protocols.

The third feature of the LLC layer, something that is implemented in most network layer and transport layer protocols as well, is some method of flow control. Flow control at layer 2 is a way to control the transfer of frames between the source and destination systems that are communicating. In most cases, this function is used by the destination station to inform the source station that it needs to slow down or speed up. The idea is to prevent the loss of data when a destination system's receive buffer is filling up, or to tell the source to speed up because of excess capacity in the destination station's receive buffer. This function can be accomplished in several ways. We will explore a very basic approach, called stop and wait, after we describe the services that a data link layer protocol can provide to a network layer protocol. These services are accessed at the layer boundary between the LLC layer and the network layer at specific service access point (SAP) addresses.

The three primary service classifications that the data link layer provides to the network layer are listed below (sometimes these three services are known as LLC types):

1. Connection-oriented service (type one).
2. Acknowledged connectionless service (type two).
3. Unacknowledged connectionless service (type three).

The type one service, known as COS, implements a reliable service. Type one service does not use acknowledgments because it opens a connection, and all communication flows on that connection reliably. Thus, no acknowledgments are required. The second two types of service are really two subclassifications of the same connectionless type service, known as CLS. Type two provides for acknowledgments and type 3 does not. Type two service is the middle ground with respect to reliability between type 1 and type 3.

Frames are used to exchange information at layer 2. In some cases, the data being exchanged between two systems is large enough that it requires more than one frame. Each layer 2 protocol has its own maximum data field size, and this size is often exceeded by the information exchange. Therefore, most communications between source and destination systems consist of a large number of frames.

Wireless layer 2 protocols especially tend to use relatively small data field sizes in their frame protocols. The small size acts to increase the importance of the acknowledgment mechanisms that may be present in the data link layer; hence the critical nature of the classification of service type. The more frames sent from source to destination, the greater the probability that one or more will arrive garbled or not arrive at all.

Remember, protocols have a special and specific meaning for the word *service*. A layer always provides services to the layer above it and uses the services provided to it by the layer below it. In this section, these service types are discussed as only data link layer services. Recall from Chapter 3, however, that all layers provide services based upon these two models of service, namely, COS and CLS.

Just like the MAC layer, the LLC layer uses a frame structure. It is not based, however, on HDLC. That distinction is reserved for the MAC layer protocols. LLC frames are constructed by building upon the frame structure of the MAC frame. Essentially, a header is added to the MAC frame. See Figure 4-3. Recall the discussion in Chapter 3, where the two-part header of the data link was briefly mentioned. Figures 4-2 and 4-3 allow examination of this two-part header structure.

By examining Figures 4-2 and 4-3, you can see that the data link layer, with its two sublayers, actually applies two separate headers to the LLC data unit as it moves down

Figure 4-3
LLC Frame Structure

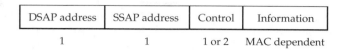

DSAP address	SSAP address	Control	Information
1	1	1 or 2	MAC dependent

the protocol stack. The first header, the LLC header, specifies the locations in memory, or SAPs, where the network layer requests services of the data link layer. The second header, the MAC header, specifies the MAC address of the source station. Thus, both the location in memory and the physical location of that memory are identified by the two-level addressing done at the data link layer.

The four fields shown in Figure 4-3 are defined as follows:

Destination service access point (DSAP): DSAP is the address representation for the SAP(s) located at the destination. While each SAP address is restricted to 1 byte, it may identify one or more SAPs for which the LLC information field is intended to be delivered.

Source service access point (SSAP): SSAP is the address representation for the SAP(s) located at the source. Just like the DSAP, each SSAP address is restricted to 1 byte, but each may identify one or more SAPs from which the LLC information field originated.

Control: The control field is where certain command and response data are contained. It is the place where sequence numbers, if needed for each frame, are located. For any connectionless service, sequence numbers are required.

Information: The MAC frame format for the particular MAC layer is placed in this location.

As mentioned above, flow control in the data link layer is a critical function. Not all stations can operate at the same speed. A slower station must have a way to tell a faster station that it is sending data to it, that it needs to slow down. A basic approach is the stop-and-wait protocol. In this approach, the source station sends one frame of data and then waits for an acknowledgment. No further data is sent until the destination station responds back to the source station.

While this approach has the merit of never overloading the destination, it does require an acknowledged service option, or LLC type 2. There is also the severe limitation of alternating or half-duplex communication. While almost all metallic and fiber channels are capable of full-duplex communication, this procedure is limited to half-duplex communication. The stop-and-wait protocol requires each station to wait for receipt of a message from the other before it can transmit one of its own. Finally, it provides no means to recover from a network the occasional lost frame. Clearly, if any frame is lost in this procedure, the communication stops as each station is locked into a procedure that requires it to wait for a response from the other station before transmitting again.

There are many advanced protocols for implementing the concept of flow control. Exploration of these approaches is outside the scope of this book. Many modern protocols use a variation of an approach called a sliding window protocol. These approaches to flow control allow full-duplex communication, include a procedure for recovering a lost frame, and other advantages. The next several sections describe the evolution of the CSMA/CA access methodology used in the IEEE 802.11 wireless LAN standard.

Most wireless systems use a variation of CSMA in their implementation. For example, the IS-95 cellular standard uses a slotted ALOHA channel for certain handshaking tasks between the mobile station and the base station. The IEEE 802.11 standard for wireless LANs uses a carrier sense mulitple access/collision avoidance (CSMA/CA) access methodology in the radio channel.

■ ALOHA

The core element of the widely implemented access methodology, carrier sense multiple access (CSMA), grew from the early access methodology called ALOHA, developed by researchers at the University of Hawaii. The problem was that Hawaii was a bunch of islands, with only one big computer on one island. They wanted to share access to the computer, and they needed a way to allow all the islands to access it through a satellite system. The satellite had one channel used for communication to the big computer. Everyone had to share this channel in some fair way. The ALOHA system was what emerged from this dilemma.

ALOHA works by letting each station or island transmit whenever it needs to. When only one island transmitted, and no one else did for the entire message length, this system worked well. In other words, each station is free to transmit whenever it has data to send. Occasionally two islands wanted to use the big computer at the same time, and a collision occurred. In a collision, both messages were corrupted. Then each station had to wait for some random amount of time and then try again. Whether they actually waited was not determined or required in a pure ALOHA environment.

This type of access works well only if the channel is not used very often. The fewer messages sent, the fewer collisions. In high-load conditions, few messages get through. ALOHA works well only if the channel utilization is less than about 20 percent, which means that the big computer can be "talked" to only about 20 percent of the time. Still, for that time, about 1970, ALOHA was a revolution in access protocols.

What is a collision, and what does it have to do with communications? Think about how ALOHA actually works: "each station is free to transmit whenever it has data to send." That situation sounds a little like anarchy. Imagine a room full of people, each with permission to speak freely and whenever they choose to. It is very likely that some will be talking at the same time others are. In some social situations, this overlap makes no difference. But on a network with only one logical channel, only one message can be transported at a time. What happens when a second station starts talking before the first is finished? The answer is, A collision occurs.

Slotted ALOHA

Slotted ALOHA is quite similar to ALOHA except that the opportunity to transmit is not totally random. Instead, the time is divided into a number of time slots, hence the name slotted ALOHA. When a station has information to send, it waits for a time slot and then transmits, instead of all of them attempting to send at once. This refinement significantly increases the channel utilization to approximately 35 percent. Just as in ALOHA, collisions do occur. The channel is divided up into several time slots, and each station must wait for the next slot to transmit. Collisions always occur at the beginning of the transmission and not later during the time slot. Therefore, if the station begins transmitting and does not immediately encounter a collision, the probability of the message getting through increases dramatically.

Spread ALOHA Multiple Access (SAMA)

This technique is a hybrid of ALOHA and a specific form of code division multiple access that may be used in some satellite architectures. This topic will be discussed in Chapter 10 because it is specific to satellite applications.

■ CARRIER SENSE MULTIPLE ACCESS (CSMA)

CSMA protocols refine ALOHA because they "listen" (carrier sense) prior to transmission. Slotted ALOHA uses time slots, which require each station to have the same sense of time and, thus, some kind of synchronization signal or external timekeeping agency. In CSMA, each station listens to the network channel to determine if any other station is transmitting. If there is a transmission, the station backs off for a random amount of time and checks again. If the channel is idle, the station goes ahead and transmits its frame. Sometimes, of course, there will be a collision. For this reason, CSMA should be used only with an acknowledging transport or network layer. Since the transmitting station has no way of ensuring a clear channel when it begins its transmission, it must rely on an acknowledgment from the destination.

Because any network, wireless or wireline, has some propagation delay, there will be occasions when two stations listen, find the channel idle, and begin transmitting approximately simultaneously. The second station, because of the propagation delay, does not hear the first station's transmission. It assumes that the channel is idle. Due to the effects of propagation delay, these two transmissions will collide. This problem increases in direct proportion to the propagation delay in the channel. The carrier sense approach still allows collisions because of channel propagation delay, but it is an improvement over ALOHA.

The propagation delay is normally very short compared to the message length, so most of the time a station will hear another transmission before it begins its own. The actual propagation delay depends on the average distance between the source and the destination, because the velocity of the transmission is constant. It is possible to measure the maximum propagation time from one station to any other station. When a station begins its transmission and no other station begins transmitting during this time period, the transmission will get through with no collisions. By the time this period has passed, every station has heard the transmission and will wait its turn.

As you might imagine, the effectiveness of CSMA increases as the frame length increases compared to the propagation time or maximum distance between stations. Therefore, the best environment for such an access methodology would be where the ratio of the distance between the stations and the frame size is fairly large, which is exactly the situation in both wireless LANs and most cellular systems. See Example 4-1.

■ **EXAMPLE 4-1**   A particular wireless LAN system has a cell radius of 1 km. This fact allows computation of the maximum propagation time. Using the traditional relationship between rate, time, and distance and the speed of light for the rate, the maximum propagation time for a cell with a diameter of 2 km is:

$$t_p = \frac{d}{r} = \frac{d}{c} = \frac{2 \times 10^3}{3 \times 10^8} \approx 7 \; \mu s$$

If the frame size is set to 1,000 bytes and the channel capacity to 11 Mbps, the maximum transmission time for any frame can be found:

$$t_t = \frac{\text{frame length}}{\text{channel capacity}} = \frac{8,000}{11 \times 10^6} \approx 700 \; \mu s$$

Note that even for a frame size as small as 100 bytes, the transmission time is still 10 times the maximum propagation time. Decreasing the cell size or increasing the frame size just reduces the chance of a collision and improves performance. As wireless systems increase bit rate, the performance of CSMA will degrade unless compensatory actions, such as limiting the frame length, are taken. This simple step will present potential interoperability problems down the road.

Carrier Sense Multiple Access/Collision Detection (CSMA/CD)

In carrier sense multiple access/collision detection, each station listens for a collision with another station's transmission (carrier sense/collision detection). If no transmission is heard, the station is free to transmit whenever it has data to send (multiple access). When a collision is sensed, each station backs off for a random period of time and then attempts to retransmit its message. This access methodology is widely used in wired LANs. Simply, it is a listen-before-talk protocol.

In wired Ethernet, both stations are on the same LAN, so they both hear pretty quickly that another is transmitting at the same time they are. Because each knows that simultaneous transmission will result in a garbled message, they both back off, or cease transmitting, for some random amount of time. Then each station listens again. If it doesn't hear activity on the channel, it starts transmitting again. Each station is at a different physical location on the network; thus, it takes different amounts of time for each station to hear another station's transmission(s). As a result, collisions are not uncommon. If the network is lightly loaded (in other words, if few stations want to transmit or use the channel at the same time), CSMA/CD does work well. The average load for this access method to work well is less than 50 percent.

Sometimes the station is successful in its attempt at communication, sometimes not. This situation is at the heart of why CSMA/CD has unbounded message delivery time under high-load conditions. Think about a network where every station wants to transmit at the same time. Only very infrequently will a station have a clear channel to transmit. The constant cycle of starting to send, only to hear a collision and then be forced to wait and retransmit, makes it very likely that the time to send a message can grow enormously. How long a particular station will have to wait to transmit grows exponentially with the amount of message traffic.

As an access protocol, CSMA/CD is a good choice as long as three conditions exist. The first is that the number of messages on the network is small relative to the bandwidth of the network. A good value to keep in mind is less than 50 percent of the bandwidth of the network is in use at any one time. This constraint will avoid excessive message delivery times that can be a problem in heavily loaded networks using the CSMA/CD access protocol.

The second condition is that message priority is not required. If every message has the same priority, then CSMA/CD works fine. Because of how the bandwidth of the network is allocated (transmit first and listen later), there is no way to identify any message as most important. Third, and this condition is related to the first, it is not important that a station knows if a message has been sent. Circumstances can arise where a message might get blocked for some amount of time. There is no way to guarantee how long a message will take to get through from source to destination. For certain systems where it is critical to know if a message has been sent, CSMA/CD is not a good choice.

Before discussing the particular approach of CSMA used in many wireless systems, a more detailed examination of CSMA/CD is appropriate. CD stands for "collision detection." This protocol element adds collision detection to CSMA, giving it the ability to determine if a collision has occured in the channel. If a collision has occurred, then both transmitting stations become aware of it and stop transmission. Without this capability, the channel is useless for the entire length of the original transmission. As we have already shown, the length of time that the channel is used increases as frame length increases. Frames thus have a tendency to grow large, increasing the desirabilty of the CD function.

In a wireline CSMA/CD access methodology, once the collision is detected, usually through a special jamming signal sent by the first station to detect the collision, both stations back off for a random amount of time and try again. In fact, the Ethernet standard requires frames to be of a minimum length so that it is certain a collision will be detected by all stations on the segment before that transmission can be completed. The functioning of the collision detection algorithm is thus guaranteed. This approach depends on one key aspect of a wired Ethernet environment that is not present in most wireless systems: unlimited bandwidth. In a wired environment, total control over the spectrum is easy. A guided channel is usually dedicated to that transmission. The atmosphere is an unguided channel shared among many users and natural phenomena. The bandwidth that is functionally free on a wired channel is very expensive on an unguided, shared one. Any transmissions in use effectively steal bandwidth from another source, usually in the same communication system.

Thus, the collision detection signal would consume bandwidth from every other station. This detection signal would somehow have to be guaranteed to reach every other deployed station in the local network. This second requirement is very difficult to achieve in an unguided channel. Given that fact, note that even if a collison detection system was in place, the bandwidth had been allocated, and a collision occurred only where the destination station could sense it, the collision detection algorithm would fail. Failure would result in reducing the channel utilization gain that the collision detection algorithm was designed to achieve. The critical insight to understanding this dilemma is realizing that the amount of wasted bandwidth in an Ethernet network is directly related to the time it takes for all potential transmitters to detect a collision. In an Ethernet environment, this time reaches a maximum for the two most distant stations both beginning to transmit at roughly the same time. As an approximation, the longer the Ethernet segment, the longer it would potentially take for a collision to propagate the entire length of the segment. Hence, twice the length of an Ethernet segment corresponds to the maximum propagation time.

This fact is the fundamental reason why Etherent segments have maximum lengths. It also explains why these lengths can be exceeded and yet the system does not shut down immediately. The channel utilization improvement of collision detection is reduced. This reduction leads to reduced maximum utilization before collisions become so numerous that few transmissions get through. Initially, this degradation is small, but it grows exponentially and can have a large impact quite quickly in networks that already feature high utilization factors.

Carrier Sense Multiple Access/Collision Avoidance (CSMA/CA)

CSMA/CA is the access technique used in the IEEE 802.11 wireless LAN standard. Probably the most important difference between wired LANs and wireless LANs is the inability to detect collisions in a wireless environment, where the transmit and receive antennas are one and the same in many implementations or, at a minimum, adjacent to each other. During any transmission, the transmitted signal dominates any potential collision signal such as is used in wired Ethernet. In general, the entire frame will be transmitted and an ACK indicating an invalid checksum will be sent when a collision occurs during transmission. This use of the channel is potentially very inefficient, and a way was needed to avoid the collision rather than detect it and back off. This collision avoidance was accomplished by modifying the CSMA algorithm with the CA module.

As a result of that evolution, CSMA/CA is also a listen-before-talk protocol, except that its reaction to a sensed collision is different than that of CSMA/CD. Before describing that process in detail, it is worth asking the question, "When is a collision most likely to occur?" There are three possibilities:

1. During an idle period
2. During a transmission
3. During the transition between the first two in this list

During an idle period, no other station is transmitting and some collisions will occur. The important observation here is that any collisions will be random in nature. There is no queue of stations waiting for a transmission to end to begin their transmission. During a transmission, after a very short period of time, all stations hear the transmission and thus wait. Collisions could occur during this period, but it is a very short one and, again, there is no queue of stations waiting to transmit.

Only during the transition between another station's transmission ending and an idle period will there be a large likelihood of a collision. During the original station's transmission, other stations will have a significant period of time to get data transmission requests from a higher layer and to wait for the transmission to end to begin theirs. A station ending its transmission is called releasing the channel. When this occurs, all stations start timers and wait for some random period of time. Each station then senses the medium and begins transmitting when the timer expires and no transmission is sensed.

The CA module used in CSMA/CA is designed to address this short period of transition described above. Since the CSMA core of the protocol is present, the random backoff described above remains unchanged from the wireline CSMA case. Stations will still collide when their timers expire relatively simultaneously. In CSMA/CA, however, instead of starting the timer again, as in CSMA/CD, every station holds its timer until the medium is sensed to be free if the medium is originally sensed to be busy.

The station then enters what is known as a backoff state. During this state, the station continually senses the medium and when the channel is clear, restarts the counter from its held position. This action significantly reduces the probabilities of repeated collisions as the timers expire. Equation 4-1 shows how the backoff time is calculated and illustrates its direct relationship to the slot time and collision window, the former of which can be modified through the simple network management protocol (SNMP) management information base (MIB). The collision window is a variable that increases as the number of retransmissions increases. This concept is discussed more fully in Chapter 13, which explores in more detail the effect of the transport layer on wireless systems.

$$\text{Backoff} = \text{random } (n) \text{ slot time}$$

$$0 \leq \text{random } (n) \leq \text{collision window} \qquad \textbf{(4-1)}$$

Figure 4-4 illustrates this algorithm of collision avoidance. It shows a block diagram flow structure of the CSMA/CA protocol.

Figure 4-4
CSMA/CA Block Diagram

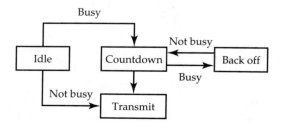

Like any probabilistic access mechanism, CSMA/CA works best if the number of stations is small and those stations have little to transmit. If the number of stations is small, then the probability of two stations transmitting at the same time is small. If the stations have little to transmit, the channel is idle more often.

MAC ARQ Protocol

CSMA/CA also incorporates a classic automatic repeat request (ARQ) protocol handshaking procedure that is initialized when the channel is clear and a station is preparing to transmit its data. This procedure is intended to ensure that collisions are recognized when they occur. Recall that the collision signal cannot be monitored by the transmitting station during transmission. Its own signal completely dominates any collision that might occur. The only way the transmitting station can know that a collision has occured is if the destination station sends back an NAK, usually indicating that the CRC check failed.

Hidden Node Problem   This ARQ protocol in the MAC also addresses one of the classic problems with linking multiple wireless stations into a network: the hidden node problem. This problem arises from the fact that, in general, all stations will not be able to hear each other because of the distance between them. As this distance increases, the attenuation of the radio signal also increases as the square of the distance. As a result, a node close to the edge of the network geographic area of transmission may not be heard by a station at the other edge.

In many ways this feature is advantageous because, without this attenuation, the whole concept of frequency reuse, which is fundamental to all cellular systems as well as 802.11, would not work. In the situation cited above, both stations could transmit on the same frequency without interference due to the range between them. However, consider what this situation means to a station in the middle, which hears both of the other stations. It could conceivably be buried under collisions from the two previous stations. These two situations illustrate the fundamental problem with carrier sense (the CS in CSMA): in a radio environment, it depends on local information. The use of this ARQ protocol at the MAC layer provides a way around this dilemma.

In this procedure, the source station first transmits a request to send (RTS) frame. If the destination station receives the RTS frame, it responds with a clear to send (CTS) frame. Note that every station in the range of either the source or destination hears one of these handshaking messages. Because all stations are in the range of one of the two communicating stations, the hidden node problem is resolved. If the original source station hears this frame, it then transmits its data frame and awaits an ACK frame from the original destination station. Figure 4-5 illustrates the content of the RTS and CTS frames and shows how the duration field is modified to keep the time slots synchronized.

At any point in this process, a collision might occur. Due to how the CA module works, however, it is most likely to occur during the original RTS, CTS exchange. These frames are short. If a collision occurs, the frame is retransmitted quickly and the channel is not held by a relatively long data transmission doomed to require a retransmission because of a CRC check failure. By the time the RTS, CTS exchange is completed, and other stations should be in their backoff state. They will hold there until the data transmission is completed and the channel is released. This procedure is illustrated in Figure 4-6.

Figure 4-5
RTS and CTS

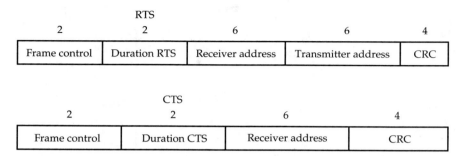

The duration RTS calculation (in ms) is determined by adding the times to transmit:

1. The information frame itself.
2. 1 CTS frame.
3. 1 ACK frame.
4. 3 SIFS frames.

The duration CTS calculation (in ms) is determined by taking the duration frame transmission time and subtracting:

1. 1 CTS frame.
2. 1 SIFS frame.

Figure 4-6
CSMA/CA Handshaking Procedure

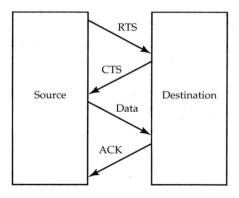

Interframe Spacing   The final piece to the CSMA/CA algorithm as used in the IEEE 802.11 wireless LAN protocol is the concept of a time slot, reminiscent of slotted ALOHA or TDMA. This protocol can be seen as combining slotted time division multiplexing (TDM) with CSMA/CA to avoid collisions occurring a *second* time. The standard describes a minimum slot time called an interframe space (IFS), which is the minimum period that can elapse between any two steps in the handshaking process described above.

Figure 4-7
SIFS and ARQ

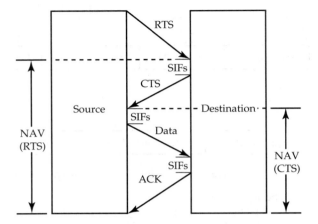

Four intervals are actually described in the standard, but only the two most important will be examined here. The first, and shortest, is the short interframe space (SIFS). This slot time is the smallest defined by the standard. The value of the SIFS varies with the physical layer chosen, but it is in the range of 20 to 40 μs. Figure 4-7 illustrates how the SIFS is used in the ARQ protocol. This figure also introduces a new variable called the network allocation vector (NAV). NAV timers work in conjuction with the duration field in the MAC frame to indicate a clear channel. Functionally, the NAV timers work by counting down from the value in the duration field of the last data frame transmitted on the channel until they reach zero, when the channel is assumed to be clear. This clear channel check is in addition to the check that the physical layer makes by listening to the channel. In almost all cases, the physical layer will indicate a clear channel prior to the applicable NAV timer expiring. When both layers agree that the channel is clear, a transmission can occur.

The other interval of importance is the distributed IFS (DIFS) period. This period is the time that elapses between data and contention or backoff process. (The contention interval is the interval between the end of the last ACK and the beginning of the next RTS.) Figure 4-8 amends Figure 4-7 to show the DIFS interval, and Figure 4-9 shows how the DIFS interval fits into the CSMA/CA block diagram.

IEEE 802.11 also implements an alternate access methodology called priority based access. Switching to this alternate access methodology disables the capability of ad-hoc networking and is used only on those links where quality-of-service parameters must be met. It is implemented with a polling scheme. Virtually all multiple-user 802.11 networks use the CSMA/CA access procedure, and both methodologies offer similar performance.

Polling To further increase the reliability of the wireless channel, the concept of polling was applied to the MAC layer. Polling can be viewed as an access methodology on its

Figure 4-8
DIFS Amendment to SIFS and ARQ

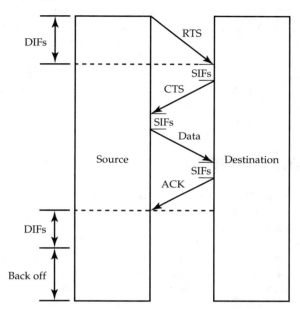

Figure 4-9
CSMA/CA with DIFS Interval

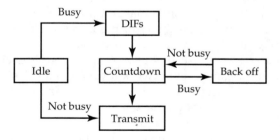

own, lying somewhere between pure TDMA and CSMA/CA. Polling requries that there be a master station to conduct the polling and that the master be an access point. The access point has total control over the channel. If polling is chosen as the access methodology, ad-hoc networking is disabled. By definition, ad-hoc networking is the establishment of a LAN without an access point. As implemented in IEEE 802.11, polling is different from pure TDMA. The frame size is not fixed but is under the control of the mobile station, depending on the amount of data to transmit.

Polling works by having the access point send a special frame, called a poll, to the mobile station. When the mobile station receives a poll addressed to its MAC address, it goes ahead and transmits its data frame. There is no chance for a collision in this case because the mobile stations transmit only under the control of the access point. Like the CSMA/CA mechanism, there is a time slot, DIFS, that must expire after the access point receives the data transmission from the mobile station before it can poll another mobile station.

Polling can be implemented in a fixed way, with the access point just moving through the mobile nodes in turn, or through a reservation procedure, where the mobile stations can request a poll when they have data to transmit. In general, a registration procedure is preferred over a registration slot mechanism when this access methodology is selected. The reason is that the implementation and management of such a procedure is independent of the number of mobile stations communicating with the access point. When the number of mobile stations is very small, then a slot mechanism may be easier to implement and manage.

■ LAYER 3 PROTOCOL

Since almost all networks now use, or are being transitioned to, the TCP/IP protocol model, this section will cover IP exclusively. Both IP and TCP functionality and limitations in a wireless environment are explained in Chapter 16.

The Internet uses an unacknowledged connectionless network layer familiar to many as Internet protocol (IP). The transport layer protocol, transport control protocol (TCP), is connection oriented and is responsible for performing any error recovery needed when using the Internet. Given the wide adoption of the TCP/IP model, the rest of this section will examine IP in a packet-switched environment, like that of the Internet. Before that discussion, however, it is important to point out how the terms *internet* and *Internet* are related. Any interconnected set of networks can be referred to as an internet. Only the internet called the Internet features World Wide Web (WWW) pages and is familiar to millions of people around the globe.

The major responsibility of the IP network layer is to route messages through the network. Just like any layer, the network layer uses service primitives and SAP addresses to accomplish this task. Again, there is a virtual communication between the source and destination network layers and a physical connection using the entire network. The packets the network layer exchanges are often referred to as datagrams because the service is a connectionless one. When the IP protocol is being discussed, the terms *packets* and *datagrams* are interchangeable.

It is important to emphasize that the network layer is a source-to-destination layer. It maintains a virtual communication from the original source to the ultimate destination. Contrast this setup with the data link layer, which maintains a virtual communication only between network systems on the same segment. See Figure 4-10. For example, the underlying network between system A and system B might be a low error-rate system, while the underlying network between system B and system C might be very error prone. Recall that, in an unacknowledged CLS environment, only the destination system can know if the message gets through, and also recall that the data link layer can communicate only between systems on the same network segment. You can see the importance of the network layer.

Ask yourself the following question, If there is a need for retransmission of a portion of the data received by system C, how does it communicate that to system A, the original source system, without some kind of virtual communication from A to C? The

Figure 4-10
Network Layer Virtual Communication Versus Data Link Layer Virtual Communication

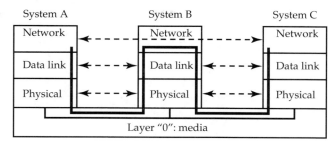

answer is, it can't without a network layer. The primary task of the network layer is to route the message through many hops or segments to the destination. If there are many alternate routes from the source to the destination, how does the network layer decide which to use? To perform routing, a routing algorithm is needed. An algorithm is just a specific approach to solving a problem. Algorithms used by network layers can plan how to route packets from source to destination.

There are two broad classes for routing. First, if the source network layer already has a virtual communication open to the destination network layer, then it just uses that one. Second, if the destination is a new one, then a new virtual circuit is set up. A router keeps tables of virtual circuits so that once it determines a path, it can use it again and again. To understand routing, we will explore routers and the algorithms used.

Router

A router can also be used to connect together two networks running different protocols or to extend a single network so it has all the functionality of a bridge. However, its real power comes from its additional intelligence and switching capabilities. In fact, one definition of the term *router* is "a device that selects the correct path for a message." Routers are layer 3 devices. They connect networks with a common network layer. Therefore, a router performs two activities:

1. Finds the best path, or least link cost, to route packets using a routing algorithm
2. Switches packets through an internetwork using the MAC address (bridging functionality)

This section will focus on the first of these tasks because it is unique to a router. The task of how to choose a path to pass a packet to, thus to route it to its destination in the shortest time or most reliable manner, can be very complex. Imagine that you are at your PC in San Francisco sending an e-mail to your friend in Hawaii. As you send your message, a router must decide to send it across the undersea trunk to Hawaii or perhaps route it through New York, London, and Tokyo to get to Hawaii. Obviously, it is going to make a difference which path your message takes. Routers examine every message they receive and decide which pathway to the destination is best or has the lowest link cost.

Routers are used in almost every network that features both LAN and WAN segments. Figure 4-11 will help illustrate how they are attached. Each router has several

Figure 4-11
Router LAN-WAN-LAN Hookup

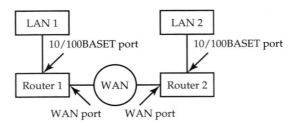

interface ports. These ports always include at least one WAN port and one LAN port, as shown in the figure. Here the LAN port is assumed to be Ethernet 10/100BASET. In principle, any LAN can be attached. Usually several ports of both types are available, or card slot(s) are provided for user configurability.

Often, the router also provides for a serial interface to link serial devices such as a channel service unit/digital service unit (CSU/DSU) to the router. The router stands between the address space on the LAN and the address space on the WAN. Its routing table contains the information on how to route messages from one LAN through the internetwork, or WAN, to the other LAN. As discussed below, this routing decision can be made in several ways.

There are many fundamental ways that a router can make a routing decision on which path to take. However, all the decisions can be sorted according to four basic algorithmic approaches:

1. Static
2. Dynamic
3. Flat
4. Hierarchical

Routers also feature extensive network management capabilities. *Network management* is a term used to describe the statistics of the traffic flow on the network and the status of the components that make it up. The router gathers this information and can present it to a user through a PC or terminal interface. By examining this data, skilled practitioners can identify trouble spots and, with sophisticated approaches, predict potential trouble spots before they occur. Routers can also be used to facilitate network security by blocking certain routes from users. Network management is critical to the reliable operation of a network.

Figure 4-12 illustrates the operation of a router, an operation similar to those used in Figure 4-10, where the bold line shows the actual path of the packet as it moves from one system to another through a router. Figure 4-12 also indicates that a router can be used to facilitate communications between systems that use two different protocol stacks at layers 1 and 2. In this case, the two protocol stacks must have a common protocol architecture at layer 3 and above. Layers 1 and 2 can be different. A good example might be an Ethernet and token ring network, where both are running the TCP/IP protocol stack over the different LAN technologies.

Figure 4-12
Layer Perspective of a Router

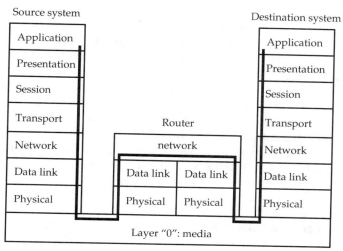

Layer "0": media

Routing Algorithms

To perform routing, a routing algorithm is required. An algorithm is just a specific approach to solving a problem. For example, to add two numbers, you usually add the second number to the first. Addition can be done the opposite way, but this is the algorithm that we will use. Similarly, routers use algorithms to plan how to route packets from source to destination.

A routing algorithm is used to build a routing table that lists the paths to all LAN and WAN network segments connected to the network. It constructs this table by determining a path to a particular destination and then noting several features about that path. The two most important features are the link cost and the address of the "hop" to the next network on the way to the destination. The "hop" address is the address of the next router between source and ultimate destination. There may be many hops between the source and destination. At each hop, the packet is routed to the next hop until it ultimately reaches its destination. The many ways of determining link cost will be discussed later.

As mentioned, all routing algorithms can be sorted according to four categories. We will examine these four operational categories in pairs because the first two are closely related, as are the second two. The first category pair, static or dynamic, is very straightforward. All four categories define possible algorithmic approaches that the router uses to construct its routing table.

Static or Dynamic Algorithms   Static algorithms always use the same path, independent of any special situations. Dynamic algorithms take into account the time-varying nature of the path and try different paths in an attempt to find the best one in real time. In an analogy with automobile traffic, dynamic algorithms try alternate routes, sometimes cutting through subdivisions, etc. Static algorithms get on the expressway like always, no matter how congested it is. In a static algorithm, the routing never changes once the routing table is built. Once a path is found for a message, it never changes until the routing table is changed, usually manually.

A table constructed with a static algorithm is predetermined, and the router does not attempt to change the table or react to changes in the network. Typically, this table of routes is computed once, when the network is first turned on, and then ignored.

The dynamic approach requires the router to figure out on its own which path is best, which path reacts to the real-time status of the network. The router does this job by periodically sending out messages to other routers and asking them for any "good" paths they might have found to any destinations that this router needs. The word *good* is usually defined by some concrete parameter or combination of parameters, such as best quality, least cost, shortest path, etc. (The general way to refer to any practical group of these parameters is link cost.) It also sends out test messages, times how long different paths take, and stores this information in a routing table.

Static algorithms do not respond to changes in the network; dynamic algorithms do. Both need a way to determine the paths, and this method depends on how the network is set up. This decision is the domain of the second pair of algorithmic categories: flat or hierarchical algorithms.

Flat or Hierarchical Algorithms   A router can assume that networks are interconnected in one of two ways: flat, like lines on a piece of paper, or hierarchical, like in a pyramid. In a flat architecture, all routers must be aware of all other routers, because all routers are considered part of one domain. *Domain* is a term describing a group of nodes. Flat algorithms assume that every domain can communicate directly with every other domain. Because of the complexity of the paths determined by this approach, most algorithms today use a hierarchical approach.

In a hierarchical approach, routers are grouped into domains hierarchically; that is, not every router can communicate with every other router. Instead each must follow a route constructed of a group of backbone routers that interconnect domains. Think of the backbone routers as wide and fast expressways and domains as exit ramps from the expressways. At each exit ramp is a router that stands between the backbone and the nodes in the domain.

The biggest advantage of hierarchical routing is obvious: it minimizes the traffic that occurs inside a domain. Since all traffic not destined for a node inside a particular domain does not pass through it, "cross-town" traffic is minimized. Similarly, most traffic in a domain never leaves that domain, so the traffic on the backbone is minimized as well. Several layers of domains may be constructed hierarchically. Additionally, in a hierarchical approach, the routers connected in that domain need to know only about the other routers inside that domain, making the routing tables relatively small and simple. This simplicity is another major advantage of hierarchical routing.

A domain is usually an organization, for example, the network that connects the workstations of everyone in a particular office or school. Most of the traffic generated, like e-mail to associates or fellow students, stays inside the domain. The congestion on that network is due to the traffic generated by users of the network if the routing is done hierarchically. In the case of a flat algorithm, messages going to some external domain destination would also be present.

Hierarchical routing is further subdivided into two subsections, called interdomain and intradomain algorithms. As you might imagine, the routing that must be accomplished by the backbone routers is of a different order of complexity than the routing accomplished inside a domain. Backbone routing is usually done with an algorithm most efficient in routing interdomain packets. Inside a domain, routing is accomplished with an algorithm most efficient in routing intradomain packets.

Link State or Distance Vector Algorithms

The vast majority of routers use dynamic hierarchical algorithms. Two families of these algorithms are widely used when IP is the layer 3 protocol. This section describes these two different approaches, the link state family of algorithms and the distance vector family of algorithms, to implementing such algorithms. In each family are specific product standards that will be used to illustrate the two approaches.

Link state routing is the more widely used today, but many networks, including major portions of the Internet, still use distance vector routing. A widely used distance vector algorithm is called routing information protocol (RIP). A widely used link state algorithm is open shortest path first (OSPF). Since these protocols are widely known and are often used as the definition of the two approaches, the discussion below refers to the two implementations, RIP or OSPF.

In a nutshell, OSPF floods routing information to *all routers* on the internetwork sending that portion of its routing table describing the link cost to every other router it knows about. RIP does the same but only to *routers it is connected to,* typically those in its domain. Before continuing, we will say a few words about flooding, a basic concept of most routing protocols.

Flooding   Flooding is just what you might expect from the name. It is a way to find paths on the internetwork. Each router floods the network with test messages, and each message counts the number of hops. Flooding is a brute force method of determining the shortest path between any particular source and destination. Flooding always finds the shortest path because it tests each and every path due to the flooding of the network.

As the message packet is received at each router, each router then sends out the message again to all its attached routers, except the one from which it received the message. This technique has the potential for creating a large amount of excessive traffic for two reasons: first, because of the geometrically increasing messages retransmitted after each hop and, second, because many packets will find their way back to the retransmitting node along another path.

A disadvantage of this approach is that a lot of excess traffic that has no value to any user is produced. Each router on the network sends out test messages on every line it has to every other router. It is possible to limit the excessive nature of flooding by applying the random walk idea from statistics. In this refinement, each router sends a *single random destination message* to search out unknown paths at random. This method can be effective and limits the number of messages sent by each router in the network, thus mitigating the main problem with flooding.

RIP and OSPF

Routing Information Protocol (RIP)   RIP uses the algorithmic approach known as distance vector routing to find the shortest path to each attached network. Each attached router maintains three vectors. The first vector represents a link cost value to every other inter-mediate system (IS) (router) that it is connected to. Thus, each router maintains a cost to reach every other attached network. The second vector represents the distance or delay from the router to the attached network, and a third keeps track of the current minimum delay route from the router to the attached network. Thus, the router has three pieces of information called vectors:

1. Cost to any attached network
2. Delay to any attached network
3. Current best path to any attached network

About every minute, each router exchanges its distance/delay vector with only its directly attached neighbors. On the basis of the incoming vectors that each node re-ceives, each router updates its distance/delay and current link cost vectors, thus mini-mizing the distance/delay to those directly attached neighbors. These updates are then used to update the cost vector.

This approach has a problem when it is used with IP as the layer 3 protocol because IP operates asynchronously. Implicit in the idea of RIP is that all incoming vectors arrive almost simultaneously. In an asynchronous network such as TCP/IP, simultaneous ar-rival is not possible. Additionally, some vectors may not arrive at all because they are lost in transmit. RIP does not do any error recovery at the transport layer.

If a router does not respond after three exchanges of vectors, RIP assumes that path is gone. Vectors arriving at different times, or not at all, can cause the protocol not to keep an accurate least distance/delay map due to the changing conditions on the network. By adding additional complexity to the basic distance vector approach used to keep its vec-tor tables current, these difficulties are minimized.

The real problem with RIP is that it responds very slowly to changes in the network topology because the routing table is distributed throughout the attached networks. This wide distribution results in messages not being sent over the least link cost (typically least delay) path. Additionally, RIP allows only sixteen levels of path cost, resulting in a granularity inappropriate to the highly interconnected and large networks of today. These two concerns led to the development of OSPF, described next.

Open Shortest Path First (OSPF)   OSPF uses a simple method of determining the delay to any other attached network. This method is called link state routing. In this approach, each router is responsible for finding out the location and delay to each of its neighbor-ing routers. Once this data is assembled, each router sends out a link state message packet to each of its identified neighbors, informing them of what it has found out. Each router that receives such an information packet then forwards it to all of its neighbors, and so on. In this manner, a table indicating the shortest path to every other router on any attached network is gradually built up *in each router*. Each router in the configura-tion knows the shortest path to any other router.

When a router is first turned on, it determines a cost/delay value to each attached net-work by sending out the link state message packet to all attached routers and measuring the round trip delay. This delay is divided by two, and the result provides a good measure of path length to the router. By performing this task periodically, the router provides a way to measure variations in path delay. As we saw in the RIP situation, both implementations use flooding. RIP used messages called vectors; OSPF uses messages called link state messages.

This traffic raises the issue of when to update, similar to that experienced in RIP. Link state routing labels each link state message packet with an age value. At each router where a message packet arrives, the router decrements the age value by 1, then sends it on in the manner described above. Once the age value reaches 0, the link state message packet is discarded. In this manner, old routing information, contained in young link state packets, is automatically discarded. This method both limits the number of packets being generated and eliminates the problem of when to update the routing table because old information never replaces newer information.

Once the router has constructed its table, it sends out to all routers on any attached network, not just those it is directly attached to, its set of costs. This information is sent out again when the periodic measurements it makes reveal a significant change in path delay or cost, which typically happens only when the topology changes because a router goes down or a new path is established in the topology. This feature tends to minimize excess traffic on the network. Therefore, each router has a table that lists the cost to every other router in any attached network. Additionally, each router must respond to each update it receives.

The cost linked to each path is usually defaulted to the delay the message experiences on that path. This time delay is represented to a first order well by the number of hops. Each time a message passes through a router, it is called a hop. This simple approximation allows the network administrator to set the cost on any specific link by configuring the router appropriately. Essentially, a cost number is associated with each port on the router. These link costs, also known as types of service (TOS), are maintained in several different ways, each representing cost from a different perspective. Each router maintains at least one of these cost structures and may maintain all five. The message packet sent out to determine the path cost indicates which TOS it is requesting.

TOS 0: minimum number of hops to destination; the default approach

TOS 2: minimum monetary cost; usually funny money used for internal administration

TOS 4: maximum reliability of network segments; usually error-rate driven

TOS 8: maximum throughput; data rate of network segments

TOS 16: minimum propagation delay

Other protocols route packets through the internetwork. In addition to the two already discussed, other protocols include:

1. Interior gateway routing protocol (IGRP)
2. Exterior gateway protocol (EGP)
3. Border gateway protocol (BGP)
4. Intermediate system to intermediate system (IS–IS)

Link Cost

Basically, link cost is what you would think it would be from the name. It is defined between source and destination. This link cost is measured and then sorted among all other paths between the same source and destination. The smallest link cost wins, and all messages from a particular source route that path to a particular destination.

Now a good question is, "What defines link cost?" There are at least three ways that link cost is measured. Probably the most primitive technique would be to go to a map and note the number of miles between source and destination and enter that as the link cost. For example, a path from Detroit to Chicago that routes through Atlanta would have a larger link cost than one that proceeds directly between the two cities.

Another way would be to count the number of hops between source and destination. Each hop is a router node and can be easily counted. This technique assumes that the number of routers between the source and destination is representative of the magnitude of the link cost. This assumption is not always the case, but it can work a first approximation.

Another way would be to time a packet to completing the path. If timing is done only once, then it is an example of static routing. If regular tests are carried out and the routing table is updated on the completion of each test, this type of algorithm can become a dynamic approach. In the dynamic approach, the smallest link cost is defined to be the shortest time. A nice consequence of the dynamic approach is that the table tends to reflect actual conditions on the network if the testing is carried out on a regular basis. To many users, especially on the Internet, the shortest time is the measure of interest.

The most sophisticated approaches to link cost computations combine elements of two or more of the above basic approaches. Some are very complex, but all strive to achieve some measure of smallest link cost and use that idea alone to complete the router table.

Congestion and Fairness

Once the choice of algorithm is made, a routing table is constructed according to that algorithm. Then two situations can arise. First, if the source network layer already has a virtual connection open to the destination network layer, then it just uses that connection. Second, if the destination is a new one, then a new virtual circuit is set up. Routers keep tables of virtual connections so that once they determine a path, they can use it until, in a dynamic environment, a better path is found. In a static environment, the router uses the identified path for every message to that destination.

If a router table becomes corrupted, the entire communication between source and destination can be corrupted. This predicament explains why some prefer static routing. If it cannot be updated, the routing table cannot get corrupted. In a corrupted environment, the router keeps trying the same path to send the messages. If that path is wrong, it can take some time to discover it. This situation is further complicated by the fact that only rarely does the source to destination communication take place directly. Usually there are several intermediate hops taken. For example, if you send an e-mail to a friend on the other side of the world, the message goes through several internetworks that connect to others, which finally connect to the LAN your friend's station is connected to. If one of the routers were to get confused about how to route to the next router in line, not just your message but all others going through that switching point would be disrupted.

All routing protocols are rated according to how fair they are. Fair routing protocols consider not only the end-to-end efficiency of the routing path but also the message traffic between hops, or congestion. The more congested a path is, the less desirable it is for routing traffic. For example, when an expressway is congested, most would like to find an alternate path if they can. The expressway, when not congested, is the shortest time path from source to destination; however, when it is congested, sometimes by local traffic, it is no longer the quickest path. If you have ever tried to take your regular path home from the office when there is a special local event, like a ball game, the additional local traffic overwhelms the efficiency of that path. If you imagine yourself as a router, your job is to find the optimum route for any traffic that you are responsible for forwarding, taking into account the local situation.

Local traffic isn't the only factor that can alter the efficiency of a particular path. For example, long-distance traffic patterns change dramatically on a holiday weekend. Again, trying to take your regular path home from work may not be desirable on the Friday before a long weekend.

This congestion on a network, just like on a highway, increases the time the message takes to get from source to destination. In communications, we say that the performance of the network degrades. Again, this expression is not dissimilar from what you would say of a jammed expressway, although perhaps you would say it with less frustration.

■ SUMMARY

This chapter has described aspects of several critical layer protocols widely used in both wired and wireless architectures. In many cases, later chapters of the text will assume that the material in this chapter has already been learned. While some protocols covered in this chapter may be familiar to many, others may prove to be new.

REVIEW QUESTIONS

1. What are the two primary pieces of the data link layer?

2. Give a brief description of how CSMA/CD works.

3. What is a hidden node problem?

4. What are short interframe space (SIFS) and distributed IFS (DIFS)?

5. State the difference between CSMA/CD and CSMA/CA.

6. Explain polling.

7. Describe static and dynamic routing algorithms.

8. Describe flat and hierarchical routing algorithms.

9. Describe link state and distance vector routing algorithms.

10. Explain flooding.

11. Which factor must be considered when choosing a routing algorithm to route a packet from source to destination?

12. Describe the role of IP and TCP.

The Wireless Physical Layer

Digital Modulation

■ INTRODUCTION

In this chapter, we will explore several effective modulation techniques for communicating digital signals. These digital techniques are in wide use today and are used almost universally in wireless voice and data communications. A brief summary of the concepts of bandwidth efficiency, symbol and baud rate, and M-ary encoding will begin the chapter. After an introduction to basic analog AM modulation, the three basic types of digital modulation, ASK, FSK, and PSK, are introduced. Specific attention is paid to the methods used in wireless systems: GMSK and DQPSK. Emphasis is on bandwidth and spectral issues. Constellation diagrams, error distance high-density modulation techniques, and a brief discussion of demodulation in a wireless environment follow. The chapter closes with several representative BER graphs and how to use them to gauge performance of a modem. Specific development of the C/N and E_b/N_o relationships will be covered in Chapter 9.

Communications systems that use the principles of digital modulation go by the general name of digital radio. Digital radios, like modems, use pulsed waveforms as the modulating signals. They use analog modulation techniques, sometimes implemented digitally, that result in the information being carried on a carrier signal. This book classifies all modulator techniques into three classes. These classes specifically ignore the implementation technique used to construct the device, which may be analog or digital. Instead, the focus is on the form of the input and output signals. Figure 5-1 clarifies this terminology.

Part (a) of the figure shows a classic analog modulator with an analog input and an analog output. A good example of this type of modulator is the one used in the original

Figure 5-1
Classification of Modulation Techniques: (a) Analog Modulator; (b) Digital Modulator; (c) Pulse Modulator

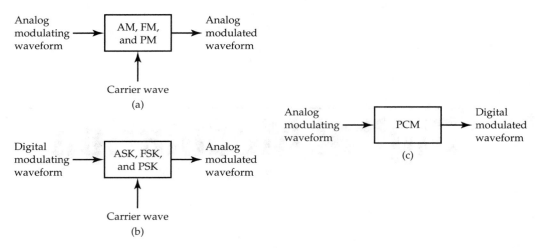

AMPs cellular network. Part (b) shows a digital modulator, which uses a digital form of the input signal. A good example of this type of modulator is the one used in the GSM cellular network. Part (c) shows a pulse modulator. This modulator accepts an analog signal and generates a digital output signal. A good example is a PCM coder, which is not considered digital modulation. For that reason, it will be discussed in Chapter 6 as being representative of pulse modulation.

■ BANDWIDTH EFFICIENCY

All modulation techniques can be characterized by their bandwidth efficiency, which is a measure of how efficient the modulation technique is in terms of how much bandwidth it requires. For example, a modulation technique with a bandwidth efficiency of 1 would mean that, for every 1 Hz of bandwidth occupied by the double-sided modulated signal, 1 bps was transmitted. Many modern modulation schemes feature bandwidth efficiencies much greater than 1. To determine the bandwidth efficiency of a modulation technique, simply divide the input data rate by the bandwidth of the output signal. This process is shown in Equation 5-1 and illustrated in Example 5-1.

$$BW_e = \frac{DR_b}{BW}$$ (5-1)

BW_e = bandwidth efficiency
DR_b = binary input data rate
BW = transmission bandwidth

■ **EXAMPLE 5-1**   Find the bandwidth efficiency of a particular modulator that has a transmission bandwidth of 1 MHz and an input binary data rate of 1 Mbps.

$$BW_e = \frac{DR_b}{BW} = \frac{1\ Mbps}{1\ MHz} = 1$$

It is also possible to talk about the concept of bandwidth efficiency by using the concept of the number of input changes required to produce a single output change. This topic is explored next.

■ M-ARY ENCODING

M-ary encoding (M is short for multiple) is directly adapted from the word *binary*. It can be defined as the power of 2 of the number of input changes required to produce a single output change. It can also be defined as the number of states of the output modulated signal. For example, in BPSK, there are two phase states, 0 and 180 degrees, so they are said to be 2-ary systems. A system with four amplitude levels would be called 4-ary. A system with four phase states in its output spectrum would also be called 4-ary.

The essential idea here is to capture a way of speaking about how many input signal changes are required for a single output change. Look at the mathematical definition, expressed by Equation 5-2, for M-ary in terms of the number of input signal changes required for a single output change:

$$M = 2^N \tag{5-2}$$

M = measure of the density of signal states
N = number of input changes to produce one output change

For BPSK, which is 2-ary, a single input change, from a logical 0 to a logical 1, results in a modulation signal that changes once. For QPSK, which is 4-ary, two changes, $N = 2$, are required for each output change. QPSK is twice as efficient as BPSK.

Bandwidth efficiency, sometimes called information density, is a measure of the density of signal states. It is not the same as the concept of information capacity discussed in Chapter 1. The higher the density, the higher the efficiency of the modulation technique. Additionally, the higher the density, the less bandwidth required for a given input signal rate. Therefore, large values of M mean less transmit bandwidth for a given input data rate and hence, high bandwidth efficiency.

■ DI-BITS, TRI-BITS, AND M-ARY ENCODING

When levels of modulation where M grows greater than 2 are discussed, a couple of new terms are used. It will prove useful to describe them in terms of M-ary encoding. For 4-ary systems, the binary bit streams are encoded as di-bits. They are accepted by the modulator in pairs. For 8-ary systems, the binary bit streams are encoded as tri-bits. They are

Figure 5-2
4-ary Encoding and Di-bits Illustrated

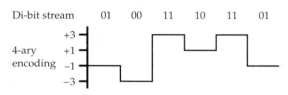

accepted by the modulator in threes. To illustrate the concept of how di-bits and 4-ary encoding are coupled, see Figure 5-2.

Tri-bits and 8-ary encoding proceeds in exactly the same way, except there are eight amplitude levels instead of four. Note also that all the encoding values have the same voltage difference. Here, it is 2 volts. This convention ensures that the decision threshold always works with the same minimum noise margin for any di-bit code transition. In the following sections, the modulating waveform will always be referred back to its binary bit rate. It is traditional, however, to show a single bit stream into the modulators, and Figure 5-2 shows how they are distinguished.

■ SYMBOL AND BAUD RATE

The traditional way to talk about bandwidth efficiency has been the baud, which is used to define the number of symbols/second in the communications channel. Baud rate is defined as the rate at which the modulated carrier changes in response to a change in the input signal. Baud rate is defined only for the modulated carrier signal in the communications channel. For example, a BFSK system that has a modulating data rate of 1 kbps and a QPSK system that has twice the modulating data rate, or 2 kbps, would both have the same baud rate, or 1 kilobaud. QPSK has twice the bandwidth efficiency of BPSK; therefore, it can represent twice the input signal rate with the same number of changes in the modulated carrier. Most modern system descriptions have dispensed with the term *baud* and instead use either the M-ary designation or talk about symbols/second explicitly. We will follow this trend.

Although this chapter covers digital modulation, it is appropriate to introduce briefly several basic concepts of modulation before discussing the various digital modulation techniques. We will start with the straightforward analog modulation technique, amplitude modulation.

Three signals are shown in Figure 5-3. The first is the carrier frequency. This wave is where the information is "carried"; hence, the name *carrier frequency*. The second signal is the modulating signal; it modifies or adjusts the carrier wave and represents the information to be transmitted. The terms *information signal* and *modulating signal* are used interchangeably. The modulating signal is typically a baseband signal. Some examples include voice and music, as in AM radio, or binary data streams when sending

Figure 5-3
AM Signals

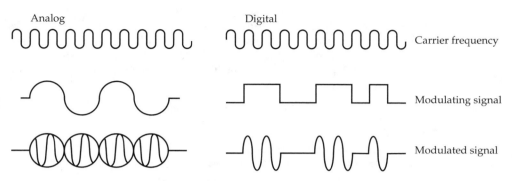

data. The final waveform is the modulated signal. Note the terminology: the carrier frequency is modulated by the modulating signal, which results in the modulated signal.

◼ DOUBLE SIDEBAND, SUPPRESSED CARRIER AMPLITUDE MODULATION (DSB-SC AM)

The amplitude modulation process can be represented by Equation 5-3:

$$m(t) = a(t)\cos(\omega_c t) \tag{5-3}$$

> $m(t)$ = modulated signal
> $a(t)$ = modulating signal
> $\omega_c = 2\pi f_c$, where f_c = carrier frequency

Note that this equation represents a cosine wave modulated by an information or modulating signal given by $a(t)$; $m(t)$ is the modulated signal that results from a cosine wave being multiplied (or modulated) by a modulating baseband signal containing the information to be transmitted, or $a(t)$.

So that the period of the waveform can be read directly from the argument of the cosine function, it is expressed in radians/s. To demonstrate, ask yourself, In which form below is it clearer that the cosine wave is completing 100 full cycles?

$$\cos(100t) \quad or \quad \cos(628.3t)$$

Obviously, either method of expression is equivalent, but the first more readily yields the information about the number of full cycles. In this chapter and the next, radians will be used almost exclusively. Convert if you wish by recalling that $2\pi f = \omega$.

The exploration of Equation 5-3 will prove very useful and will introduce terminology that will form the basis of much of this chapter. To begin, examine $a(t)$ more closely. Now $a(t)$ is the modulating signal. It is multiplying the carrier frequency, represented as the cosine function. The modulating signal is the signal where all the information to be sent is contained.

For example, in AM broadcast radio, the modulating signal is the announcer's voice or the program material you are listening to. The program material can be anything: music, sports, commentary, advertisements, etc. It is possible to write a representative equation for the modulating or information signal. Equation 5-4 represents the information signal as one cosine wave oscillating the modulating signal frequency, ω_m, with a peak voltage of E_m. Instead of a complex signal, like that for the program material above, we have simplified the modulating signal, representing it as a single frequency.

$$a(t) = E_m\cos(\omega_m t) \tag{5-4}$$

E_m = peak voltage of modulating signal
ω_m = frequency of modulating signal, measured in radians per second
$\omega_m = 2\pi f_m$, where f_m = modulation frequency

By inserting the expression from Equation 5-4 for $a(t)$ into Equation 5-3, $m(t)$ becomes:

$$m(t) = E_m\cos(\omega_m t)\cos(\omega_c t) \tag{5-5}$$

By a very useful mathematical relationship, you can always write the product of two cosine waves as the sum of half the amplitude of their difference, shown in Equation 5-6:

$$m(t) = \frac{E_m}{2}\{\cos[(\omega_c + \omega_m)t] + \cos[(\omega_c - \omega_m)t]\} \tag{5-6}$$

The result in Equation 5-6 is important and needs to be examined closely for what it will tell about most forms of modulation, including AM. Note first that the original modulating signal, at the modulating frequency, is not present in the modulated signal. Note also that there is no sinusoid oscillating at ω_m. Instead, it has split into two pieces, each of half amplitude, each shifted in frequency by the modulating frequency. Therefore, when you modulate, or multiply, one frequency by another, AM, the result is that both of the original frequencies disappear, and two new frequencies result.

These new frequencies are closely related to the two original ones: the modulating frequency shifts the carrier frequency in both a positive and a negative direction. Examine Figure 5-4, which illustrates this phenomenon. The figure shows the frequencies in frequency space, or the way they would be viewed with a spectrum analyzer. It will be very useful to be able to analyze modulation practices with this type of graph. It uses frequency, rather than time, as the horizontal axis. The farther to the right the signal is, the higher in frequency it is.

You should be able to make two observations from Figure 5-4. First, there are two components, which are called sidebands in communications. This type of AM is called double sideband (DSB) modulation. Second, because the carrier frequency is absent from the modulated signal, this type of AM is called suppressed carrier (SC) modulation. Because the carrier frequency is not present, it is suppressed. These two naming

Figure 5-4
DSB-SC AM Spectrum Analyzer View: (a) Premodulation Signals; (b) Postmodulation Signals

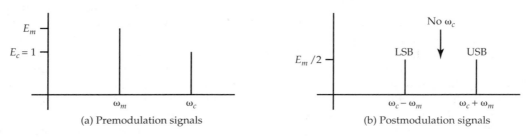

(a) Premodulation signals

(b) Postmodulation signals

Figure 5-5
DSB-SC AM with Multiple Frequencies, Spectrum Analyzer View: (a) Modulating Signal; (b) Modulated Signal

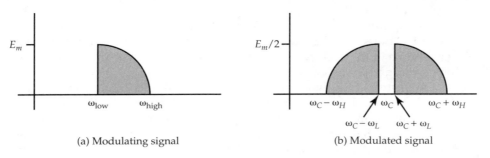

(a) Modulating signal

(b) Modulated signal

conventions are combined, and the shorthand designation that results for this type of modulation is double sideband, suppressed carrier amplitude modulation (DSB-SC AM). Often the AM is not written because DSB-SC implies AM modulation. The two components, or sidebands, also have special names, the upper sideband (USB) and the lower sideband (LSB). The term *sideband* is used because they are be*side* the original location of the carrier signal in frequency space. (Sideband will be defined carefully in Figure 5-5.)

In addition, note carefully what happened to the amplitudes. In the first part of Figure 5-4, the carrier frequency is assumed to have an amplitude of one, and the modulating frequency is shown to have the amplitude E_m. When these two frequencies are multiplied or modulated, the amplitudes of the two resulting signals are the same and are half of the product of the two individual amplitudes, or $E_m/2$.

Next, note that the carrier frequency is shown as a higher frequency than the modulating frequency. This depiction is always the case. Usually, it is much higher, typically a factor of at least 100. The purpose of Figure 5-4 is to highlight the changes in frequency location that occur as a result of the modulation process, not to indicate precisely the relative frequency locations of ω_m and ω_c. Just be aware that the carrier frequency is much higher than the modulating signal frequency in most real systems. For example, in broadcast AM, the carrier frequency is about 1,000 kHz, and the modulating signal is about 10 kHz.

Now, most of the time the modulating signal will not be just a single frequency. Not many would want to listen to an AM broadcast radio station that transmitted a single tone or frequency! So it is important to examine how the picture would change if the modulating signal were composed of a group of different frequencies rather than just one. If you think about it, the frequencies used would be in some bandwidth, and there would be a lowest frequency and a highest frequency. Figure 5-5 shows what would result. The filled curve shown in (a) illustrates all the frequencies that would exist between the lowest and the highest frequency of the modulating signal.

What would be expected from our analysis of a single frequency modulating signal in Figure 5-5 is that two groups of frequencies result. There would be one above the carrier frequency and one below, resulting in what is called the modulated signal. In Figure 5-5(b), these two groups of frequencies are shown by the two shaded areas, one above the carrier frequency, one below. Also, note that the spectrum does not extend right up to the carrier frequency. It stops a little short of it on both sides because, in this diagram, the assumption is that the lowest frequency in the modulating signal is not DC but some slightly higher frequency. This feature results in the exaggerated gap shown in the spectrum in Figure 5-5. Note also that since this is DSB-SC, there is no carrier present at ω_c.

In the rest of the diagrams in this text, the modulating signal will be assumed to extend down to 0 Hz. In other words, the group of frequencies that is modulating the carrier extends down to 0 Hz, or DC. Signals that have this characteristic are called baseband signals, and it makes the exaggerated gap in the center of Figure 5-5 disappear. In the rest of the spectral diagrams in the chapter, it will be assumed that the modulating signal is always baseband.

Another characteristic that you may be wondering about is why the shapes are shown as curved. In most applications, the highest frequencies in the modulating signal occur less frequently, so there is less energy at those frequencies. If all frequencies were equally probable, then the shape would be rectangular. But there is more energy at low frequencies than at high frequencies. The spectrum is shown as smooth because most practical systems also apply something called scrambling to smooth it. If scrambling were not done, the spectral shape would be very ragged, with peaks for several common data patterns.

Finally, the diagrams throughout the book will do away with the high and low frequency markers and assume that, as stated above, the low frequency mark will be DC and the high frequency mark will be labeled ω_m. ω_m is the maximum modulating frequency.

■ SIDEBANDS

Figure 5-6 shows the three important signals of DSB-SC AM. Part (a) of the figure shows the input modulating baseband signal; part (b), the carrier frequency; and part (c), the modulated output signal. The dip has disappeared because the modulating signal is shown extending down to zero frequency. Note that the carrier voltage, E_c, has no contribution to the output signal.

Figure 5-6
DSB-SC Showing Sidebands, Spectrum Analyzer View: (a) Input Modulating Baseband Signal; (b) Carrier Frequency; (c) Modulated Output Signal

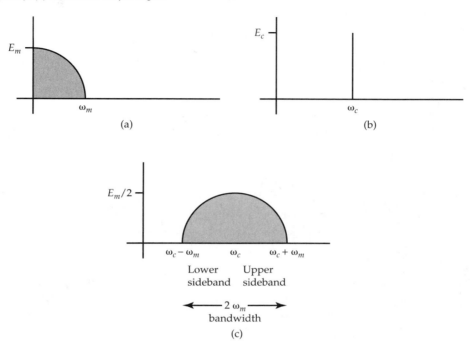

Sideband generation is characteristic of all DSB types of amplitude modulation. It is important to note what the bandwidth of the output signal is. Because there are two sidebands and each is exactly the bandwidth of the original modulating signal, as shown in Figure 5-6, the resultant bandwidth is twice the original modulating signal's bandwidth. This relationship is expressed in Equation 5-7:

$$BW = 2\omega_m \qquad or \qquad BW = 2f_m \qquad \text{(5-7)}$$

Additionally, the sidebands are always symmetric about the carrier frequency:

$$\omega_{usb} = \omega_c + \omega_m \qquad or \qquad f_{usb} = f_c + f_m \qquad \text{(5-8)}$$

$$\omega_{lsb} = \omega_c - \omega_m \qquad or \qquad f_{lsb} = f_c - f_m \qquad \text{(5-9)}$$

The last item that is important to note about the sidebands is that they have exactly the same peak voltage, given by $E_m/2$. Since they have the same peak voltage, the average powers will also be the same. Power can be found by just squaring the RMS voltage and dividing by the load resistance. Figure 5-7 shows how the signals in Figure 5-6

Figure 5-7
DSB-SC Oscilloscope View

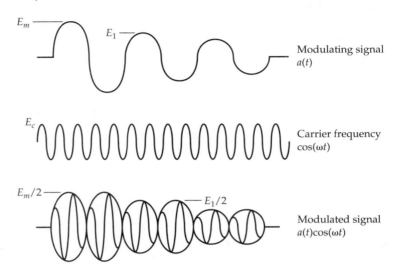

would look on an oscilloscope. Pay special attention to the timing and voltage relationships between the modulating signal and the modulated signal. Note especially that the shape of the modulating signal defines the shape of the modulated signal's envelope. If the first changes, so does the second.

■ **EXAMPLE 5-2**   This example will show sample calculations for a particular DSB-SC modulator with a carrier frequency of 1,000 radians/s and a modulating signal $a(t)$ given by the following expression:

$$a(t) = 10 \cos(20t)$$

Find the following:

1. The peak voltage, E_m
2. The carrier frequency, ω_c
3. The modulation frequency, ω_m
4. The DSB-SC bandwidth
5. The USB bandwidth and frequencies
6. The LSB bandwidth and frequencies

Since the carrier signal amplitude is assumed to be unity, the peak voltage is just the peak voltage of the modulating signal, $a(t)$, or 10 volts. The carrier frequency is 1,000 radians/s. The modulation frequency is 20 radians/s. The DSB-SC bandwidth is twice ω_m, or 40 radians. The USB and LSB are each 20 radians in width. The LSB starts at 980 radians and extends to 1,000 radians. The USB starts at 1,000 radians and extends to 1,020 radians.

Figure 5-8
ASK, Oscilloscope View

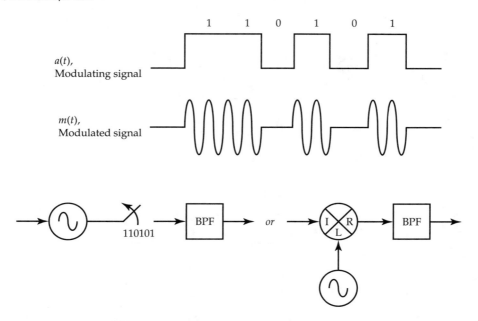

◼ AMPLITUDE SHIFT KEYING (ASK)

Amplitude shift keying (ASK) is a modulation technique where the amplitude of the carrier wave is switched between two or more amplitude states. It is the digital modulation form of DSB-SC AM. In the most basic implementation, this modulation technique is referred to as on off keying (OOK). OOK is a binary-modulated waveform, with the presence of the carrier corresponding to a binary state of 1, and the absence of the carrier corresponding to a binary state of 0. See Figure 5-8 for an example waveform, the digital modulating signal that produced it, the resultant modulated signal after filtering, and two sample modulators. Note that these sample modulators are equivalent. In the second, the mixer is just acting like a switch controlled by the baseband-modulating signal.

As you can see in Figure 5-8, ASK is very straightforward. An implementation consists of an oscillator, a switch controlled by a binary-modulating signal, and a bandpass filter to restrict the bandwidth prior to transmission. This OOK implementation is just the old telegraph system sending dots and dashes. The 0 binary input represents the spaces between dots and dashes, while the dots are single binary ones and the dashes are two adjacent binary ones. In the figure, it would be read dah-di-dit; if using the international Morse code, it would represent the letter D.

For each basic modulation technique, two figures will be shown. The first is a time domain view, or oscilloscope view. It emphasizes the timing relationships of the input and output signals and, for ASK, was presented in Figure 5-8. The second is a frequency domain view. It emphasizes the spectral relationships of the input and output signals. See Figure 5-9.

Figure 5-9
ASK, Spectrum Analyzer View: (a) Modulating Signal; (b) Carrier Frequency; (c) Modulated Signal

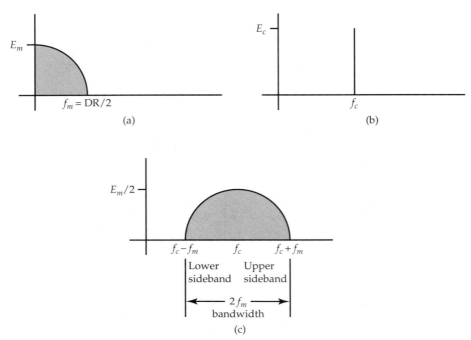

Part (a) of the figure shows the modulating signal; part (b), the carrier frequency; and part (c), the modulated signal. Note that the power spectrum of ASK is centered at the carrier frequency and, on each side of the carrier frequency, has a shape exactly that of the baseband modulating signal. Thus, the resultant bandwidth will be twice that of the modulating signal bandwidth. This result is the same as that found in the previous section on analog AM. You should not be surprised because nothing has changed except the form of the modulating signal. ASK is still AM, but it is driven by a digital waveform rather than an analog one.

Note that, to make comparisons between digital modulation techniques easy, the peak voltage of the modulated signal will always be defined as E_m. This convention will allow power comparisons without reference to the peak voltage of the modulating waveform. A DSB spectrum is centered on the carrier frequency. The bandwidth of the double sideband spectrum is twice that of the baseband modulating signal. Note that the bandwidth of the modulating signal is half the binary bit rate.

■ FREQUENCY SHIFT KEYING (FSK)

Frequency shift keying (FSK) is a modulation technique where the instantaneous frequency of the carrier wave is shifted, or switched, between two frequency states. Binary FSK (BFSK) implies two frequency states in the modulated output.

The same situation that we saw in the ASK case holds with regard to the bandwidth of FSK modulators. The baseband bandwidth will be half the Nyquist binary data rate. The Nyquist bandwidth is the minimum bandwidth that the modulated signal has while in the channel. In a binary modulation environment, the frequency deviation is equal to the frequency shift caused by the modulating signal. It is equal to the difference in frequency of the two signaling frequencies. Each signaling frequency is offset from the carrier rest frequency by the carrier swing.

BFSK has special names for these two signaling frequencies: the mark and space frequencies. Mark frequency is the signaling frequency corresponding to a modulating signal value of a logical 1, and space frequency corresponds to a modulating signal value of a logical 0. It will always be assumed that the mark frequency is the lower frequency of the two.

Recall from Chapter 1 that a modulating frequency is equal to half a binary modulating signal's bit rate. If you consider only those systems where the minimum bandwidth required to pass the fundamental component of any data pattern is used, comparisons between various modulation schemes and density of modulation become straightforward.

There are two broad classifications of FSK techniques: coherent modulation and demodulation, and noncoherent modulation and demodulation. Coherency has implications for both the modulator and demodulator in FSK. For the modulator to be coherent, the switching between two frequencies is accomplished in a very specific way. Noncoherent modulators do not have this restriction; thus, they are far more widely implemented. Coherency for the demodulator means demodulation of the modulated signal using knowledge of the phase of the carrier wave. Noncoherent detection would not require knowledge of the phase of the carrier wave and hence, requires less complex circuitry and consequently, lower cost.

Traditionally, many commercial and military systems utilized FSK techniques because of the relative simplicity of implementing noncoherent modems, most significantly the demodulator: no carrier synchronization is required. Additionally, at least conceptually, the modulator design is simpler. In practice, almost all FSK modulators are constructed using the coherent approach, even if the demodulator is constructed following the noncoherent approach. The noncoherent method produces all kinds of undesirable transients, significantly complicating the transmitter. For reasons that will be explored throughout this chapter, however, FSK is not the preferred technique for most new designs, although many systems in the field today still use FSK techniques.

It is useful to summarize the concept of coherency versus noncoherency and to illustrate it. A binary frequency shift keyed (BFSK) modulator switches between two frequencies based on a digital modulating signal. See Figure 5-10. Both a coherent modulated waveform and noncoherent modulated waveform are shown.

In both cases, you can see clearly that one output frequency is produced for one binary state and another frequency is produced for the other binary state. Note that only the noncoherent modulated waveform has the identified phase discontinuities.

Noncoherent or Incoherent
FSK Modulation

Some discuss these techniques as incoherent instead of noncoherent, but the choice is a matter of language preference. Use whichever term you are more comfortable with. Both the modulators and demodulators are different in the two approaches.

Figure 5-10
BFSK Waveforms, Oscilloscope View

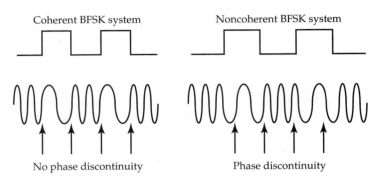

Figure 5-11
Noncoherent FSK, Oscilloscope View

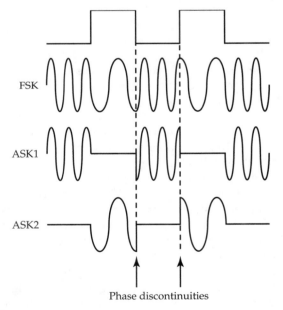

Any BFSK system shifts the carrier frequency between two distinct frequencies. A noncoherent approach does it in the most straightforward manner. The waveforms are shown in Figure 5-11. In the figure, the first signal is the binary-modulating signal. The second signal, FSK, is the binary FSK modulated signal. Note that it is just shifting between two frequencies. Look at the next pair of signals, ASK1 and ASK2. These signals are just the two frequencies of the FSK modulated signal broken apart. If you added ASK1 and ASK2, you would get FSK. Also, note the phase discontinuities shown.

ASK1 and ASK2 are just individual ASK, or OOK, modulated signals. So binary FSK modulation can be viewed as just two ASK modulated signals added together. The ASK

Figure 5-12
BFSK Modulated Waveform, Spectrum Analyzer View

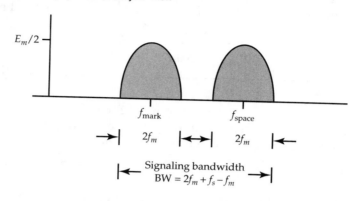

modulator design illustrated in Figure 5-3 is also part of a BFSK modulator, with a few additions to generate two ASK signals and then add the signals together before transmission.

A good question would be, "What would the spectrum of such a signal look like?" Since it is just the addition of two ASK signals, that fact describes the spectrum. Each individual ASK waveform looks just like a DSB spectrum centered at its carrier frequency. In this case, however, there is no carrier frequency, just the mark and space frequencies. Therefore, the FSK signal would just be two double sideband spectrums centered at the mark and space frequencies. The frequency difference of the two frequencies that the modulator switches between is determined by the application. In principle, any two frequencies could be used. Figure 5-12 illustrates what the signal would look like in frequency space.

Only one of these spectral envelopes exists at any one time because the modulator is switching between the two of them. Only one piece of the total bandwidth that the modulator requires is in use at any one time! This feature points out the major disadvantage to this technique. These systems require more bandwidth than equivalent phase shift keying (PSK) systems, which limits their application. Most systems designed today would use PSK or a related form of modulation. (PSK will be introduced after completing the discussion on FSK.)

■ **EXAMPLE 5-3**   The total bandwidth of the BFSK waveform is very dependent on the choice of the signaling frequencies. In Figure 5-12, if the baseband signal data rate is 1 kbps, what is the bandwidth of *each* envelope?

$$f_m = \frac{DR_b}{2} = \frac{1 \text{ kbps}}{2} = 500 \text{ Hz} \Rightarrow BW = 2f_m = (2)(500) = 1 \text{ kHz}$$

The answer is twice the baseband bandwidth, or 1 kHz. Further, what is the total spectrum bandwidth shown in Figure 5-12? The answer depends on the choice of the signaling frequencies and cannot be determined without more information.

■ **EXAMPLE 5-4**   If in Figures 5-11 and 5-12, the signaling frequencies are 20 kHz for the space frequency and 10 kHz for the mark frequency, what is the signaling bandwidth?

$$BW_{BFSK} = |f_{space} - f_{mark}| + 2f_m \tag{5-10}$$

$$BW_{BFSK} = |20 \text{ kHz} - 10 \text{ kHz}| + 1 \text{ kHz} = 11 \text{ kHz}$$

Examples 5-3 and 5-4 and Figure 5-12 point out that the minimum bandwidth required by an FSK system is four times the modulating frequency! This value is twice the bandwidth of ASK and is expressed in the relationship below:

$$BW_{BFSK}(\min) = 4f_m \tag{5-11}$$

A modulator could be constructed using the ASK approach described above, but this method is not a good idea. As noted above, noncoherent modulators produce unwanted transients that complicate the transmitter and add cost. When the switch between two frequencies occurs abruptly, there is no guarantee that the two waves will be synchronized in phase. This noncoherent shifting occurs unless the frequencies are chosen carefully and they are aligned with the bit transition times exactly. These switch points are shown as the phase discontinuities in the figures above. Go back and review the oscilloscope views. Notice that the noncoherent waveforms are switching at points where the signaling frequencies are not crossing the axis at zero.

Coherent modulators implemented by adding together two ASK signals are possible but require specific pairs of frequencies to be used as the mark and space choices. In a nutshell, it is conceivable that a noncoherent modulator using two distinct frequency sources and switching between the two could be implemented. In general, however, this technique is impractical because of the transients generated from the lack of phase coherence between the two frequency sources. Coherent FSK modulators, discussed next, get around this limitation of having to pick special frequencies to avoid this kind of "spectral splash."

Noncoherent FSK demodulators demodulate the received signal directly, much like that of the simple envelope AM demodulator. It is hard to imagine a simpler or less expensive demodulation circuit, hence the attractiveness and wide adoption of incoherent FSK systems.

Coherent FSK Modulation or Minimum Shift Keying (MSK)

Coherent modulation is similar to noncoherent when you consider what it needs to accomplish, but it has the additional requirement that the phase change remain continuous when switching between frequencies. Signals that exhibit this property are called continuous phase waveforms. Coherent FSK modulation is often referred to as minimum shift keying (MSK). The idea is that the frequency is shifted the minimum amount required to guarantee a continuous phase shift when shifting frequencies. It is also sometimes referred to as fast FSK (FFSK), the word *fast* coming from the early days of digital modulation where, due to the implementation choices available, the minimum shift in frequency was also faster, allowing higher data rate systems. Of course, this situation is no longer true and use of the name FFSK is fading away.

The requirement that the two output frequencies be phase coherent when switching implies two possible approaches. The first would be to assume that the only way an MSK system could be implemented is by having the two signaling frequencies synchronized to the data rate. This implementation uses the two ASK modulator approach and was demonstrated in the last section. The other possibility is that the requirement for phase coherency can be met if the frequency switches that represent the input signal are not always switched by the same amount. In the second approach, each frequency shift must depend on the previous frequency shift. Each shift of the frequency must be just enough, the minimum, until the new frequency phase matches exactly that of the existing one.

The key idea here is that, because the frequency shift is not always the same, it is not easy to tell where the data shifts occurred by just examining the modulated waveform. This situation has some important consequences; the most important is that MSK performs much better than noncoherent FSK. This performance advantage is due to the continuous phase shift when moving between mark and space frequencies as compared to the abrupt phase shifts that occur in noncoherent FSK at the same transitions. It is also interesting to note that there is some amplitude modulation on the MSK waveform because of these continuous phase shifts and that the modulation index of MSK is always approximately 0.5. An MSK signal must have the following characteristics to be considered MSK:

1. The frequency deviation is exactly $\pm\frac{1}{2}\,\omega_m$, where ω_m is the frequency version of the bit rate.
2. As already mentioned, the modulated signal is phase coherent at the switching instant.

The main downfall for MSK in many wireless applications is that it is a wideband system and thus requires wide channel spacings. For wireline applications, this feature is not a concern; however, for wireless applications, it is a real problem. A refinement to MSK discussed in the next section addresses this important issue.

Gaussian MSK (GMSK)   This modulation technique is a refinement of the general form of MSK. The refinement makes it better suited for mobile wireless applications. In a cellular application like GSM, where this form of modulation is used, the frequency channels are tightly packed. The goal is to place as many adjacent channels as possible. If a way to restrict the output power spectrum could be found, it would enhance the overall performance of the cellular system by allowing more channels in a particular spectrum allocation. Gaussian MSK does just that by prefiltering the baseband signal prior to the MSK modulation. The modulation technique gets its name from the form of the response of this filter: a Gaussian response low-pass filter. Since MSK is usually generated by direct modulation, the output spectrum can be narrowed by placing this filter in the baseband signal path. See Figure 5-13.

Figure 5-13
GMSK Modulator Block Diagram

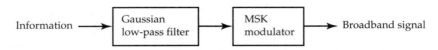

Figure 5-14
MSK and GMSK Output Spectra

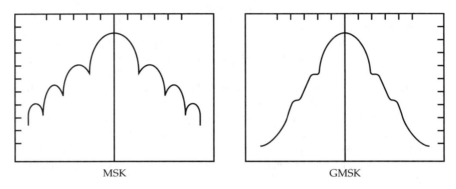

The central problem with MSK is the very large sidelobes produced. They assume a classic $\sin x/x$ fall-off response. With the addition of the Gaussian prefilter, these sidelobes are dramatically reduced. To illustrate the difference that the prefiltering makes, two transmit spectra are shown in Figure 5-14. Note the dramatically reduced sidelobe energy that results. This reduction makes it clear why GMSK is preferred over MSK in most wireless environments.

■ PHASE SHIFT KEYING (PSK)

Phase shift keying (PSK) is a modulation technique where the phase of the carrier wave is switched between two or more phase states. It is essentially an analog angle modulation system where the phase was changed, except PSK requires the modulating signal to be pulsed in two or more amplitude states, bin-ary or M-ary. PSK is the most dominant modulation technique in use today for implementing digital communications.

It can be shown that frequency and phase modulation are closely related and, when the input signal is analog in nature, are often grouped together and studied as angle modulation. When the input signal is digital in nature, as in the case of PSK, there will be analogy to amplitude modulation systems rather than frequency modulation systems. This comparison stems from the fact that the performance of several PSK systems is identical to that found in similar AM systems.

Figure 5-15
BPSK Waveforms, Oscilloscope View

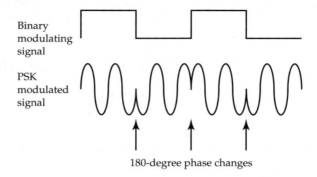

Binary
modulating
signal

PSK
modulated
signal

180-degree phase changes

Figure 5-16
BPSK Modulated Waveform, Spectrum Analyzer View

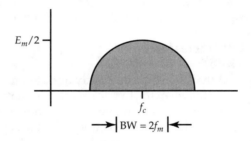

$E_m/2$

f_c

$\text{BW} = 2f_m$

Binary PSK (BPSK)

BPSK modulators take a binary modulating signal, switch between two amplitude states, and output a modulated signal where a carrier switches between two phase states, usually 0 degrees and 180 degrees. The waveforms are shown in Figure 5-15.

Mathematically, the bandwidth of a BPSK modulated waveform is found by the following expression:

$$\text{BW}_{\text{BPSK}} = 2f_m \tag{5-12}$$

The spectrum is shown in Figure 5-16. This spectrum is the same that resulted from DSB-SC AM! Look again at Figure 5-16. Notice how the spectrum looks exactly like that of Figure 5-6. In theory, the waveforms of DSB-SC AM and BPSK are identical. There is no carrier signal present in the output waveform, a surprising result given that the modulated signal is just the carrier wave switching between two phase states.

All the results that we obtained in the analysis of DSB-SC will apply here. The bandwidth is just twice the modulating frequency, and there is no carrier present. Also, just like the ASK situation, the modulating frequency equals half the bit rate. Example 5-5 illustrates these ideas.

■ **EXAMPLE 5-5**   For a BPSK modulator with a 1 Mbps binary modulating signal and a 10 MHz carrier frequency, find the upper and lower sidebands and the bandwidth of the resulting modulated signal. Remember to convert from a binary data rate to a baseband modulating frequency.

$$f_m = \frac{DR_b}{2} = \frac{1 \text{ Mbps}}{2} = 500 \text{ kHz}$$

$$f_{USB} = 10.5 \text{ MHz}$$

$$f_{LSB} = 9.5 \text{ MHz}$$

BPSK Modulator

An illustrative BPSK modulator design is shown in Figure 5-17. Its design is very straightforward. It consists of a binary modulating signal applied to a mixer, a carrier oscillator at the carrier frequency, and a BPF to filter out any unwanted harmonics generated by the mixing process. The spectrum would look just like that shown in Figure 5-16.

Bandwidth Considerations of BFSK and BPSK

Earlier we stated that BFSK is not widely adopted in the field, while PSK is. The bandwidth relationship between the two forms of modulation is the fundamental reason. The bandwidth of BFSK is often dominated by the choice of the mark and space frequencies. This choice led to a calculation of bandwidth that depended on two factors: the signaling frequency choices and the modulating signal data rate. This calculation was expressed in Equation 5-10. For BPSK, we found that the bandwidth of the modulated signal depended only on the modulating signal data rate, as expressed in Equation 5-7. Both equations are reproduced below for your reference.

$$BW_{BFSK} = |f_{space} - f_{mark}| + 2f_m \qquad BW_{BPSK} = 2f_m$$

As you can see, for any given data rate, the bandwidth of the BPSK modulated waveform will be less than the BFSK modulated waveform. Even if the chosen mark and

Figure 5-17
BPSK Modulator

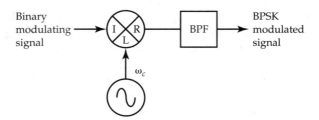

space frequencies minimize the bandwidth, the difference must be at least twice the modulating frequency. This calculation is shown in Equation 5-13. Note the similarity to Equation 5-11.

$$\text{BW}_{\text{BFSK}}(\min) \geq 2f_m + 2f_m = 4f_m \qquad \textbf{(5-13)}$$

BFSK is always at least half as efficient a user of bandwidth as BPSK. Realistically, the above minimum could never be reached because some kind of BPF must separate the two spectra centered on the mark and space frequencies. Allowing for filter roll-off, for all real systems, BPSK is likely to be at least three times as efficient a user of bandwidth as BFSK. This result is at the heart of why FSK systems are disappearing from the marketplace.

■ QUADRATURE PHASE SHIFT KEYING (QPSK)

QPSK is a 4-ary modulation technique. It is very similar to BPSK; in fact, it is generally made by just adding two BPSK modulated signals 90 degrees out of phase. Since it is 4-ary, the number of changes in the input signal required to produce a change in the output signal is two. As you might expect, the bandwidth efficiency of QPSK is also two.

QPSK Modulator

QPSK modulators are implemented in practice by adding together two BPSK modulators 90 degrees out of phase. The 90-degree phase shift is accomplished by simply driving one BPSK modulator with a cosine wave at the carrier frequency and the other with a sine wave. A sample QPSK modulator is shown in Figure 5-18.

You may notice a few unfamiliar items in the figure. First, the serial-to-parallel converter takes the input data, which is running at the bit rate of the modulator, and breaks it into two parallel streams of data, each running at half the bit rate. This rate is called

Figure 5-18
QPSK Modulator

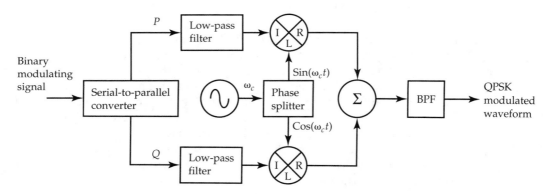

the symbol rate. For QPSK, its value is two. The symbol rate always expresses the band-width efficiency and signal state density as well (again, another way to talk about the same idea). The labels P and Q in the figure represent the in-phase and quadrature-phase symbol pairs. (It might make more sense to label them I and Q, which is some-times done. Both practices are in use.) These labeled items are the two quadrature sig-nals. They are combined in a way sometimes referred to as quadrature mixing. The mathematical expression of the signal at the output of the summer (shown in Equa-tion 5-14) is easy to find. $P(t)$ represents the in-phase data line and $Q(t)$ represents the quadrature-phase data line.

$$\text{Summer output} = P(t)\sin(\omega_c t) + Q(t)\cos(\omega_c t) \tag{5-14}$$

For the second time, you can see an LPF in line with the binary symbol data streams. Like in the GMSK situation, these filters are used to control the bandwidth of the base-band symbol lines. In advanced systems, they are also used to shape the spectrum of the modulating symbol lines to improve performance. After this preconditioning, the sig-nals are applied to mixers, with each LO port being driven by a cosine wave or a sine wave at the carrier frequency. The signals are then summed together and passed through a BPF to eliminate unwanted harmonics generated by the mixing process.

The bandwidth required by such a modulator is less than you might expect. What is important to this calculation is not the input bit rate but the symbol rate! This feature is always true of M-ary modulators. For BPSK and BFSK, the symbol rate and bit rates were the same, so this relationship was not apparent. Now that we are considering modulation other than binary, this relationship becomes clear. To see why the bandwidth is less than twice the bit rate, think about the symbol rate, which is really 2 bps/Hz. From the figure, the baseband signal modulated by the mixers is half the input bit rate. Since the baseband bandwidth of the symbol rate is half, the bandwidth of the resultant sig-nal is also half.

A look at the constellation diagrams in the next section should make clear why the word *quadrature* is used to describe what we see. Note that by adding the second BPSK pair of phasors, we take a constellation diagram that was "flat" and make it square!

■ CONSTELLATION DIAGRAM

A constellation diagram is the most commonly used diagram to discuss M-ary modula-tion. It can be produced on an oscilloscope by taking the two outputs of the modulation circuit and attaching them to the x and y inputs of the oscilloscope. This process must be done before the signals are added together to produce the modulated signal at the same relative place in the demodulator. It is helpful to think of these diagrams as phasor dia-grams. They are very similar except there is no vector representation, just a dot on the screen where the typical phasor arrowhead would terminate. They do show exactly where in amplitude and angle the modulation energy is placed. See Figures 5-19 and 5-20.

In actual practice, both of the BPSK signals produced by the modulator are also shifted 45 degrees during the process of adding them together. Therefore, the actual con-

Figure 5-19
BPSK Constellation Diagram

Figure 5-20
QPSK Constellation Diagram

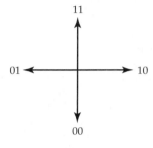

Figure 5-21
Actual QPSK Constellation Diagram

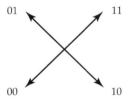

stellation diagram becomes like the one in Figure 5-21. The reason is a result of the mathematics and is worked out in Example 5-6. Note that for both Figure 5-20 and Figure 5-21, to confuse one state with another, an error of ±45 degrees must be made in the demodulator, which is the difference between each state. To illustrate how the signals are combined in a QPSK system, see Example 5-6. This example will demonstrate how the extra 45 degrees of phase shift come about during the process of combining the signals. Note also that the values of $P(t)$ and $Q(t)$ are now explicitly broken out.

■ **EXAMPLE 5-6**   To make this example easier, refer to Figure 5-18, where a QPSK modulator design is shown. The binary data input to the serial-to-parallel converter splits pairs of the binary modulating signal into two paths. Each possible 2-bit pair, or di-bit, will be examined and the signal phase at the output of the summer calculated. The mathematics of adding a sine wave and cosine wave are not instructive and so will not be presented here; only the results will be shown. For those of you who are interested, consult the function sum and function difference formulas used in trigonometry.

$$00: P = -1, Q = -1 \Rightarrow P = (-1)\sin(\omega_c t), Q = (-1)\cos(\omega_c t)$$

$$P + Q = -1.414\cos\left(\omega_c t - \frac{\pi}{4}\right) = +1.414\sin\left(\omega_c t - \frac{3\pi}{4}\right)$$

$$01: P = -1, Q = +1 \Rightarrow P = (-1)\sin(\omega_c t), Q = (+1)\cos(\omega_c t)$$

$$P + Q = -1.414\sin\left(\omega_c t - \frac{\pi}{4}\right) = +1.414\sin\left(\omega_c t + \frac{3\pi}{4}\right)$$

$$10: P = +1, Q = -1 \Rightarrow P = (+1)\sin(\omega_c t), Q = (-1)\cos(\omega_c t)$$

$$P + Q = +1.414\sin\left(\omega_c t - \frac{\pi}{4}\right)$$

$$11: P = +1, Q = +1 \Rightarrow P = (+1)\sin(\omega_c t), Q = (+1)\cos(\omega_c t)$$

$$P + Q = +1.414\cos\left(\omega_c t - \frac{\pi}{4}\right) = -1.414\sin\left(\omega_c t - \frac{3\pi}{4}\right) = +1.414\sin\left(\omega_c t + \frac{\pi}{4}\right)$$

Each term in these equations is expressed as a positive function of sine to illustrate exactly how the constellation diagram is tilted by 45 degrees in the process of summing. Figure 5-22 shows the di-bit locations.

From this figure and the equations above, a QPSK truth table can be constructed. See Table 5-1.

Figure 5-22
QPSK with Di-Bit Locations Shown

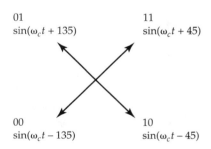

01
$\sin(\omega_c t + 135)$

11
$\sin(\omega_c t + 45)$

00
$\sin(\omega_c t - 135)$

10
$\sin(\omega_c t - 45)$

Table 5-1
QPSK Output Phase Truth Table

Binary Signal P bit	Binary Signal Q bit	QPSK Output Phase
0	0	−135
0	1	+135
1	0	−45
1	1	+45

■ **EXAMPLE 5-7**   What is the bandwidth required for a QPSK modulated signal? The bandwidth required by such a modulator is just half of the bit rate, or the symbol rate.

The bandwidth efficiency for any binary input was given by Equation 5-1, which is reproduced here:

$$BW_e = \frac{DR_b}{BW} \tag{5-1}$$

Using this relationship, the bandwidth efficiency can be calculated. When this equation was examined earlier, however, it was assumed that the input bit rate was applied to the mixers. In M-ary modulation schemes where M is not equal to two, what is actually in the numerator of this equation is the symbol rate. Equation 5-1 must be rewritten with this new information in mind:

$$BW_e = \frac{SR}{BW} \tag{5-15}$$
$$SR = \text{symbol rate}$$

Now, for QPSK only, the symbol rate is half the binary data rate, so Equation 5-15 can be evaluated as:

$$BW_e = \frac{DR/2}{BW} = \frac{DR}{2BW} \tag{5-16} \text{ (QPSK only)}$$

Therefore, the bandwidth efficiency is twice that of a 2-ary modulator. We can apply this to an actual example, for a bit rate of 1 kbps, and obtain the following results:

$$BW = \frac{BW_{\text{binary}}}{2BW_e} = \frac{1 \text{ kbps}}{(2)(1)} = 500 \text{ Hz} \qquad (\text{QPSK only})$$

This computation illustrates that, for a QPSK modulator, the bandwidth required is just half of the input bit rate. Rearranging the above results (keep in mind that the symbol rate is always the same numerical result as the transmit bandwidth) gives us a general relationship among the bandwidth or symbol rate, binary data rate, and bandwidth efficiency. Use Equation 5-17 for all calculations relating bandwidth efficiency and symbol rate for the various modulation types.

$$SR = BW = \frac{BW_{\text{binary}}}{BW_e} \tag{5-17}$$

■ **8-ARY PSK**

The next level of signal state density from QPSK is 8-ary PSK. 8-ary PSK uses eight phase states, each $\pi/4$ apart, which results in a bandwidth efficiency of 3 bps/Hz. Also, the symbol rate will be one-third the bit rate. This process can go on and on and is limited only by the effectiveness of the result in a noisy environment. Systems using 16-ary and higher are in common use today.

Differential QPSK (DQPSK)

DQPSK is very similar to QPSK, with a small change in both the modulator and demodulator. The basic problem that differential encoding solves is that QPSK and many variations of it are rotationally invariant. "Rotationally invariant" means that if the constellation is rotated in some multiple of $\pi/2$, or 90 degrees, it is impossible to tell the unrotated constellation from the rotated one. When a constellation satisfies this property, it is called rotationally invariant. A rotationally invariant constellation is a disadvantage in wireless systems because the atmospheric channel can rotate the constellation during transmission. The data is thus received in error because the receiver has no way of detecting a constellation that has been rotated a multiple of $\frac{\pi}{2}$. A QPSK constellation rotated 90 degrees looks fine and the symbol(s) represented by the new phase state are well within the error distance of the modulation technique.

The solution to this problem is differential encoding of the information input. The information is encoded by using the *change* in the constellation symbol position rather than the *absolute* symbol position. The constellation is thus no longer rotationally invariant because the received symbol state now depends on the previous symbol state. If the constellation is rotated 90 degrees, the symbol will be recognized as a rotational error and will be repaired by the receiver. Repair can occur because a rotation of 90 degrees will change the differentially encoded symbol state.

As shown below, a differential encoder has one output when the information bits are the same and another when they are different. A 90-degree shift will violate this coding, and the receiver will be able to recover from a single bit error with no additional information. After reading the following discussion, demonstrate this process to yourself by changing a bit during transmission, which represents a 90-degree phase shift. When decoding, note that it is possible to predict what the correct phase state should have been. The decoding process is the exact inverse of the differential encoding process.

A form of DQPSK modulation is used in almost all variations of the DAMPS cellular systems. Instead of the output phase being determined only by the state of the current input symbol, the phase is determined by the state of the current symbol and the previous symbol transmitted. Fundamentally, a comparison is made between the state of the two symbols individually, and that comparison determines the output state.

The differential state of the incoming bit streams determines the output phase of the modulator. This approach can be applied to any signal state density form of PSK. Figure 5-23 illustrates the encoding process for a DBPSK modulator. A DQPSK modulator is constructed from two offset DBPSK modulators, just as a QPSK modulator is constructed of two BPSK modulators offset by 90 degrees. Specifically, if the previously encoded bit and the next information bit are the same, the encoder output is a 1; if they are different, the output is a 0. Hence, the name, differential encoding. Figure 5-24 illustrates the encoding process for one channel of a DQPSK encoder.

Figure 5-23
DBPSK Encoding

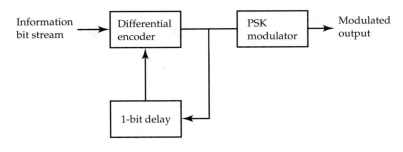

Figure 5-24
One Channel of a Differential QPSK Encoder

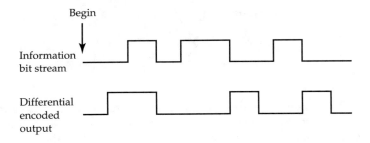

$\frac{\pi}{4}$-DQPSK

While the modulation technique in Chapter 12 for the DAMPS or IS-136 cellular systems is listed as DQPSK, the actual implementation is $\frac{\pi}{4}$-DQPSK. These systems transmit 2 bits per symbol, or di-bits. Rather than these 2 bits mapping into a constellation with four phase states, as would be the case with QPSK or DQPSK, $\frac{\pi}{4}$-DQPSK maps the 2 bits into a constellation with eight phase states. The constellation is illustrated in Figure 5-25. Note how similar it looks to an 8PSK constellation.

This modulation technique works by separating the incoming symbols into even and odd symbols. The output phase state is determined by using Table 5-2. Even and odd symbols each have a separate look-up table for establishing the output phase. For convenience, these phase states are listed in absolute degrees in Table 5-2. Thus, as each successive bit is passed to the last stage of modulation, the modulator must shift by either 45 degrees or 135 degrees. To illustrate, if a particular bit is output as 45 degrees, it must be an even bit.

The next bit can take on only the odd values in Table 5-2, which means that the output phase must be one of those listed in the table. The first two listed, 0 and 90 degrees, represent a 45-degree phase shift from the previous state, while the second two, 180 and 270, represent a 135-degree shift from the previous state. When the differential encoding is factored in and the possible bit values are calculated and once these values are applied to the phase state from Table 5-2, the phase shift is always 135 degrees.

Figure 5-25
$\frac{\pi}{4}$-DQPSK Constellation

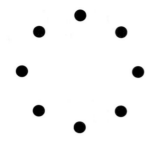

Table 5-2
$\frac{\pi}{4}$-DQPSK Output Phase Selection

Phase State				
Even	45	135	225	315
Odd	0	90	180	270

This fact makes this modulation technique practical from multiple viewpoints. First, the error distance is now 135 degrees, greater than the 90 degrees of standard QPSK. Second, the data is differentially encoded, so the receiver can always calculate what the next phase shift should be given accurate demodulation of the previous symbol. Single bit errors are corrected automatically without any additional coding techniques, which consume bandwidth.

■ QUADRATURE AMPLITUDE MODULATION (QAM)

It doesn't make much sense to keep adding additional phase states in PSK. The error distance or difference between them gets smaller and smaller, and the performance in a noisy environment doesn't get any better for the additional cost of implementation. Eventually, the requirement that all the phasors have the same amplitude must be dropped. Most modern modems designed today use a combination of amplitude and phase modulation. For example, there might be eight phase states, each one with two different amplitudes, yielding a 16-ary modulation technique.

Before we discuss systems that are more complex, the simplest form of modulation, where both amplitude and phase are combined, is examined. This simplest form is quadrature amplitude modulation (QAM). QAM uses four amplitude states. Note that four amplitude states are exactly like four phase states. Refer back to Figure 5-21. Why not regard the phasor values as amplitudes instead of phases? Moving clockwise from the top, you would read (1, 1) (1, −1) (−1, −1) (−1, 1). This perspective is just as valid as thinking of these values as angles of 45 degrees, 315 degrees, 225 degrees, and 135 degrees, respectively.

If the constellation diagrams representing two different ways of talking about modulation are the same, then the two different modulation techniques are the same. Therefore, QPSK is identical to QAM. This similarity is a special case and relies on the fact that the baseband symbol streams of both modems are shaped in the same way. Additional discussion of this situation is beyond the scope of this book.

In conclusion, we obtain two interesting results with phase modulation: BPSK is identical to DSB-SC AM, and QPSK is identical to QAM, which is just the combination of two DSB-SC AM signals. As mentioned in the paragraph above, this pattern does not hold true for higher M-ary rates. Once higher values of M, other than two or four, are considered, this equivalence of PSK to DSB-SC breaks down. The last modulation technique that we will examine will illustrate why this breakdown occurs.

■ 16-ARY QAM

A popular implementation choice, 16-ary QAM can be implemented in a couple of different ways. The approach examined here will illustrate how advanced M-ary modulators arrange the amplitude and phase states so that they achieve maximum performance in noise. One obvious way to implement this modulator would be to generalize from 8-ary PSK and add another phase state between each existing one. That method would reduce the phase difference between states from $\frac{\pi}{4}$ to $\frac{\pi}{8}$, and would represent a maximum phase difference or error distance of only about 45 degrees between phasors.

This difference might seem like a lot, and it can be implemented and work effectively. However, is there a different way of maximizing the difference between phasors? As will be shown in Chapter 7, the goal of source coding is to make the states different enough to tell them apart. Can something be done to make these states more different from each other?

The answer is to combine amplitude modulation and phase modulation, thus obtaining a constellation diagram that has larger error distances between each constellation point. Look at Figure 5-26 for a comparison of what can be accomplished using amplitude and phase as compared to phase only. The distances between the constellation

Figure 5-26
16-ary QAM and 16-ary PSK

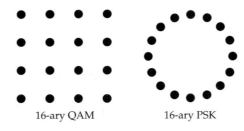

16-ary QAM   16-ary PSK

Figure 5-27
Noise Added to Constellation Diagrams from Figure 5-26

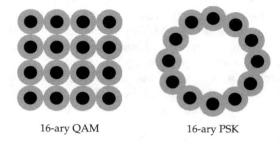

16-ary QAM   16-ary PSK

points are important because as noise is added to the signals, the effect is to render the sharp points represented in the figure into noisy "smudges." For this reason, the regions centered about each point are called decision regions. The effect of noise on this process is shown in Figure 5-27. The noise is shown by the light gray regions surrounding the signal state. As you can see, when noise is added to the figure, errors in determining which state is which become more probable. Clearly, QAM is more practical than PSK as the number of signal states grows.

■ ERROR DISTANCE

The constellation diagram should be familar and the concept of how noise affects the ability to determine the correct phase state should be clear. Now it is possible to introduce a way to characterize the performance of any PSK system based on the distance between any two decision regions, called the error distance. The larger the error distance, the more practical is the performance of the modulation technique. In a pure PSK system, the decision regions all lie on a circle. It is possible to find a simple relationship to describe the distance between any two points. A circle is described by a sin function rotating through an entire 360 degrees, or 2π radians. Since the signal density tells us directly how many signal points are placed on this circle, a relative measure of the error distance is easily found. Examine Figure 5-28, where the error distance for QPSK is illustrated.

Figure 5-28
PSK Error Distance

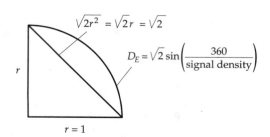

$$\sqrt{2r^2} = \sqrt{2}r = \sqrt{2}$$

$$D_E = \sqrt{2} \sin\left(\frac{360}{\text{signal density}}\right)$$

r

$r = 1$

Table 5-3
PSK Relationship Table

Modulation Technique	Baseband Bandwidth (W_m)	Transmit Bandwidth (BW)	Symbol Rate	Signal State Density	Bandwidth Efficiency (BW_e)	Error Distance
BPSK	500 Hz	1 kHz	1000	2-ary	1	2.8
QPSK	250 Hz	500 Hz	500	4-ary	2	1.4
8-PSK	166.7 Hz	333.3 Hz	333.3	8-ary	3	1.0
16-PSK	125 Hz	250 Hz	250	16-ary	4	0.55
32-PSK	100 Hz	200 Hz	200	32-ary	5	0.28
64-PSK	83.3 Hz	166.7 Hz	167	64-ary	6	0.14

At first glance, it would seem that the error distance is given by the straight line shown in the figure and depends on the amplitude of the input signals, here, taken to be unity. However, this error distance is not quite correct because the decision points follow the circle. The error distance is the distance between the two points shown but not the straight line distance; it is the distance as traveled on the arc of the circle defined by those two points. This distance is expressed in Equation 5-18, which is the product of the radius of the circle and the value of the sin function over that arc.

$$D_E = \text{error distance} = \sqrt{2}\left(\sin\left(\frac{360}{\text{signal density}}\right)\right) \qquad \textbf{(5-18)}$$

■ **EXAMPLE 5-8**   This example illustrates the caclulation of the error distance for QPSK.

$$D_E = \text{sqrt}(2)*\sin(360/4) = 1.414$$

Note that this relationship breaks down for the BPSK case, but it takes little imagination to see that the error distance in that case is just twice the error distance for QPSK.

To summarize the relationships between the various concepts introduced, Table 5-3 compares several PSK modulation techniques. It relates the quantities of baseband bandwidth, transmit bandwidth, symbol rate, signal state density, bandwidth efficiency, and error distance. For ease of comparison, the binary modulation data rate is assumed to be 1 kbps for all cases shown in the table.

■ **HIGHER MODULATION DENSITY SYSTEMS**

High density wireless systems use modulation densities that are at least 64-ary. Many systems feature modulation densities of up to 256-ary. These systems are

Figure 5-29
64-QAM Constellation

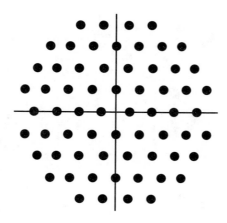

QAM, with constellation patterns similar to the 16-QAM pattern shown in Figure 5-26. For illustrative purposes, a 64-QAM constellation is shown in Figure 5-29. As you can see, the 64-QAM constellation is essentially four 16-QAM constellations, one in each quadrant.

Traditionally, QPSK or QAM has been used for so long because there is no amplitude information in the output constellation; thus, very inexpensive receivers are possible because the signal has a constant energy envelope. Each symbol is represented by a phase state with the same amplitude. Very inexpensive digital receivers are possible because no amplitude information is needed during the demodulation process. From an implementation point of view, the input signal at the receiver can immediately be hard-limited. Hard-limiting preserves the zero-crossing information, which is all that is important to determine the phase, but it discards all amplitude information. A hard-limited signal is processed easily by a simple digital receiver. (A simple digital receiver has no expensive A/D conversion and analog circuitry.)

In a microwave system where there is only one receiver for each transmitter, such a cost tradeoff does not make sense. In this case, performance improvement is the critical objective. Since power into the receiver is an important indicator of error rate, any way to increase the power at the receiver for a given total cost of the system is desirable. Lower power amplifiers cost less. If a more efficient constellation could feature the same modulation density, this find would be an improvement.

Such an approach makes use of hexagonal constellations. A hexagonal constellation is so named because the decision regions are hexagonal in shape. This type of constellation places reduced power requirements on the transmitter because it minimizes the largest amplitude state for a given decision region. Figure 5-30 illustrates a 64-QAM constellation configured as a hexagonal constellation.

Figure 5-30
Hexagonal 64-QAM Constellation

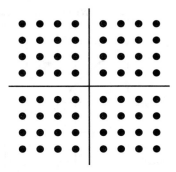

DEMODULATION IN A RADIO ENVIRONMENT

In general there are three methods for detecting a signal: coherent, differential coherent, and noncoherent.[*] Coherent detection requires generation of a frequency and phase coherent waveform at the demodulator. Differential coherent does not require this absolute phase information in the demodulator because it obtains the signal itself. Fundamentally, instead of comparing the received signal to some reference standard, it compares the last symbol's phase to the current one and maintains a phase reference between successive symbols. Noncoherent detection, of course, requires no phase reference at all.

In general, noncoherent detection is always preferred in a mobile radio environment such as cellular and wireless LAN applications, but the performance penalty of noncoherent detection is severe. Coherent detection performs better but it is expensive. Its cost tends to rule it out in commercial applications. Differential detection is the middle of the road: not quite the performance of coherent detection but significantly less expensive to implement. The main problem when using differential detection is that if an error is made in one symbol, the error carries over to the next because the previous symbol is used to detect the subsequent symbol.

BIT ERROR RATE (BER) CALCULATIONS

Bit error rate calculations tell how many bits will be in error, on average, for a given E_b/N_o ratio. Carrier to noise (C/N) and energy per bit relationships (E_b/N_o ratios) are explored in detail in Chapter 9. In this section, we discuss the use of graphs to determine experimentally the performance of a modem.

---

[*]For more detail on each of these approaches, see Thurwachter, Charles N. 2000. *Data and Telecommunications: Systems and Applications* Upper Saddle River: Prentice Hall, Inc. (ISBN: 0-13-793910-8).

Figure 5-31
BER Graph for ASK Systems

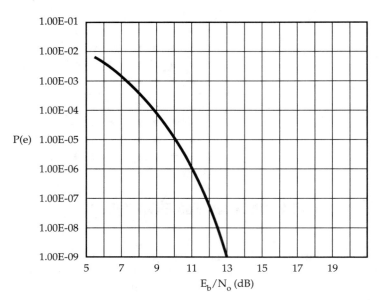

BER is a probability function and makes assumptions on the channel characteristics. The primary assumption is that the noise in the channel is just thermal noise. Thermal noise is the only type considered here. Other noise sources, especially those with spikes in their spectrum, will result in lower BER numbers for a given E_b/N_o ratio. The mathematical analysis of BER requires knowledge of statistics that is outside the scope of this text. However, BER is drawn as a curve and these diagrams will be reproduced to allow calculations to proceed.

The first graph, shown in Figure 5-31, is for ASK or OOK systems. There is only one curve because all ASK systems work in the same way. For any ASK system, the probability of error in a transmission is given by the curve in the figure. To illustrate how this type of graph would be used, follow Example 5-9.

■ **EXAMPLE 5-9**   Find the BER for an ASK system operating at an E_b/N_o ratio of 11 dB.

In Figure 5-31, the E_b/N_o ratio corresponds to a BER of 1E-6. For every million bits sent, on average, one will be in error.

The next graph, shown in Figure 5-32, is for FSK systems. As you know, all FSK systems can be divided into two classes: those that use coherent detection and those that use noncoherent detection. In the figure, the curve to the right is the one for noncoherent detection. As you might expect, that curve represents a higher error rate for the same received power or E_b/N_o.

Figure 5-32
BER Graph for FSK Systems

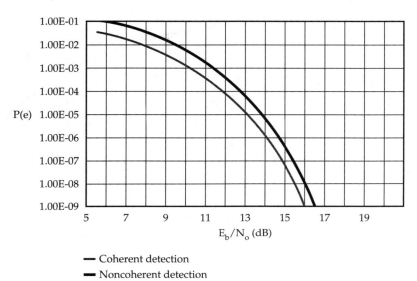

— Coherent detection
— Noncoherent detection

■ **EXAMPLE 5-10**   Compare the error rate performance of coherent and noncoherent FSK systems. For a BER of 1E-4, what are the E_b/N_o requirements of the two systems?

For the coherent FSK system, the E_b/N_o required is about 11.9 dB. For the noncoherent system, the E_b/N_o required is about 12.8 dB. For this error rate, an additional almost 1 dB of power is required at the receiver for the same performance. Note that this additional power gives nothing in terms of increased message or information density. On the other hand, for the additional expense of adding coherent recovery to the receiver, the transmit power requirement can be reduced by almost 1 dB. Unless there are a lot of receivers, this tradeoff probably does not make sense.

The next graph, shown in Figure 5-33, is for M-ary PSK systems. The line on the left is used for both BPSK and QPSK, which have almost identical error rates. The next line to the right is for 8-ary PSK, and the line to the far right is for 16-ary PSK. As you can see, more power per bit is required for the same error rate as the bit density grows.

■ **EXAMPLE 5-11**   Compare the BER performance of QPSK and 16-PSK systems for an E_b/N_o received power of 11 dB. For QPSK, a BER of 1E-6, or 1 bit in a million, is read. For 16-PSK, a BER of about 8E-2, or about 1 bit in 12, is read. Although more bits per symbol are achieved by using a 16-PSK system, the BER soars for this power level. How much power must be received for the 16-PSK system to perform like the QPSK system?

Figure 5-33
BER Graph for PSK Systems

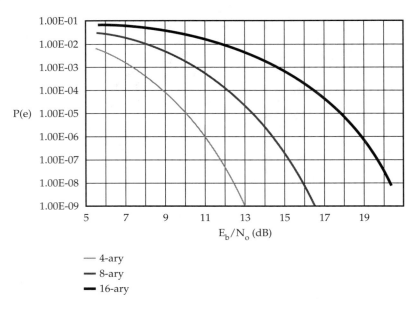

Look at the graph in Figure 5-33. For a BER of 1E-6, a 16-PSK system requires an E_b/N_o value of 19 dB. A value of 8 dB more power is required! That represents a power ratio of about 2.5:1! For this extra power requirement, what do you get? The BER is the same, so no increased performance will be had in that area. E_b/N_o graphs like the one in the figure remove any dependency on data rate, so no more data is passing through the system.

The only consideration left is bandwidth. A 16-PSK system is much more bandwidth efficient. It features a bandwidth efficiency of 4 bps/Hz. Stated another way, four input state changes are required for one output change. A QPSK system has a bandwidth efficiency of 2 bps/Hz. Therefore, for this increase in power requirements, transmission bandwidth is reduced.

A 16-PSK system uses half the bandwidth of a QPSK system. Look carefully at Figure 5-33. If you halve the bandwidth, you double the power. (Recall that 3 dB is double the power.) As you move to the right, each curve uses half the bandwidth of the previous one. Additionally, each curve is roughly 3 dB higher than the previous one for the same error rate. For a reduction in bandwidth, an increase in power is required.

The net result is that the bandwidth requirement is halved, but at a large price. The received power must be at least twice as great, which is generally the case with higher-ary transmissions. The bandwidth savings always grow more slowly than the power requirements. In many cases, bandwidth is so valuable that this penalty is gladly paid.

Figure 5-34
BER Graph for QAM Systems

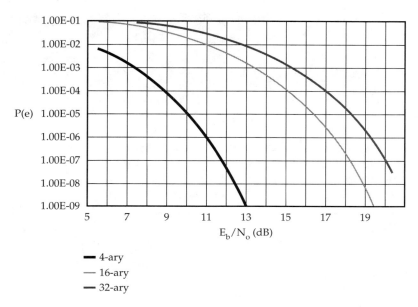

The last BER graph is illustrated in Figure 5-34 for M-ary QAM systems. This graph shows three curves, one each for 4-ary QAM, 16-ary QAM, and 32-ary QAM. Most new designs utilize the QAM approach for the reasons shown below. The error rate performance is better than any other technique considered here.

■ **EXAMPLE 5-12**   To illustrate that QAM and QPSK are equivalent systems, find and compare the BER of the two modulation techniques for an E_b/N_o of 11 dB.

For the QPSK system, the BER is 1E-6, and the QAM curve shows the identical result. QAM and QPSK are equivalent systems. Under the same conditions, they will perform identically. If you examine the constellation diagrams, the distance between each signal state is identical.

Compare the performance of 16-PSK and 16-QAM. Note how the difference in the constellation diagrams affects the error performance. Since both have the same bandwidth efficiencies, both will take up the same bandwidth. This feature is a clear indication of how the distances between each amplitude/phase state affect the performance of a modem in the presence of noise. Here, the comparison is dramatic. The performance at an E_b/N_o value of 18 dB is very different. For 16-PSK, an error rate of about 1E-5 results. For 16-QAM, the error rate is only about 8E-7! That difference represents almost two orders of magnitude better performance for the same received energy per bit.

The final example for this chapter will illustrate how these graphs are used in industry to characterize the performance of a particular modem. In any manufacturing

Table 5-4
QAM Performance Measurements

Modulation	E_b/N_o	Measured P(e)	Pass/Fail/Comment
QAM	11 dB	1^5	Pass
QAM	11 dB	1^4	Fail
QAM	11 dB	1^7	Not possible!

environment, the actual performance of any individual manufactured system will vary due to several factors. What is important is that the device meet the published performance specifications prior to shipment to a customer. BER graphs are often used to characterize the performance of modems prior to shipment.

■ **EXAMPLE 5-13**   Determine if a standard QAM modem is functioning within tolerance or not. For this product, the tolerance is defined to be within 1 dB of theoretical. The theoretical curve is shown in the BER graphs. All real modems operated to the right of, or above, the curve for that modulation type. If any modem ever shows a BER to the left of, or below, the curve, something is wrong. This situation would represent better performance than theory predicts, an impossibility unless the theory is wrong.

Table 5-4 lists three points in Figure 5-33. Examine the points yourself and verify that the entries in the table are correct.

Note that the first data point is to the right of the curve, but within 1 dB. The second data point is also to the right of the curve, but 2 dB off, which fails the test of 1 dB. The third data point is to the left of the curve, so something is wrong: Either the modem is mislabeled or the test was performed incorrectly.

■ **SUMMARY**

The relationship between the bandwidth required for a particular modulation method and the tradeoff in increased power for the same error rate has important implications for the types of modulation that are preferred for certain channel types. Modulation techniques to be used in wireless channels are particularly sensitive to this relationship. If the application is a dedicated wired one, like a LAN, then it does not make much sense to increase the signal density and add complexity to the modem. The only advantage to this approach is less bandwidth used for any bit rate. If the channel is a dedicated one, it makes little sense to pay the complexity costs of the design and the cost of a higher power signal as well.

On the other hand, for channels where the bandwidth must be shared among competing services, like a cellular, wireless LAN or satellite system, or where the bandwidth is strictly limited by the technology, like the telephone system, it can make a large difference. In these situations, one is ready to pay for complexity and amplification to get increased throughput from an existing limited channel allocation. Subscribers to a service will usually pay more for increased throughput. The only effective means of increasing throughput when the bandwidth is limited is increased signal density, and since subscribers usually pay more for increased throughput, the project often pays for itself.

REVIEW QUESTIONS

1. Describe briefly the difference between traditional FSK and MSK modulation techniques.

2. Find the bandwidth required for a BPSK modulator operating at 1 kbps.

3. Find the bandwidth required for a QPSK modulator operating at 1 kbps.

4. Find the bandwidth required for a QAM modulator operating at 1 kbps.

5. A certain modem has a transmit bandwidth of 10 kHz. The binary data rate is 20 kbps. What is the symbol rate?

6. Another modem has a transmit bandwidth of 20 kHz. Without knowing the binary data rate, can you find the symbol rate? Assume that the modem is PSK.

7. For an 8-PSK modem, what is the transmit bandwidth if the binary data rate is 2 kbps?

8. For each of the modulation types in Table 5-5, determine from the measurements

taken if the modem is functioning within tolerance or is not. Also check to see if the data you were supplied on the modulation type is correct. Assume that the modem is operating correctly if the performance is within 1 dB from theoretical. Use the BER graphs to make these determinations.

9. A channel has a lower limit of 1 kHz and a BFSK modulator. Find the signaling frequencies (mark and space) that would minimize the upper limit of the bandwidth of the transmission. Also find the transmission bandwidth. The modulating signal is 10 kbps. Find these values for each of the guardband options below:
   a. No guardband
   b. Guardband of 10 kHz

10. Draw a tri-bit encoding diagram and relate it to 8-ary amplitude assignments, like we did in the chapter discussion for di-bit and 4-ary assignments.

Table 5-5

Modulation	E_b/N_o	Measured P(e)	Pass/Fail/Comment
ASK	11 dB	1exp-4	
ASK	11 dB	1exp-5	
QPSK	11 dB	5exp-6	
8-PSK	11 dB	1exp-4	

6

Pulse Modulation

■ INTRODUCTION

This chapter will introduce pulse modulation, the variation of a digital pulse in response to an information signal. In Chapter 5, modulation techniques varied a continuous sinusoid, in amplitude, frequency, or phase, in response to an information signal. Those techniques are called continuous wave (CW) modulation because the waveform varied was continuous in nature.

In the study of pulse modulation, ways to vary a pulsed waveform in amplitude, width, or position will be discussed. The focus of the chapter will be pulse amplitude modulation (PAM) and its close cousin, pulse code modulation (PCM). Pulse width modulation (PWM) and pulse position modulation (PPM) will be surveyed only briefly because the bulk of economic interest is in PCM. Note that almost every telephone call, whether you are using the wireline telephone system or any of the digital cellular systems, is digitally encoded using PCM techniques. Before going into detail about these techniques, fundamental concepts of sampling theory will be reviewed.

■ SAMPLING THEORY

The first thing to realize about pulse modulation is that most time analog waveforms need to be represented in a digital manner. The best examples are telephone calls. One way digital representation can be done is by using traditional analog-to-digital (A/D)

Figure 6-1
Sampling an Analog Signal: (a) Before; (b) After (*Reprinted by permission of Pearson Education, Inc., Upper Saddle River, NJ*)

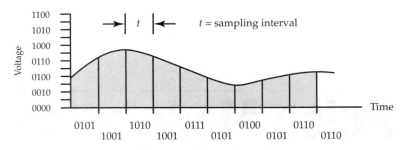

The series of binary data points.
Note that sample value is taken at the start of the sampling interval.
(a)

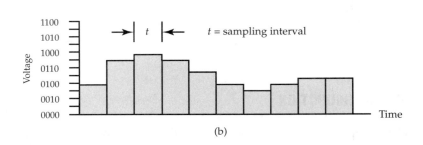

(b)

circuits at the source and digital-to-analog (D/A) circuits at the destination. Most of you have already had a course that has introduced these circuits, so some important points will be only briefly reviewed. An A/D converter digitizes or samples an analog signal at regular intervals and represents the original analog or continuously varying signal as a series of binary data points that represent the amplitude of the original signal. Part (a) of Figure 6-1 illustrates the analog signal. Below, on the horizontal axis, are the binary values used to represent each sampling interval. Part (b) of Figure 6-1 shows that once the sampling is performed, the original information in the analog signal is lost and only the binary values taken in the sampling interval remain.

The series of binary data points in the figure is an accurate representation of the original signal if certain criteria are met. The first is that the sampling must be regular, which means that the samples must be evenly spaced. The sampling interval t, shown in Figure 6-1, is always the same. Note that the spacing must remain constant, which is a requirement of sampling theory. The second criterion is that the signal must be sampled often enough to represent the signal accurately. If it is too slow, the binary data points, when reconstructed, will not look anything like the original signal. Figure 6-2 illustrates the effect of varying the sample interval.

The third criterion is the number of bits used for each binary value. The more bits used to represent each sample, the more accurate are the sampled quantities. Any error

Figure 6-2

Effect of Varying the Sampling Interval (*Reprinted by permission of Pearson Education, Inc., Upper Saddle River, NJ*)

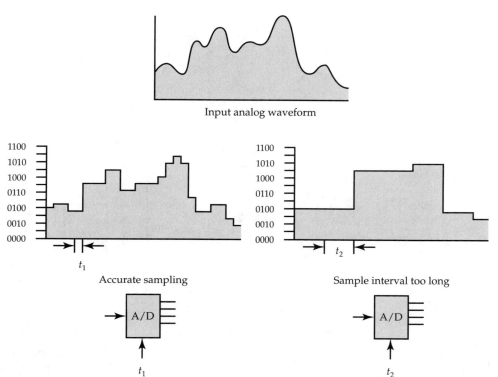

due to this effect is called quantization error. Suppose that you had an analog signal that varied from 0 to 10 volts, and that you tried to sample it with a 2 bit A/D. Two bits gives just four possible amplitude values, each about 2.5 volts apart. The quantization error here could be as much as 1.25 volts, or half the step size. On the other hand, if you used an 8 bit A/D, 8 bits gives 256 possible amplitude values, each only about 0.25 volts apart. It should be clear that the latter approach gives a lower quantization error and smaller step size. See Figure 6-3.

Note that in Figure 6-2, the number of bits was held constant and the sampling interval was varied. In Figure 6-3, the sampling interval was held constant and the number of bits was varied. Also, note that in Figure 6-3, the amplitude steps accommodated the same dynamic range. Both comparisons were made to help you understand how these factors interact. The sampling interval and the dynamic range will be held constant when we talk about the number of bits needed for a specific performance level. In this way, the comparisons can be kept real and will be representative of tradeoffs that occur in practice.

Figures 6-2 and 6-3 illustrate the effects of not following either criterion 2 or criterion 3. As you can see, skimping on bits or the sampling clock speed will have negative

Figure 6-3
Effect of the Number of Amplitude Bits Used in Sampling (*Reprinted by permission of Pearson Education, Inc., Upper Saddle River, NJ*)

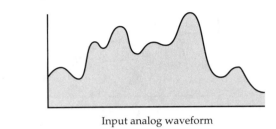

effects. Remember, once the signal is digitized (or sampled), only the samples represent the information in the original analog signal. If the signal is not represented well, information is lost and can never be recovered. It is crucial to set these parameters correctly. To summarize, three criteria must be obeyed to obtain an accurate sampling of analog data:

1. The sampling interval must be held constant.
2. The signal must be sampled often enough.
3. Enough bits must be used to represent each sample so that the amplitude accuracy is a close approximation.

Now that you understand the practical issues, it is time to examine the second criterion more carefully. That point is at the real heart of sampling theory. The Nyquist theorem defines sampling theory. Usually theorems are part of mathematics studies, which most of us would rather forget, but here it has a very practical and easy-to-understand application. For any sampled data system, a minimum sampling rate must be observed if the results are to be accurate. This minimum sampling rate is defined by the maximum frequency component present in the original analog or continuous waveform to be sampled. The sampling rate must be at least twice that frequency.

So criterion 2 is very easy to follow. Measure the input signal to determine the maximum frequency present. Multiply it by two, and the minimum frequency of the sample clock is determined. This criterion is expressed in Equation 6-1.

$$f_{sample} \geq 2 * f_{signal} \tag{6-1}$$

It is possible to sample more often, but all that is necessary to recover the original signal is twice the highest frequency present in the original signal. Stated another way, the result of this criterion is to make sure that *each* cycle of any frequency present in the original signal is sampled at least twice. This results in a mathematical guarantee (a theorem and proof) that the original waveform can be reproduced from the sampled data.

■ ALIASING

In practice, each waveform is usually sampled slightly more than two times to make sure that the second criterion is not violated. If it is violated and the original waveform is not sampled at the rate of at least twice its highest frequency, then a phenomenon known as aliasing will arise. To understand what aliasing is, take apart the word. The first part of the word is *alias*. Alias is often used in sentences like "His alias is Clark Kent, but secretly he is named Jorel." So *alias* means "an assumed name." Now the subject here is not comic book characters but rather frequencies, and aliasing means that a frequency has folded over and is an alias, or false image, of the correct frequency. Aliasing always results in making the frequencies appear lower in frequency than they actually are.

If the signal is not sampled fast enough and hence the fundamental frequency isn't high enough, then due to the sampling process, the negative frequency image folds over and appears in the output spectrum. The sampling process itself is essentially amplitude modulation of the sampling waveform by the original analog waveform. (Please verify the above sentence to yourself. It is an important insight about how aliasing actually arises. Pay special attention to which waveform is modulating which.) This signal that appears is an *alias* of the original analog frequency that should be in the output spectrum. Remember, any alias is always lower in frequency than what it should be because it is always the sum of some multiple of the sampling waveform and the negative image of the analog waveform.

Figure 6-4 illustrates how aliasing occurs. As stated above, if the signal is not sampled often enough, the negative frequency image of the sampling clock folds over and appears in the output spectrum. As you can see from Figure 6-4, the alias frequency is of a lower frequency than the signal frequency. It has resulted from the sample clock not sampling each cycle of the input signal twice. Note that for the first cycle of the input signal, the sample clock did sample twice, but it did not for the second cycle. Criterion 2

Figure 6-4
Illustration of Aliasing (*Reprinted by permission of Pearson Education, Inc., Upper Saddle River, NJ*)

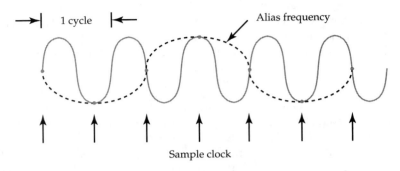

says that, for every cycle, the sample clock must be fast enough to sample it twice. Example 6-1 illustrates the effects of aliasing and how to calculate the alias frequency.

■ **EXAMPLE 6-1**   Sample an 8 kHz signal with a 12 kHz sample clock. This sampling clearly violates criterion 2 expressed in Equation 6-1.

$$f_{sample} \geq 2*f_{signal}$$

$$12 \text{ kHz} \geq 2*8 \text{ kHz} = 16 \text{ kHz}$$

Unless the sample frequency is at least 16 kHz, criterion 2 has been violated. The sampling has not occurred often enough. Determine the maximum frequency that could be sampled with the 12 kHz sample clock.

$$f_{sample} \geq 2*f_{signal}$$

$$12 \text{ kHz} \geq 2*6 \text{ kHz} \Rightarrow 6 \text{ kHz} = f_{signal, max}$$

To be sampled accurately by the 12 kHz sample clock, the maximum input frequency would have to be 6 kHz. Determine the alias frequency in the original problem.

$$f_{alias} = f_{sample} - f_{input} \tag{6-2}$$

$$f_{alias} = 12 \text{ kHz} - 8 \text{ kHz} = 4 \text{ kHz}$$

Equation 6-2 illustrates that you determine the alias frequency by subtracting the input frequency from the sample frequency. As an additional test, always check the answer to make sure that the alias frequency determined from Equation 6-2 is always less than the maximum input sample frequency that would be sampled accurately with the sample clock. This test can be written mathematically as:

$$f_{alias} \leq \frac{f_{sample}}{2} \qquad \qquad \text{(6-3)}$$

$$4 \text{ kHz} \leq \frac{12 \text{ kHz}}{2} = 6 \text{ kHz}$$

As you can see, the answer checks.

If the reasons for aliasing are not clear, do not worry. If you always sample fast enough, there is no need to be concerned. Many everyday events make it possible to see the aliasing that results from insufficient sampling. One of the most common is watching a fan blade rotating. Once the blade is rotating fast enough, it appears to be rotating in the opposite direction but much more slowly. This effect is also often seen on stationary exercise bikes, etc. You know that this effect cannot be true and correctly assume that it is some kind of optical illusion. The cause of that optical illusion is aliasing. Since the item is rotating faster than you can actually see it, your brain tries to "sample" the image and reconstruct it. It makes a mistake, however, because your brain can only sample so fast. It violates the Nyquist theorem, and you observe an alias of the actual image. Note that the alias is always slower than the rotating blade or spoke, which emphasizes the statement that any alias frequency will be lower in frequency than the frequency it masks.

We reiterate here that aliasing, except for rare occasions, must be avoided. The easiest and surest way is to pass the input signal through a low-pass filter (LPF) before sampling the waveform. This low-pass filter will attenuate any frequencies above its cutoff frequency. The cutoff frequency of the filter is set to less than half of the sampling rate on the A/D, thus preventing aliasing.

Of course, the corner and cutoff frequency and the sampling rate must be chosen so that you get all the information needed from the original analog signal. This information is a system design problem and is usually defined by the bandwidth of the signal. For example, in a voice frequency design, the upper frequency of interest is 3,400 Hz. For an audiophile design, the upper frequency of interest is 20,000 Hz. When used in this way, low-pass filters have a special name and are referred to as anti-aliasing filters.

■ **EXAMPLE 6-2** Determine the cutoff frequency of an LPF for a sampled data system. Also determine the sampling rate. The input signal is from a high-quality stereo system and you are to determine the anti-alias filter's cutoff frequency for a new type of compact disk product.

Although you search for a specification of the upper frequency, you cannot find any numbers. You conclude that the information is not in the document and try to find another way to set the frequency. You ask your friend, who is an audiophile, the maximum frequency of her stereo system. She tells you that her system response is flat up to 20,000 Hz. Since she has a reputation for excessive zeal about her hobby, you imagine that if that frequency satisfies her, it will probably satisfy whoever wrote the document as well.

(Perhaps part of the assignment was to determine what that frequency should be, as well as to find the cutoff frequency!)

Now that a maximum input signal frequency is determined, it is an easy matter to find the cutoff frequency for the LPF. You specify an LPF that is flat to 20 kHz, has a corner frequency of 22 kHz, and rolls off by 40 dB by 30 kHz. These specifications set the sampling rate to 60 kHz. You have no idea if such a filter is feasible at this point, but it should satisfy the sampling theory restrictions given to you.

■ ANALOG AND DIGITAL COMPARISON

Modulating analog waveforms into digital pulses, recovering them, and reconverting them into analog waveforms is the main focus of this chapter. In many communication systems, this activity takes place. Remember, all real-life systems are analog in nature. Digital techniques may be used to move them from one place to another because digital techniques offer several advantages over analog techniques of transmission and reception, but in the end and at the beginning, analog waveforms are required. As you might have guessed from the above discussion on sampling theory, A/D converters and D/A converters will be key components. However, a simple approach of doing just the conversion does not yield the best overall system.

The first point to make here is that when analog and digital are compared, the subject is *not* analog inputs versus digital inputs. AM is analog modulation with analog inputs. ASK, FSK, and PSK are analog modulation techniques with digital inputs. It is sometimes confusing that these techniques are often referred to as digital modulation; however, digital modulation refers to the form of the input signal, not the form of the output modulated signal in the channel. Remember, in pulse modulation, the input is an analog signal but the modulation is digital in form. It uses pulses to represent information.

This difference is of interest when comparing digital modulation schemes such as ASK, FSK, and PSK to the pulse modulation schemes studied in this chapter. Either set of modulation schemes can be used to transmit digital or analog input data from source to destination. The key difference is what form the signal takes while in the channel. If the channel is wireless in nature, digital modulation schemes will be used. If it is wireline in nature, pulse modulation schemes will often be used. The key items are the relative advantages and disadvantages of these two basic approaches.

Analog modulation/demodulation has only one real advantage today: it consumes less bandwidth for a given input bandwidth than do digital techniques. This advantage is the core reason why all wireless systems use either analog or digital modulation techniques for transmission. Digital transmission techniques using pulse modulation waveforms claim almost all other advantages because of the technology available today. Thirty years ago, the story was much different, and it may be quite different again in thirty years if optical techniques evolve in certain ways. One key technology that makes such a difference is the availability of digital ICs in large numbers and at low cost. These devices are very reliable compared to analog components, and they don't require the periodic tuning that many analog designs inherently require for optimum performance.

However, the availability of cheap flip-flops, and the like, isn't the only reason. Coding techniques work to help digital schemes and, because the data is already in binary form, applying coding theory to improve signal-to-noise ratios is not cumbersome and the designs mesh nicely. Coding can also be used for encrypting and making the transmission secure. Thus, it becomes a critical advantage for some applications. Coding will be discussed in Chapter 7.

Another major advantage to digital techniques is that much larger dynamic ranges are possible when using large numbers of bits to store amplitude values. Anyone who owns a CD player and an old cassette player can compare the specifications between the two for dynamic range. The CD player, utilizing 16-bit PCM, far outperforms the analog cassette player.

The only real disadvantage to digital implementations, one that analog implementations avoid, is the requirement that the demodulator must accurately sample the signal for digital signals to be recovered accurately at the demodulator. This accuracy means that the demodulator must sample the signal at the same time for each sample, relative to its position. Since the time delay between each sample can vary during transmission, synchronization is required at the demodulator. Synchronization generally requires some kind of PLL circuit.

Analog systems may use some form of synchronous demodulation that also requires this type of circuitry, but for digital modulation systems, the circuitry is a requirement. This requirement adds cost and complexity to the receiver. On the other hand, cost is becoming less of an issue because these circuits can now be implemented routinely in digital form. It seems clear that analog and digital modulation will predominate in channels where the signal is not confined to a wired or optical media, for example, atmospheric channels. Because bandwidth is a shared resource in this environment, bandwidth utilization is a critical issue.

In most modern wireless systems, pulse modulation is reserved to convert the fundamentally analog input, usually voice, into a series of digital pulses. These digital pulses are then modulated using some digital modulation scheme, such as those introduced in Chapter 5. This fundamentally analog signal, carrying analog information rendered into digital form though pulse modulation techniques, is used in the wireless channel.

Take an example from cellular telephony. Your analog voice is first converted into a series of pulses using a pulse modulation technique. These digital pulses are applied to a digital modulation scheme like those discussed in Chapter 5. The output of this modulator is what actually passes through the atmosphere from source to destination. This signal is analog in nature. At the destination, the demodulator converts the modulated signal received into a series of pulses. These pulses are then pulse demodulated back into analog signals that represent your voice.

■ PULSE AMPLITUDE MODULATION

Probably the simplest implementation of pulse modulation is pulse amplitude modulation (PAM). In this scheme, the amplitude of the pulse train produced carries the analog

Figure 6-5
Pulse Amplitude Modulation (PAM): (a) Input Analog Waveform; (b) Sample Interval; (c) PAM Waveform
(*Reprinted by permission of Pearson Education, Inc., Upper Saddle River, NJ*)

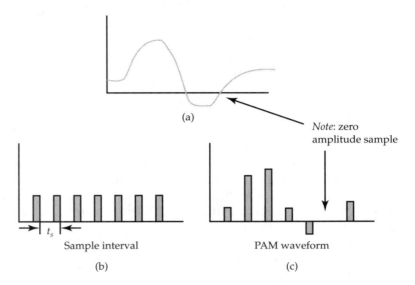

information. Essentially, a sample-and-hold circuit samples the input analog waveform at regular intervals and holds that value for the sample interval. Then each pulse is an (almost) instantaneous sample of the input analog waveform's amplitude. PAM is illustrated in Figure 6-5.

The first waveform, shown in Figure 6-5(a), is the input analog waveform that is sampled. The second, shown in Figure 6-5(b), is the pulse clock that drives the sample-and-hold circuit. Figure 6-5(c) is the resultant PAM waveform; the amplitudes of the pulses match the amplitude of the input analog waveform at the moment of sampling. As you can see from the figure, connecting the samples of the PAM waveform would reproduce the original input waveform. Note the zero amplitude sample and the negative amplitude sample. A typical circuit implementation of this type of modulator is shown in Figure 6-6.

Note that only two components, a sample-and-hold module and an oscillator, produce the sample timing. Since the sample-and-hold module uses a capacitor to store the voltage, there is no issue of amplitude quantization as with the A/D discussed earlier. This feature is a plus for this inherently analog approach to measuring magnitude. Will this circuit work so simply to produce pulse amplitude modulation? The answer is yes if the maximum frequency of the input analog waveform is less than twice the oscillator frequency to ensure a sample twice each period.

Of course, there is always Murphy's Law to suggest that everything won't go as anticipated. So, be a good engineer and use a little precaution to make sure that no frequencies that are higher than expected get into the sampled data system. As suggested earlier, one simple method is to put in a low-pass filter just prior to sampling the wave-

Figure 6-6

PAM Circuit Implementation (*Reprinted by permission of Pearson Education, Inc., Upper Saddle River, NJ*)

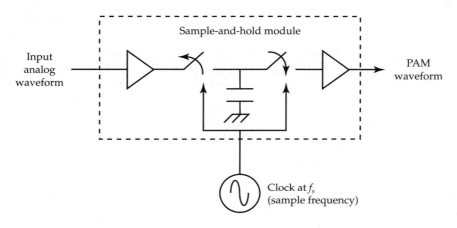

Figure 6-7

PAM Circuit with Anti-aliasing Filter Added (*Reprinted by permission of Pearson Education, Inc., Upper Saddle River, NJ*)

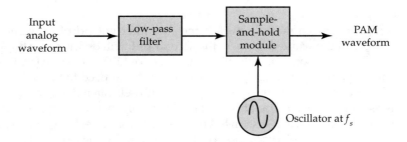

form. Figure 6-7 shows this modification. When used in this manner, an LPF is often called an anti-aliasing filter because that is the function it is performing.

The next question is, Now that a PAM waveform is captured, ready to send, how is the original signal recovered at the other end? The answer is, With a certain curious symmetry, with another low-pass filter! The second filter will remove all the high-frequency harmonics of the pulse train, essentially connecting the dots or, in this case, the tops of the pulses, to form a smoothly varying, continuous waveform that closely represents the original. Getting the waveform as close as possible is important for best performance, but nothing is absolutely perfect. Just approximate it as closely as possible, within the constraints of insight, time, and money. A PAM modulator and demodulator is shown in Figure 6-8. Note that the communications channel is not considered to contribute any noise or distortion to the signal as it passes through it from source to destination.

The only remaining question is, How is the cutoff frequency for the second LPF chosen? It is selected to be identical to the first LPF, to a first approximation. If you think about

Figure 6-8

PAM Modulator and Demodulator (*Reprinted by permission of Pearson Education, Inc., Upper Saddle River, NJ*)

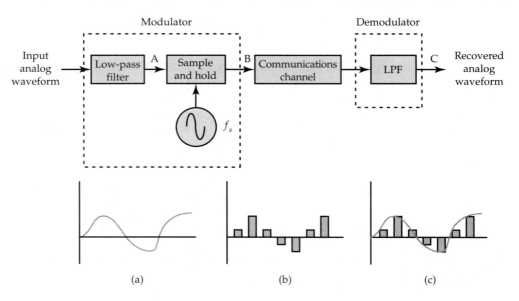

it, this approach makes sense. If the first filter got rid of unwanted stuff, it must have passed all the wanted stuff. So the second LPF will do the same thing on the other end.

There are some caveats. First, the choice of LPFs is not so easy. The passband characteristics and corner frequency are also of interest and need to be chosen carefully for best performance. Sometimes these features are purposely altered to compensate for modulation effects and channel degradations. Almost certainly in any real system, they will not be exactly identical. Second, this system has no provision for noise. It assumes a perfect channel, which is okay for textbook discussions but not so good for the real world. Third, the design of the sample-and-hold circuitry also has an impact. For example, just how instantaneous is the sampling? Maybe more important, how instantaneous does it have to be to meet the specifications? Finally, you may choose to change the sample rate to improve the circuit. More points allow greater frequency coverage, but this alteration usually costs money.

Where is the tradeoff point where people will pay more for better performance, compared to that point where the system already sounds okay so keep the cost down? This last concern is present in every system, not just PAM. It is important to point this out often: cost frequently determines success or failure in the marketplace. You might ask yourself, If the product line fails, how secure will my paycheck be?

■ PULSE CODE MODULATION

PAM, while simple in concept, is not the best choice for modern implementations. An important step in understanding a very good system is a widely used technique called

Table 6-1
PAM Pulse Amplitude and PCM Code Word Representation

Input Analog Amplitude	PAM Pulse Shape	PCM Code Word
−3		011
−2		010
−1		001
−0		000
+0		100
+1		101
+2		110
+3		111

The sign bit is first. It takes the value 1 for positive amplitudes, 0 for negative amplitudes.

pulse code modulation (PCM). PCM uses PAM as a first stage; it makes use of PAM techniques to convert an analog waveform into a series of pulses. But PCM goes further: it actually sends a serial code word, a group of bits, for each sample. This group of bits is the output of an A/D that is sampling the waveform in just the way that the sample-and-hold circuit did for the PAM case. Now it is clear why the number of bits determining the amplitude resolution becomes important.

Did you wonder what happened to criterion 3 in the discussion where we used a sample-and-hold circuit instead of an A/D? Recall that in the sample-and-hold case, a capacitor was used to store the sampled voltage. To convert to an all digital design, the number of bits needed to sample the amplitude accurately enough for the system objectives must be carefully determined. PCM sends this group of bits, called a PCM code word, at every sampling interval. Typically, it first serializes the PCM code word and sends a stream of data bits at every sampling interval.

Furthermore, this group of bits, which is sent for every sample, provides a way of accurately representing the amplitude of the pulse while sending pulses that are all the same amplitude. This point is critical for understanding why PCM works so well when it is used to transmit information in wired channels. Most communications channels work best for signals in a relatively narrow amplitude range. Certainly there are preferred amplitudes that minimize distortion. So a big advantage of PCM over PAM, and for that matter over almost any analog system, is that all the signals sent are of the same amplitude. The varying amplitudes of the samples are represented as different PCM code words instead of pulses of different amplitudes. See Table 6-1.

■ QUANTIZATION ERROR

Any PCM system that uses PAM techniques suffers from the same source of error, or quantization error. This source of error results from the fact that an analog waveform can take on an infinite number of values, while a digital approximation can take on only some fixed number of values. The number of values a digital approximation can take on is determined by the number of bits in the A/D converter.

As you may recall from your classes on digital systems, A/D converters give a stairstep approximation of the input waveform. The smaller the stairsteps, the better, and the more expensive the A/D. Quantization error is defined as half the step size of the converter and is represented mathematically in Equation 6-4.

$$\text{Quantization error} = \frac{\text{step size of A/D converter}}{2} \tag{6-4}$$

One always wants to minimize the quantization error in any system design. The quantization error is a key component in determining the signal-to-noise ratio of both linear and nonlinear PCM systems. A linear PCM system is one where the step size always stays the same. A nonlinear PCM system is one where the step size varies according to some rule. For now, assume that all PCM systems are linear. Later it will be discovered that most real implementations are not linear. These nonlinear systems are referred to as companded PCM systems. For any PCM system, better potential performance in the presence of noise is obtained as the quantization error is reduced. These calculations for companded PCM systems are complex and outside the scope of this text. However, to illustrate this concept, a calculation for a linear PCM system will be performed in Examples 6-3 and 6-4.

■ **EXAMPLE 6-3**   The number of bits used in the A/D converter indicates the number of bits used for the magnitude portion of the PCM code word. It also indicates the amplitude resolution or step size. Any step voltage used is an approximation of the actual input to the A/D. That error is called quantization error. For a certain PCM system, an 8-bit A/D converter with a step size of 0.2 V is used. What is the quantization error?

$$\text{Quantization error} = \frac{\text{step size}}{2} = \frac{0.2}{2} = 0.1 \text{ V}$$

■ **EXAMPLE 6-4**   To illustrate further the relationship between step size and quantization error, suppose a 2 V peak analog waveform is to be sampled. The waveform has a 0 V DC offset, so a sign bit to represent the negative digital magnitude values will be needed. Assuming a 4-bit code word, determine the number of bits needed in the A/D converter, and compute the step size and quantization error.

Since 1 bit is used for the sign, 3 bits are used for the magnitude. A 3-bit A/D converter is required. The sign bit will be detected in parallel with the converter. Three bits means $2^3 - 1 = 7$ steps. Recognize that one step is always reserved for 0 volts. A 2 V peak signal means a peak-to-peak voltage swing of 4 V. These two quantities are divided to obtain the step size. The quantization error is then easily obtained.

$$\text{Step size} = \frac{\text{peak-to-peak voltage}}{\text{number of steps}} \tag{6-5}$$

$$\text{Step size} = \frac{4}{2^3 - 1} = \frac{4}{7} = 0.57$$

$$\text{Quantization error} = \frac{\text{step size}}{2} = \frac{0.57}{2} = 0.285$$

Before going any further, examine the typical PCM modulator shown in Figure 6-9(a). As you can see, this modulator is similar to the PAM design discussed earlier. The input anti-aliasing filter is still included, but instead of a sample-and-hold circuit, an A/D converter is present. Additionally, a parallel-to-serial converter takes the parallel output of the A/D and serializes it. Again, the more bits in the A/D, the better the performance over any fixed range of input voltage values. More bits means lower quantization error.

Part (b) of Figure 6-9 illustrates how the PAM sampling of the input analog waveform converts the analog values to 4-bit digital output words. Once these values are serialized, they become PCM code words. Essentially, the PCM process takes the digitized amplitudes of the input signal and outputs them serially as digital words.

Figure 6-9
PCM Modulator (*Reprinted by permission of Pearson Education, Inc., Upper Saddle River, NJ*)

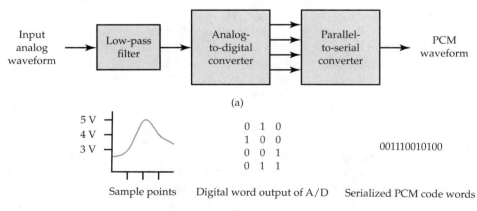

■ NONUNIFORM QUANTIZATION

While the design in Figure 6-9 is straightforward, it is not complete. Consider how costs could be reduced while still accommodating a wide range of input voltage values. One way to keep the quantization error constant while accommodating a wider input voltage range would be to add bits to the A/D. However, this method gets expensive quickly. A better way would be to compress the input signal in some way and then uncompress it on the other end. But how can this process be done? It does not seem possible to compress an analog signal and still keep all the information necessary to recover it.

The answer lies in the application definition. In all digital telephony, the voice signal is first PCM encoded. Depending on whether the system is wireline or wireless, the PCM code words either are transmitted directly or form the input to a digital modulation step, respectively. Therefore, perhaps knowledge of how humans vary their voices can be used to reduce the number of bits necessary for acceptable accuracy. It turns out that human beings generally talk in low voices on the telephone. Although some people scream, most telephone conversations are conducted using only about half of the dynamic range of human speech. Furthermore, people tend to expect that screaming voices will be distorted, so they tolerate easily a little amplitude distortion for people who talk loudly.

Therefore, people will apparently tolerate some distortion of the input analog waveform at high amplitudes. We can take advantage of that fact and increase the quantization error for large amplitudes. This increase is the central idea behind nonuniform quantizing. Traditional A/D devices always use uniform quantization: each stairstep is the same size. For the nonuniform case, the steps are set small for small input voltages and large for large input voltages. This process is illustrated in Figure 6-10.

Examine how the idea of moving to a nonlinear quantization can directly affect the number of bits that must be sent in each code word. In the figure, the two curves are approximately similar. But in the uniform case, the number of states is 8, while in the nonlinear case, the number of states is only 4. Note that each series of stairsteps covers the entire dynamic range of the input signal. But in the uniform case, 8 binary states are required, while in the nonuniform case, only 4 binary states are required. This change results in a savings of 1 bit for each sample.

Notice that, in the nonuniform case, 3 of the 4 bits are used for amplitudes of less than 70 percent of the total dynamic range of the input signal. This result gives good amplitude resolution for normal speech, close to the number of states in the uniform case. Notice that the entire upper 30 percent of dynamic range is represented by only 1 bit, or 25 percent of its states. This feature highlights the difference between the nonuniform quantization approach and the uniform approach, which requires 50 percent of its states, and hence its bits, to lie in that range.

Therefore, good amplitude resolution or quantization is obtained for that portion of the dynamic range of the input that concerns us, and at a significant cost savings. This savings is accomplished by changing the word size of the A/D. It is reduced by 1 bit or, in this example, we achieve a 33 percent reduction in the number of bits transmitted. Also, on the other end, we save that same bit in the D/A required, another cost savings with little impact on the system.

Figure 6-10
Uniform and Nonuniform Quantization (*Reprinted by permission of Pearson Education, Inc., Upper Saddle River, NJ*)

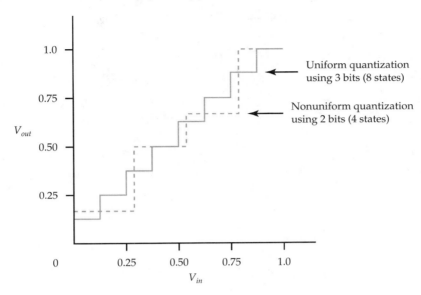

■ **EXAMPLE 6-5**   Suppose you use companding to reduce the PCM word size by 2 bits. What is the information flow impact on a system that originally used 8 bits but now requires only 6 bits? Assume an original data rate of 1 Mbps.

$$\text{Original number of words transmitted} = \frac{\text{data rate}}{\text{word size}} = \frac{1\text{ Mbps}}{8\text{ bits}} = 125,000 \text{ words/s}$$

$$\text{New number of words transmitted} = \frac{\text{data rate}}{\text{word size}} = \frac{1\text{ Mbps}}{6\text{ bits}} = 166,666 \text{ words/s}$$

As you can see, a significant improvement in the amount of information flow occurs for the same data rate by using a shorter PCM code word. Clearly, companding is an important technique.

Table 6-2 and Figure 6-11 show the quantization in both the linear and nonlinear cases, with the number of bits held at 2 in both cases. The table and the figure will allow you to compare how nonlinear quantization works when the number of bits stays the same.

It is clear that one way to optimize a PCM system is to employ nonuniform quantization. You might ask, Are there other ways to optimize? The answer is yes. A choice was skipped earlier and was not addressed. The A/D converter output was simply converted to a serial stream of bits. In that process, it was assumed that no further optimization

Table 6-2

V_{in}	V_{out} Linear Stairstep	V_{out} Nonlinear Stairstep
0	0.15	0.15
0.1	0.15	0.45
0.2	0.15	0.7
0.3	0.5	1.0
0.4	0.5	1.0
0.5	0.5	1.0
0.6	0.7	1.0
0.7	0.7	1.0
0.8	1.0	1.0
0.9	1.0	1.0
1.0	1.0	1.0

Figure 6-11
2-Bit Quantization

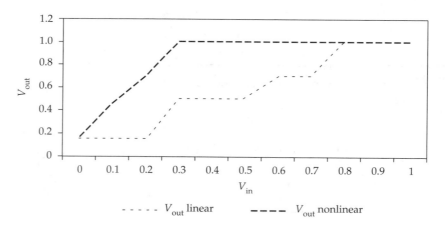

would be exercised on the assumed binary output of the A/D. The binary sequence would be directly serialized. Now this process will work, of course, but in doing it, a fundamental part of what makes PCM work so well is overlooked.

Why not explore applying an alternate binary code? Many transformations could be made to that binary code that would improve the overall performance of the PCM system. Two that will be discussed are the folded Gray code, and the code typically used in commercial implementations of CODECs: the folded binary code Table 6-3 summarizes how these codes work.

It is important to make a couple of points about Table 6-3. The first column shows the input analog voltage thresholds where the digital output code will change. Note that there are two values for 0: $+0$ and -0. Every code must be able to handle both positive

Table 6-3
PCM Codes

A/D Input Voltage	3 Bit, Sign-Magnitude, PCM Code Words[*]		
	Binary Code	**Folded Gray Code**	**Folded Binary Code**
+3	**1**11 (7)	**1**10 (+3)	**1**11 (+3)
+2	**1**10 (6)	**1**11 (+2)	**1**10 (+2)
+1	**1**01 (5)	**1**01 (+1)	**1**01 (+1)
+0	**1**00 (4)	**1**00 (+0)	**1**00 (+0)
−0	**0**11 (3)	**0**00 (−0)	**0**00 (−0)
−1	**0**10 (2)	**0**01 (−1)	**0**01 (−1)
−2	**0**01 (1)	**0**11 (−2)	**0**10 (−2)
−3	**0**00 (0)	**0**10 (−3)	**0**11 (−3)

[*]The sign bit appears in bold.

and negative input voltages, so every code needs a sign bit. Symmetry requires that if there is any output for zero input, then there must be two versions of it, one with a positive sign bit and one with a negative sign bit. In all the codes shown, the first bit position is a sign bit: 1 for positive values and 0 for negative values.

The first code column, a traditional binary code, will output 000 for its most negative input and 111 for its most positive. The second column, the folded Gray code, has the property of changing only 1 bit between adjacent code positions. So 100 and 000 represent the smallest input magnitudes: 0 volts. The codes change by 1 bit position with each input magnitude step change. The third column, the folded binary code, has the same code outputs as the Gray code for the smallest input magnitude, and just counts binary up as the magnitude increases.

Note that both the folded Gray code and the folded binary code are 100 percent symmetric about 0, while the binary code is not. (By the way, this symmetry about 0 is what gives rise to the adjective *folded*. Any code that exhibits this property is known as a folded code.) This symmetry offers some cost savings not only when encoding the code in the transmitter, but also when decoding the code at the receiver. The idea is to design a simple circuit to detect sign, strip it, and remember it, followed by a single circuit for decoding the magnitude portion. The magnitude is then inverted or not, depending on the state of the sign bit. Almost all real implementations use some form of folded code.

By not requiring the D/A converter to interpret the sign bit, you can obtain a reduction in the complexity of the D/A converter required for each receiver. Simply by using a folded code and absorbing the expense of a simple circuit to detect the status of the sign bit, the cost of the D/A in every receiver is reduced.

The number of bits chosen for encoding the A/D output has several important consequences. It determines the resolution of the PCM system, its performance in noise, and its bandwidth. So the number of bits used (or, as it is referred to, the length of the PCM code word, or even just PCM word) has a big impact on system performance. Generally, the longer the code word, the better the system performance both in precision and

Figure 6-12
Quantization Error (*Reprinted by permission of Pearson Education, Inc., Upper Saddle River, NJ*)

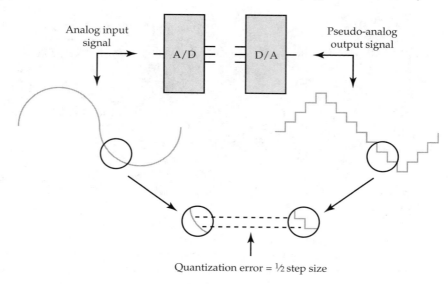

Quantization error = ½ step size

performance in noise, but the wider the bandwidth required and also the greater the cost. So understanding how these features change with the code word is an important step in understanding how PCM systems work.

Although quantization error has already been discussed, it is prudent to include a short discussion here in the context of all the factors that influence the choice of the number of bits in a PCM code word. First, the number of bits used in the code word is determined from the number of bits output by the A/D converter. As has already been discussed, the number of bits output from an A/D determine its amplitude resolution, or step size. Typically, the same output step voltage results from a range of values of the input signal. So any step voltage used is an approximation to the actual value input to the A/D. The error that results from this approximation is, again, called quantization error, and the process of assigning the code to the input value is called quantizing. See Figure 6-12.

As you can see from the figure, quantization error does not appear until the A/D in the receiver converts the digital code back to an analog signal. Therefore, this error is not an effect of the channel but of the actual PCM encoding and decoding process. This kind of error will add to whatever other sources of error that might creep in through filtering, transmission, etc. It is a fundamental component of error in any PCM system.

■ QUANTIZATION NOISE

The second impact that a code word has is a result of its length: the longer the code word, the better the performance in the absence of noise. Every PCM system has one source of error, the quantization error. The other main source of error *not* due to the channel environment (which introduces bit errors in the PCM code word and hence amplitude error at the output) is the relationship between S/N and word length. This second source of error is known as quantization noise.

A special rule called the 6 dB rule applies to PCM systems. For each additional bit you add to the code word, you get an additional 6 dB signal-to-noise (S/N) performance improvement at the reproduced analog output. For a 2-bit code word, the S/N would be 12 dB; for a 3-bit word, 18 dB; etc. So by using only a small number of bits, good S/N performance results.

Some of you are no doubt wondering why S/N is being discussed here when it has been stated that this effect is not due to channel effects. Usually, the S/N ratio is discussed in the context of noise. For a given signal level, more noise results in a lower S/N ratio, and hence a lower level of performance. S/N ratios have theoretical maximums, and if you do not provide for a sufficiently large theoretical S/N ratio, you can never get them big enough once they start being degraded by the noisy channel. The 6 dB rule allows calculation of the maximum S/N that can be expected in a perfect environment. Channel effects will always reduce the S/N ratio.

■ PCM BANDWIDTH

The third major effect that code word length has on PCM systems is the relationship between word length and bandwidth required for transmission. This relationship is actually one between bit rate and bandwidth. Since the longer the code word, the faster the bit rate, and if everything else is held equal, the relationship can be discussed in either context. The bit rate is equal to the number of bits in the code word times the sampling rate of the sample-and-hold because each sample will produce one code word output from the A/D encoder circuit. From the discussion on sampling theory, we know that the sample rate must be at least twice the maximum frequency component in the input signal. So the minimum bandwidth required by a PCM signal is defined as shown below:

$$BW_{PCM} = \text{bit rate} \geq N*f_{sample} \qquad \textbf{(6-6)}$$

$$N = \text{number of bits in PCM code word}$$
$$f_{sample} = \text{sampling rate}$$
$$f_{sample} \leq 2*BW_1$$
$$BW_1 = \text{bandwidth of input analog signal}$$

$$BW_{PCM} \geq N*2*BW_1 \qquad \textbf{(6-7)}$$

Equation 6-7 shows that the bandwidth of the PCM signal is always at least twice, and usually more, than the bandwidth of the input analog signal. This disadvantage is one of the only reasons not to use digital transmission in wireless systems. It is always true that a digital transmission takes up more bandwidth in the channel than an analog one if the modulating signal is held constant. However, the other benefits of digital

Table 6-4
Bandwidth Comparison of Modulation Techniques

Modulation Technique	Bandwidth Comparison
AM-DSB	20 kHz
Narrowband FM (two sidebands)	40 kHz
PCM	160 kHz

transmission (the cost saving by using digital components, performance improvements possible through coding techniques, and no accumulation of noise because of analog repeaters) usually outweigh this single disadvantage for wired media.

■ **EXAMPLE 6-6**   Find the minimum PCM bandwidth for a system with an analog signal maximum frequency of 10 kHz and using an 8-bit PCM word. Then determine the maximum theoretical signal-to-noise ratio using the 6 dB rule.

$$BW_{PCM} \geq N*2*BW_1 = (8)(2)(10 \text{ kHz}) = 160 \text{ kHz}$$

Before the signal-to-noise ratio is computed, think a moment about the result obtained in Example 6-6. Recall our earlier discussion of digital transmissions and their comparison to analog. At that time, the only real disadvantage of digital modulation techniques was the amount of bandwidth they required as compared to analog modulation techniques.

For a DSB AM modulation, the bandwidth required would be only 20 kHz! For any reasonable modulation index, an angle modulated waveform would also require less bandwidth. In some applications, this bandwidth advantage is critical and should not be overlooked. It is always true that bandwidth is a critical issue for systems designed to work in wireless channels, especially those like cellular telephone or satellite television. On the other hand, for a system designed to work on a guided channel, like twisted pair wiring, it does not make so much difference. The telephone system is an example. These results are summarized in Table 6-4. The modulation frequency is taken to be 10 kHz. The signal-to-noise ratio is given by the number of bits in the PCM code word:

$$S/N_{max} = N*6 \text{ dB} = (8)(6 \text{ dB}) = 48 \text{ dB}$$

■ **COMPANDING**

The next topic to discuss about PCM systems is companding. Actually, this topic was touched on briefly when nonuniform quantization and its usefulness in maintaining resolution across wide dynamic ranges was pointed out. The United States, Canada, and Japan have agreed to use one rule for this nonuniform quantization, called the μ-law. In Europe

Figure 6-13
(a) μ-law and (b) A-law (*Reprinted by permission of Pearson Education, Inc., Upper Saddle River, NJ*)

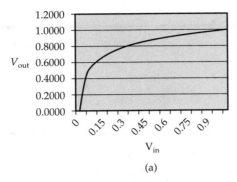

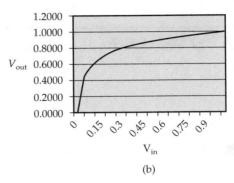

(a)   (b)

and most of the rest of the world, standardization is based on a slightly different rule called the A-law. Both are very similar and the practice of applying them is known as companding.

Companding is nonuniform quantization and it results in the compression and then expansion of the input signal. The word is formed from *compressing* and then ex*panding* (yet another term formed from an expression of the physical processes that it describes). Essentially, larger amplitude signals are "amplified" less than those of lower amplitude. This variable "amplification" effect is accomplished by applying fewer code decision points in the large amplitude input range and making use of the extra resolution offered for low amplitude signals. Figure 6-13 shows how it works. This diagram illustrates both the μ-law and A-law standards, which are very close.

The charts in the figure are just simple V_{in} versus V_{out} types of graphs. The values of μ and A are set by agreement but, of course, can take on any value desired. In the United States, Canada, and Japan, μ is set at 255. In Europe and most everywhere else, A is set at 100. (This difference has always pointed out to me just how profound is the difference in unit preferences between the States and Europe.) Note that the number used in the United States is odd and has no relation to anything except that it is implemented easily in an 8-bit converter. On the other hand, the European quantity is set at 100, an even, metriclike number, and the equation was adjusted so that good performance would result with A set equal to 100. The three equations are shown below. Both approaches closely emulate a logarithmic response curve.

μ -law $\qquad |V_o| = \dfrac{\ln[1 + \mu|V_i|]}{\ln[1 + \mu]} \qquad 0 \le |V_i| \le 1 \qquad$ **(6-8)**

A -law $\qquad |V_o| = \dfrac{A|V_i|}{1 + \ln A} \qquad 0 \le |V_i| \le \dfrac{1}{A} \qquad$ **(6-9)**

A -law $\qquad |V_o| = \dfrac{1 + \ln[A|V_i|]}{1 + \ln A} \qquad \dfrac{1}{A} \le |V_i| \le 1 \qquad$ **(6-10)**

Note that the input voltage is always taken as its absolute value, or is always positive. The only concern is with magnitude, so the sign bit is always added after the companding to the code word. This practice retains the sign of the input signal. Additionally, note that the crossover point for the right expression to use for the A-law is V_{in} equal to 0.1 V.

■ **EXAMPLE 6-7** Find the output voltage for a μ-law companding if the input voltage is 0.5 V and $\mu = 255$.

$$|V_o| = \frac{\ln[1 + \mu|V_i|]}{\ln[1 + \mu]} = \frac{\ln[1 + 255|(0.5)|]}{\ln[1 + 255]} = \frac{4.86}{5.54} = 0.875 \text{ V}$$

■ **EXAMPLE 6-8** Repeat Example 6-7 but set the input voltage to -0.5 V.

$$|V_o| = \frac{\ln[1 + \mu|V_i|]}{\ln[1 + \mu]} = \frac{\ln[1 + 255|(-0.5)|]}{\ln[1 + 255]} = \frac{4.86}{5.54} = 0.875 \text{ V}$$

Note that the result is the same. The values given by these equations are always of a positive magnitude. Negative values are generated by applying a negative sign bit to the positive magnitude.

Figure 6-14 illustrates what the μ-law curves look like for various values of μ. These curves were generated by using Equations 6-8, 6-9, and 6-10. The linear "curve" that is

Figure 6-14
Curves for μ-law (*Reprinted by permission of Pearson Education, Inc., Upper Saddle River, NJ*)

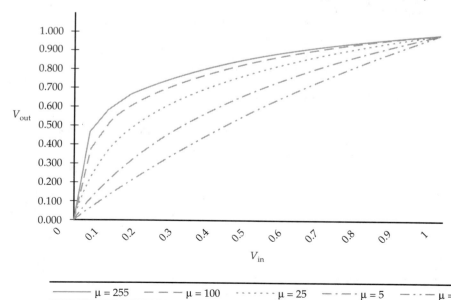

lowest in the figure is for $\mu = 1$, the next is for $\mu = 5$, the next for $\mu = 25$, then for $\mu = 100$ and for $\mu = 255$. Note the large output values for small input voltages, and the little change in the output for larger input values.

■ DYNAMIC RANGE

Dynamic range for PCM systems will be discussed next. In particular, we will cover how the number of bits is selected for a given input dynamic range. Recall that the dynamic range of a signal is the logarithmic ratio of the maximum amplitude level divided by the minimum amplitude level that the signal can obtain. Since dynamic range is an easily measured quantity, it would be convenient to be able to calculate how many bits the minimum PCM code word would require to accommodate it. This calculation would let us estimate the bandwidth and bit rate from the input dynamic range specification, typically supplied or easily measured.

The calculation is almost too easy, but it does depend on recognizing three items. First, if you double the dynamic range and want to keep the quantization error constant, you must double the number of "steps." This procedure adds 1 to the number of bits in the code word. Second, the minimum code word always represents the first amplitude point after 0. In other words, for any sampled data system like a PCM system, the first "step" in the quantization curve is always above 0 volts. Some small magnitude, positive output voltage value is always selected to correspond to 0 input volts. These two observations combine to yield the simple formula shown in Equation 6-11:

$$V_{DR} = 2^N - 1 \tag{6-11}$$

V_{DR} = voltage dynamic range of input analog signal

In Equation 6-11, the 2^n term arises from the first observation and the -1 comes from the second. The third critical observation is that the formula above will yield the number of bits required for the peak magnitude variation of the signal. If the use of a sign bit is needed, it must be added. Sometimes you may see dynamic range defined as a logarithm of a ratio of voltages. The voltage dynamic range is related to the dynamic range in the following way:

$$DR(dB) = 20 \log \frac{V_{max}}{V_{min}} = 20 \log V_{DR} \Rightarrow V_{DR} = \frac{V_{max}}{V_{min}} \tag{6-12}$$

Be careful when applying these formulas. Make sure you know whether the dynamic range is expressed as a ratio of voltages or as a decibel value. If one expresses the dynamic range as a ratio of voltages, a relationship for the number of bits required can be developed from Equation 6-12:

$$V_{DR} = 2^N - 1$$

$$\log(V_{DR} + 1) = (2^N) = N\log(2)$$

$$N = \frac{\log(V_{DR} + 1)}{\log(2)} \tag{6-13}$$

■ **EXAMPLE 6-9**   For a dynamic range of 15 dB, how many magnitude bits are needed in a PCM code word? First, find the voltage dynamic range by applying Equation 6-12 then apply Equation 6-13.

$$DR(dB) = 20 \log (V_{DR}) \Rightarrow V_{DR} = 10^{\left(\frac{15}{20}\right)} = 5.62$$

$$N = \frac{\log(V_{DR} + 1)}{\log(2)} = \frac{\log(6.62)}{\log(2)} = 2.73 \Rightarrow 3 \text{ bits}$$

It should make sense that you always round up in this type of calculation. Ask yourself the question: If there are between three and four cups of water in a container, how many cups are required to hold all the water in the container? The answer is four.

■ **EXAMPLE 6-10**   Continuing Example 6-9, how many quantization levels would be needed for 3 bits? The answer is just the power of 2 for both linear and nonlinear quantization.

$$\text{Number of steps} = 2^{(\text{number of bits})} = 2^N = 2^3 = 8$$

What would be the step size of the A/D if we assume linear quantization?

$$\text{Step size} = \frac{V_{DR}}{2^N - 1} = \frac{5.62}{(8 - 1)} = 0.8 \text{ V}$$

An interesting observation results when one realizes that the V_{min} in Equation 6-12 is actually the step size or resolution of the A/D converter. When this fact is combined with the observation that the V_{max} in the same equation is the peak-to-peak voltage of the input analog signal, the following equation can be found by rewriting Equation 6-12 (the result is not so useful for calculations, but it illustrates a couple of important results):

$$V_{DR} = \frac{V_{max}}{V_{min}} = \frac{\text{peak-to-peak voltage of input analog waveform}}{\text{step size or resolution of the A/D converter}} \quad \textbf{(6-14)}$$

Equation 6-14 tells us that if the peak voltage doubles, the voltage dynamic range is also doubled. If the step size is halved to obtain more resolution, the voltage dynamic range is again doubled. The voltage dynamic range is related to the number of bits needed for the code word size. A direct relationship exists between the peak voltage of the input waveform and the number of bits needed for a particular code word size. The relationship also indicates an indirect relationship between the step size and the number of bits needed for a particular code word size.

■ PCM POWER

PCM is unique among the analog and digital modulation techniques studied thus far because the power is not linearly related to the spectrum, or bandwidth, of the input analog signal. There is no direct relationship between the input signal and the output pulse waveform. Before, the input analog signal could always be traced through the circuitry and represented in the output waveform. For PCM, the analog input is first changed to PAM, the last place where the relationship to the input signal is clear. When the encoding and compression are performed, all track of the input waveform's "shape" is lost and only a digital representation of the signal remains. The information is in a digital code that has no relationship to the frequencies that generated it.

For example, a PAM outputs a pulse of amplitude 5 for an input sample of the analog waveform that had amplitude 5. After the additional steps in a PCM, the code might be 1101, or 101101, or almost anything, depending on the code choice and the number of bits used in the code word. The frequencies produced by these bit sequences have no relationship to the input analog signal; they contain only the information. Therefore, the shape of the waveforms produced depend only on that information, not on the original shape of the input analog signal.

Where does this leave us in determining the power? The power does not depend on the analog waveshape of the input signal, but it must depend on its own pulseshape and, not unreasonably, on the bit rate. As the bit rate increases, the bandwidth required should also increase because more information is contained in the signal. Unless the representation of that information changes, more information means more bandwidth is required. Remember that PCM uses code words, so for each sample, the number of bits used in the code word must also be considered.

Some of you may be wondering where the dependency of the pulseshape comes in because so many different kinds of pulses could be used. The result presented in Equation 6-15 is valid only for rectangular, polar NRZ pulses. These are common and will be assumed in all pulse modulation analyses.

The power in a PCM signal depends on the same factors as it does with the CW modulation techniques we have already discussed, like voltage and load resistance. After all, it is just a signal like any other, and Ohm's law holds for square waves or pulses just as it does for analog waves. There is one interesting distinction when determining the power of a pulsed waveform, however, and that is the duty cycle of the pulses themselves.

Clearly, a pulse train with a 90 percent duty cycle has a higher average power than a pulse train with only a 10 percent duty cycle. Now in PCM signals, the duty cycle is influenced by two factors: the actual shape of the pulses, and how many of them are 1s and how many are 0s. The first factor is illustrated in Figure 6-15 and expressed in Equation 6-15:

$$P_{PCM} = (\text{pulse duty cycle})\left(\frac{V^2}{R}\right) \qquad \textbf{(6-15)}$$

$$P_{PCM} = \frac{t}{T} * \frac{V^2}{R}$$

Where $\quad \dfrac{t}{T} = $ duty cycle of individual pulse

Figure 6-15
50 Percent Duty Cycle Pulse (*Reprinted by permission of Pearson Education, Inc., Upper Saddle River, NJ*)

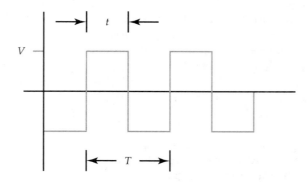

The expression for power in a PCM waveform is then given by the relationship shown in Equation 6-15. If the pulses are all 50 percent duty cycle, then:

$$\text{At 50 percent duty cycle:} \quad T = 2*t \text{ and } P_{\text{PCM}} = \frac{V^2}{2R}$$

The next factor, how many 1s and how many 0s, will be accounted for in the same way as the pulse shape factor. It is shown in Equation 6-16. The use of Equations 6-15 and 6-16 is illustrated in the next two examples.

$$P_{\text{PCM}} = \left(\frac{\text{total number of 1 bits}}{\text{total number of bits}} \right)(\text{pulse duty cycle})\left(\frac{V^2}{R} \right) \quad \textbf{(6-16)}$$

■ **EXAMPLE 6-11**   A PCM pulse train uses an 80 percent duty cycle pulse. The bit pattern of the pulse train is 50 percent 1 bits and 50 percent 0 bits. The peak voltage of the waveform is 5 volts and the load impedance is 100 Ω. What is the power?

$$P_{\text{PCM}} = \left(\frac{t}{T} \right)\left(\frac{V^2}{R} \right) = (0.80)\left(\frac{5^2}{100} \right) = 0.2 \text{ W}$$

■ **EXAMPLE 6-12**   A certain PCM pulse train has been analyzed and found to contain, on average, three 1 bits for every single 0 bit. The duty cycle of each bit is 40 percent. The peak voltage of the pulse train is 5 volts and the load impedance is 200 Ω. Find the power in the PCM waveform.

$$P_{\text{PCM}} = \left(\frac{\text{total no. of 1 bits}}{\text{total no. of bits}} \right)(\text{pulse duty cycle})\left(\frac{V^2}{2R} \right) = \left(\frac{3}{4} \right)(40\%)\left(\frac{5^2}{200} \right) = 37.5 \text{ mW}$$

where the shape factor of $\frac{3}{4}$ results from the fact that three out of every four bits is a 1 bit.

Figure 6-16
Duty Cycle and Shape (*Reprinted by permission of Pearson Education, Inc., Upper Saddle River, NJ*)

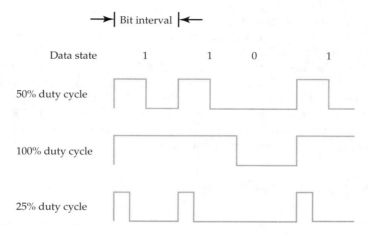

Example 6-12 illustrates how to combine the two effects on power. Find the product of the two shape factors and insert this result into the appropriate equation.

To be absolutely clear about the pulse duty cycle factors, examine Figure 6-16, which illustrates various shape factors or duty cycles for the bits themselves. Each of the waveforms was generated from the same logical pattern of bits.

■ DPCM AND ADPCM

An interesting refinement of the PCM method is called differential pulse code modulation (DPCM). This scheme makes use of the fact that many types of analog signals do not change rapidly from sample interval to sample interval. As a result, the amplitude change must be small and the code words very "close." When this case is true, as it is for most voice and some video applications, then a lot of redundant information is present in adjacent PCM code words.

If there was a way to represent only the differences between the analog values at the sample points, the amplitude resolution and hence the number of bits could be reduced by sending only the difference from the last sample. DPCM sends only the differences between successive samples. Because *differential* is defined as "showing a difference," the name differential pulse code modulation results.

If the range of amplitude values needed to accommodate the PCM code word shrinks, then the number of bits needed to represent it accurately also shrinks. The same amount of information can be sent with fewer bits in DPCM as compared to PCM. (Reread the discussion on how the step size or resolution is related to code word size.) The implementation of such a system is very similar to the PCM systems already described, but the A/D conversion process outputs only a code word that represents the difference in amplitude from the last sample. Note that the circuitry must be able to sense a positive or negative difference independent of the actual magnitude of the sample.

Adaptive differential pulse code modulation (ADPCM) is a further refinement of DPCM. ADPCM is used in many telecommunications speech applications. It works by calculating the difference between consecutive voice samples. The adaptive part comes in with the use of a filter, called an adaptive filter, that allows transmission of the data at a lower rate. Typical voice systems use a 64 kbps data rate. This data rate is calculated by taking the product of the sample rate, 8 kHz, and the number of bits sent. ADPCM systems have similar performance at half that data rate, or 32 kbps. Instead of sending 7 bits of amplitude information, they send only 3 or 4. The savings in bandwidth explains the popularity of this method in many telecommunications voice applications.

■ DM, ADM, AND CVSDM

Three variations of delta modulation, or DM, will be briefly discussed here. The first of this family is delta modulation. This technique yields good results for slowly changing signals. In DPCM, the modulator outputted a code word that depended only on the difference in amplitude from the previous sample. Although the code word shrunk due to the decreased dynamic range it was required to represent, several bits were still sent.

DM takes this same idea a step backward. It ignores the actual amplitude difference and sends just a single bit, a 1 or a 0, depending on whether the current sample is larger or smaller, respectively. This technique shrinks the PCM code word to a single bit. For that reason, this approach is sometimes discussed not as a variant of PCM but rather as a completely different technique. This categorization makes some sense because there is no attempt to gauge the absolute amplitude or, for that matter, the relative amplitude differences between successive samples. DM just outputs a 1 if the input signal is larger than the previous sample and a 0 if it is smaller. You can think of this system as sending only a special sort of sign bit that indicates the sign of the slope but contains no magnitude information.

Another reason why it is sometimes viewed as an alternate technique is that delta modulation, in all its forms, samples at the bit rate, not the Nyquist rate. The sample clock used to sample the input is running at the same clock rate as the D/A. Because there is no sampling theory limitation, many view DM as a completely different class of pulse modulation. Figure 6-17 illustrates a typical DM signal.

This system relies on the fact that the changes are slow and quite regular from sample instant to sample instant. If the input analog signal changes too fast, the stairstep approximation cannot keep up and will seem to lag the input analog waveform. This lagging is viewed as a distortion component and has a special name, slope overload distortion. Its name comes from the observation that if the slope of the input analog waveform is too large, DM will not be able to keep up and will distort. Figure 6-18 illustrates what happens when the signal is changing too rapidly to be a good candidate for representation using DM.

As you can see, the output does not track the input well. It misses the fast rising and falling components of the input analog waveform. It also does not do a very good job of

Figure 6-17
Delta Modulation (*Reprinted by permission of Pearson Education, Inc., Upper Saddle River, NJ*)

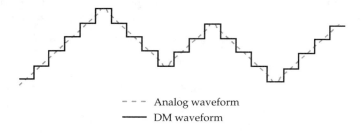

— — — Analog waveform
———— DM waveform

Figure 6-18
Delta Modulation Showing Slope Overload Distortion (*Reprinted by permission of Pearson Education, Inc., Upper Saddle River, NJ*)

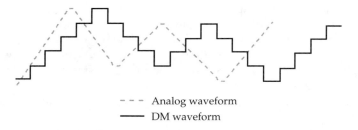

— — — Analog waveform
———— DM waveform

tracking the peaks and valleys of the amplitude variation. A good question is, What does an input signal with too large a slope actually represent? It represents a signal with higher frequency components than the sampled system was designed to handle. For DM, the sampling rule needs to be modified a little because the rate of output amplitude change is limited.

Depending on the dynamic range of the input analog signal, the input frequencies must be restricted to less than half of the sample rate. It should be clear to you that, for large input amplitude variations when using DM, the sampling frequency must increase above the minimum specified by sampling theory to avoid slope distortion. Let's generate a simple equation that estimates when the step size will be too small for the input signal to delta modulate. To begin this analysis, recall how slope is defined:

$$\text{Slope} = \frac{y}{x}$$

$y = \text{step size}$

$x = \text{sample interval} = \dfrac{1}{\text{sample frequency}}$

$$\text{Slope} = \frac{\text{step size}}{\text{sample interval}} = (\text{step size})(\text{sample frequency})$$

Now, assume that the input analog waveform to be sampled is a sine wave of amplitude A, $A\sin(\omega t)$. Then the slope is just the first derivative, or $A\omega\cos(\omega t)$. This function has a maximum value of $A\omega$ because the maximum value a sinusoid can have is unity. Inserting this value into the equation developed above, you will find:

$$\text{Slope} = (\text{step size})(\text{sample frequency})$$
$$A\omega_i \leq (\text{step size})(\omega_s)$$
$$\text{Step size} \geq \frac{A\omega_i}{\omega_s} \qquad \textbf{(6-17)}$$

ω_i = input signal frequency
ω_s = sample frequency

For best results, keep the step size close to equality in Equation 6-17. Also note that, with DM and its variations discussed below, the first few samples are usually garbage. Because no absolute amplitude information is ever sent, the code is not valid until the input analog signal hits one amplitude rail or the other. Once it hits a rail, then the code, or bit, in this case, typically syncs up and you have an accurate tracking.

Adaptive delta modulation (ADM) tries to overcome this disadvantage by changing the rules about the operation of an A/D. Instead of keeping the amplitude step size constant, the step size is varied in amplitude to track the input analog waveform. Note that only 1 bit is still sent for each change, but the value of this bit is changed. For example, when a larger step size is required to track the input signal accurately, the code word output is still just a single bit but it represents a larger step.

Most implementations keep track of the most recent few code word outputs. If they stay all 1s or all 0s for a predetermined number of samples, then the step size is increased or decreased, respectively, by about 50 percent. The number of samples before a step change is implemented and the amount of the step size changes varies from implementation to implementation. The typical number of consecutive samples increasing or decreasing, before a step size change would be made, varies from 3 to 5. And 30 percent to 50 percent is a typical percentage adjustment.

Continuously variable slope delta modulation (CVSDM) is a further refinement of ADM. It is exactly similar to ADM, except that there is no predetermined number of in-

Table 6-5
PCM and DM Techniques Compared

Type	Concept	Restriction
DPCM	Send difference, arbitrary number of bits	Slowly changing signals
ADPCM	Send difference, smaller number of bits	Slowly changing signals
DM	Send difference, only 1 bit	Slowly and smoothly changing signals
ADM	Send difference, arbitrary step size	Slowly changing signals
CVSDM	Send difference, arbitrary step size	Depends on algorithm

tervals before the step size is varied, and there is no fixed amount by which the step size is varied in amplitude. These exceptions allow the use of sophisticated algorithms to track input signal variations accurately when the input signal is accurately characterized. Due to the commercial importance of digitizing speech, this characterization analysis has been widely performed.

Table 6-5 summarizes the concepts and limitations of the various forms of PCM and DM discussed in this chapter.

■ SUMMARY

This chapter introduced pulse modulation from a practical perspective. It began with an overview of sampling theory and aliasing, two general concepts that are applicable to all pulse modulation techniques. PAM was introduced as a fundamental building block in what is the most widely used form of pulse modulation, PCM. The effects of quantization and companding were explored.

Almost every digital wireless device that can be used for voice communications digitizes that voice with PCM or a closely related technique. The sample rate may change with technological progress, but PCM will be with us for the foreseeable future as a way of digitizing speech.

REVIEW QUESTIONS

1. For a signal with a maximum frequency component of 10 kHz, determine the minimum sampling period.

2. For an amplitude resolution of 1 percent, how many bits would each of the following A/D converters have to be?
   a. A sign-magnitude converter
   b. A magnitude-only converter

3. For a sample rate of 8 kHz, determine the maximum input frequency that a sampled data system could support.

4. For the following input frequencies, determine if aliasing is occurring and at what frequency. Assume a sampling clock speed of 20 kHz.
   a. $f_{signal} = 8$ kHz
   b. $f_{signal} = 10$ kHz
   c. $f_{signal} = 12$ kHz
   d. $f_{signal} = 16$ kHz

5. List three advantages of digital systems compared to analog systems.

6. Sketch a 1 kHz sinusoid and identify the sample points. Assume a 6 kHz sample rate. Draw two cycles.

7. For the waveform in problem 6, sketch the PAM waveform that would result. Draw two cycles.

8. What is the dynamic range for an 8-bit sign magnitude A/D converter?

9. For an input dynamic range of 60 dB, find the number of bits needed in the PCM code word.

10. What is the dynamic range capability for a 16-bit sign-magnitude PCM code?

11. If the resolution of a PCM system is 1.0 volts, determine the output voltages for

the following 4-bit sign-magnitude PCM codes. The codes are all expressed as folded binary.
a. 1100   c. 0111
b. 0101   d. 1000

12. If the resolution of a PCM system is 0.5 volts, determine the output voltages for the following 3-bit sign-magnitude PCM codes. The codes are all expressed as folded Gray codes.
a. 110   c. 011
b. 001   d. 100

13. Determine the output voltage of a μ-law compander with $\mu = 255$ for the following input voltages. Assume a maximum input voltage of 1.0 volts.
a. 0.01   c. 0.2
b. 0.1   d. 0.6

14. Determine the output voltage of an A-law compander with $A = 100$ for the following input voltages. Assume a maximum input voltage of 1.0 volts.
a. 0.01   c. 0.2
b. 0.1   d. 0.6

15. Repeat problem 13 with $\mu = 100$.

16. For a PCM system using a 6-bit code word and a sample frequency of 10 kHz, find the bandwidth of the PCM signal.

17. For a PCM system designed to accommodate an input analog signal that has a dynamic range of 45 dB and a maximum frequency component of 8 kHz, find the bandwidth of the PCM signal.

18. For a PCM code word length of 6 bits, find the power in a PCM signal for a high voltage level of 5 V and a low voltage level of 0 V. The load resistance is 50 ohms. Assume a 50 percent duty cycle for the pulses and a bit pattern of 50 percent 1 bits.

19. Repeat problem 19 assuming a 25 percent duty cycle for the pulses.

Error-Correcting Codes

■ INTRODUCTION

This chapter examines a variety of error-correcting codes used in wireless systems. The chapter begins with a definition of channel encoding. After a brief review of Galois field arithmetic for the $n = 2$ case only, codes based on parity are introduced. These codes include simple parity and Hamming codes. The chapter then turns to cyclic codes, specifically, CRC codes and their operation. This section concludes with a brief discussion of BCH codes. The last major section of the chapter is devoted to convolutional coding. The specific instances chosen are trellis code generation and decoding using a maximum likelihood technique. Here, the Viterbi algorithm is covered.

■ WHY USE CODES?

The first point to understand about error-correction coding is that any time a code is used to improve the error rate of transmission, overhead is added to the transmission and a cost is paid. Every extra bit transmitted that does not actually contain information lowers the system's efficiency. Therefore, improving the error rate must be traded off with the extra bandwidth taken up by the code. Any modulation scheme can be made more effective by improving its performance in noise, so channel coding is widely used in many communication systems.

Figure 7-1
Block Diagram of Coding and Decoding Elements

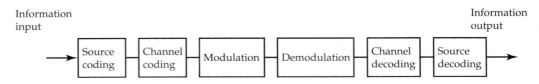

In all digital communications systems, the information is sent as a series of binary digits. These digits are first converted to an analog form through modulation, sent through the channel, and converted back to a digital form at the demodulator. Ideally, the communications channel would not corrupt the information in any way. This ideal, however, is not the case. Any real channel introduces noise and distortion. Through the use of codes, these channel effects can be addressed.

In general, any error-correction coding scheme is the addition of redundant bits into the information stream so that any errors can be first detected and then corrected. This task requires one of two events: either the actual transmission rate in the channel must be increased to accommodate the additional bits, or the information rate must be reduced.

As you can see in Figure 7-1, two functional components of coding are illustrated. All coding can be broken into two broad groups: source coding and channel coding. This chapter will discuss only channel coding. Also observe that all coding and decoding necessary precede the modulation step and follow the demodulation step.

■ CHANNEL ENCODING

Channel encoding involves decisions about how to represent the data transmitted from a source so the data are "different enough" that, in the presence of small variations of noise, any errors caused by the noise can be detected and corrected. It is called channel encoding because it compensates for imperfections in the channel. Channel encoding is concerned with error detection and error correction. Error detection is the process of detecting errors in the received transmission; error correction is the process of fixing those detected errors. There are many ways to approach channel encoding, and in some approaches, all the information necessary, or available, is contained in the encoding itself. Therefore, to correct an error, two steps must take place:

1. Error detection
2. Error correction

There are two main subgroups to channel coding: those where all the information is contained in the message itself, forward error coding (FEC) codes, and the other main group of channel codes known as automatic repeat request (ARQ) protocols. All FEC codes are channel codes, but not all channel codes are FEC codes. For example, ARQ pro-

Figure 7-2
Channel Codes

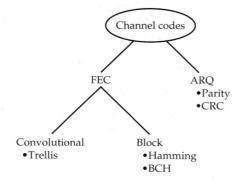

tocols are not considered FEC codes. The channel codes that will be explored in this chapter are illustrated in Figure 7-2, which shows how they are classified.

ARQ techniques require the receiver to request retransmission of any frame of data where it detects errors. Therefore, the distinction between these two groups is *where* the information to correct the errors is located. In the first case, FEC codes, it is contained in the code itself. In the latter case, ARQ codes, it is held by the transmitter in the sense that the receiver requests the frame of data be sent again.

Many layer 2 (data link layer) protocols use some form of the latter type, usually accomplished by the ARQ protocol utilizing CRC codes to detect errors. Once the CRC code detects the error, the ARQ protocol requests a retransmission. In this chapter, we will see how a CRC code works to detect errors. In Chapters 3 and 4, the protocol elements necessary to support the requests for retransmission where error(s) were detected were described.

Be sure to recognize the distinction between those channel codes that stand by themselves (FEC codes) and those approaches that require retransmission from the source to operate (ARQ protocols). The first group requires no additional support from the transmitter once the message is sent. The second group requires coordination between the source and destination and a way of messaging between them, from either direction. Channel encoding of the type known as ARQ protocols are referred to as protocols because a protocol must exist to communicate about errors that are detected and request retransmissions of data already sent. These protocols are more complex than simple FEC codes, where once you have sent it, you can forget it. ARQ protocols are often encountered in LAN/WAN environments.

This distinction between FEC codes and ARQ protocols can also be cast in another way. Sometimes ARQ protocols are called feedback error correction. In feedback error correction (as contrasted with forward error correction), the code only detects the errors and relies on a "feedback" or return path to correct them. Any modulation technique can be used to transmit either type of channel code just as they can be used to transmit any type of data. The codes are just a few additional bits sent with each message.

FEC codes have an additional subclassification: they may be a block or convolutional code. The general definition of block codes are codes where the length of the code stays fixed. However, a more fundamental distinction exists between block codes and convolutional codes. While both codes have fixed output lengths, block codes do not use the previous bits to determine the output state. In other words, for block codes, the output is determined only by the bits applied to the input of the decoder. All convolutional codes use some number of previously received bits to determine the output.

All error-correction codes are classified by whether they are systematic or not. If the input bits included in coder output are unmodified, the code is described as systematic. In this case, the additional bits introduced by the coder are always known as parity bits. There are many examples of these types of codes and two of the most important are discussed below. Unsystematic codes do exist but are not used often so they will not be discussed further.

■ MODULO-2 ARITHMETIC

To perform any analysis on error-correcting codes, a specialized form of arithmetic is required. This arithmetic is known in general as Galois field arithmetic and is abbreviated by the symbol GF(n), where n is always a prime number. It is outside the scope of this text to explore this field in general. However, when $n = 2$, the case for binary arithmetic, Galois field arithmetic reduces to modulo-2 arithmetic. Modulo-2 arithmetic is quite simple and is briefly summarized below:

Addition: the exclusive OR (XOR) function

$0 + 0 = 0$

$0 + 1 = 1$

$1 + 0 = 1$

$1 + 1 = 0$

Multiplication: the AND function

$0 \times 0 = 0$

$0 \times 1 = 0$

$1 \times 0 = 0$

$1 \times 1 = 1$

■ BLOCK CODES

A block code is one where the number of bits of the output is constant for any input. Another way of describing it is that a block code is a mapping of j input bits onto k output bits, where $j < k$. Parity is an example of a block code. In the next section, this simple example of an FEC block code, parity, is explored.

When using block codes, the input bits are divided into blocks of j bits. The coder then produces a block of k output bits for transmission. Thus, each block code can be described by the phrase, "This code is a (j, k) code." Each block is always coded and each decodes separately from all other blocks.

■ PARITY

While parity is properly classified as an ARQ code, it forms the foundation for many useful block codes. Parity is one of the oldest and simplest methods of error detection. It is the simplest block code possible. It is classified as a systematic $(j, j + 1)$ code. There are two types of parity: even and odd. They are differentiated from each other by whether the parity bit is the modulo-2 sum of the information bits or 1 minus that sum. When the sum of the $j + 1$ bits of the output of the coder, known as the codeword, is 0, the parity is described as odd parity.

Note that parity is independent of the number of 0s and not sensitive to the number of paired 1s, a direct result of the property of modulo-2 addition. Zeros have no impact in the sum. Therefore, for even parity, the parity bit is a 0 if the number of 1s is even. It would be a 1 for an odd number of 1s. Similarly, for odd parity, the parity bit would be a 1 if the number of 1s is even. It would be a 0 for an odd number of 1s. Typically, a single XOR gate can be used to detect errors once the number of 1s has been determined.

Note that parity is a very primitive error-detection scheme. One parity bit can detect one error and will not detect two errors. Parity will detect only an odd number of errors. Parity should never be used unless there is a very low probability of error in the channel for any single block of information bits. The only real advantage of parity is that it is very simple to use and implement. The value of the parity bit is determined by modulo-2 addition of all bits in the block. Note that this process is the same as adding all the 1 bits in the block, again due to the properties of modulo-2 addition.

For even parity, a parity bit of 1 is added if the number of 1s in the block is odd, and a parity bit of 0 is added if it is even. *After adding the parity bit, the total number of 1 bits is even for even parity.* In Example 7-1, both even and odd parity are illustrated.

If only one parity bit is added, only one single bit error can be detected. Double bit errors, such as a 10 switched to a 01, cannot be detected. In the latter case, the number of 1 bits remains the same. This fact illustrates an interesting property of FEC codes in general. FEC codes must include enough extra bits to detect the maximum number of errors expected in each group or block of data. Parity is known as a single error-detecting (SED) code.

■ EXAMPLE 7-1

Even Parity

Data: 1001010   Parity: odd number of 1 bits means add a parity bit of 1
Data: 1011010   Parity: even number of 1 bits means add a parity bit of 0

Odd Parity

Data: 1001010   Parity: odd number of 1 bits means add a parity bit of 0
Data: 1011010   Parity: even number of 1 bits means add a parity bit of 1

As is clear from the previous discussion and Example 7-1, parity as a stand-alone concept can detect errors only. There is no mechanism for determining which exact bit is corrupted and is thus correctable. Because of this inability to correct, parity (when employed alone) requires some kind of ARQ protocol that will request error-correction information from the source. For this reason, parity is classified as an ARQ code. In the next section, where parity is combined with a powerful algorithm, the power of sensing parity forms the basis for the widely employed family of Hamming codes.

■ HAMMING CODES

Single-bit parity may be a crude code, but it can be built upon to render it much more useful. This type of block FEC code is the family of Hamming codes. They are block codes where bit error(s) are not only detected at the receiver but also corrected, if the length of the code is long enough. Hamming codes give the best code for single-bit correction when performance in a white noise environment is considered. Richard Hamming developed not only the code that bears his name but also two terms used to define the performance of all block FEC codes. These terms are Hamming weight and Hamming distance.

Hamming weight is the number of binary 1 bits in a code word. The Hamming distance between two code words is the number of bit positions by which they differ. A larger Hamming distance is desirable. The more bit positions that differ, the higher the probability that the code words can be distinguished in the presence of noise (the more different they are). These two ideas are examined in Example 7-2.

■ EXAMPLE 7-2   Find the Hamming weight of the code word 11001010.
There are 4 nonzero bits, so the Hamming weight is 4.
Find the Hamming distance between the two code words c1 and c2.

$$c1 = 11001010$$
$$c2 = 11011000$$

The Hamming distance is the number of bit positions in which they differ: $d = 2$.

In Example 7-2, the two code words, c1 and c2, had a Hamming distance of d. When c1 is transmitted into the channel, d errors must occur before code word c2 is received. Therefore, if the number of errors that occur in any block is less than $d - 1$, the two code words could not be confused. Thus, the errors would be detected. Stated another way, the decoder is capable of detecting $d - 1$ errors.

A relationship exists between the Hamming distance and Hamming weight. In Example 7-2, if the two code words are added modulo-2, the Hamming weight of the sum will have the same numerical value as the Hamming distance. The distance between any two code words in any set of code words will equal the weight of some other code word in the set. From this observation, the minimum distance of a set of code words equals the minimum weight of any nonzero code word. This property is very useful when analyzing the performance of a particular set of code words.

■ **EXAMPLE 7-3**   Add the two code words, c1 and c2, from Example 7-2 and show that the sum has the same numerical value for the Hamming weight as it does for the Hamming distance.

$$
\begin{array}{ll}
\text{c1} = 11001010 & \\
+\ \text{c2} = 11011000 & \\
\hline
\phantom{+\ \text{c2} = }00010010 & \text{Hamming weight} = 2 = \text{Hamming distance of c1 and c2}
\end{array}
$$

The minimum Hamming code length is generated by adding some number of bits, m, to a source data transmission determined by the inequality below:

$$2^m \geq m + n + 1 \tag{7-1}$$

n = original data word length
m = number of Hamming bits required

So the new length of the data transmission, when you include the Hamming bits, would be $m + n + 1$ for an original data word of length n. The number of Hamming bits added is given by m. The background for this relationship is given below, but first let's do a quick example showing the application of Equation 7-1.

■ **EXAMPLE 7-4**   Determine the number of parity bits required for a 4-bit message. Because Hamming codes are groups of 1-bit parity codes, the value of m that will satisfy the inequality in Equation 7-1 must be found. Remember to add the number of 1-bit parity checks to the code word length, as shown below.

$$2^m \geq m + 4 + 1 \Rightarrow m = 3$$
$$2^3 = 8 \geq 3 + 4 + 1 = 8$$

For Example 7-4, the number of parity bits required was 3. That number was added to the original message length, so the new length of the message was $4 + 3 = 7$ bits. The next job is to place the bits in the original data word.

The Hamming code is a series of 1-bit parity checkers. There are many members of the Hamming code family. In most, the placement of the 1-bit parity bits is arbitrary, while in some, it is not. This chapter will explore a straightforward Hamming code where the placement of the series of 1-bit parity bits, or Hamming bits, is not arbitrary. This relatively simple code illustrates the Hamming approach.

Table 7-1
Binary Representation of Bit Positions

Bit Position	Binary Representation
1	0001
2	0010
3	0011
4	0100
5	0101
6	0110
7	0111
8	1000
9	1001
10	1010
11	1011
12	1100

In the Hamming approach, the placement is fixed because the resultant number generated at the receiver must indicate the positions of the errors, and the coding technique guarantees this result. The number that indicates the positions of the errors is known as the syndrome. One way to think about why the bit positions are crucial is that each parity bit must cover one and only one parity bit location. Since parity can detect only a single bit error, it is critical that only one parity bit location be covered by each parity bit.

Hamming's insight was to recognize that every binary representation that has a 1 in the last bit position must be in the first parity check. From the same reasoning, the second parity check must be for those binary representations that have a 1 in the second from the last bit position, etc. For this reasoning to work correctly, the positions of the check bits are predefined. They occupy the positions 1, 2, 4, 8, etc. In this specific Hamming code, the check bits occupy the positions:

$$2^m \qquad m = 0, 1, 2, \ldots$$

The first parity check covers bit positions 1, 3, 5, 7, 9, 11, etc. The second parity check will cover bit positions 2, 3, 6, 7, 10, 11, etc. The third parity check will cover bit positions 4, 5, 6, 7, 12, etc. The fourth parity check will cover bit positions 8, 9, 10, 11, 12, etc. Table 7-1 should help make this process clear. Notice that the first parity check catches all those binary representations with a 1 in the LSB position. The second parity check catches all those binary representations with a 1 in LSB + 1, etc. Example 7-5 continues Example 7-4 and illustrates the proper placement of the parity bits.

■ **EXAMPLE 7-5**   The original code word of length 4 bits is 0101. The parity or check bits occupy the positions 1, 2, and 4 for a word of this length. After encoding, the new data word becomes _ _0_101, where the _ represents insertion points for the check bits.

To determine the check bits to enter at which positions, compute the parity check at all open positions, here, bit positions 1, 2, and 4. The first parity check, which goes into bit position 1, checks 1, 3, 5, and 7 from Table 7-1. Looking at the new data word, _ _0_101, there is an even number of 1-bits. To keep even parity, the check bit in position 1 must be set to a 0. Therefore, a 0 is entered into bit position 1. The new data word now becomes 0_0_101.

The second parity check is over positions 2, 3, 6, and 7. Again examining the new data word, an odd number of 1-bits are in those positions. To keep even parity, the parity bit must be set to a 1. The new data word now becomes 010_101. The third parity check is over bit positions 4, 5, 6, and 7. Here, an even number of 1-bits occurs in those positions. To keep even parity, the parity bit is set to a 0. Inserting this zero into the new code word results in 0100101. This step completes the Hamming encoding process.

■ **EXAMPLE 7-6**   For an original data word length of 8 bits, how many parity checks are required? What are the bit positions that these bits occupy? Finally, is there any advantage to regarding the message as one 8-bit message rather than two 4-bit messages?

The number of parity checks required for any message length is given by Equation 7-1. For $n = 8$, the number of bits required works out to 4 bits. The bit positions are given by evaluating increasing powers of 2.

$$1:2^0 = 1 \qquad 2:2^1 = 2 \qquad 3:2^2 = 4 \qquad 4:2^3 = 8$$

Since only four bits are required, these numbers are the four locations of the parity bits.

Finally, treating the message as two 4-bit words would require 3 parity bits for each 4-bit message. The overhead in the message due to the Hamming coding would be a total of 6 bits: two messages, each with 3 parity bits. For the single 8-bit message, only 4 parity bits would be required. It seems clear that it is more desirable to send a single long message rather than two short messages. As we will see later, for this additional message overhead, however, more potential checks on the data stream for errors do occur. Nevertheless, in all but the most severe noise environments, it is better not to break up the message into smaller sections and code each individually. Usually, the primary concern is message overhead due to coding.

■ **EXAMPLE 7-7**   Now introduce a single bit error and see if the Hamming code worked out in Examples 7-4 and 7-5 can find it. If the first bit position is changed from a 0 to a 1, then the corrupted code word from the earlier example is now 1100101. Apply the parity checks in the same order at the receiver that was done at the transmitter.

Parity check one yields an odd number of 1-bits, so the LSB of the syndrome becomes a 1. Parity check two yields an even number of 1-bits, so the LSB + 1 of the syndrome becomes a 0. The last parity check counts an even number of 1-bits, so the LSB + 2 of the syndrome becomes a 0. The check is complete.

The checks performed at the receiver indicate that there was an error in bit position 1. This error is determined by taking the checks and writing down a 1 if the parity is odd and a 0 if the parity is even. Then read out the binary number and its decimal equivalent to indicate the bit position in error. The result here was 001, with only the first parity check yielding odd parity. The first parity check occupies the LSB, and subsequent checks move one bit position to the left for each check.

Therefore, by flipping back the bit in position 1, which corrects the error in transmission, the result is a corrected received word of 0100101. Strip off the Hamming bits to recover the original message. The result is 0101, the original 4-bit word.

The set of calculations in Example 7-7 is very easy for digital logic circuits to perform, and Hamming codes have found application in many integrated circuits. The entire operation is summarized in Figure 7-3. Note that the Hamming code syndrome in-

Figure 7-3

Hamming Code for Example 7-7 (*Reprinted by permission of Pearson Education, Inc., Upper Saddle River, NJ*)

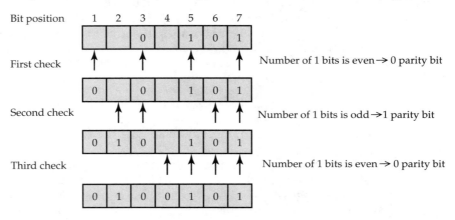

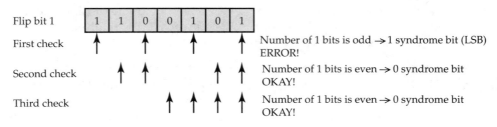

Syndrome is 001 → bit position 1 in error.
Flip bit one back.

Strip off parity check bits → 0101.
The original data sequence is recovered.

dicates the position of the bit in error, regardless of the message content. If a different original data message had been chosen, the Hamming bits might be different, but if bit 1 were flipped, the syndrome would still be 001. Also note that a syndrome of all zeros indicates a message received with no errors. The Hamming code uses a variety of 1-bit parity codes to identify the bit position of the error. This error then can be corrected at the receiver.

A code with 3 Hamming bits has a Hamming distance of 3 ($n = 3$). It detects a double bit error or corrects a single bit error. Generally, the larger the Hamming distance, the more (bits − 0) in error that are detected or the more (bits − 2) that are corrected. A Hamming distance of 2 means a single bit error detection. A Hamming distance of 4 means single bit error correction *plus* double bit error detection, or triple bit error detection, etc. A Hamming distance of 5 is the minimum for double bit error correction.

Finally, the overhead ratio for Example 7-7 was high; three additional bits were needed to find a single bit error in a 4-bit message. A 75 percent overhead is not the general case. By examining Equation 7-1, you can see that, for reasonable message lengths encountered in typical applications, the Hamming code is quite efficient. This efficiency is illustrated in the next two examples.

■ **EXAMPLE 7-8**   Find the number of Hamming bits needed for a message length of 1,000 bits.

At first glance, it is difficult to solve Equation 7-1 when the number of Hamming bits is unknown. An alternate approach is to find out how long a message can be encoded given a certain number of bits. The first guess is 9 bits. This number implies a message length of 1001 bits plus m. The value of 2 raised to the tenth power is 1,024. Inserting that into Equation 7-11, we get:

$$2^m \geq m + n + 1 \Rightarrow 2^9 \geq 9 + 1,000 + 1 \Rightarrow 512 \geq 1,010$$

Not quite enough bits for a message length of 1,000 bits, but by adding one Hamming bit, the message length that can be covered is 1,024 bits. Therefore, 10 bits are required for an original message length of 1,000 bits. Note that after adding the Hamming bits, the new message length is 1,011 bits.

$$2^{10} \geq 10 + n + 1 \Rightarrow 2^{10} \geq 10 + 1,000 + 1 \Rightarrow 1,024 \geq 1,011$$

■ **EXAMPLE 7-9**   Find the number of detection and correction bits possible for 10 parity bits.

Since 10 bits are added, 10 errors could be detected at 1 bit per error. It is also possible to correct some bits. Recall that each bit corrected takes 2 additional bits over the detection bit. Any bit to be corrected must first be detected. Therefore, the following combinations are possible:

10 detection bits and 0 detection + correction *or*

7 detection bits and 1 detection + correction *or*

4 detection bits and 2 detection + correction *or*

1 detection bit and 3 detection + correction

Another way to look at this is to take the Hamming distance, which is just the number of parity check bits applied to the original data word, and calculate the possible checks:

Hamming Distance	Possible Checks
1	1 bit error detection
2	2 bit error detection
3	1 bit error detected and corrected
4	1 bit error detection plus 1 bit error detected and corrected
5	2 bit error detection plus 1 bit error detected and corrected
6	2 bit error detected and corrected

In Examples 7-8 and 7-9, an original data word length of 4 bits was used and it required 3 Hamming bits. As you can see from the list in Example 7-9, 1 bit in error could both be detected and corrected in the message. This bit could be part of the original message or a Hamming bit itself; it makes no difference. With longer data words, more Hamming bits are required, so increasing numbers of errors can be detected and potentially corrected. The overhead of a Hamming approach to error detection and correction grows in a predictable fashion with the data word length.

Summary of Hamming Method

The four basic steps in constructing a Hamming code are summarized below:

1. Use Equation 7-1 to determine the number of Hamming bits required for the message length.
2. Determine the location of the Hamming bits using the 2^m rule.
3. Write the new message, including the blank spaces, for placement of the Hamming bits.
4. Perform parity checks to determine the value of the Hamming bits.

■ INTERLEAVING TECHNIQUES

Interleaving is a technique used to combat burst errors. While it could be debated that interleaving is not actually a coding technique, in its application to wireless systems, it functions as one in at least two ways. The most important aspect is that, through the various forms of interleaving, the information content of any particular message is spread over many symbols when they are placed into the radio channel. Since radio channels are characterized by short bursts of errors, spreading out of any one information word among many symbols means that the loss of any individual symbol or closely grouped symbols does not result in the loss of the information word. In general, a sudden burst of errors will result in only one character in error in any information word.

The second way that interleaving functions as a coding technique is that, due to the nonconsecutive nature of the information in any single transmission, many transmissions must be monitored by any unauthorized listener to capture any particular information word. This process increases the cost and complexity of monitoring others' conversations, and thus acts to increase the security of the transmission in the same sense that the application of a scrambler would. The practice of interleaving is demonstrated in Example 7-10.

■ **EXAMPLE 7-10**   A group of five 7-bit Hamming coded words are to be sent in a noisy radio channel characterized by regular bursts of errors 2-bit periods long. Without interleaving, the five blocks of data would be sent end-to-end continuously. For any single burst of errors, unless the burst happened to occur exactly at a word boundary, one data word would be lost in transmission because the Hamming coded words can detect and correct only one bit in error.

```
1234567  1234567  1234567  1234567  1234567
```

If interleaving were performed, however, the data sent would appear as follows:

```
11111  22222  33333  44444  55555  66666  77777
```

When a burst of errors occurs, it would tend to distort only 1 bit in any data word. If a method such as the Hamming code were applied, this error could then be both detected and corrected at the destination.

The next line of data shows a burst of errors that corrupts the fourth block of data sent:

```
11111  22222  33333  56612  55555  66666  77777
```

As you can see, the fourth block of data is completely corrupted. If possible it would require a large overhead to both detect and correct this entire block of data. After applying the inverse of the interleaving at the destination, however, the received bits reordered into the original data look much better:

```
1235567  1236567  1236567  1231567  1232567
```

Now, just 1 bit in each block of data is corrupted. Therefore, a relatively small overhead could both detect and correct this single bit error. Hamming codes are often used for this purpose.

■ **CYCLIC CODES**

All cyclic codes are examples of block codes. Block codes are those codes where the number of bits of the output is constant for any input. A block code is also a mapping of j input bits onto k output bits, where $j < k$. But cyclic codes have another special characteristic: every cyclic shift of the code word results in a new code word that is also a member of the code family.

A cyclic shift is accomplished by shifting each symbol of the code word some number of symbols to the left or right. Leaving out a lot of mathematical development, this technique has one interesting property. All code words of a cyclic code are multiples of the generator polynomial. Before exploring two cyclic codes in some detail, a short section on the mathematics of cyclic codes is appropriate.

Mathematics of Cyclic Codes

Like the CRC code and the BCH code (both introduced later), cyclic codes use a polynomial notation. In this approach, the j input bits of the information word are treated as coefficients of a polynomial $p(j)$. The polynomial associated with the information word is always of the same length as the information word. The order of polynomial is always $j - 1$, and the variable x in the polynomials used below never needs to be evaluated because its only function is as placeholder. Example 7-11 illustrates this calculation for a simple 4-bit information word.

■ **EXAMPLE 7-11**   The information word is 1011. The polynomial associated with this information, or information polynomial, is:

$$M(x) = 1x^3 + 0x^2 + 1x^1 + 1x^0$$

The arithmetic of polynomials is always done using Galois field arithmetic for binary values (GF(2)), also known as modulo-2, for binary data. To illustrate, the multiplication of a simple generator polynomial, represented as $G(x)$, and the information polynomial will be performed:

$$G(x) = 1x^1 + 1x^0$$
$$G(x)M(x) = 1x^4 + 1x^3 + 1x^2 + (1 + 1)x^1 + 1x^0$$

However, the coefficient of the first power term must be evaluated using modulo-2 arithmetic. In that case, $1 + 1 = 0$! The correct result is shown below:

$$G(x)M(x) = 1x^4 + 1x^3 + 1x^2 + 0x^1 + 1x^0$$

We introduced the term *generator polynomial* above without any explanation. A generator polynomial defines any (k, j) cyclic code. The generator polynomial's order is determined by the value r and has $r + 1$ coefficients:

$$r = k - j$$

For example, the generator polynomial used in this example is of order 1. Thus, the cyclic code for which it is used would be characterized as a (4, 3) cyclic code.

In systematic cyclic codes, the first j bits of the code word are the information word exactly. The last r bits are the parity bits. Recall, $r = k - j$, hence $k = j + r$! In the next section, where a CRC code is explicitly calculated, note that after the CRC encoding, the code word always consists of the original message bits plus r parity bits.

■ ARQ PROTOCOL ERROR-DETECTION CODES: CRC CODES

This section will examine cyclic redundancy check (CRC) codes, which are commonly implemented as error-detection codes in ARQ protocols. Here we will explore how the CRC code mechanism works.

Every data communication system is concerned with moving blocks of data from one location to another without error. CRC codes are a popular way to identify if an error has occurred in the block of the data being sent. A group of bits is added to the end of a block of data. These bits are added so that the received block of data, including the CRC code, will be exactly divisible by some number. If this division is exact, with no remainder, the message has no errors. If the division is not exact and some remainder is evident, then the transmission is assumed to be in error.

The division described above is done in a special way: modulo-2 arithmetic. If you know how an exclusive-OR gate works, you know how modulo-2 arithmetic works. To summarize, modulo-2 arithmetic is binary addition *with no carries*. For example, if we add 1101 and 1001 in modulo-2, we obtain the following answer:

$$
\begin{array}{r}
1101 \\
+\ 1001 \\
\hline
0100
\end{array}
$$

The same result would be found if both numbers were applied, bit by bit, to an exclusive-OR gate.

CRC codes are block codes that are implemented by performing the following steps:

1. Shift the original data block the same number of bits left as the highest order exponent of the generating polynomial, P. In Example 7-12 below, the generating polynomial is given by $P = 110101$.
2. Divide the shifted data block by the generating polynomial. Discard the quotient, Q (the result), and keep the remainder, R. The remainder should have the same number of bits or less than the highest order exponent of the generating polynomial. This remainder is the CRC code for the original data word.
3. To form the transmitted data block, take the original data word and append the CRC code found in step 2. This data block is transmitted over the communications channel.
4. At the receiver, divide the received data block by the generating polynomial. If the remainder is zero, then no error occurred. If the remainder is not zero, then error occurred in the transmission.
5. When no error occurred, strip the bits appended in step 3 and pass the correctly recovered data block out of the receiver. If an error occurred, either request retransmission or apply a method of error correction.

■ **EXAMPLE 7-12**   The original message to be sent is given by $M = 10101100$:

$$
P = x^5 + x^4 + x^2 + 1
$$

The generating polynomial is represented by $P = 110101$. A 1 is in the binary representation of P when the generating polynomial is not zero and a 0 when it is zero. This particular generating polynomial is known as a CRC-5. (The highest power of the generating polynomial gives it its name.) The remainder to be calculated in this example will be 5 bits or less because the highest power of the generating polynomial is 5.

To illustrate how CRC encoding works in detail, this message will be worked through step by step. For your reference, table versions of the long divisions shown below can be found in the appendix at the end of this chapter. The appendix allows an easy check of the bit positions, useful when first learning about this type of modulo-2 division.

First, shift the message by the highest power of the generating polynomial, P. This shift is 5 bits:

$$\text{New message} = 1010110000000$$

Next, divide the new message by the generating polynomial and discard the quotient. Keep the remainder. Pay careful attention to how the division is carried out. Remember, it is modulo-2 arithmetic, which means that you ignore which number is bigger and apply the two numbers as if they were connected to an exclusive-OR gate:

```
                   11011110
       110101 )1010110000000
            − 1101010000000
              0111100000000
             − 110101000000
               001001000000
               − 1101010000
                 0100010000
                − 110101000
                  010111000
                 − 11010100
                   01101100
                  − 1101010
                    0000110
```

Therefore, the quotient is 11011110 and is discarded. The remainder is 110 and is retained. The next step is to add the remainder to the new message:

```
       1010110000000
             + 110
       1010110000110
```

This sum is the message that is actually transmitted.

Two cases will be explored: first, no errors in transmission. In this case, when the message above is divided by the generating polynomial, P, the remainder should be zero:

```
                   11011110
       110101 )1010110000110
            − 1101010000000
              0111100000110
             − 110101000000
               001001000110
```

```
                                  − 1101010000
                                    0100010110
                                  − 110101000
                                    010111110
                                  − 11010100
                                    1101010
                                  − 1101010
                                          0
```

The answer is exactly what one would expect. Now, flip one bit and see if the answer changes. The LSB is the bit that is flipped from a 0 to a 1:

```
                             11011110
                   110101 )1010110000111
                          − 1101010000000
                            0111100000111
                          − 110101000000
                            001001000111
                          − 1101010000
                            0100010111
                          − 110101000
                            010111111
                          − 11010100
                            1101011
                          − 1101010
                                  1
```

Note where the last bit was flipped that, again, exactly what one would have anticipated is found: a nonzero remainder indicating an error in transmission. Note also that the CRC method does not give any indication of where the error occurred, only that it did occur.

For a better idea of where the bit positions line up, see the appendix at the end of the chapter. These three modulo-2 divisions are shown in table form. The appendix shows exactly the alignment for each step in the division. To illustrate how the number of bits in the remainder is determined by the order of the CRC polynomial, a fourth-order division is also illustrated in the appendix.

The increased number of data bits required to be transmitted when using any CRC code depends on the length of the data word. Like any code, how the length increases with the length of the data word is an important subject in determining the applicability and efficiency of the code. The CRC codes are used with long data words to minimize the overall bandwidth increase required to use them, especially when the communication channel is well known and does not contribute too many errors.

Many generating polynomials have been defined. Table 7-2 gives four of these CRC polynomials. The CRC-5 is the least used in practice, but the most used to illustrate the concept.

Table 7-2
CRC Generating Polynomials

CRC-5	$x^5 + x^4 + x^2 + 1$
CRC-12	$x^{12} + x^{11} + x^3 + x^2 + x^1 + 1$
CRC-16	$x^{16} + x^{15} + x^2 + 1$
CRC-CCITT	$x^{16} + x^{12} + x^5 + 1$

Figure 7-4
Implementation of the CRC-5 Polynomial (*Reprinted by permission of Pearson Education, Inc., Upper Saddle River, NJ*)

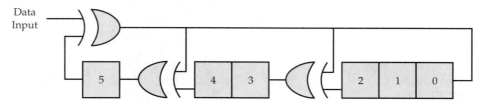

Finally, it is of interest to note how this algorithm is implemented in circuitry. Figure 7-4 illustrates the implementation. As you can see, only two types of components are required. Each of the rectangular boxes represents 1 bit of delay, usually implemented with a D-type flip flop. The other is the exclusive OR gate, which performs the modulo-2 addition required.

The chief distinction of CRC codes is that they are used widely with ARQ protocols as the mechanism for detecting errors. ARQ protocols require a full-duplex channel for this type of approach to work. Remember, CRC codes cannot correct an error, only detect it. A protocol must be in place for the destination to request retransmission when a message is received in error. CRC codes are widely used in LAN environments where a full-duplex channel exists naturally. CRC codes are preferred over Hamming or any type of parity code in a full-duplex environment where an ARQ protocol can operate because they are concerned only with error detection. While both Hamming and parity codes can be used as error-detection codes only, they are not the most efficient approach because most full-duplex LAN environments experience very few errors. A typical error rate for a LAN environment is one error in 1 billion bits of transmission. In many cases, there will be no errors at all in a message.

On the other hand, when errors occur, they are likely to occur in a group, known as a burst. An error burst begins and ends with an error, but all bits between are not necessarily in error. The bottom line is that parity, or any scheme based on parity such as a Hamming code, does not do a good job of detecting error bursts. CRC codes are the preferred choice when the overall error rate is low and the errors are likely to occur in bursts.

■ BINARY BCH CODE

Bose-Chaudhuri-Hocquenghem (BCH) codes are some of the most widely applied error-correcting cyclic block channel codes. All code word symbols considered in this section will be binary. The code word bits are coefficients of the generating polynomial, as was the case for the CRC codes.

There are two classes of BCH codes: primitive, which has a code word block length of

$$k = 2^k - 1$$

$$k = 3, 4, \ldots$$

and nonprimitive, which features a block length k, where k is some factor of the above expression. These codes work like the CRC codes examined above in some detail. BCH codes require the use of a look-up table to compute the actual code word. For that reason, an example will not be explored here. However, the implementation of the coder and decoder is very similar to that shown in the section on CRC codes.

■ CONVOLUTIONAL CODING

A convolutional code is also a form of FEC coding. Convolutional codes are error-correction codes and also block codes that rely not only on the bits in the current block but also some number of bits in previous blocks to determine the output. *Convolution* means "the folding of one part over another." Here, the values of the previous bits are used to determine the current input sequence of bits: they are "folded over." At the receiver, each bit in the data block is compared with some number of previous bits. Since the convolutional encoder sets the state of the nth bit depending on the state of the previous few bits, the previous bits can sometimes be used to correct the bit in error if an error occurs due to noise in the channel.

The convolutional code that will be explored in this section uses the concept of a trellis. The name *trellis* comes from the diagrams used to determine the value of the additional bit(s) inserted. The diagrams appear similar to trellis fences, which are exemplified by an overlay or lattice of boards. A brief example will illustrate the concept.

Typically, these systems are implemented as shift registers. The value of bits sent is determined through a truth table of the value of the input bits. In this example, a shift register of 4 bits will be used, and the value of the shift register contents will be applied to two exclusive OR gates to obtain a modulo-2 value of the shift register contents. Both output bits will be sent. Figure 7-5 illustrates the circuit diagram. Assuming that the shift register is loaded with all zeros to begin, a bit pattern such as the one shown in Table 7-3 would result.

■ MAXIMUM LIKELIHOOD DETECTION

In general, maximum likelihood detection uses the Euclidean distance between potential received signals as a method of detection. However, since parity and Hamming distance

Figure 7-5
Convolutional Encoder Used for TCM

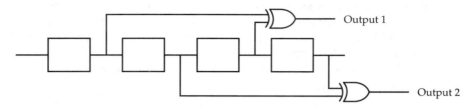

Output 1

Output 2

Table 7-3
Input and Output Sequences for Figure 7-5

Input Data	SR1	SR2	SR3	SR4	Output Bit 1	Output Bit 2
0	0	0	0	0	0	0
1	1	0	0	0	1	0
0	0	1	0	0	0	1
1	1	0	1	0	0	0
1	1	1	0	1	1	0
1	1	1	1	0	0	1
0	0	1	1	1	1	0
1	1	0	1	1	0	1
0	0	1	0	1	0	0
1	1	0	1	0	0	0

have already been introduced, a specific form of this algorithm will be explored instead. Maximum likelihood detection is another name for minimizing the Hamming distance between all possible locations in the trellis.

The fact that a decoding algorithm uses Hamming distance to characterize its performance does not ensure that the Hamming distance used is minimized in all cases. Maximum likelihood detection methods ensure that the Hamming distance is minimized. In other words, if the performance of a decoder does not degrade the performance of a code below that predicted by the Hamming distance calculation, then that detection method is a maximum likelihood detection method.

Detection methods are the other half of the coding process. Advanced coding techniques can be very complex and processor-intensive to decode. Because of this fact, it is important to characterize decoding algorithms according to their effectiveness, independent of the data block input to them. A maximum likelihood detection method is implemented by calculating all possible Hamming distances that could result from the coding process and by selecting the one with the minimum Hamming distance for each step of the decoding process.

Employing a maximum likelihood detection algorithm will guarantee that the possible received sequence selected, which will be used to construct the trellis, represents those sequences that feature the minimum Hamming distance possible for that received sequence of bits from the original bits sent. Recall that Hamming distance is a measure of how many bits are different between two potential word blocks. Clearly, word blocks with the fewest differences are most likely to be the correctly decoded result.

Utilizing a trellis structure for the decoder eliminates many redundant paths through the decoder that would normally have to be considered for optimal decoding. This efficiency is why the convolutional coding and decoding techniques are so popular. They offer a close approach to optimal decoding, thus implementing the theoretical optimum closely, while at the same time reducing the computational complexity of such an optimal solution. The Viterbi algorithm, in particular, rejects some paths before the complete received sequence has to be evaluated. With each rejection, the number of computational steps required for optimal decoding is reduced.

■ VITERBI ALGORITHM

Viterbi algorithm decoding is an implementation of an efficient optimal decoding approach based on a trellis structure. It is widely used in terrestrial and satellite-based microwave links. Example 7-13 illustrates the application of the Viterbi algorithm. Remember, the key to this algorithm and its implementation is the reduction of computations needed to detect and correct any errors resulting from transmission.

■ **EXAMPLE 7-13**   The information bits = 10110, followed by a tail of two 0 bits to clear the decoder for the next information bit. All convolutional codes make use of this approach, and the number of 0 bits following the data stream is always one less than the inverse of the rate. Here the rate is $\frac{1}{3}$ and the inverse is 3. Subtracting 1 yields the fact that 2 zero bits are added. Hence, the actual bit stream applied to the decoder is 1011000. A rate $\frac{1}{3}$ convolutional code means that each information bit will be represented in the channel as a three-bit sequence. Figure 7-6 illustrates the shift register for the convolutional encoder.

The trellis representation of the state is shown in Figure 7-7. Use the bit pairs on the vertical margin to the left of the figure to calculate how the information bits propagate through the trellis. This is shown in a heavy line. The idea is to take each state, here represented by a pair of bits, and move to the next state value corresponding to the horizontal line represented by those 2 bits.

In Figures 7-6 and 7-7, the first initial state is 00. They are the last two bits of the shift register, the tail of the previous information sequence. Reading across from the 00 bit pair in Figure 7-7, the starting point of navigating through the trellis is established. The next state is 10. Reading horizontally across from 10 establishes the first link in the path through the trellis and the first code output of 111. The code output is the three-bit sequence written next to the trellis link.

The next state is 01, so the next link is found to be 001, etc. Work the rest of the way through the trellis yourself. For convenience, the bit pairs are written above the trellis in

Figure 7-6
Conditional Encoder for Example 7-13

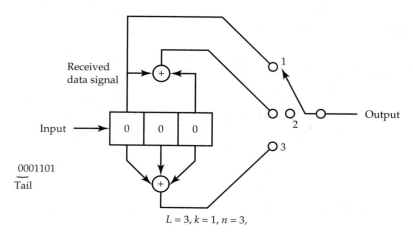

Figure 7-7
Viterbi Trellis Diagram 1

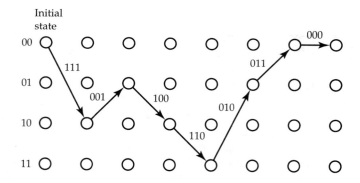

Figure 7-7. Only the path taken by this particular code word is shown. The final transmitting sequence is:

```
111 001 100 110 010 011 000
```

To illustrate how a hard, decoded trellis reduces computation, suppose that several bits were received in error. Specifically, suppose that bits 3, 9, 10, 14, and 18 were received in error. What effect would this have on the number of paths through the trellis that need to be compared to find the correct path?

The received bit sequence would now be:

```
110 001 101 010 000 010 000
```

Figure 7-8
Viterbi Trellis Diagram 2

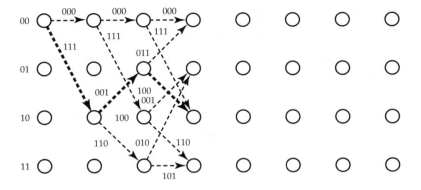

Table 7-4

Paths	Actual Data	Trellis Estimated Data	Hamming Distance
00–00–00–00	110 001 101	000 000 000	2 + 1 + 2 = 5
00–00–00–01	110 001 101	000 000 111	2 + 1 + 1 = 4
00–00–10–01	110 001 101	000 111 001	2 + 2 + 1 = 5
00–00–10–11	110 001 101	000 111 110	2 + 2 + 2 = 6
00–10–01–00	110 001 101	111 001 011	1 + 0 + 2 = 3
00–10–01–10	**110 001 101**	**111 001 100**	**1 + 0 + 1 = 2**
00–10–11–01	110 001 101	111 110 010	1 + 3 + 3 = 7
00–10–11–11	110 001 101	111 110 101	1 + 3 + 0 = 4

After evaluating the first three links in the trellis diagram, some paths have been discarded, or only some paths "survive." The rule for determining which path survives is that the path with the smallest Hamming distance of the pair is kept. Figure 7-8 shows this rule in action for the first three links. The dotted line paths have not survived; hence, they do not need to be calculated, which results in faster throughput for the decoder.

Figure 7-8 shows several potential decoding paths that make it through at least three links. These paths are shown as either a dotted line or a solid line. Any path that makes it this far has its Hamming distance calculated. Then, using the results of this calculation, the maximum likelihood path is chosen. In this example, the solid line path resulted in the minimum Hamming distance; therefore, it is the chosen path. Table 7-4 tabulates the paths shown in Figure 7-8, and the path set in bold type indicates the maximum likelihood path.

The procedure shown in Example 7-13 should be continued with the Hamming distance computed after each link is completed. Valid paths, paths that reach through all links made so far, are not discarded, until one path reaches the final link. Then any paths with Hamming distances larger than the one for the final path are discarded. Incomplete paths are computed until their Hamming distance exceeds the completed path. Once all paths with Hamming distances equal to or less than the original completed path are also completed, then the path with the lowest Hamming distance is kept. Every time a path does not survive, the corresponding links do not need to be computed, which saves computational resources. The Viterbi algorithm changes the number of computations from

$$2^k - 1 \quad \text{to} \quad k2^{l-1}$$

where k is the state and l is the inverse of the rate. The second value is much smaller than the first for typical rates, but for large values of l, this is not the case. Typical values of l are less than 8. This number implies an increase of 8 times the information being sent through the channel; this is not a practical result. Most Viterbi coders use values of 3 to 5 for l.

■ SUMMARY

This chapter provided an overview of the error-detection and error-correction methods widely used in wireless systems. Several error-correcting codes were worked through in detail to demonstrate algorithmic principles and to provide a high level of confidence to the reader. Channel coding is critical to wireless networking because the performance of these channels is notoriously low.

Coding is usually required to get any reasonable throughput in most wireless channels, but it comes at a cost. This cost comes not only from the significant number of bits required to detect the relatively large number of errors caused by wireless channels, but also from the complexity and power implications of the decoding process, coupled with the implementation of any coding process. This chapter provided a guide for determining when to apply certain techniques as long as the performance of the channel is known in general terms. The efficiency of the decoding process is also critical in many mobile wireless systems because battery power is limited. By reducing the processor cycles required to decode communications received, the experience of the subscriber using the mobile device is enhanced.

REVIEW QUESTIONS

1. Determine the Hamming code length for the data word lengths below:
   a. 10 bits
   b. 8 bits
   c. 16 bits
   d. 4 bits

2. Draw code trees for each of the following alphabets. Make sure that no code word has the prefix of any other code word. Compute the average length of each code.
   a. A   33 percent
      B   33 percent
      C   33 percent
   b. A   10 percent
      B   20 percent
      C   30 percent
      D   40 percent
   c. A   25 percent
      B   25 percent
      C   25 percent
      D   25 percent
   d. A   10 percent
      B   45 percent
      C   45 percent

3. Explain why 1-bit parity codes are useless for double bit errors.

4. Generate the Hamming code words of 7 bits for each of the following 4-bit data words. Assume that the Hamming bits are inserted in the correct positions, like the example in the chapter.
   a. 1010   c. 0001
   b. 1100   d. 1110

5. For each of the 7-bit code words developed in problem 4, flip bit 3 and verify that the syndrome predicts the bit error location correctly.

6. Write a software program to take an arbitrary data word input of 8 bits and find the CRC code. Use the generating polynomial below:

$$x^5 + x^4 + x^1 + x^0$$

7. Write a software program to check received data words that have been CRC encoded in problem 6. Verify that it recovers the original transmitted data word correctly when no errors in transmission occur and that it identifies when an error occurred.

8. Explain why channel encoding is needed for digital communciations.

9. Classify channel encoding schemes and list some examples for each classification.

10. Define FEC and an error-detecting code with an ARQ protocol. Discuss the difference between them.

APPENDIX

This appendix gives the table versions of the modulo-2 division performed in the section about CRC codes. The data and polynomial are given below:

$$D = 10101100 \qquad P = 110101$$

															1	1	0	1	1	1	1	0
		1	1	0	1	0	1		1	0	1	0	1	1	0	0	0	0	0	0	0	0
									1	1	0	1	0	1								
									0	1	1	1	1	0	0							
										1	1	0	1	0	1							
										0	0	1	0	0	1	0	0					
												1	1	0	1	0	1					
												0	1	0	0	0	1	0				
													1	1	0	1	0	1				
													0	1	0	1	1	1	0			
														1	1	0	1	0	1			
														0	1	1	0	1	1	0		
															1	1	0	1	0	1		
															0	0	0	0	1	1	0	

As you can see, the remainder is 00110. Note that the number of bits is always the same as the order of the polynomial. The remainder must be added to the original data message and sent, so the transmitted message is:

$$D = 1010110000110$$

At the receiver end, the message is again divided by the polynomial. If there is no remainder, the message is assumed to contain no errors.

											1	1	0	1	1	1	1	0
1	1	0	1	0	1	)	1	0	1	0	1	1	0	0	0	0	1	1	0
							1	1	0	1	0	1							
							0	1	1	1	1	0	0						
								1	1	0	1	0	1						
								0	0	1	0	0	1	0	0				
										1	1	0	1	0	1				
										0	1	0	0	0	1	0			
											1	1	0	1	0	1			
											0	1	0	1	1	1	1		
												1	1	0	1	0	1		
												0	1	1	0	1	0	1	
													1	1	0	1	0	1	
													0	0	0	0	0	0	0

As you can see, there is no error for the above division. To demonstrate what happens if an error occurs, we will change 1 bit. The division now predicts this error because it results in a nonzero remainder. The bit that has been changed is the last bit in the message.

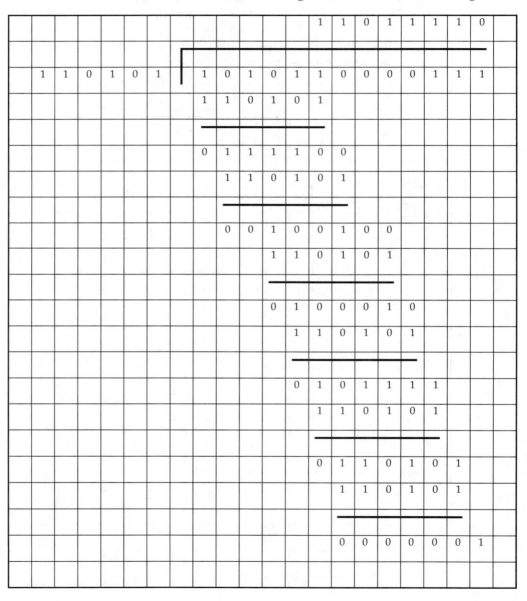

This example also illustrates modulo-2 division as applied to CRC coding. The data and polynomial are given below. Note that this polynomial is of the fourth order, so the remainder is 4 bits.

$$D = 1101101 \qquad P = 10011$$

```
                        1 1 0 0 1 1 1
              ┌─────────────────────────────
1 0 0 1 1  │  1 1 0 1 1 0 1 0 0 0 0
              1 0 0 1 1
              ─────────
              0 1 0 0 0 0
                1 0 0 1 1
                ─────────
                0 0 0 1 1 1 0 0
                    1 0 0 1 1
                    ─────────
                    0 1 1 1 1 0
                      1 0 0 1 1
                      ─────────
                      0 1 1 0 1 0
                        1 0 0 1 1
                        ─────────
                        0 1 0 0 1
```

As you can see, the remainder is 1001. Note that the number of bits is always the same as the order of the polynomial. This remainder must be added to the original data message and sent, so the transmitted message is:

$$D = 11011011001$$

You can proceed as you did for the first example in this appendix.

8

Antennas

■ INTRODUCTION

The subject of antennas is long and still growing. Since the very early days of radio, antenna design has been an important issue. Without an antenna, how can a broadcast be transmitted or received? Early antennas were developed as the most straightforward ways of solving a problem, namely, proving that radio worked. A comprehensive study of antenna design is outside the scope of this book, but many excellent texts cover the topic thoroughly. For those of you who would like a practical introduction that goes beyond this textbook, explore the literature on amateur radio, or ham radio.

This chapter will begin with several sections summarizing the basic ideas that are necessary to understand antennas. Link analysis is a theme that will tie together topics covered in Chapters 8 to 10. We will begin in this chapter with those components specific to antenna design. Finally, the chapter closes with an examination of several common antenna types found in wireless communications systems.

Note that most of the discussion below is about transmitting antennas only. It can be applied to receiving antennas in exactly the same way. For many of you, it will be more convenient to talk about power being transmitted from an antenna than to talk about power being received by an antenna.

■ TERMS AND DEFINITIONS

The whole idea of antenna design is to get the maximum power transferred to the transmitting antenna and pick up the maximum power at the receiving antenna. These issues drive the entire subject and are always critical issues in communication design. Since antennas radiate electromagnetic energy into free space, this energy travels in one of the three modes of propagation that was described in Chapter 1: ground wave (GW), space wave (SW), and line of sight (LOS).

What we assumed in Chapter 1 was that the antenna would radiate power in a particular pattern, specifically an isotropic pattern. *Isotropic* means "identical in all directions." Thus, we assumed earlier that the power was radiated equally in all directions. This assumption resulted in simple equations that relied on the geometry of a sphere to determine the power radiated and how it would dissipate with distance from the transmitting antenna. That discussion also assumed that the receiving antenna was ideal and its effects were neglected. In this chapter, nonideal antennas and their effects will be addressed.

The first broad classification of antennas concerns how they radiate power. Many types of patterns are used, but they all fall into one of two classifications: isotropic or anisotropic. *Anisotropic* means "different in one or more directions." Therefore, an anisotropic antenna is an antenna that does not radiate power equally in all directions. Of course, this condition is equally true for receiving antennas. An isotropic antenna radiates power equally in all directions.

Almost all real antennas are anisotropic, not isotropic, because the whole point is to get power to the receiver most efficiently. Most of the time, a particular direction is more desirable for transmitting power than another. A radio or television station positioned on the coastline rather than inland comes to mind, or perhaps a radio or television station at the edge of a city rather than at its center. There are many examples, some driven by technical considerations, some by regulatory requirements. The terms *isotropic* and *anisotropic* are scientific and not easily remembered by most people, so for the rest of this discussion, the terms *unidirectional* or *directional* will be used instead. Antennas that radiate power isotropically are commonly called unidirectional, or all directional. Antennas that radiate power preferentially in one or more directions are commonly called directional antennas. A good example of a directional receiving antenna is the television antenna you may have on your roof.

■ RADIATION PATTERN

The radiation pattern of an antenna is the shape or pattern taken by the electromagnetic signals radiated by the antenna. These patterns are usually plotted on graphs like those shown in Figure 8-1. They are called polar coordinate graphs because they use the polar coordinate scheme, r and θ, rather than the typical rectangular coordinate scheme, x and y. The first pattern, shown in Figure 8-1(a), is for a unidirectional antenna. The pattern shown in Figure 8-1(b) is for a very common and basic directional antenna called a di-

Figure 8-1
Radiation Patterns (Top View): (a) Unidirectional Antenna; (b) Directional Antenna (*Reprinted by permission of Pearson Education, Inc., Upper Saddle River, NJ.*)

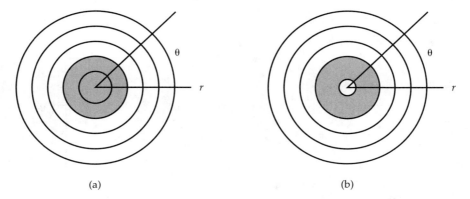

(a)   (b)

pole. The perspective in the figure is what you would see if you were looking down onto each antenna from above.

As you can see in Figure 8-1, the power radiated from the first antenna is isotropic or unidirectional. The second is radiating power in an anisotropic or directional pattern. In part (a), the power radiates equally in all directions (unidirectional). In part (b), you can see the power is the same in all horizontal directions. From the top and bottom or vertical directions, no power is radiated. Figure 8-1 shows the horizontal radiation pattern of the two antennas. This perspective comes from looking down at the power radiating from this antenna.

The horizontal radiation pattern is seen by looking from a vertical perspective. The vertical radiation pattern is seen by looking from a horizontal perspective. This terminology is often confusing when students are first introduced to antennas. It should be clear that you can view the horizontal pattern only by looking at something from a vertical perspective. For example, you are now reading a pattern of letters and spaces, a horizontal pattern, from a vertical perspective.

A natural question to ask is, "What does the pattern look like in the vertical direction?" From the side, the unidirectional antenna looks just the same. If it is radiating equally in all directions, it must fill a spherical shape, like a balloon. From the side, the dipole antenna radiation pattern looks like a doughnut with a very small hole. The doughnut is placed so that the hole of the doughnut is placed at the middle of the antenna. The antenna element would be placed vertically through the center of the doughnut. Figure 8-2 shows the doughnut on its side, cut in half. (If you like, imagine that you have eaten half of the doughnut while trying to picture this concept.)

The different perspectives shown in Figures 8-1 and 8-2 should give you a good idea of how power radiates from two simple antennas. In general, the patterns that result are not so regular. In most cases, antennas radiate power in different amounts in different directions, as we will see a little later in the chapter.

Figure 8-2
Radiation Patterns (Side View): (a) Unidirectional; (b) Directional (Dipole) (*Reprinted by permission of Pearson Education, Inc., Upper Saddle River, NJ.*)

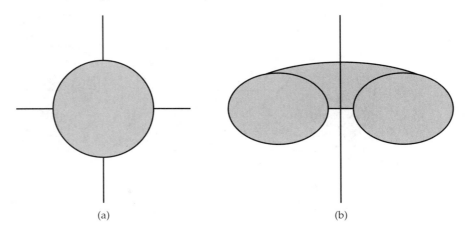

(a)   (b)

■ POLARIZATION

The polarization of a radiated electromagnetic signal is always considered to be in the direction of the electric field of the antenna. Antennas designed for different purposes use different polarizations, depending on the propagation mode used by the electromagnetic signals being transmitted. Only two polarizations, vertical and horizontal, are commonly used.

For example, waves below 3 MHz use GW propagation, and almost all signals transmitted in that frequency range use vertically polarized waves. All AM broadcast stations use vertically polarized transmitters. Hence, best results come from using vertically polarized antennas, like your car antenna.

In the 3 to 30 MHz band, both vertically and horizontally polarized antennas are used, depending on transmission distance and environmental conditions. For waves above 30 MHz, where LOS propagation occurs, most signals are naturally vertically polarized. Modern communication system designs at microwave frequencies often transmit both a vertically polarized and a horizontally polarized signal from a single antenna with two active elements. Broadcast television and radio would naturally lend themselves to transmission that is vertically polarized, but early systems were implemented to transmit horizontally. As a result, they still transmit horizontally today.

Now, what type of antenna would best receive a vertically or horizontally polarized transmitted signal? The answer is the same type used to transmit! Vertical antennas receive vertically polarized transmissions best, and vice versa. For those transmissions below 30 MHz, however, the ionosphere sometimes twists ground waves or sky waves out of polarization due to reflection effects. Therefore, better reception occasionally results from using the opposite polarized antenna. Sometimes the best results are obtained somewhere between the two. Like the reflection of electromagnetic waves from the ion-

Figure 8-3
Television Transmitter Tower (*Reprinted by permission of Pearson Education, Inc., Upper Saddle River, NJ.*)

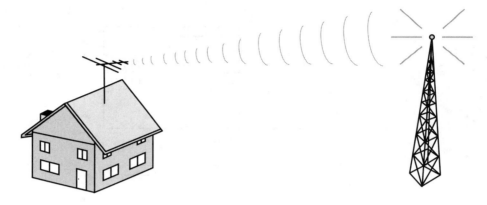

osphere, a lot depends on the frequency, time of day, angle of impact, chemical content of the ionosphere, etc.

Another good question many people have is, "How can you tell what the polarization of an antenna is?" If the antenna is simple, like a wire or aerial, the direction of the wire is the polarization. If it is horizontal with respect to the surface of the planet, it is horizontally polarized, and vice versa. Antenna towers are often erected vertically for height, but the actual antenna elements in those towers can be oriented horizontally *or* vertically, depending on which polarization is intended. This construction explains why radio and television towers are very tall and yet transmit horizontally polarized waves. It also explains why television antennas are polarized horizontally on top of a mast. The mast takes advantage of height and the antenna elements are held horizontally. Figure 8-3 shows the horizontal placement of a transmitting antenna placed on top of a tower and the horizontally placed antenna on a mast at the receiving end.

■ ANTENNA RADIATION

Before going any further, it is appropriate to describe how a very simple antenna really works and how simple antennas are made. The most basic antenna, called a half-wave dipole, is a piece of wire with an insulator at each end. It is suspended above the ground either horizontally or vertically. If the wire is cut to equal half of the wavelength of the signal, it is designed to radiate or receive; it acts like a simple LC oscillator. This type of analog oscillator is covered in every textbook on basic electronics; however, a brief review is presented here. The paragraphs below analyze the circuit with a classic mechanical analogy for oscillators: the flywheel analogy.

Figure 8-4(a) shows a simple LC circuit with a battery in series and a switch attached across the capacitor. As the switch is closed instantaneously and then opened

Figure 8-4
Antenna Flywheel Analogy (*Reprinted by permission of Pearson Education, Inc., Upper Saddle River, NJ.*)

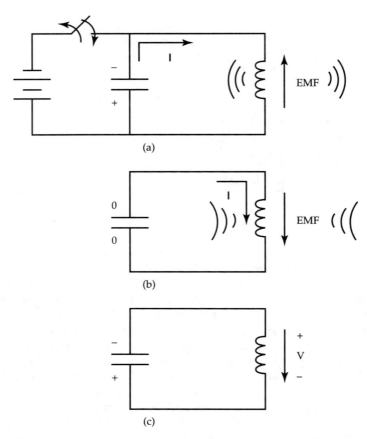

again, electrons flow from the battery and deposit themselves on the top plate of the capacitor, charging it negatively. Simultaneously, electrons are pulled from the bottom plate of the capacitor, charging it positively. Since the coil has inductance and inductance resists the flow of current through its electromotive force, or EMF, no significant current flow occurs in the brief instant that the switch is closed. When the switch is opened, the electrons on the top plate start to move through the inductor toward the bottom plate of the capacitor, thus balancing the charge across the capacitor. In essence, the switch operation has started the current moving using the power from the battery.

Examine the following sequence. As current flows through the inductor in part (a) of the figure, it generates a magnetic field, which generates a *counter EMF* in the inductor. The counter EMF resists the flow of current and prevents the capacitor from discharging immediately. The important part here is the generation of the magnetic field that expands outward from the inductor. As the capacitor discharges in Figure 8-4(b), both plates of the capacitor eventually have the same number of electrons, and the current flow slows and stops. The magnetic field generated in the inductor begins to collapse inward, which induces a voltage in the inductor. This voltage is shown in Figure 8-4(c). It charges the bottom plate of the capacitor slightly negatively and the top plate slightly positively.

If this process is repeated but the polarity of the battery is switched before briefly closing the switch, the same event takes place except in the opposite direction. For an ideal circuit with no resistance, the charging/discharging cycle would continue and a generated ac voltage results. This oscillation of electrons is where the analogy to a flywheel comes in. Many traditional designs of sinusoidal oscillators use this or some variation of this effect to produce ac oscillators.

The same process occurs in simple antennas. An antenna is cut to the proper length, and once it is excited by an ac voltage at that frequency, the free electrons in the antenna will oscillate back and forth along it. Most antennas have very low resistance. They are made of conductors, so this oscillation can continue for some time. But what really steals the energy from the system is the radiation of the electromagnetic field. This energy lost by the antenna while it is oscillating is what generates the electromagnetic signals that propagate through the atmosphere.

As the electrons propagate from one end of the antenna to the other (just like the electrons propagated from one plate of the capacitor to the other), they pile up at one end or the other and create an electric field. One end of the antenna is full of electrons, the other has none. Of course, the electrons move immediately to correct this disparity and a short time later, the other end is full of electrons and the original one is empty.

As the electrons move back and forth, they produce changing electromagnetic fields that expand and then contract, like the magnetic field produced by the inductor in the above example. Some of these fields travel so far that they are not coupled back by the reverse action. This lost energy is transmitted out from the antenna as an electromagnetic wave, which represents the radiated power of the antenna.

■ RADIATION RESISTANCE

When an antenna is excited into oscillation, the antenna radiates power. Conceptually, the antenna is acting as a voltage or current source and, like any such source, has some impedance associated with it. By applying Ohm's law, you can compute the radiation resistance if you know the power and current using the formula $R = P/(I*I)$. The current is measured at the center of the antenna, or at the designated feedpoint. The resistance computed is the radiation resistance. Each antenna type has a different radiation resistance, depending on its geometry and the construction materials used.

Radiation resistance is important because it defines the matching impedance necessary to couple maximum power either into or out of the antenna. In an antenna application, a cable connects the antenna to the transmitter or receiver. It is important to choose the impedance of that cable so that it matches, as closely as possible, the radiation resistance exhibited by the antenna.

Every cable has an associated characteristic impedance. This impedance should match, as closely as possible, the radiation resistance of the antenna for maximum power transfer between the antenna and the cable that connects it to your transmitter or receiver. Of course, the output impedance of the transmitter must also match this impedance for maximum power transfer to the load. When these impedance values are not the same, standing waves result and the standing wave ratio (SWR) gives a direct reading of how much power is transferred to the antenna. Sometimes the impedance calculated from the radiation resistance is called the input impedance of the antenna. When used this way, the term usually refers to the input impedance of the feedpoint of the antenna. The feedpoint of the antenna is that point where the cable is attached to the antenna.

Radiation resistance can also be used to calculate the efficiency of an antenna. An antenna's efficiency is defined as the ratio of the actual power radiated or received to that of an ideal antenna. For most basic antennas, the efficiency is very high because the loss in the antenna itself is very small. The loss in the antenna depends on the impedance matching. This efficiency is expressed in Equation 8-1. Note that radiation resistance can be defined for a transmitting or receiving antenna.

$$\eta = \frac{P_{tx}}{P_{\text{input}}} = \frac{P_{rx}}{P_{\text{output}}} = \frac{R_r}{R_r + R_a} \tag{8-1}$$

P_{tx} = transmitted power
P_{rx} = received power
P_{input} = applied power
P_{output} = extracted power
R_r = radiation resistance
R_a = antenna resistance

As we explore different antenna types, you will find that a particular type has an impedance associated with it. For simple dipole antennas, this dependence is based on the length of the antenna. For example, half-wave dipole antennas have a radiation resistance of 73 Ω, while quarter-wave antennas have a radiation resistance of half that, or about 36 Ω. You can see that smaller values for radiation resistance lead to less efficient antennas. This phenomenon is somewhat counterintuitive, so Example 8-1 demonstrates it.

■ **EXAMPLE 8-1** Determine the antenna efficiency for the two antennas mentioned above. Assume that the antenna resistance of both is 1 Ω. This number is reasonable for an antenna made of a metallic conductor.

$$\text{Half-wave dipole: } \eta = \frac{R_r}{R_r + R_a} = \frac{73}{73 + 1} = 0.986 = 98.6\%$$

$$\text{Quarter-wave dipole: } \eta = \frac{R_r}{R_r + R_a} = \frac{36}{36 + 1} = 0.973 = 97.3\%$$

■ EFFECTIVE RADIATED POWER (ERP)

Antennas must be classified according to how much power they radiate compared to how much power is used to excite them. This comparison is called the gain of an antenna. Of course, an antenna is a passive device, so it cannot actually supply gain to the signal, but the term is a good one for classifying antennas according to how well they work. Antenna gain is typically expressed in decibels. Usually, manufacturers give the gain of an antenna in dBi. The number of dBi is the number of dB gain that the antenna features over that of the ideal isotropic unidirectional antenna.

Precise definitions of the effective radiated power (ERP) of an antenna are complex and are always referenced to an equivalent perfect antenna that radiates isotropically. Thus, the directivity of the antenna plays a role in determining its power gain. For most applications, except actual antenna design, it is sufficient to be able to measure the input power and output power of an antenna and to determine if the antenna is operating effectively at the frequencies and directions of interest.

Most antenna configurations are also specified by their directive gain and transmitted or radiated power. These two components are used to determine the ERP for any antenna.

$$\text{ERP(dB)} = 10\log(G_{d,tx}) + 10\log(P_{tx})$$

G_d = directive gain of transmitting antenna
P_{tx} = transmitted power **(8-2)**

The directive gain component of Equation 8-2 is a measure of how much power gain a particular antenna has compared to the equivalent perfect isotropic radiating antenna. The equation developed in Chapter 1 for power density from an antenna at any distance is used to compute P_{tx} and then this result is modified by the directive gain component, which accounts for the directivity of the antenna. Note that, because the ERP is being calculated for a particular direction, the result applies only to the direction where the directive gain component applies. Example 8-2 illustrates the use of Equation 8-2.

■ **EXAMPLE 8-2**   An FCC specification states that the maximum ERP of a transmitter and antenna combination must be no greater than 100 W. You have selected a 6 dB gain antenna to use. What is the maximum power (the radiated or actual transmitted power) that you can apply to the antenna?

$$\text{ERP(W)} = 100 \text{ W}$$
$$\text{ERP(db)} = 10\log(100) = 20 \text{ db}$$
$$20 \text{ dB} = 6 \text{ dB} + P_{tx}$$
$$P_{tx}(\text{dB}) = 14 \text{ dB}$$
$$P_{tx}(\text{W}) = 10^{14/10} = 25 \text{ W}$$

In Chapter 1, we assumed that the antenna radiated unidirectionally so that the power expanded like a balloon when inflated. The directive gain will modify this assumption in the same way that an entertainer can when he or she blows up balloons and twists them into interesting and diverse shapes. From now on, in this and following chapters, gain will be assumed to be directional. Thus, the antenna directional gain must be calculated when transmitted or received power is expressed.

Multiply the isotropic gain by the directive gain to obtain the anisotropic gain and the power density at any point. Note that the directive gain changes as the direction changes; the directive gain will be different in different directions. Figures 8-1 and 8-2 illustrating the radiative patterns are important because they show the directive gain shapes. The definition of power density is shown in Equation 8-3. Note that power density has units W/m^2, while power has units W.

$$P_d = \frac{P_{tx}G_{d,tx}}{4\pi r^2} \left(\frac{W}{m^2} \right) \tag{8-3}$$

P_d = power density

Another way to express the relationship in Equation 8-3 uses the idea of the effective area of an antenna. This concept can lead to the same conclusion by defining the effective area of an antenna. The antenna effective area has units ^{2}.

$$A_{\text{eff}} = \frac{G_d\lambda^2}{4\pi}(m^2) = \text{antenna effective area} \tag{8-4}$$

The power into a receiver can then be written as the product of the power density and the antenna effective area. This concept is shown in Equation 8-5:

$$P_{rx}(W) = P_d\left(\frac{W}{m^2} \right)A_{\text{eff}}(m^2) \tag{8-5}$$

It is possible to write an equation for received power as the product of the power density and the antenna effectiveness. Note that the transmitted power is now modified by adding a directive gain, like we did in Equation 8-3 for the received power. Equation 8-6 is used to calculate the power received at one antenna from another. Note that the directive gain for both the transmitting and receiving antennas is explicitly noted.

$$P_{rx}(W) = P_dA_{\text{eff}} = \left(\frac{P_{tx}G_{d,tx}}{4\pi r^2} \right)\left(\frac{G_{d,rx}\lambda^2}{4\pi} \right) = \frac{P_{tx}G_{d,tx}G_{d,rx}\lambda^2}{16\pi^2 r^2} \tag{8-6}$$

One important physical idea captured in Equation 8-6 is that the power received by such an antenna is directly proportional to the wavelength squared and inversely proportional to the radius or distance squared. Obviously, it is also directly proportional to the transmitted power. The wavelength and distance variables are expressed in the proportionality relationship shown in Equation 8-7. Note that Equation 8-7 is only meant to convey a proportionality relationship and equality is not implied.

$$\text{Power received} \propto \frac{\lambda^2}{r^2} \tag{8-7}$$

This fundamental result will always give you a rough idea of how effective an antenna pair will be. The longer the wavelength, or the lower the frequency, the more effective the antenna. Communications systems utilizing higher frequencies will have to pay this penalty, namely, with the use of higher frequencies, antenna effectiveness generally declines. This tradeoff requires additional transmit power for any given receive power requirement. This situation clearly has an impact on wireless mobile devices operating at high frequencies, where additional power requirements affect battery life. Second, the farther the distance between the antennas, the more transmit power that is required for any given receive power requirement. The latter result is just an expression of the inverse square law.

■ **EXAMPLE 8-3**   Find the ERP of an antenna with a directive gain of 6 in the direction of interest and a transmit power of 10 W. Further, at a distance of 2 km, find the power density.

$$\text{ERP(dB)} = 10\log(G_{d,tx}) + 10\log(P_{tx}) = 10\log(6) + 10\log(10) = 17.8 \text{ db}$$

$$P_d = \frac{P_{tx}G_{d,tx}}{4\pi r^2} = \frac{(10)(6)}{4\pi(2 \times 10^3)^2} = 1.19 \ \mu W/m^2$$

The approach in Example 8-3 can also be used to calculate the power received at one antenna from another. The idea is that if one antenna radiates so much power and another absorbs so much power, the product of the two gains must yield the ratio of power transferred.

■ **EXAMPLE 8-4**   Assume that two antennas are half-wave dipoles and each has a directive gain of 3 dB. If the transmitted power is 1 W and the two antennas are separated by a distance of 10 km, what is the received power? Assume that the antennas are aligned so that the directive gain numbers are correct and that the wavelength used is 100 MHz.

$$P_{rx} = \frac{P_{tx}G_{d,tx}G_{d,rx}\lambda^2}{16\pi^2 r^2} = \frac{(1)(3)(3)\left(\dfrac{3 \times 10^8}{100 \times 10^6}\right)}{16\pi^2(10 \times 10^3)^2} = 1.71 \times 10^{-9} \text{ W}$$

In practical situations, it is convenient to know the power, but what really is of interest is the voltage applied to the input of the receiver. Most receivers specify their sensitivity as a voltage. This sensitivity specification tells how much signal the receiver needs to supply an output other than noise. Assuming that the receiver is matched to the radiation resistance of the antenna, here, 73 Ω, the following sensitivity is found:

$$P = IR = \frac{V^2}{R} \Rightarrow V = \sqrt{PR} = \sqrt{(1.71 \times 10^{-9})(73)} = 3.53 \times 10^{-4} = 353 \ \mu V$$

To explore further, check the specification sheet of your stereo receiver or radio and see how the voltage applied to the antenna input compares with what your receiver expects as a minimum signal strength. It should be more than enough. Note that the frequency used in this example is close to the center of the FM broadcast band.

■ BEAMWIDTH

Another important quantity for applying any antenna is the beamwidth of that antenna. Just as we had a directive gain, where more of the gain of the antenna was concentrated in a preferred direction, the beamwidth is a measure of the area toward which the gain is concentrated. In general, the greater the size of the antenna, the narrower the beamwidth. A good example of this effect are the headlights of your automobile. Behind each lamp is a reflector that concentrates the light beam into a specific direction and area. That area is called the beamwidth of the headlight, and the location of maximum gain is called the directive gain or directivity of the antenna.

The beamwidth of an antenna is always specified as the number of degrees to either side of the location where the gain falls off by 3 dB or half power. Alternately, this value can be specified as twice the angle of deviation from the center of the beam where the signal strength drops by 3 dB. Beamwidth is sometimes specified in two numbers, the azimuth and elevation. For most purposes, these numbers can be approximated to a single radius value in a circular cross-section. Figure 8-5 illustrates this concept.

At most frequencies above about 10 GHz, beamwidths are usually less than 10 degrees. Antennas can be manufactured with beamwidths as low as 1 degree. Narrow beamwidths are desirable because interference from adjacent antennas as sources is minimized. On the other hand, the narrower the beamwidth of an antenna, the more precisely it must be mounted and the more stable that mounting must be. In general, an inverse relationship exists between the beamwidth and the gain of an antenna. As the gain and size of an antenna grow, so also does its beamwidth narrow.

Figure 8-5
Antenna Beamwidth

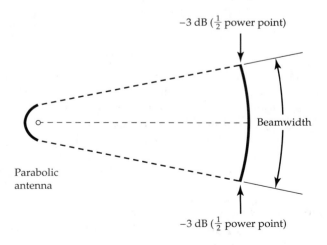

Beamwidth can be calculated for a parabolic antenna using Equation 8-8. Note that the number in the numerator is a constant appropriate for *parabolic antennas* only.

$$B_w = \frac{21 \times 10^9}{fD} \tag{8-8}$$

B_w = beamwidth in degrees
D = diameter of parabolic antenna

■ **EXAMPLE 8-5**   Find the beamwidth of a parabolic antenna that has a diameter of 3 meters and is designed to receive a wavelength of 10 cm.

First, convert 10 cm wavelength to a frequency:

$$c = \lambda f \rightarrow f = \frac{c}{\lambda} = \frac{3 \times 10^8}{0.10} = 3 \times 10^9 \text{ Hz}$$

Now apply Equation 8-8:

$$B_w = \frac{21 \times 10^9}{(3 \times 10^9)(3)} = 2.33°$$

■ FRONT-TO-BACK RATIO

The front-to-back ratio of an antenna describes the ratio of the gain in the forward direction to the gain in the backward direction. Sidelobes on the back of the antenna are generated in typical radiation patterns for directional antennas. A typical value is in the range of 20 to 30 dB. This parameter is critical when selecting an antenna for a repeater application because, in such an application, the same frequencies are used in both directions. If there isn't adequate isolation between the two antennas, which are placed back to back in such an application, the signal can feed back on itself and oscillate between the two antennas. This oscillation can be avoided with proper physical design of the mounting and careful selection of the antennas. Figure 8-6(a) shows a typical installation that is sensitive to the front-to-back ratio of the antennas. Note how the antennas are placed very close to each other. Figure 8-6(b) is a line drawing illustrating the oscillation that this physical arrangement can engender. Figure 8-6(c) shows the desired situation: no oscillation between the two antennas being used as a repeater.

■ CHANNEL ATTENUATION

In Chapter 1, the concept of power radiation loss was introduced, as was an equation to determine that loss. In Equation 1-14, there was no wavelength dependence but rather a simple inverse square law dependence. In actual practice, there is a dependency on wavelength due to the directive gain of the antenna. This concept was introduced in Equation 8-3.

Figure 8-6
Front-to-Back Ratio: (a) Installation Sensitive to Front-to-Back Ratio; (b) No Oscillation; (c) Resulting Oscillation
(*Digital Imagery* © 2001, *PhotoDisk, Inc.*)

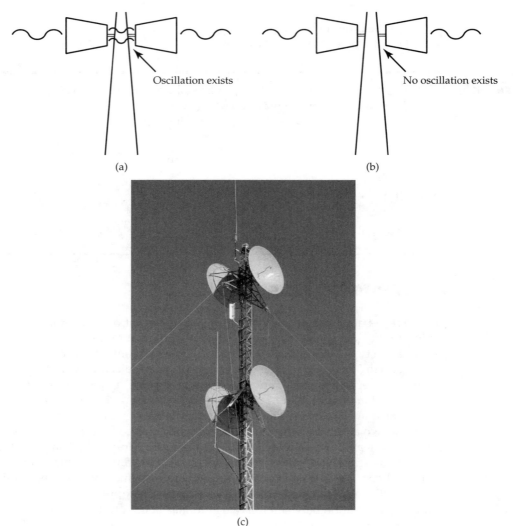

To describe this dependency better, we will define a new variable called channel at-tenuation. Channel attenuation is defined as the attenuation in decibels that a signal will experience as it propagates through the atmosphere. It captures the wavelength de-pendence and distance that the signal propagates into a single term. Note the inversion of how these terms appear in Equations 8-3 and 8-4. Channel attenuation should dimin-ish or subtract from the power that is transmitted. This property will prove helpful when writing link equations.

$$A_c = \left(\frac{4\pi r}{\lambda}\right)^2 \tag{8-9}$$

$$or \qquad A_c(\text{dB}) = 10 \log\left(\frac{4\pi r}{\lambda}\right)^2$$

A_c = channel attenuation

Equation 8-9 leads us to the following important relationships. Equation 8-10 is Equation 8-6 written in logarithmic form. Equations 8-11 and 8-12 both express received power in logarithmic form.

$$P_{rx} = \frac{P_{tx}G_{d,tx}G_{d,rx}\lambda^2}{16\pi^2 r^2} = (P_{tx})(G_{d,tx})(G_{d,rx})\left(\frac{\lambda}{4\pi r}\right)^2$$

$$P_{rx}(\text{dB}) = 10\log(P_{tx}) + 10\log(G_{d,tx}) + 10\log(G_{d,rx}) - 10\log(A_c) \tag{8-10}$$

$$P_{rx}(\text{dB}) = P_{tx}(\text{dB}) + G_{d,tx}(\text{dB}) + G_{d,rx}(\text{dB}) - A_c(\text{dB}) \tag{8-11}$$

$$P_{rx}(\text{dB}) = \text{ERP}(\text{dB}) + 10\log(G_{d,rx}) - 10\log(A_c) \tag{8-12}$$

Equations 8-11 and 8-12 are important. They will be used later when we discuss link calculations and how to connect them to performance criteria. The next several examples will illustrate the concepts introduced in these equations.

■ **EXAMPLE 8-6** A transmitting station has a transmitter capable of applying 100 W to an antenna with a gain of 30 dB. Find the ERP in decibel form.

$$\text{ERP}(\text{dB}) = 10\log(100) + 30\text{ dB} = 50\text{ dB}$$

■ **EXAMPLE 8-7** Find the attenuation, in decibel form, of a channel for a satellite-to-earth-based transmission. Take the frequency to be 4 GHz and the distance to be 40 million meters. (This distance is typical for a television satellite-to-earth transmission link.)

$$A_c(\text{dB}) = 10\log\left(\frac{4\pi(4\times10^7)}{\dfrac{3\times10^8}{4\times10^9}}\right)^2 = 196.5\text{ dB}$$

■ **EXAMPLE 8-8** Find the received power for the channel attenuation calculated in Example 8-7 and the ERP calculated in Example 8-6. Assume that the directive gain of the receiving antenna is 20 dB.

$$P_{rx}(\text{dB}) = \text{ERP} + 10\log(G_{d,rx}) - 10\log(A_c)$$

$$P_{rx}(\text{dB}) = 50\text{ dB} + 20\text{ dB} - 196.5\text{ dB} = -126.5\text{ dB}$$

■ SOME COMMON ANTENNA TYPES

The next portion of this discussion is an examination and classification of several common antennas. We will begin with a review of the fundamentals of a classic half-wave dipole. Following that, we will explore a close relative of the half-wave antenna: the folded dipole.

Half-Wave Dipole

The half-wave antenna is also known as the Hertz antenna. Any antenna with a length equal to half the wavelength of operation, as shown in Equation 8-13, is a Hertz antenna. These antennas are used for frequencies above 3 MHz or in any band from the HF band upward. The half-wave dipole antenna is constructed of two conductors arranged end-to-end and tied to some kind of metallic conductor, for example, coaxial or twin lead wire. Figure 8-7(a) shows a traditional half-wave dipole antenna; part (b) of the figure is a line drawing showing how the wave length of such an antenna is measured. The length shown in the figure defines the wavelengths or frequencies that the antenna is designed to radiate or receive. To calculate this value, set the length equal to half the wavelength, as shown in Equation 8-13. Of course, one can also relate the length to frequency by recalling $\lambda = \frac{f}{c}$.

$$l = \frac{\lambda}{2} = \frac{f}{2c} \tag{8-13}$$

These antennas are characterized by effective operation at only a narrow range of frequencies around the design choices. This choice is determined by applying Equation 8-13, as demonstrated in Example 8-9.

■ EXAMPLE 8-9   Find the optimum wavelength and frequency for a half-wave dipole that is 10 meters long.

$$10 = \frac{\lambda}{2} \Rightarrow \lambda = (2)(10) = 20 \text{ m}$$

$$l = \frac{f}{2c} \Rightarrow f = 2lc = (2)(10)(3 \times 10^8) = 6 \times 10^9 = 6 \text{ GHz}$$

Figure 8-7
(a) Traditional Half-Wave Dipole Antenna; (b) Measurement of its wavelength (*Reprinted by permission of Pearson Education, Inc., Upper Saddle River, NJ.*)

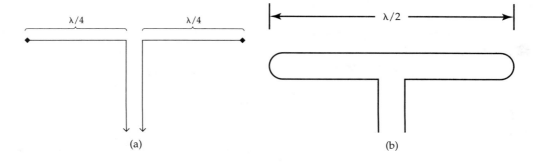

(a)   (b)

Antenna impedance is determined by the geometry of its construction. For the half-wave dipole antenna that is traditionally constructed, the impedance is always about 73 Ω.

Folded Dipole

The folded dipole antenna is a very simple antenna design based on the half-wave dipole. Essentially, it consists of two conductors, instead of a single conductor, connected together at each end. One of the conductors is split in the middle and connected to the antenna leads attached to the receiver. To picture it, recall the free antenna received with your last clock radio or receiver. In all likelihood, it was a T-shaped piece of 300 Ω twin lead wire. This antenna is a folded dipole. The top of the T is the horizontal segment and the two conductors are connected at each end. Since the active element is a half-wave dipole, this type of antenna should really be referred to as a half-wave folded dipole antenna.

One of the conductors in the folded dipole is broken in the middle and the vertical part leads down to attach to the television or radio. All simple folded dipole antennas have a characteristic impedance of 73 Ω, so let's examine the match such an antenna offers. The quality of a match is determined by the ratio of the two impedances, called the standing wave ratio (SWR). For close impedances, a very good match results. The SWR calculation for such a system is shown in Equation 8-14:

$$\text{SWR} = \frac{Z_{\text{larger}}}{Z_{\text{smaller}}} = \frac{300}{73} = 4.1 \qquad \textbf{(8-14)}$$

As you can see from this calculation, the match is terrible! The closer the characteristic impedance of the antenna is to the impedance of the cable that connects the antenna to the transmitter or receiver, the better the performance. An ideal SWR is 1.0. As the value of SWR increases above this value, performance worsens. One might imagine that a matching transformer would be needed. However, due to the special way such an antenna is manufactured, as shown in Figure 8-8(b), the actual impedance of the antenna becomes about 292 Ω. This construction allows a direct connection without any matching circuitry required, a real cost savings!

Figure 8-8
(a) Half-Wave Folded Dipole Antenna; (b) Measurement of Its Wavelength (*Reprinted by permission of Pearson Education, Inc., Upper Saddle River, NJ.*)

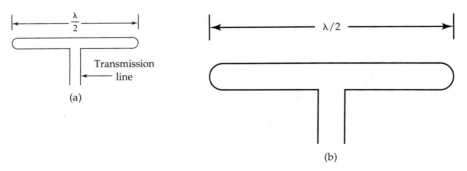

The reason that this type of design yields such a high impedance is the construction techniques used. Because the two wires that make up the loop are so close together, the relative phase of the currents flowing in them are the same. This condition results in an impedance that is about four times the expected impedance of a half-wave dipole. It also features wide bandwidth operation, unlike the traditional dipole, which operates well only at a particular design frequency. The impedance that results is:

$$Z = 4Z = (4)(73) = 292 \ \Omega$$

This equation can be applied only to the special situation where the two conductors that make up the antenna are very close together, which is the case with the antennas commonly shipped with clock radios and inexpensive receivers. If you recalculate the SWR using Equation 8-14, a very good match now results:

$$\text{SWR} = \frac{Z_{\text{larger}}}{Z_{\text{smaller}}} = \frac{300}{292} = 1.03$$

Folded dipole antenna elements are used in many other antenna designs. The popular Yagi and log-periodic both use half-wave folded dipoles. Since these designs are used almost exclusively for terrestrial reception of broadcast television and radio signals, they will not be explored in depth. Figure 8-8(a) shows a half-wave folded dipole antenna. Figure 8-8(b) shows how the wavelength of such an antenna configuration is measured. As mentioned above, each element in many antenna types consist of half-wave folded dipoles. A comparison of Figures 8-7(b) and 8-8(b) is instructive. The physical size of these antennas is the same, but because of the construction of the folded dipole antenna, its impedance is four times the traditional dipole or Hertz antenna.

■ **EXAMPLE 8-10**   Calculate the optimum wavelength for a half-wave folded dipole like the one in Figure 8-8. Assume a length of 1.5 meters. Find the optimum frequency.

$$l = \frac{\lambda}{2} = \frac{f}{2c} \Rightarrow 1.5 = \frac{f}{2c} \Rightarrow f = (1.5)(2)(3 \times 10^8) = 9.0 \times 10^8 = 900 \ \text{MHz}$$

Quarter-Wave Antenna

The quarter-wave antenna is commonly substituted for the half-wave antenna. It is often mounted vertically and receives or transmits effectively the same bands as the half-wave antenna, namely, LF and MF. It works by using the ground plane as at least one additional quarter wavelength of the desired frequency. Its physical length is one-quarter the wavelength with which it is designed to operate. Its radiation resistance is about half the half-wave antenna, or about 36 ohms. This antenna is the type used in automobiles for AM, FM, and CB as well as many cellular radio antennas. Example 8-11 shows how tall a quarter-wave antenna would be for two common cellular frequency bands, 900 and 1900 MHz.

■ **EXAMPLE 8-11**   Find the length of a cellular antenna using a quarter wavelength design at the frequency bands of 900 and 1900 MHz.

$$900 \text{ MHz: } \lambda = \frac{c}{f} = \frac{3 \times 10^8}{900 \times 10^6} = 0.34 \text{ m} \rightarrow \frac{\lambda}{4} = 0.083 \text{ m} = 3.45''$$

$$1900 \text{ MHz: } \lambda = \frac{c}{f} = \frac{3 \times 10^8}{1900 \times 10^6} = 0.16 \text{ m} \rightarrow \frac{\lambda}{4} = 0.04 \text{ m} = 1.67''$$

While each value is the total length of the antenna, it is not the actual height of the antenna as implemented. If you divide the above values by two (since each antenna is folded), you will arrive at the actual physical height of the antennas for this application, approximately $1\frac{3}{4}''$ for the 900 MHz system and approximately $\frac{3}{4}''$ for the 1900 MHz system.

When an antenna is mounted on an automobile, the metal surface of the car acts as the ground plane, sometimes extending into the surface of the road. Note that the little metal ball on the top of the antenna is a safety device. If the tip of the antenna were not rounded off and was instead a sharp point, the electrical field concentration would be increased dramatically.

Since an antenna oscillates or resonates, the electrical field becomes very concentrated every half cycle of the oscillation. Given the right atmospheric conditions (those that enhance static electric charge), the tip of the antenna could actually discharge a spark, just like you experience with static electricity. This discharge, in and of itself, is not a danger because there is virtually no current flow, but around gas fumes, the spark could ignite the gas and an explosion would occur. Can you imagine if it was considered dangerous to turn on the radio when working on your car in the garage or when pulling into a gas station? Maybe when electric cars appear, the antenna design will change!

Helical Antenna

Everyone today is familiar with the helical antenna used for cellular telephones mounted in a car. These short, helically wound antennas are effective when used in the VHF and UHF bands. The distinguishing characteristic of this type of antenna is that it radiates or receives symmetrically throughout the arc of a circle. Therefore, it is especially effective when the polarization of the transmitted wave may be altered by reflections, such as in a city environment.

A helical antenna is constructed by taking a short piece of wire and loosely winding it three to six times so that the number of turns times their diameter approximately equals the total length of the wire. The circumference of the turns should approximately equal the wavelength of the frequency selected.

When used as a transmitting antenna, the helix emits electromagnetic waves in two polarizations: circularly (as described above) and axially (that is, out one end). The axial

Figure 8-9
(a) Helical Antenna; (b) Helical Radiation Pattern (*Reprinted by permission of Pearson Education, Inc., Upper Saddle River, NJ.*); (c) Side View; (d) Radiation Pattern as Seen from Top View

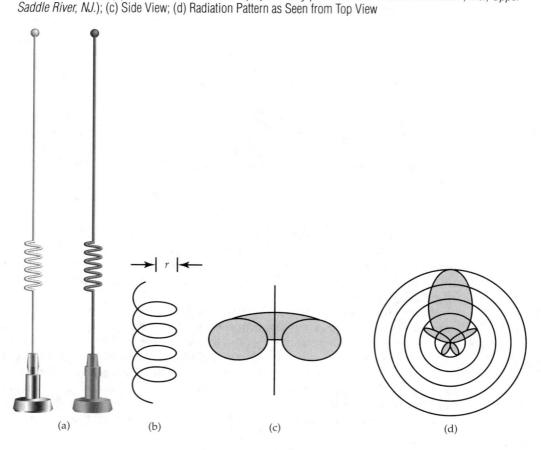

(a)   (b)   (c)   (d)

polarization is fairly broadband and directional. The radiation pattern for this type of antenna is shown in Figure 8-9. Note that the last view is the field radiating out from the top of the helix.

As already discussed, any antenna also receives the same way it radiates. Here, the helical antenna pulls signals in primarily from the vertical direction. This construction makes sense in designing an antenna for use in a city environment with lots of blocking buildings. The only LOS path available is often up.

■ **EXAMPLE 8-12**   Design a helical antenna for use at 1 GHz. Assume five turns for the wire. First, determine the wavelength for that frequency.

$$\lambda = \frac{c}{f} = \frac{3 \times 10^8}{1 \times 10^9} = 0.3 \text{ m}$$

Next, apply the criterion that the circumference of the turns should equal the wavelength. Since the circumference of a circle is given by $2\pi r$, where r is the radius:

$$r = \frac{0.3 \text{ m}}{2\pi} = 4.8 \text{ cm}$$

Finally, apply the criterion that the number of turns times their diameter equals the length of wire. Since the diameter of a circle is just twice its radius and five turns of wire are necessary, the length of the wire is:

$$(\text{Number of turns})(2)(\text{radius}) = (5)(2)(4.8) = 4.8 \text{ cm}$$

Parabolic Antenna

Parabolic antennas are used in many fields, as discussed below, but they are almost universally used for short-haul terrestrial microwave links. Parabolic antennas are also commonly seen as satellite television receiving antennas. These antennas come in at least two sizes now, each designed for a different frequency range. The larger ones are approximately 1.5 meters in diameter and are designed for use in the 4 GHz to 6 GHz range, the low end of the SHF band. The more recently introduced parabolic antennas are designed for use in the 14 GHz to 16 GHz band and are significantly smaller, approximately 18 inches in diameter. These types of antennas are designed for use in the upper UHF band and throughout the SHF band. Certain specialized applications, namely, radio astronomy, extend the use of these antennas down into the VHF band as well. The largest such antenna installation is in Arecibo, Puerto Rico, and is 305 meters in diameter!

Parabolic antennas (or "dish" antennas, as they are sometimes called) rely on the special shape of the parabola (hence the name *parabolic*) to reflect energy absorbed back to a focus that lies above the surface of the parabola. Recall that a parabola has two foci. As the electromagnetic waves reach the surface of the antenna, they are reflected back to a point somewhere above the surface and centered. This point is where the energy is gathered if the antenna is used as a receiver and is where the antenna is fed energy if it is used as a transmitter. In parabolic antennas, geometry rules the day. The better the surface follows the arc of a parabola, the more effective the antenna will be in gathering or transmitting power.

Since the shape of the parabolic antenna, and not the surface, per se, is used to focus the electromagnetic energy, the parabolic antenna can be made of a mesh that will operate equally well, as long as the shape is preserved. The requirement for rigidity in shape often means that these antennas are made out of a solid material, although designs with rigid meshes are in use. The rule of thumb for choosing any mesh is that it is as good as a solid if the hole size in the mesh is less than a quarter of the wavelength of the frequency being used. Figure 8-10(a) shows a parabolic antenna used in communications; Figure 8-10(b) is a line drawing of a parabolic antenna with a representative radiation pattern.

The gain of a parabolic antenna is given by the ratio of the diameter of the parabola defining the antenna and the wavelength of the signal. As the antenna gets larger, more

Figure 8-10
(a) Parabolic Antenna; (b) Radiation Pattern (*Reprinted by permission of Pearson Education, Inc., Upper Saddle River, NJ.*)

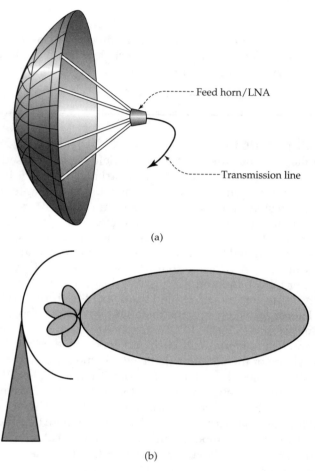

(a)

(b)

gain results for a given wavelength. Since the parabolic antenna has this relationship, the directive gain is given by the formula in Equation 8-15:

$$G_d = 6\left(\frac{D}{\lambda}\right)^2 \qquad \textbf{(8-15)}$$

The constant 6 in Equation 8-15 is a proportionality factor.

■ **EXAMPLE 8-13**   Determine the gain for a parabolic antenna that has a 3 m diameter and is designed to receive wavelengths of 10 cm. Make sure that both D and λ are in the same units (here, meters).

$$G_d = 6\left(\frac{D}{\lambda}\right)^2 = 6\left(\frac{3}{0.10}\right)^2 = 5400 \Rightarrow G_d(\text{dB}) = 10\log(5400) = 37.3 \text{ dB}$$

For the case of a parabolic reflector antenna, it is possible to find a relationship to define the directivity of the antenna, measured in degrees. Equation 8-16 is used to calculate the directivity:

$$d = \frac{70\lambda}{D} \qquad \qquad \textbf{(8-16)}$$

d = directivity

Directivity and beamwidth are closely related concepts; to see the close relationship between the two, review Equation 8-8, which was used to calculate beamwidth. Note that both relationships feature an inversely proportional relationship to the diameter of the antenna and a proportional relationship to the wavelength used (although in Equation 8-8, they are expressed in terms of the frequency used). Therefore, both beamwidth and directivity decrease as the diameter of the parabolic antenna increases. Also, both decrease as the frequency increases.

■ **EXAMPLE 8-14**   Find the directivity of a parabolic reflector antenna 3 meters in diameter and operating at a frequency of 4 GHz.

$$c = \lambda f \rightarrow \lambda = \frac{c}{f} = \frac{3 \times 10^8}{4 \times 10^9} = 7.5 \times 10^{-2}$$

$$d = \frac{70\lambda}{D} = \frac{(70)(7.5 \times 10^{-2})}{3} = 1.75 \text{ degrees}$$

Horn Reflector Antennas

Horn reflector antennas are used for almost all long-haul terrestrial links. These antennas feature high gains and narrow beamwidths with relatively wide bandwidths. Table 8-1 illustrates typical values.

Table 8-1
Typical Horn Reflector Antenna Characteristics

Gain (dB)	40–50
Beamwidth (degrees)	1–2
Bandwidth (MHz)	20–40

Figure 8-11
(a) Horn Antennas; (b) Critical Dimensions

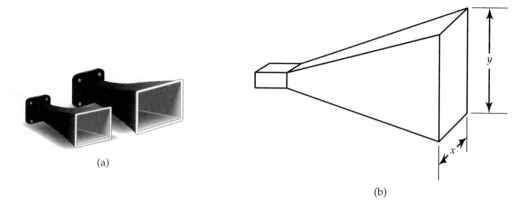

(a)

(b)

Horn antennas are widely used to send and receive data in the microwave frequency bands. (Refer back to Figure 8-6 to see how antennas are used for this purpose.) The shape of the horn helps focus the energy, like any antenna. Horn antennas are available in various shapes and are generally intended to be connected directly to a waveguide. The shape of the horn determines its primary use. For example, optimizing for gain results in one shape, optimizing for a particular radiation pattern, another. Sometimes the need is to match optimally the impedance of the waveguide feeding the antenna. In this case, the shape would be designed with this characteristic in mind.

Because of the great dependency on shape, any detailed analysis of horn antennas is outside the scope of this text. However, it is possible to obtain a few general relationships. Figure 8-11(a) shows two rectangular horn antennas, and Figure 8-11(b) is a line drawing of a rectangular horn antenna showing the critical dimensions. For rectangular horns such as the one shown in Figure 8-11(b), it is possible to write a proportionality equation for the gain. The gain for such an antenna is proportional to the size of the opening and indirectly proportional to the square of the wavelength used. These relationships are shown in Equation 8-17:

$$G \propto \frac{xy}{\pi\lambda^2}$$ **(8-17)**

At microwave frequencies, where horn antennas are used, as the size of the antenna decreases, the gain decreases. Because the gain is inversely proportional to the wavelength squared, the gain of the antenna of any given size also increases as the frequency band used by the antenna increases. This relationship explains why these antennas are so well suited to very high frequency applications, which include satellite and terrestrial microwave-based systems.

■ SIMPLIFIED LINK ANALYSIS

Now that central concepts and equations have been developed, it is possible to perform a simplified link analysis for a two parabolic antenna system, like one used in a link between a terrestrial transmitting location and a satellite in orbit. This situation occurs every day when the network studios in New York uplink the evening news. This satellite, in turn, will relay this signal to other terrestrial locations such as local television stations, which will then transmit them to local viewers. We will assume the channel characteristics are completely defined by the channel attenuation formula defined in Equation 8-9 and focus only on the antenna issues discussed in this chapter. (The results of this section and a similar section in Chapter 9 will be combined in Chapter 10 to illustrate many important aspects of a link analysis.)

Several effects are not considered here, however; primarily, we will assume that the antenna is precisely aligned. This precision is rarely the case so some loss will occur. Also, an antenna is not composed of a single component; there are subsystems. Each of these subsystems will also add attenuation to the signal. The most common for a parabolic antenna is referred to as the feeder losses. The feeder assembly is a series of waveguides, filters, and couplers that connect the amplifier to the antenna dish itself. We have chosen to lump these factors into antenna directivity and ignore them.

Also, atmospheric losses are not considered. Atmospheric losses are of two types: fixed and time dependent. The fixed are a consequence of the composition of the atmosphere; the time dependent depend on the weather. Specifically, rain and snow/ice fades are common. Finally, the sun itself can be a large contributor to noise if the receiving antenna is pointed in the direction of the sun. This position is to be avoided because the sun effect is large and will swamp the receiving signal in most cases. Adding all these factors and excluding the sun effect, most applications will experience about a 1 to 2 dB variable performance penalty from them. The solution is easy, if not inexpensive: add more overhead in your design or learn more about these effects and design around them. In Chapter 9, these effects will be discussed more fully.

■ **EXAMPLE 8-15** You have an orbital satellite transmission from earth to the satellite. Take the frequency to be 6 GHz and the distance to be 40 million meters. Assume that the transmit power applied to the antenna is 200 W and the parabolic transmitting antenna is 3 meters. For the satellite receiving antenna, assume a 1 meter parabolic antenna. First, find the directive gain of the terrestrially located transmitting antenna.

$$G_d(\text{dB}) = 10 \log\left(6\left(\frac{D}{\lambda}\right)^2 \right) = 20 \log\left(\frac{6D}{\lambda}\right) = 20 \log\left(\frac{6(3)}{\dfrac{3 \times 10^8}{6 \times 10^9}} \right) = 20 \log(360) = 51 \text{ dB}$$

Next find the ERP. Convert the power to dB and add it to the directive gain of the antenna.

$$P_{tx} = 10 \log\left(\frac{200}{1}\right) = 23 \text{ dB}$$

$$\text{ERP(dB)} = P_{rx}(\text{dB}) + G_d(\text{dB}) = 23 + 51 = 74 \text{ dB}$$

Then find the channel loss in decibels and subtract this power from the result. This calculation will yield the power arriving at the space based receiving antenna.

$$A_c = 10 \log\left(\frac{4\pi r}{\lambda}\right)^2 = 10 \log\left(\frac{4\pi(4 \times 10^7)}{\dfrac{3 \times 10^8}{6 \times 10^9}}\right)^2 = 10 \log(1.01 \times 10^{20}) = 200 \text{ dB}$$

$$P_{rxA} = 74 \text{ dB} - 200 \text{ dB} = -126 \text{ dB}$$

Finally, to find the power delivered to the satellite receiver stage, calculate the directive gain of the receiving antenna and add it to the power determined in the last step.

$$G_d(\text{dB}) = 20 \log\left(\frac{6D}{\lambda}\right) = 20 \log\left(\frac{6(1)}{\dfrac{3 \times 10^8}{6 \times 10^9}}\right) = 20 \log(120) = 41.6 \text{ dB}$$

$$P_{rx}\,(\text{dB}) = -126 \text{ dB} + 41.6 \text{ dB} = -84.4 \text{ dB}$$

$$P_{rx}(\text{W}) = 10^{\frac{-84.4}{10}} \text{ W} = 3.63 \times 10^{-9} \text{ W} = 3.63 \text{ nW}$$

Example 8-15 is a simplified link analysis for a typical commercial satellite application with a focus on the antenna portion. In Chapter 9, a more detailed calculation for a terrestrial LOS link will be presented, and in Chapter 10, a comprehensive link calculation for a satellite will be covered.

■ SUMMARY

This chapter introduced the basic concepts and designs of antennas in wide use today for wireless communications. Several basic concepts were introduced, common antenna types were described, and a simple link analysis was performed. The ideas in the link analysis will be used to tie together common aspects of the topic in this chapter and Chapters 9 and 10.

REVIEW QUESTIONS

1. Find the antenna efficiency for a half-wave dipole. Assume that the antenna resistance is 1 Ω.

2. Find the antenna efficiency for a quarter-wave dipole. Assume that the antenna resistance is 1 Ω.

3. Find the ERP of an antenna with a directive gain of 10 in the direction of interest and a radiated power of 1 W. Assume a half-wave dipole antenna. (You will need this information for problem 5.)

4. For problem 3, find the power density emitted by the transmitting antenna.

5. For problem 3, find the voltage coupled to the receiver if the receiving antenna has a directive gain of 3 and is 10 km from the transmitting antenna. Assume that the antennas are aligned so that the directive gain numbers are correct and that the wavelength used is 50 MHz.

6. How large would a half-wave dipole antenna be if the desired resonance frequency is 100 MHz?

7. How large would a quarter-wave dipole antenna be if the desired resonance frequency is 1 MHz?

8. How large would a half-wave folded dipole antenna be if the desired resonance frequency is 10 MHz?

9. Assuming five turns of wire, design a helical antenna for resonance at 2 GHz. Specifically, what is the total length of the wire used? Follow the rules discussed in the chapter.

10. What is the gain for a parabolic antenna 6 m in diameter and designed to receive wavelengths of 10 cm? Compare the results with Example 8-13 in the chapter.

11. Compare the gains of two horn antennas designed to operate at 10 GHz. The first is 1 cm wide and 10 cm tall; the second is 10 cm wide and 10 cm tall. Repeat at 1 GHz. Contrast your results.

12. Perform a simplified link analysis for a satellite-to-satellite transmission. Follow Example 8-15. Assume that both antennas have a diameter of 1 meter and the distance is 10 million meters. The transmit power is 100 W. Perform the calculation at carrier frequencies of 1 GHz and 10 GHz. Contrast your results.

13. Using the results from problem 12, determine the minimum transmit power at each carrier frequency for a receiving power of 1 pW, 1 nW.

9

Terrestrial Broadband Wireless

■ INTRODUCTION

After a short introduction to line of sight systems, this chapter introduces several aspects of line of sight links from the perspective of the performance in the channel. (Modulation, demodulation, and multiplexing techniques are not discussed here; refer to Chapters 2 and 5.) As pointed out in Chapter 8, this installment is the second of a link analysis theme across three chapters. In these three chapters, the main elements of a comprehensive link calculation are presented; however, many minor factors are not addressed. The chapter provides some background on a typical terrestrial link, followed by an introduction to the most promising microwave link band for future high bandwidth communications, LMDS. The rest of the chapter focuses on the details of link analysis.

This discussion of link analysis begins with a comprehensive introduction of the fundamental concepts of noise, which are important for understanding its use in channel analysis. Subsequent sections relate these concepts to the fundamental measurements of a channel and the performance in it: carrier-to-noise ratio (C/N), bit error rate (BER), and E_b/N_o curves. A detailed example illustrates the second installment of this introduction to link calculations.

Finally, several sections discuss modifications of the example, including degradations such as rain fades, and the use of Fresnel zones to calculate required antenna tower height. These considerations are applied to another example worked out in detail. Elements of this analysis are carried over to a satellite link calculation in Chapter 10, where appropriate.

Figure 9-1
Basic Line of Sight Link

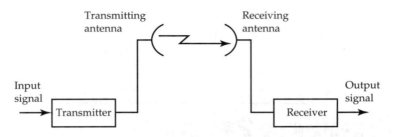

■ LINE OF SIGHT LINKS

Line of sight links are widely used by various organizations to pass broadband data from one point to another without the need for any physical link between them. Traditionally, a long-distance carrier would use these links to pass telephone trunk calls across areas where it was either economically undesirable or physically impossible to establish a land line, like no-right-of-way areas or swampland. Today, line of sight links are often used to place broadband links in cities where the cost to enter a building and run a line up several floors is prohibitive. Sometimes a reluctant landlord will not wish to have any new lines run inside the building because of the possibility of disturbing other leases.

Line of sight systems get their name from the fact that the transmission path must be line of sight (LOS) because they use signal frequencies above 30 MHz that propagate in this mode. The transmitter and receiver must have a clear path, with no obstructions, for the signal to travel. The most basic line of sight link has four components, shown in Figure 9-1.

Although the transmission in this figure is simplex, in fact, duplex systems are also common. In a duplex arrangement, the transmit and receive signals are horizontally and vertically polarized. In that case, the transmitter and receiver shown become transceivers and the antennas become dual function, one polarization for transmission in one direction, the other for transmission in the opposite direction.

The next major enhancement to the basic line of sight link is the addition of repeaters. Repeaters extend the distance over which the link may be used. The distance between repeaters varies, depending on several factors that will be explored in this chapter. As a rule of thumb, however, the maximum distance a single link may extend is between 35 and 40 miles. Typical links are rarely longer than 30 miles. This more general description of a line of sight link is illustrated in Figure 9-2. In this figure, the system is assumed to be full duplex and towers are used for antenna placement.

Almost all line of sight systems use one of two intermediate frequencies (IFs) in the modulator and demodulator: 70 MHz and 140 MHz. These IF frequencies are also widely used in the satellite industry because much of the technology is similar. The carrier frequency used in the channel is much higher in frequency, often in the high UHF or low SHF bands. Figure 9-3 shows a block diagram of a typical transceiver, like what

Figure 9-2
General Line of Sight Link

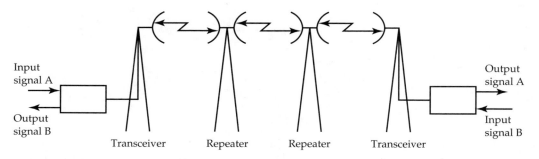

Figure 9-3
Block Diagram of Transceiver: (a) Transmitter; (b) Receiver

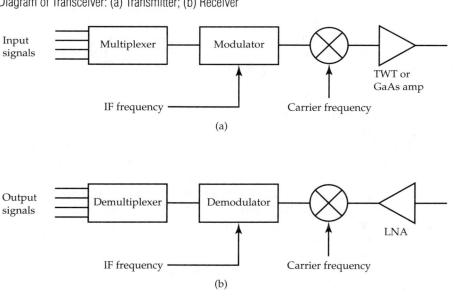

would be used in Figure 9-2. Figure 9-3(a) shows the transmitter half of a transceiver; Figure 9-3(b) shows the receiver half.

In part (a), the data rate that this type of system can support is usually much higher than any one information source; therefore, a multiplexer combines several information sources. This combined information flow is then modulated at the IF frequency used. The next step is to upconvert the IF signal to the RF signal used in the channel and then amplify this RF signal with either a traveling wave tube (TWT) or GaAs microwave amplifier.

In part (b) of the figure, which shows the receiver portion of the transceiver, the RF signal is again amplified by a low noise amplifier (LNA). This amplification and the

Table 9-1
FDM Multiplexing Hierarchy

Group Name	Number of Voice Channels
Basic	12
Super	60
Master	600

noise performance of the amplifier is critical because the incoming signal is often at a very low power. The RF signal is then down-converted to the same IF frequency used in the transmitter portion of the transceiver and the information signal is demultiplexed out appropriately.

All line of sight links are fundamentally analog communication links because the form of the signal is analog while in the channel. However, communications systems are generally referred to as analog or digital according to either the form of the information being carried or the method used to multiplex that data into a single signal for transmission. Analog systems usually mean analog information signals multiplexed in a frequency division multiplexing (FDM) manner, and digital systems use digital information signals and often multiplex them in a time division multiplexing (TDM) manner.

Analog systems use FM modulation with pre-emphasis and de-emphasis to improve performance. A standardized multiplexing hierarchy is used in the United States and in many locations throughout the world. These standardized multiplexing techniques are referred to as groups and are based on the number of 4 kHz voice channels modulated onto a single carrier. See Table 9-1.

Digital line of sight systems are available in a much wider range of bandwidths. Typical offerings include DS-1, DS-3, OC-1, OC-3, and 10 Mbps and 100 Mbps systems. The latter two provide an easy interface to 10BaseT and 100BaseT Ethernet LANs. Much higher data rates are possible. The emerging local multipoint distribution service (LMDS) system, due to its very large spectral allocation, is a good example. For this reason, LMDS will be discussed in the next section, before we cover the engineering of microwave link systems.

■ LMDS

Local multipoint distribution service (LMDS) has much promise for delivering broadband service over a long distance to a wide audience. It has the potential to provide a single broadband pipe for voice, video, and data in those areas where wireline solutions are not feasible. It is well suited to the types of bursty digital transmissions that characterize the nonsymmetric access characteristics of many consumer services, including Internet access, because of the very large bandwidth available. While it is engineered in a cellular fashion, which makes full use of the frequency reuse concept, it is not intended

Table 9-2
LMDS Bands

Carrier Frequency (GHz)	Lower Frequency (GHz)	Upper Frequency (GHz)	Maximum Bandwidth (MHz)
LMDS: 28	27.5	28.35	850
LMDS: 29	29.1	29.25	150
LMDS: 31	31.0	31.3	300

for mobile applications because no provision is made for handing off calls from one location to another, a critical element in any mobile system.

As indicated in Table 9-2, LMDS is composed of three noncontiguous blocks of spectrum totaling 1300 MHz in bandwidth. This spectrum was designated by the Federal Communications Commission (FCC), and licenses were auctioned in 1998 for a total of just under $600 million. In contrast to the third-generation (3G) cellular spectrum auctions, most of the spectrum was purchased by entrepreneurs and venture capital firms rather than the traditional operating companies. The purchase price was also much less. Many expect these groups to lease the spectrum back to competitive local exchange carriers (CLECs) and the mobile cellular companies. The winners of the bidding have ten years to offer substantial services, so many expect initial service to enter the market in the 2002–2005 time frame.

As its name suggests, LMDS is intended to be a local access facility. It uses low power transmissions that limit the range of any cell to approximately 5 km in diameter. LMDS is multipoint. Signals are transmitted in a point to multipoint, while the return path is point to point. The LMDS architecture is the only broadband terrestrial wireless system specifically designed with that function in mind. Essentially, a central hub is connected to several other networks with traditional wireline or satellite links. This site then is wireline-connected to several hubs or towers, each of which communicates simultaneously with numerous distribution locations. This concept is illustrated in Figure 9-4.

As shown in the figure, each multipoint site is equipped with a directional antenna that must have line of sight to the central hub or tower. The equipment located at the multipoint site or central hub is often referred to as the network node equipment (NNE). It provides the gateway function between the central hub information links and the LMDS wireless equipment. The transceiver subsystem of the NNE is similar to the transceiver in a cellular telephony base station. It performs the multiplexing/demultiplexing, modulation/demodulation, and other functions necessary for this conversion. Figure 9-5 illustrates a block diagram of this module.

At the subscriber site are three units: the antenna, outdoor unit (ODU), and indoor unit (IDU). The ODU will hold the radio frequency electronics, and services will be accessed and controlled by the IDU. The IDU will function like a gateway between the transceiver subsystem located in the ODU and the wireline links at the subscriber location. The IDU is likely to include a 10/100BaseT Ethernet connection for Internet

Figure 9-4
LMDS Configuration

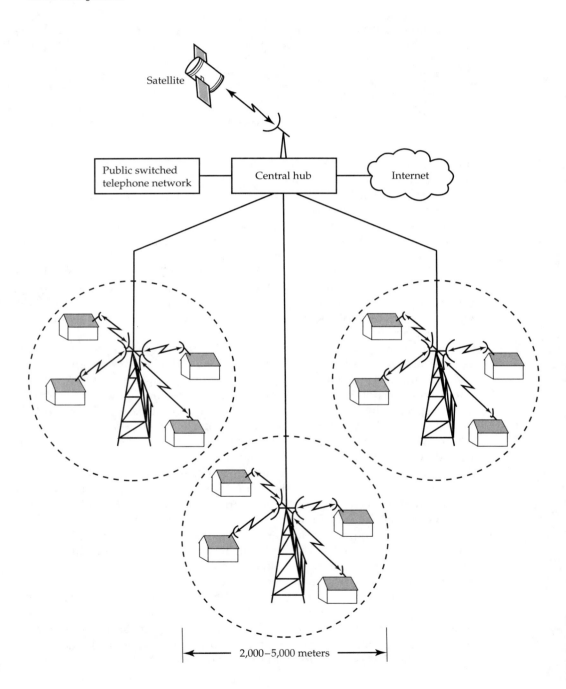

Figure 9-5
NNE Block Diagram

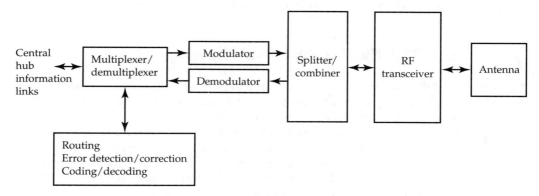

access. It may also include a private branch exchange (PBX) for circuit-switched voice service.

The indoor unit is expected to be available initially in two versions. The first will be a scalable unit intended for large business subscribers who must manage growth. It will consist of some kind of chassis with card slots that, when occupied, provide interfaces to subscriber equipment. As the subscribers' bandwidth and service needs evolve, the unit will evolve with them through the addition of cards. This scalable device will be fully manageable and configurable from a remote interface.

The indoor unit for small business and consumer applications will not be scalable. It will be a stand-alone piece of equipment with a fixed set of interfaces, probably variable when initially ordered. It will offer connectivity at up to DS-1 rates and will feature a 10/100BaseT Ethernet connection.

The network architecture is not a constraint when using LMDS. LMDS services are expected to include voice, video, and data. A wireless digital subscriber line (DSL) seems a natural fit. Both synchronous (such as the traditional T line access) and nonsynchronous (such as internet protocol [IP]) networks will exist. LMDS does not constrain the access methodology used on the link. Frequency division multiple access (FDMA), time division multiple access (TDMA), and frequency division multiplexed/time division multiple access (FDM/TDMA) schemes are all possible and all will probably exist, depending on the service offered. The modulation method is also not specified.

Because of the high carrier frequency used in this system, building blockage, terrain blockage, rain fades, and foliage fades will be real problems. EHF band signals of 28 GHz do not penetrate trees, and during heavy rain conditions connectivity will be lost. Experimental trials suggest that fades of 40 dB will not be uncommon. Recall that signals at this frequency use LOS propagation. Regions with hilly terrain or frequent heavy rains will not be good candidates for this technology.

The link engineering of these systems will be critical to their economic success. It may be possible to utilize low power repeaters to eliminate some of the foliage attenuation inside any particular cell. It is also possible to create overlapping cells so that each distribution point is served by several hubs. This setup will increase the problems associated with co-channel interference and will have to be optimized carefully for each situation. The rest of this chapter will focus on frequency planning and link engineering.

■ FUNDAMENTAL NOISE CONCEPTS

Basically, two types of noise are of concern: internal noise and external noise. Internal noise is electrical or thermal noise generated inside the components used to construct various communications subsystems. All components generate some internal noise, with active, or powered, components such as transistors and diodes generating far more than passive ones, like resistors and inductors. At the most fundamental level, this noise is generated by the motion of electrons that make up any material. At any temperature above absolute zero, there is motion of the electrons and hence thermal noise. External noise is contributed by the atmospheric path in which the signal travels while in the communications channel. There are many sources of external noise, including other communications systems.

Noise contributions to communication limit the ability to communicate. For example, think of the last time you were at a party or concert and the music was so loud that you had trouble speaking to your friend. In this situation, whatever you might think of the musical qualities of the performing artist, the music is a noise source relative to your conversation with a friend. The same idea applies to communications links: any external or internal energy that is not part of the desired signal is viewed as noise and degrades the link.

Reference Noise Power

To begin, we will define a reference noise power, N_o. It is used to standardize a thermal noise value for all components at a specific noise temperature. This temperature is taken to be 290 degrees Kelvin.

$$\text{Reference noise power} = N_o = kT_0B \qquad \textbf{(9-1)}$$

k = Boltzman's constant = 1.38×10^{-23}
T_0 = 290° K
B = bandwith of channel (Hz)

Generalizing this equation to find the thermal noise power at the noise temperature actually specified, we define a quantity called thermal noise power because, in any fixed bandwidth system, the only actual variable in Equation 9-1 is the temperature.

Thermal Noise Power and Noise Temperature

The thermal noise power in any component is given by the following relationship:

$$\text{Thermal noise power} = N = kT_NB \qquad \textbf{(9-2)}$$

T_N = noise temperature

In Chapter 1, when we introduced signal-to-noise ratios, noise was expressed as a power, and it can be seen here that in communications, noise power is talked about as temperature. This concept has a solid basis in physics. Recall that the hotter a gas is, the more rapid the molecules move. This increasing movement is indicative of more energy, and the energy comes from the additional heat.

In communications studies, amplifiers and antennas specifically will often have their noise contributions characterized by a noise temperature in Kelvin. Noise temperature is related to the temperature of the component but is generally not equal to it. Every component in the link will have some contributing noise that will be characterized by its noise temperature. You must add together all these noise contributions to find the total noise.

Equation 9-2 relates the noise power to the bandwidth of the channel because the wider the bandwidth, the more noise that can enter; just like a hose: the thicker the hose, the more water can flow for any given pressure. Temperature is included for the reasons just stated, and Boltzman's constant is a universal physical constant.

It is also possible to define an RMS noise voltage across some resistance by applying Ohm's law to Equation 9-2. The factor 4 in the relationship is due to the specific circuit configuration for which this concept is defined.

$$\text{Since} \quad P = \frac{E^2}{R} \quad \text{and} \quad N = kT_NB \rightarrow E_N = \sqrt{4RkT_NB} \quad \textbf{(9-3)}$$

■ **EXAMPLE 9-1** Find the noise temperature of an amplifier with a noise power of 0.1 picowatt in a bandwidth of 20 MHz.

$$N = 0.1 \times 10^{-12}\,\text{W}$$
$$B = 20 \times 10^6\,\text{Hz}$$
$$N = kT_NB \rightarrow T_N = \frac{N}{kB} = \frac{1 \times 10^{-13}}{(1.38 \times 10^{-23})(2 \times 10^7)} = 362.3 \text{ degrees Kelvin}$$

■ **EXAMPLE 9-2** Find the internal noise and output noise of an amplifier if the noise temperature is given as 150 degrees K, and the amplifier has a gain of 45 dB and operates in a bandwidth of 500 MHz. The output noise is just the internal noise multiplied by the gain of the amplifier.

$$N_i = kT_NB = (1.38 \times 10^{-23})(150)(500 \times 10^6) = 1.035 \times 10^{-12}\,\text{W}$$
$$G = 45\,\text{dB} = 10^{45/10} = 31,623$$
$$N_o = N_iG = (1.035 \times 10^{-12})(31,623) = 3.2 \times 10^{-8}\,\text{W}$$

Noise Figure

Noise figure references the internal noise generated by a component back to what a component noise contribution at a temperature of 290 degrees Kelvin would be, the already introduced reference noise. Noise figure is also defined as the signal-to-noise ratio at the input divided by the signal-to-noise ratio at the output. Note that noise figure is a unitless value, as shown in Equation 9-4:

$$\text{Noise figure} = F = \frac{N}{N_o} + 1 \qquad \textbf{(9-4)}$$

Another commonly used term is noise temperature. Noise temperature is related to noise figure by the relationship shown in Equation 9-5:

$$T_N = T_0 (F - 1) = 290(F - 1) \qquad \textbf{(9-5)}$$

T_N = noise temperature

Example 9-3 illustrates the calculation to obtain the reference noise and thermal noise power of a particular component.

■ **EXAMPLE 9-3**   Find the reference noise of a 1 MHz channel and then the thermal noise. Assume an actual noise temperature of 100 degrees Kelvin.

$$N_o = kT_0B = (1.38 \times 10^{-23})(290)(1 \times 10^6) = 4.0 \times 10^{-15}$$
$$N = kT_NB = (1.38 \times 10^{-23})(100)(1 \times 10^6) = 1.38 \times 10^{-15}$$

As you can see from the result in Example 9-3, if the temperature of the component is lowered, the thermal noise power will decrease. As you can imagine, better components and systems have lower thermal noise powers. This value is a key figure to examine when specifying these components for use in a communications system.

■ **EXAMPLE 9-4**   Calculate the noise figure for Example 9-3.

$$F = \frac{N}{N_o} + 1 = \frac{1.38 \times 10^{-15}}{4.0 \times 10^{-15}} + 1 = 1.345$$

■ **EXAMPLE 9-5**   Calculate the noise figure of an amplifier with an internal noise of 0.1 picowatt and a bandwidth of 30 MHz.

$$N_o = kT_0B = (1.38 \times 10^{-23})(290)(30 \times 10^6) = 1.2 \times 10^{-13}$$

$$F = \frac{N}{N_o} + 1 = \frac{0.1 \times 10^{-12}}{1.2 \times 10^{-13}} + 1 = 1.833$$

It is important to remember that any of these quantities are used for specifying internal noise contributions from discrete components within a subsystem or often for the subsystem as a whole. By adding noise contributions from each subsystem, you can obtain an estimate of the total noise contribution of the communications hardware. This topic is the subject of the next section.

Noise Calculations

Each subsystem adds some internal noise to the signal as it passes through that subsystem. For example, how would the noise temperature of two individual amplifiers contribute to the total noise when they are placed in series? The answer is that the total noise temperature is the sum of all the individual noise temperatures modified by the gain of each subsequent stage of amplification. Mathematically, the total noise temperature is given by the relationship shown in Equation 9-6. Note that G is in absolute value form, not decibel.

$$T_{N,total} = T_1 + \frac{T_1}{G_1} + \frac{T_3}{G_1 G_2} + \ldots + \frac{T_n}{G_1 G_2 \ldots G_{n-1}} \tag{9-6}$$

It is also possible to write this relationship in terms of the noise figures of each individual component, as is shown in Equation 9-7:

$$F_{total} = F_1 + \frac{F_2 - 1}{G_1} + \frac{F_3 - 1}{G_1 G_2} + \ldots + \frac{F_n - 1}{G_1 G_2 \ldots G_{n-1}} \tag{9-7}$$

While both expressions are useful, which you apply usually depends on how the internal noise is specified by the manufacturer of the subsystem. We will illustrate the application of Equation 9-6 using noise temperatures because it is usually the more common way of expressing this concept. Examples 9-6, 9-7, and 9-8 indicate three different amplifier arrangements. These examples illustrate an important rule in designing amplifier chains, a common problem in any link analysis. Refer to Figure 9-6 when reading these examples.

Figure 9-6
Examples 9-6, 9-7, and 9-8

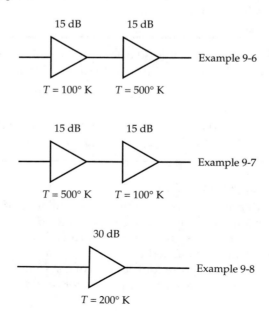

■ **EXAMPLE 9-6**   Given a two amplifier cascade with the following characteristics, find the total noise temperature.

>   Amplifier 1: $G = 15$ dB, $T = 100$ degrees Kelvin
>
>   Amplifier 2: $G = 15$ dB, $T = 500$ degrees Kelvin

First, express the gain in normal numeric form.

$$15 \text{ dB} = 10 \log(G_1) \rightarrow G_1 = 10^{\frac{15}{10}} = 31.6$$

Now substitute into Equation 9-6.

$$T_{N, \text{ total}} = T_1 + \frac{T_2}{G_1} = 100° \text{ K} + \frac{500° \text{ K}}{31.6} = 115.8° \text{ K}$$

■ **EXAMPLE 9-7**   For this example, reverse the two amplifiers. Is there any difference in the noise temperature?

$$T_{N, \text{ total}} = T_1 + \frac{T_2}{G_1} = 500° \text{ K} + \frac{100° \text{ K}}{31.6} = 503.2° \text{ K}$$

It is much better to place the lower noise temperature device first because the noise temperature of subsequent stages is reduced by the gain of the previous stage(s). For this reason, the first stage in any amplifier stage is the most critical and should be made as low-noise as is economically feasible.

■ **EXAMPLE 9-8**   What if a single 30 dB amplifier with a noise temperature of 200 degrees Kelvin were used instead of a pair of 15 dB amplifiers, as in Examples 9-6 and 9-7?

$$T_{N, \text{total}} = T_1 = 200° \text{ K}$$

It is always better to do multiple-stage amplification with a low-noise temperature amplifier as the first stage.

Carrier-to-Noise Ratio (*C/N*)

The carrier-to-noise ratio is a commonly used method of describing the amount of power that the transmitter is outputting compared to the amount of noise in the channel. It is one key measure used to relate the performance of communications systems. The carrier-to-noise ratio is calculated most easily by measuring the carrier power directly and using the fact that the noise contribution of a system is related to the noise temperature. This calculation is expressed in Equation 9-8, and the use of that equation is shown in Example 9-9.

$$\frac{C}{N} = \frac{C}{kT_N B} \tag{9-8}$$

■ **EXAMPLE 9-9**   Find the C/N in decibels for a system with a carrier power of 0.1 pW and operating at a noise power of 150 degrees Kelvin in a bandwidth of 10 MHz. Use Equation 9-8.

$$\frac{C}{N} = \frac{0.1 \times 10^{-12}}{kT_N B} = \frac{1 \times 10^{-13}}{(1.38 \times 10^{-23})(150)(10 \times 10^6)} = 4.83$$

$$\frac{C}{N}(\text{dB}) = 10 \log(4.83) = 6.8 \text{ dB}$$

As you can see from Example 9-9, increasing the power will increase the carrier component and hence the C/N ratio. Note also that increasing the bandwidth decreases the C/N ratio. Therefore, for optimum values of C/N, high power in narrow bandwidths at as low a noise temperature as possible is desired.

Bit Energy to Noise Ratio (E_b/N_o)

In digital communications systems, the bit energy (E_b) is used instead of the carrier power. Once this value is found, then an important ratio called E_b/N_o can be found. This ratio allows calculations that characterize all different kinds of digital modulation techniques by relating the energy per bit (E_b) transmitted to the noise power density. With this ratio, curves can be drawn for performance for each type of modem without worrying how fast the modem is, how powerful the transmitter is, or how noisy the channel is. E_b/N_o calculations extend the C/N ratio so that it is independent of bandwidth. Imagine that you are asked to recommend which type of modulation technique is best applied to a particular problem. It would be convenient to talk about different modulation techniques without worrying about how big the transmitter is or how fast the input data rate is. Use of the E_b/N_o ratio makes these comparisons straightforward.

The energy per bit transmitted is given by the product of the carrier power times the bit duration. This calculation can be written as shown in Equation 9-9:

$$E_b = C t_b \tag{9-9}$$

E_b = energy per bit (J)
C = carrier power (W)
t_b = bit duration (s)

It should make sense that the energy per bit transmitted is just the carrier power, which is what is modulated, multiplied by the amount of time it is modulated. This equation is just like saying that the power used by a system is the amount of power multiplied by the time it is applied.

The E_b/N_o ratio is found by just dividing the above expression by the noise power density, N_o. Note that *the noise power density is not the same as the reference noise power developed earlier.* The reference noise power subscript is a zero; the reference noise power subscript is an o, as shown in Equation 9-10.

$$\frac{E_b}{N_o} = \frac{C}{N_o} t_b \tag{9-10}$$

N_o = noise power density

To make this distinction clear, it is important to relate this quantity to the noise power, as shown in Equation 9-11. Since the noise power density of any system is measured in the same bandwidth as the receiver, the noise power of the system can be related to the noise power density exhibited by the receiver in the bandwidth of that receiver.

$$N_o = \frac{N}{B} \tag{9-11}$$

Note that when the bandwidth is 1 Hz, $N = N_o$. Hence, by using the concept of noise power density, a bandwidth-independent measurement is found. N_o, the noise power density, effectively relates the noise power in any bandwidth to that noise power found in the reference bandwidth of 1 Hz. So the noise power density can be found by dividing the noise power measured by the bandwidth of the measurement. It is critically important always to measure the noise power in some bandwidth, and that bandwidth should be the bandwidth of the receiver.

To relate the noise power density to the noise temperature analytically, substitute for N, as shown in Equation 9-12:

$$N_o = \frac{N}{B} = \frac{kT_N B}{B} = kT_N \tag{9-12}$$

The next step, shown in Equation 9-13, is to relate C/N to C/N_o:

$$\frac{C}{N} = \frac{C}{kT_N B} = \frac{C}{N_o B} \tag{9-13}$$

Finally, relate C/N to E_b/N_o and examine what Equation 9-14 reveals:

$$\frac{E_b}{N_o} = \frac{C}{N_o} t_b = \frac{CB}{N} t_b = \frac{C}{N}(Bt_b) = \left(\frac{C}{N}\right)\left(\frac{B}{f_b}\right) \tag{9-14}$$

Because the bit rate period and bit rate are the inverse of each other, they are used in Equation 9-14. If the bit rate and the bandwidth of the channel, or the noise bandwidth, are numerically equal to each other, then the two quantities cancel each other. Examples 9-10 and 9-11 illustrate the application of these equations.

■ **EXAMPLE 9-10**   Find the E_b/N_o ratio in decibels for a system operating at 2 Mbps in a bandwidth of 1 MHz. The unmodulated carrier power is 0.1 pW, and the system noise temperature is 120 degrees Kelvin.

$$\frac{E_b}{N_o} = \left(\frac{C}{N}\right)\left(\frac{B}{f_b}\right) = \left(\frac{0.1 \times 10^{-12}}{(1.38 \times 10^{-23})(120)(1 \times 10^6)}\right)\left(\frac{1 \times 10^6}{2 \times 10^6}\right) = 30.2$$

$$\frac{E_b}{N_o}(\text{dB}) = 10 \log(30.2) = 14.8 \text{ dB}$$

■ **EXAMPLE 9-11**   Change the bit rate to 1 Mbps and repeat the calculation of Example 9-10.

$$\frac{E_b}{N_o} = \left(\frac{C}{N}\right)\left(\frac{B}{f_b}\right) = \left(\frac{0.1 \times 10^{-12}}{(1.38 \times 10^{-23})(120)(1 \times 10^6)}\right)\left(\frac{1 \times 10^6}{1 \times 10^6}\right) = 60.4$$

$$\frac{E_b}{N_o}(\text{dB}) = 10 \log(60.4) = 17.8 \text{ dB}$$

Note in Example 9-11 that halving the bit rate resulted in an E_b/N_o ratio that was 3 dB better. Similarly, doubling the bandwidth and keeping the bit rate unchanged would yield the same result.

Note also that the difference between E_b/N_o and C/N at any given bit rate depends directly on the bandwidth of the receiver. Therefore, the difference between the two ratios will change if the bandwidth of the receiver is changed. In theory, the bandwidth of the channel and the bandwidth of the receiver should be identical. In actual systems, there is always some additional bandwidth in the receiver compared to what is actually used in the channel. This difference serves to increase the difference between the C/N ratio and the E_b/N_o ratio. This observation underlines why E_b/N_o is used. In actual systems, the bandwidth of the channel will vary from implementation to implementation. Not only will the actual channel change, but the input filters of the receiver may also vary. These devices determine the bandwidth from the receiver's point of view, which is important for performance measurements.

■ ANTENNA EFFECTS

We can modify Equation 9-8, the general expression for carrier-to-noise ratio, to include antenna and channel attenuation effects. In Figure 9-1, you saw several elements that contribute to a calculation of the carrier power actually received at the destination. The carrier power can be considered the power presented to the receiver.

The power presented to the receiver can be found from the transmitted power by adding any antenna gains and subtracting the channel attenuation. It is expressed in Equations 9-15, 9-16, and 9-17, first in algebraic form and then rewritten and simplified in decibel form. Refer to Chapter 8 for the source of these equations. ERP is defined as the sum of the transmitted power and the transmitting antenna's gain. Note that these equations use the terms and forms developed in Chapter 8 for parabolic antennas.

$$C = P_{tx}6\left(\frac{D}{\lambda}\right)^2_{tx}6\left(\frac{D}{\lambda}\right)^2_{rx}\left(\frac{4\pi r}{\lambda}\right)^2 \tag{9-15}$$

$$C \,(\text{dB}) = P_{tx} \,(\text{dB}) + G_{d,tx} \,(\text{dB}) + G_{d,rx} \,(\text{dB}) - A_c \,(\text{dB}) \tag{9-16}$$

$$C \,(\text{dB}) = \text{ERP} \,(\text{dB}) + G_{d,rx} \,(\text{dB}) - A_c \,(\text{dB}) \tag{9-17}$$

Note that Equation 9-17 yields the same result as that arrived at in Chapter 8 for power into the receiver. The noise power can be considered the thermal noise power. This observation yields the relationship expressed in Equation 9-18, where all values are taken in decibel form:

$$\frac{C}{N} \,(\text{dB}) = \text{ERP} \,(\text{dB}) + G_{d,rx} \,(\text{dB}) - A_C \,(\text{dB}) - N \,(\text{dB}) \tag{9-18}$$

Many times designers capture the effects of the antenna and system noise temperature, which is related to the thermal noise power, by defining a new term, called the G/T ratio or figure of merit. It is expressed in logarithmic form in Equation 9-19. This expression has a single term that includes the performance of the entire receiving antenna assembly (both the antenna itself and the performance of the first-stage amplifier), which will dominate the overall noise temperature of the system, as was demonstrated earlier in Examples 9-6, 9-7, and 9-8.

$$\frac{G_{d,rx}}{T_N} \,(\text{dB}) = G_{d,rx} \,(\text{dB}) - T_N \,(\text{dB}) \tag{9-19}$$

Note how similar in concept this equation is to the way C/N is written. With this definition, the expression for carrier-to-noise power, Equation 9-20, is:

$$\frac{C}{N} \,(\text{dB}) = \text{ERP} \,(\text{dB}) + G_{d,rx} \,(\text{dB}) - T_N \,(\text{dB}) - A_c \,(\text{dB}) - 10 \log \,(\text{k}B) \tag{9-20}$$

Now, what is really needed is the carrier-to-noise power relative to a noise power density in some uniform bandwidth. This equality will give us a measurement independent of bandwidth. See Equation 9-21:

$$\frac{C}{N} \,(\text{dB}) = \frac{C}{N_o B} \,(\text{dB}) = C \,(\text{dB}) - N_o \,(\text{dB}) - B \,(\text{dB}) \tag{9-21}$$

Inserting Equation 9-21 into Equation 9-20, we obtain Equation 9-22:

$$\frac{C}{N_o} \,(\text{dB}) = \text{ERP} \,(\text{dB}) + G_{d,rx} \,(\text{dB}) - T_N \,(\text{dB}) - A_c \,(\text{dB}) - 10 \log \,(\text{k}) \tag{9-22}$$

Note how the bandwidth term has cancelled out because a bandwidth-independent relationship is desired. Finally, it is traditional to insert a catch-all term, called losses, to account for any number of marginal effects that are outside the scope of this textbook. To address these issues a placeholder is inserted into the above relationship and Equation 9-22 becomes Equation 9-23.

$$\frac{C}{N_o} \,(\text{dB}) = \text{ERP} \,(\text{dB}) + G_{d,rx} \,(\text{dB}) - T_{N(dB)} - A_c - 10 \log \,(\text{k}) - 10 \log \,(\text{losses}) \tag{9-23}$$

Remember, the ERP term accounts for the transmitting antenna characteristics and the G/T term accounts for the receiving antenna characteristics, as well as the first-stage LNA amplification at the receiver. Finally, since Boltzman's constant never changes, it is possible to calculate the logarithmic value and insert it as a numerical constant:

$$10 \log (k) = 10 \log(1.38 \times 10^{-23}) = -228.6 \text{ dB}$$

■ **EXAMPLE 9-12** Find the C/N_o ratio and perform a link budget calculation using Equation 9-23 and the following values:

Antenna pointing loss	1 dB
Atmospheric loss	1.5 dB
Carrier frequency	—
Channel attenuation	200 dB
Receiving antenna diameter	—
Receiver G/T ratio	20 dB
Receiver sensitivity	—
Transmitting antenna diameter	—
Transmitter power	40 dB
Transmitting antenna gain	12 dB

$$\frac{C}{N_o} \text{(dB)} = \text{ERP (dB)} + G_{d,rx} \text{(dB)} - T_N \text{(dB)} - A_c \text{(dB)} - 10 \log (k) - 10 \log \text{(losses)}$$

$$\frac{C}{N_o} \text{(dB)} = 40 \text{ (dB)} + 12 \text{ (dB)} + 20 \text{ (dB)} - (-228.6 \text{ dB}) - 200 \text{ (dB)} - 2.5 \text{ dB}$$

$$\frac{C}{N_o} \text{(dB)} = 98.1 \text{(dB)}$$

■ **CARRIER FREQUENCY CHOICE AND CHANNEL ATTENUATION**

One of the most fundamental effects of link design that carries over to all cellular systems and wireless LANs as well is the relationship between the carrier frequency, or frequency band of operation, and the amount of attenuation the signal experiences while in the wireless channel. As the carrier frequency increases, the range decreases if the transmit power and antenna configuration remain constant. This concept is demonstrated in Examples 9-13 and 9-14. Example 9-15 summarizes the techniques introduced to this point in Chapters 8 and 9 and applies them to a link calculation.

■ **EXAMPLE 9-13** In Example 9-12, the channel attenuation was given as 200 dB. In this example, compare two different frequencies and recalculate the C/N_o values. Use 4 GHz and 14 GHz and the relationship from Chapter 8 to determine the channel attenuation for these two frequencies. Assume a link distance of 35,000 km.

$$A_c = 10 \log\left(\frac{4\pi r}{\lambda}\right)^2$$

$$4 \text{ GHz: } A_c \text{ (dB)} = 10 \log\left(\frac{4\pi(35{,}000)}{0.075}\right)^2 = 135.3 \text{ dB}$$

$$14 \text{ GHz: } A_c \text{ (dB)} = 10 \log\left(\frac{4\pi(35{,}000)}{0.021}\right)^2 = 146.4 \text{ dB}$$

$$4 \text{ GHz: } \frac{C}{N_o} \text{ (dB)} = 298.1 - 135.3 = 162.8 \text{ dB}$$

$$14 \text{ GHz: } \frac{C}{N_o} \text{ (dB)} = 298.1 - 146.4 = 151.7 \text{ dB}$$

You should observe from the calculation in Example 9-13 that as the frequency goes up, some additional channel attenuation occurs. This effect may seem numerically small but remember that the difference is measured in decibels, so a small numerical difference can mean a large power difference. In the example, we have a difference of about 11 dB. This numerical difference corresponds to a power difference ratio of more than 10. To give you a feel for what a difference this makes, Example 9-14 shows how much further apart the transmitter and receiver could be at 4 GHz for the same penalty in power as at 14 GHz.

■ **EXAMPLE 9-14** Use the channel attenuation equation with a carrier frequency of 14 GHz. Insert the wavelength for 4 GHz and then solve for the distance that would give this value of channel attenuation.

$$A_c = 146.4B = 10\log\left(\frac{4\pi r}{0.075}\right)^2 \rightarrow r = \frac{0.075}{4\pi}\sqrt{310^{14.64}} = 124{,}700 \text{ km}$$

Note that the range is increased by a factor of more than 3 for the 11 dB penalty incurred by increasing the frequency from 4 to 14 GHz. This example illustrates the general rule that as you move up in frequency, the range decreases if the power available remains the same. Since higher frequency components and systems generally cost more and also cover less range for a given power, the cost of moving to a higher frequency is always an issue. In Example 9-15, the maximum link distance of a 4 GHz system, given antenna diameters, carrier frequency, and a minimum receiving power, is calculated.

■ **EXAMPLE 9-15** Find the maximum link distance for the following system parameters:

Antenna pointing loss	1 dB
Atmospheric loss	1.5 dB
Carrier frequency	11 GHz
Channel attenuation	—

Receiving antenna diameter	3 m
Receiver G/T ratio	20 dB
Receiver sensitivity	10 nW
Transmitting antenna diameter	3 m
Transmitter power	40 dB
Transmitting antenna gain	12 dB

Begin with Equation 9-23 and find the value for each component. Although it may seem backward now, begin at the right-hand side and move left. The last two terms, accounting for Boltzman's constant and other losses, we can write immediately:

$$-10 \log (k) - 10 \log (\text{losses}) = +228.6 \text{ dB} - 1.0 \text{ dB} - 1.5 \text{ dB} = 226.1 \text{ dB}$$

The next term, channel attenuation, will depend on the link distance, so that term must wait. The figure of merit is given, so the next step is to find the ERP of the link. ERP is the sum of the transmitted power and the transmitting antenna gain:

$$\text{ERP (dB)} = P_{tx} \text{ (dB)} + G_{d,tx} \text{ (dB)} = 40 \text{ dB} + 12 \text{ dB} = 52 \text{ dB}$$

The only remaining quantity to be computed before solving for the link distance is the C/N_o ratio, which will correspond to the given minimum receiving power:

$$\frac{C}{N_o} \text{ (dB)} = 10 \log\left(\frac{C}{N_o}\right) = 10 \log(C) - 10 \log(N_o)$$

$$= 10 \log (10 \times 10^{-9} \text{ W}) - 10 \log (kT_N) = -80 \text{ dB} - 10 \log (k) - 10 \log (T_N)$$

$$= -80 \text{ dB} + 226.1 \text{ dB} - 10 \log (T_N)$$

We do not have a value for T_N; however, we do have a value for the receiver G/T ratio. If we can find the gain of the receiving antenna, the G/T ratio will give us T_N. To find the gain, first find the wavelength for the carrier frequency:

$$c = \lambda f \rightarrow \lambda = \frac{c}{f} = \frac{3 \times 10^8}{11 \times 10^9} = 0.027 \text{ m}$$

Next, find T_N:

$$G_{d,tx} \text{ (dB)} - T_N \text{ (dB)} = 10 \log\left(6\left(\frac{D}{\lambda}\right)^2\right) - 10 \log(T_N)$$

$$20 \text{ dB} = 10 \log\left(6\left(\frac{3}{27 \times 10^{-2}}\right)^2\right) - 10 \log(T_N) = 48.7 \text{ dB} - 10 \log(T_N)$$

$$10\log(T_N) = 48.7 \text{ dB} - 20 \text{ dB} = 28.7 \text{ dB}$$

Insert this value into the equation to find C/N_o:

$$\frac{C}{N_o} \text{ (dB)} = 146.1 \text{ dB} - 10 \log(T_N) = 146.1 - 28.7 = 117.4 \text{ dB}$$

The last step is to find the maximum link distance:

$$\frac{C}{N_o} \text{ (dB)} = \text{ERP (dB)} + G_{d,rx} \text{ (dB)} - T_N \text{ (dB)} - A_c \text{ (dB)} - 10 \log (k) - 10 \log \text{ (losses)}$$

$$117.4 \text{ dB} = 52 \text{ dB} + 20 \text{ dB} - 10 \log\left(6\left(\frac{D}{\lambda}\right)^2\right) + 226.1 \text{ dB}$$

$$10 \log\left(6\left(\frac{D}{2.7 \times 10^{-2}}\right)^2\right) = 180.7 \text{ dB}$$

$$D = 2.7 \times 10^{-2}\sqrt{\frac{10^{180.7/10}}{6}} = 1.19 \times 10^7 \text{ m} = 1.19 \times 10^4 \text{ km}$$

■ EFFECTS OF MOISTURE AND OTHER LOSS COMPONENTS IN THE ATMOSPHERE

Loss components in the atmosphere include moisture. The most important refinement to the calculations in the previous section that we want to discuss is how water droplets in the atmosphere effect the propagation of a radio wave. Water droplets can come in the form of rain, snow, or fog. These conditions are common throughout the world and vary with the season, so they are common for all link calculations where at least one component is based terrestrially. Fading or attenuation of the signal in the channel due to moisture is one of the primary disadvantages of many microwave systems.

We must take into account two primary considerations. First, the higher the frequency of the signal, the greater the sensitivity to moisture in the atmosphere. Again, we see penalties in the form of additional channel attenuation associated with moving up in frequency. Second, the more moisture, the higher the attenuation per unit distance. Above about 10 GHz, rain fades will range from 0.1 dB/km for a light mist up to 10 dB/km in a downpour. As the frequency is increased to 50 GHz, the range is from about 1 dB/km to as high as 20 dB/km. The curves shown in Figure 9-7 provide a rough guide. The top curve is for moisture accumulation of about 2.5 cm/hour; the middle curve, about 1 cm/hr; and the lowest curve, for foggy conditions.

Other significant losses besides moisture considerations include:

Foliage attenuation: This effect increases with frequency but only becomes a significant factor around 30 GHz. It is due primarily to the moisture content of the foliage. A good rule of thumb is to allow 1 to 2 dB additional attenuation when both the transmitter and receiver are not high enough to clear any intervening foliage such as trees or brush.

Reflection effects of buildings: The difference between a suburban setting and an open field varies between 15 and 20 dB of additional attenuation for the suburban setting compared to an open field. No reflections occur in an open field, and the usual problem is foliage attenuation. Generally, reflections off buildings add additional channel attenuation. Suburban settings are particularly poor because the buildings are not placed in a grid pattern. Hence, the reflections do not act to chan-

Figure 9-7
Rain Fades as a Function of Frequency and Moisture Rates of Accumulation

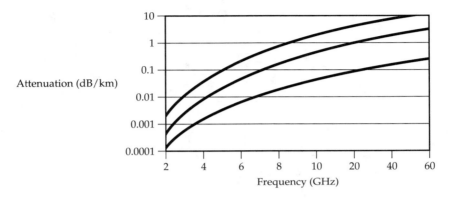

nelize the signals. The next item in this list illustrates that in metropolitan or urban areas, where the buildings are laid out in a regular grid, this pattern acts to help the signal propagation in certain preferred directions.

Channelization in an urban area: Radio waves tend to follow streets and can mean up to 20 dB difference between a location directly north, south, east, or west from a transmitter and one northeast, northwest, southeast, or southwest. The difference is that the first group of directions is assumed to be along a street directly below a transmitter, while the second group does not have this advantage in channelization. This attenuation difference happens in as little as six to ten city blocks. See Figure 9-8.

Tunnels: Long tunnels will effectively block all signal energy unless transceivers are placed inside the tunnels. Any tunnel over about 200 feet will block signal energy. Again, this effect depends on frequency: the higher the frequency, the less tunnel length that can be tolerated without special accommodation.

When both ends of the microwave link are based terrestrially, two additional factors must be taken into consideration beyond those already considered. Both are related to what is called the path profile that the communication must take to get from source to destination. Since all microwave links propagate through the atmosphere by line of sight (LOS), these factors are both related to this propagation mode. These calculations directly affect the height of the antenna towers required in a line of sight link.

Curvature of the Earth

All radio waves experience a slight curvature when propagating through the lower atmosphere. Experimental measurements indicate that, in free space, radio waves above 30 MHz will propagate in a direct line. In the earth's atmosphere, however, there is some refraction and hence curvature. The usual way to account for the earth's curvature is by introducing a constant called the effective earth radius. To find the actual path length,

Figure 9-8
Street Channelization of Radio Waves

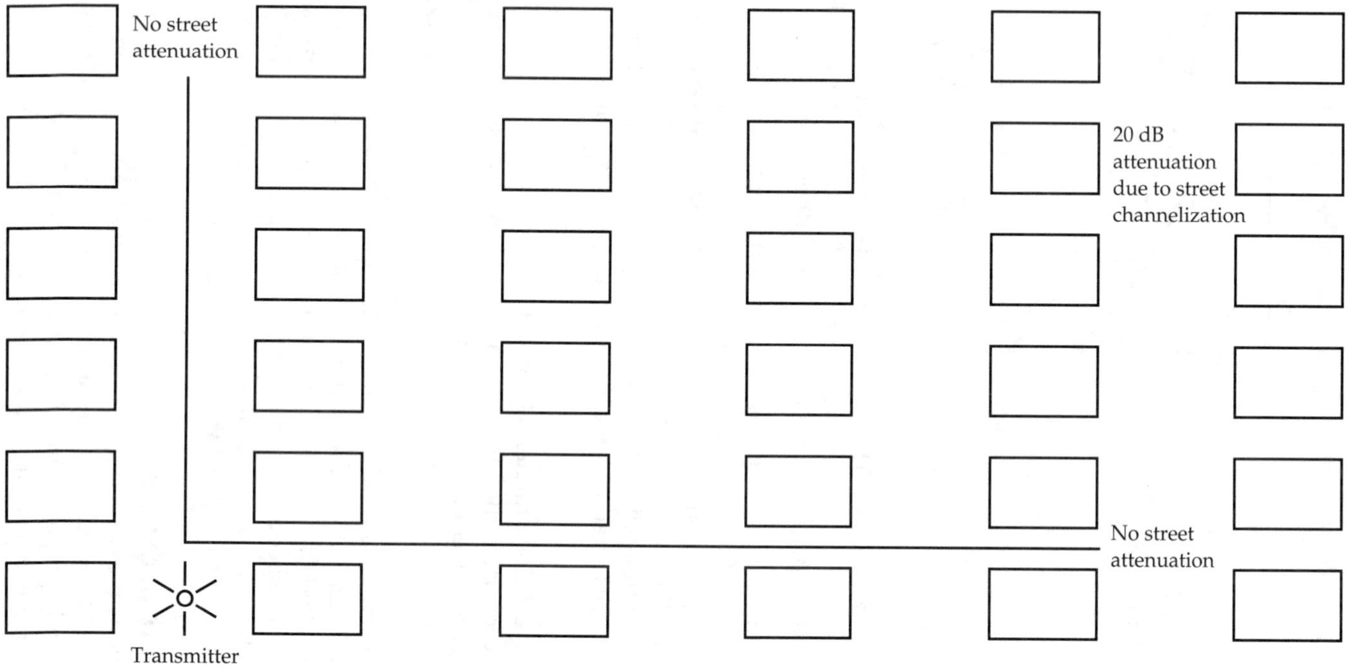

No street attenuation

20 dB attenuation due to street channelization

No street attenuation

No street attenuation

Transmitter

the true earth's radius is multiplied by this factor, which is taken to be ⅘. This value is calculated in Example 9-16 and expressed in Equation 9-24.

■ **EXAMPLE 9-16**   Find the effective earth radius, given that the true earth radius is 6380 km.

$$r_{ee} = \frac{4}{3}r_{earth} \tag{9-24}$$

$$r_{ee} = \frac{(4)(6.38 \times 10^6)}{3} = 8500 \text{ km}$$

r_{ee} = effective earth radius
r_{earth} = true earth radius

For the vast majority of applications involving microwave relay towers, the approximation in Example 9-16 is effective. If you are considering an application where the transceivers are at extreme altitudes or the weather conditions are extreme, further refinement is possible.

This curvature of the earth's radius was included in the relationship introduced in Chapter 1 for calculating antenna height: Equation 1-13, reproduced below. You may recall that this relationship was introduced as experimentally determined in Chapter 1.

$$d_t = \sqrt{2h}$$

Now, we will modify the constant factor to convert to metric units where the antenna height is measured in meters and the radio horizon is measured in kilometers. The conversion factors used are 1 mile = 1.61 km and 1 ft = 0.305 m. They are shown in Equation 9-25.

$$d_t = \sqrt{2h} \rightarrow \frac{d_t}{1.61} = \sqrt{\frac{2h}{0.305}} \rightarrow d_t = 4.12\sqrt{h} \tag{9-25}$$

d_t = LOS radio horizon (km)
h = antenna height (m)

Fresnel Zone

The key idea in applying the principle of the Fresnel zone is that if there are no obstacles inside the Fresnel ellipsoid, then the communication link can be considered interference free. Place the endpoints of the ellipsoid at the antenna locations and draw the Fresnel ellipsoid. If there are no obstructions inside the curve, you are done. If there are obstructions, you can move the antenna by using a taller mast or you can change the carrier frequency, if possible. The higher the carrier frequency, the narrower the Fresnel ellipsoid. Remember, however, that moving to a higher carrier frequency will result in greater channel attenuation and a greater sensitivity to atmospheric effects, so it is important to consider many factors.

Figure 9-9
Fresnel Ellipsoid

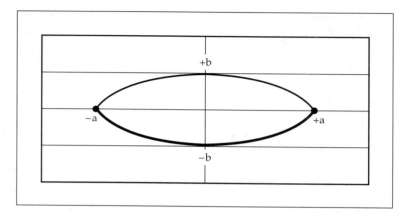

When calculating a link budget for microwave relays and when the path between the transmitting and receiving antennas is not free of obstacles, Fresnel zones are used. The mathematical construct known as a Fresnel ellipsoid defines the volume between the two antennas, which must be free of obstructions. If it is not, then further attenuation will occur. In other words, you can assume free-space conditions apply if the volume of the Fresnel ellipsoid is free from interference due to obstructions.

Basically, if this area is not obstruction-free, the radio waves reflect to some degree off the obstructions when propagating from transmitter to receiver. Whether this effect is of concern can be calculated using the concept of Fresnel zones. (Fresnel was a famous French scientist who did pioneering work on the wave theory of light in the early nineteenth century.) Note that Fresnel zones are useful when only a small number of obstructions lie in the communications path. The interference calculated using Fresnel zones and that calculated due to multipath effects, such as reflections off buildings, are not the same.

Probably the easiest way to understand this concept is to think of a Fresnel zone as a special kind of ellipse. Like all ellipses, specific conditions define the shape of it (see Figure 9-9). For example, it might be long and narrow, or almost circular. The familiar equation for an ellipse is reproduced below:

$$\frac{x^2}{a^2} + \frac{y^2}{b^2} = 1$$

a = maximum horizontal displacement from the y axis
b = maximum vertical displacement from the x axis

The antennas would be placed at $-a$ and $+a$, and b defines the maximum cross-section of the Fresnel ellipsoid. The set of points defined by this curve is called the first Fresnel zone. If the obstacle is assumed to be precisely at the midpoint of the two towers, then b will give directly the additional height required for the antennas. In this *special case*, the relationship expressed in Equation 9-26 holds.

Figure 9-10
Terrestrial Microwave Fresnel Zone, Example 9-18

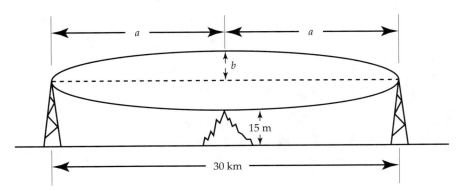

$$b = 8.65\sqrt{\frac{d_t}{f}} \qquad\qquad (9\text{-}27)$$

The value of d is measured in kilometers and the carrier frequency, f, is measured in GHz. Example 9-17 illustrates the use of these equations. This unit convention results in the 8.65 numerical factor shown in Equation 9-27.

■ **EXAMPLE 9-17**   Define and draw the Fresnel ellipsoid for a microwave-relay application where the two antennas are 40 km apart and the carrier frequency is 10 GHz.

$$a = \frac{d_t}{2} = 20 \text{ km}$$

$$b = 8.65\sqrt{\frac{40}{10}} = 17.3 \text{ km/GHz}$$

■ **EXAMPLE 9-18**   For a link distance of 30 km and a carrier frequency of 6 GHz, find the antenna height required. Calculate the first Fresnel zone assuming an obstacle of 15 m located precisely at the midpoint between the two towers. Assume both towers are the same height. Figure 9-10 illustrates a typical terrestrial microwave link and points out how to measure the values for a and b in Equation 9-26.

Since the formula given for antenna height already includes any effects of the earth's curvature, we will start with it. Both antenna towers are assumed to be the same height. We converted Equation 9-25 to metric units earlier. Remember that each tower is the same height and must be added separately. Applying Equation 9-25, we have:

$$d = 4.12(\sqrt{h_{tx}} + \sqrt{h_{rx}}) = 8.24\sqrt{h}$$

$$30 = 8.24\sqrt{h} \rightarrow h = \left(\frac{30}{8.24}\right)^2 = 13.3 \text{ m}$$

Since the obstacle has been placed conveniently at the exact midpoint of the link, b yields the exact additional height required for either antenna:

$$b = 8.65\sqrt{\frac{d}{f}} = 8.65\sqrt{\frac{30}{6}} = 43.2 \text{ m}$$

Therefore, the two antenna heights are:

$$h_{tx} = h + b + \text{obstacle height} = 13.3 \text{ m} + 43.2 \text{ m} + 15 \text{ m} = 71.5 \text{ m}$$
$$h_{rx} = h = 36 \text{ m}$$

Not all obstacles are located conveniently and precisely at the midpoint. A more general solution would be more widely applicable. Examine more closely the defining statement of a Fresnel ellipsoid. The Fresnel ellipsoid is the set of points where the sum of the distances from the two antennas is greater by half a wavelength than the direct distance between the two antennas. Again, if you take the set of points defined by a continuous cross-section through the Fresnel ellipsoid in a direction perpendicular to the direction of propagation, this set of points is called the first Fresnel zone. The key idea in applying this principle is that, if there are no obstacles inside the Fresnel ellipsoid, then the communication link can be considered interference-free. The higher the carrier frequency, the narrower the Fresnel ellipsoid.

Drawing out the Fresnel ellipsoid and then comparing it to a terrain map is fine, but it would be easier to have an analytical expression that can determine the effect on antenna height that any particular obstacle might have. This expression would depend on having a relationship similar to the one above, but it would be applicable at any point along the link. This expression is given by Equation 9-28. Since the cross-sectional distance varies along the ellipsoid, the variable r is used to represent the cross-sectional area of the Fresnel ellipsoid at the location of the obstacle. At the midpoint, $r = b$.

$$r = 17.3\sqrt{\frac{d_{tx}d_{rx}}{f(d_{tx} + d_{rx})}} \tag{9-28}$$

r = height of any point on the first Fresnel zone
d = distance from the respective towers

■ **EXAMPLE 9-19**   For a link distance of 30 km and a carrier frequency of 6 GHz, find the antenna height required. Calculate the first Fresnel zone, assuming an obstacle of 15 m located 5 km from the transmitting antenna tower. Assume both towers are the same height. Figure 9-11 illustrates the situation.

Figure 9-11
Example 9-19

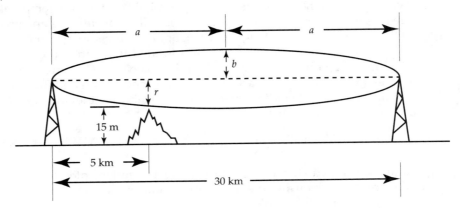

Begin as you did in Example 9-18:

$$d = \sqrt{2h_{tx}} + \sqrt{2h_{rx}} = 2\sqrt{2h}$$

$$30 \text{ km} = 2\sqrt{2h} \rightarrow h = \frac{1}{2}\left(\frac{30}{2}\right)^2 = 112.5 \text{ ft} \approx 36 \text{ m}$$

Now apply Equation 9-28 for an obstacle at an arbitrary distance between the two towers:

$$r = 17.3\sqrt{\frac{d_{tx}d_{rx}}{f(d_{tx} + d_{rx})}}$$

$$r = 17.3\sqrt{\frac{(5)(25)}{(6)(5 + 25)}} = 14.4 \text{ m}$$

The two antenna heights are found to be:

$$h_{tx} = h + r + \text{obstacle height} = 36 \text{ m} + 14.4 \text{ m} + 15 \text{ m} = 65.4 \text{ m}$$

$$h_{rx} = h = 36 \text{ m}$$

As you can see from Examples 9-18 and 9-19, the location of an obstacle at the mid-point is the worst case in determining additional antenna tower height. For that reason, it is sometimes used as a rough worst-case approximation when initially planning a terrestrial link and the exact locations of any obstacles are unknown.

■ TERRESTRIAL LINK CALCULATION

To this point, we have covered the most important loss components in a terrestrial link calculation. Example 9-20 will bring these components together with elements of Chapter 8 to illustrate how they contribute to a terrestrial link budget calculation.

■ **EXAMPLE 9-20**   Assume a 40 km link at a carrier frequency of 18 GHz. The transmit power applied to the transmitting antenna is 100 W. The noise temperature of the first-stage amplifier at the receiver is 125 degrees. The transmitting antenna height is 10 m and the receiving antenna height is 10 m. Both antennas are parabolic design, 2 m in diameter. Assume one obstacle of 20 meters located 10 km from the transmitting antenna and additional path losses of 2 dB. Find the received C/N_o ratio and power applied to the receiver. Finally, assume a 10 Mbps QPSK modulation and find the E_b/N_o ratio.

Calculate the LOS radio horizon and determine if either antenna needs to be raised due to the curvature of the earth:

$$d_t = 4.12(\sqrt{h_{tx}} + \sqrt{h_{rx}}) = 4.12(\sqrt{10} + \sqrt{10}) = 26.0 \text{ km}$$

You can see immediately that this link cannot be implemented with the antennas so low to the ground. Try raising the transmitting antenna to 20 m and determine the height needed for the receiving antenna:

$$40 \text{ km} = 4.12(\sqrt{20} + \sqrt{h_{rx}}) \rightarrow \sqrt{h_{rx}} = \frac{40}{4.12} - \sqrt{30} = 4.23 \rightarrow h_{rx} = 17.9 \text{ m} \approx 18 \text{ m}$$

Next, include the first Fresnel zone contribution from the obstacle:

$$r = 17.3\sqrt{\frac{d_{tx}d_{rx}}{f(d_{tx} + d_{rx})}} = 17.3\sqrt{\frac{(10)(30)}{(18)(10 + 30)}} = 11.2 \text{ m}$$

And sum to find the antenna heights:

$$h_{tx} = h + r + \text{obstacle height} = 20 \text{ m} + 11.2 \text{ m} + 20 \text{ m} = 51.2 \text{ m}$$
$$h_{rx} = h = 36 \text{ m}$$

Now that the antenna heights are calculated, the channel attenuation can be calculated. First, find the wavelength of 18 GHz:

$$c = \lambda f \rightarrow \lambda = \frac{c}{f} = \frac{3 \times 10^8}{18 \times 10^9} = 16.7 \times 10^{-2} \text{ m}$$

Next, find the ERP from the transmitting antenna by adding the supplied power input and the gain of the transmitting antenna. Since both antennas are identical, the directive gain found will apply to both the transmitting and receiving antennas.

$$\text{ERP (dB)} = P_{tx} \text{ (dB)} + G_{d,tx} \text{ (dB)} = 10 \log(P_{tx}) + 10 \log\left(6\left(\frac{D}{\lambda}\right)^2\right)$$

$$= 10 \log(100) + 10 \log\left(6\left(\frac{2}{1.67 \times 10^{-2}}\right)^2\right) = 20 \text{ dB} = 49.3 \text{ dB} = 69.3 \text{ dB}$$

Then find the G/T ratio:

$$G_{d,rx} \text{ (dB)} - T_N \text{ (dB)} = 10 \log\left(6\left(\frac{D}{\lambda}\right)^2 \right) - 10 \log(T_N)$$

$$= 10 \log\left(6\left(\frac{2}{1.67 \times 10^{-2}}\right)^2 \right) - 10 \log(125)$$

$$= 49.3 \text{ dB} - 20 \text{ dB} = 28.4 \text{ dB}$$

Find the channel attenuation:

$$A_c \text{ (dB)} = 10 \log\left(\frac{4\pi r}{\lambda}\right)^2 = 10 \log\left(\frac{4\pi(40 \times 10^3)}{1.67 \times 10^{-2}}\right)^2 = 149.6 \text{ dB}$$

Finally, find Boltzman's constant and the contribution of other losses:

$$10 \log (k) + 10 \log \text{ (losses)} = -228.6 + 10 \log (2) = -225.6 \text{ dB}$$

By adding these quantities together, you will find the C/N_o value for this link:

$$\frac{C}{N_o} \text{ (dB)} = 69.3 + 28.4 - 149.6 - (-225.6) - 173.7 \text{ dB}$$

To find the E_b/N_o ratio for the bit rate and bandwidth given, apply the relationship developed earlier:

$$\frac{E_b}{N_o} = \frac{C}{N_o} t_b$$

First, convert the C/N_o value to a nonlogarithmic one:

$$\frac{C}{N_o} \text{ (dB)} = 10 \log\left(\frac{C}{N_o}\right) \rightarrow \frac{C}{N_o} = 10^{\frac{C}{N_o}(dB)/10} = 10^{17.37} = 2.34 \times 10^{17}$$

Next, find the product and convert back to a decibel value:

$$\frac{E_b}{N_o} \text{ (dB)} = 10 \log[(2.34 \times 10^{17})(1 \times 10^{-7})] = 103.6 \text{ dB}$$

The answer in Example 9-20 should make sense to you. After all, the only difference between C/N_o and E_b/N_o is that the latter is the carrier power per bit. The modulation rate is 10 Mbps, and because 10 million corresponds to 70 dB, the answer found in the example could have been inferred directly from the value calculated for C/N_o. Note that it would have been just as reasonable to convert the bit time to decibels and to add the two decibel quantities: the answer would have been the same.

■ SUMMARY

This chapter focused on terrestrial microwave link analysis. This subject matter, long important for traditional point-to-point links, will become even more important when

LMDS and other high bandwidth terrestrial microwave systems become widely deployed. Note also that several fundamental concepts of link analysis (for example, the relationship of carrier frequency and channel attenuation; the effects of fades, building reflections, and street channelization) apply to cellular telephony and wireless LAN applications. Because many emerging wireless systems will operate in the same frequency bands as traditional microwave terrestrial links, the concepts explored in this chapter have wide applicability. Based on the theory and principles explored in this chapter, Chapter 10 will conclude the theme of link analysis by extending the concepts to a space-based location of one transceiver.

REVIEW QUESTIONS

1. What are the major components of a terrestrial line of sight link? What function does each perform?

2. What are the major components of a transceiver? What function does each perform?

3. Define a transceiver.

4. Discuss why an IF frequency is used.

5. Why is the noise performance of the first-stage amplification in a receiver so important?

6. Discuss the advantages of LMDS in a point-to-multipoint context.

7. What services could be accessed from a typical home using LMDS?

8. Discuss how LMDS might affect the last-mile issue.

9. Why is link engineering so important to the commercial success of LMDS?

10. Explain the difference between internal noise and external noise.

11. Why is reference noise power an important concept?

12. What is the difference between thermal noise power and noise temperature?

13. Why is the noise figure of an amplifier or other component so important?

14. Discuss what principles you would apply to an amplifier chain used in the first stage of a receiver.

15. Define the carrier-to-noise ratio and describe what it represents.

16. Define energy per bit and describe what it represents.

17. Define the E_b/N_o ratio. Explain its importance and describe what it represents.

18. Recalculate Example 9-10 using a 10 Mbps data rate in a 1 MHz channel.

19. Recalculate Example 9-10 using a noise temperature of 80 degrees Kelvin.

20. Discuss the results of Review Questions 18 and 19. What did you observe?

10

Satellite Communications Technology

■ INTRODUCTION

This chapter introduces several aspects of satellite communications. After a brief section on the history of satellite technology and a block diagram of a satellite, the chapter introduces geosynchronous earth orbit (GEO) satellites, discusses orbital issues associated with them, and compares satellites in GEO orbits with other communications systems. After a brief introduction to low earth orbit (LEO) satellites, we will compare the relative strengths of the two orbital classifications.

The chapter then turns to a discussion of satellites as communications systems, describing the major components, frequency bands of operation, and the many access methodologies or air interfaces used in the satellite industry. These sections address the technology issues of spot beams and on-board processing as associated with the air interface technique. A few sections discuss the basics of launching a satellite, how to find a satellite, and how to point the antenna. This discussion includes a tutorial on latitude and longitude and a block diagram of a typical earth station. The next sections of the chapter explore several major applications that make use of communications satellites. The chapter closes with the third installment of the three-chapter theme of link analysis calculations.

■ HISTORY OF SATELLITE COMMUNICATIONS

Traditionally, a satellite used for communications between earth and space is essentially a relay. It receives signals from the earth, amplifies them, changes the carrier frequency, and retransmits them to earth. In fact, these satellites are sometimes called repeaters because the only change they make to the signal is a physical layer one. When science fiction author and scientist Arthur C. Clarke first suggested the concept of a communications satellite in 1945, the term he used was extraterrestrial relay. As is most appropriate for a science fiction author, Clarke was ahead of his time. He called for these satellites to be placed 22,270 miles above the equator. At the time of his writing, the highest altitude that humans had achieved was about half that. The V2 rocket, the closest thing to a satellite launcher at that time, had a maximum altitude of about twice the 22,270 miles needed but with no payload.

Satellites are placed in several different orbits, depending on their primary application, and were first introduced into commercial service in the early 1960s. Although not used as a relay, the first satellite was a pure single-frequency transmitter called Sputnik. It was launched by the former Soviet Union in late 1957 as a demonstration of superior Soviet technology. The United States was caught by surprise and was quick to respond in those chilly days of the Cold War with signal communicating by orbital relay equipment (SCORE), launched in late 1958. SCORE actually worked as a relay.

Since nobody had built a satellite yet, the engineering was "by the seat of the pants." SCORE was assembled out of a modified FM pocket pager combined with a walkie-talkie and amplifier. It relayed one message at a time, copying it on one orbit and playing it back on the next. The maximum message length was 4 minutes. The uplink frequency was 150 MHz and the downlink frequency was 132 MHz. The power supply was a battery that lasted about one month. At the time, these events were revolutionary; now they are commonplace because dozens of satellites are launched into space each year. Toward the end of the Cold War, about 1985, the number of satellites approached 1,000 annually and were dominated by Soviet-launched, short-term military observation satellites intended to survey the "daily hot spot" of activity.

The first true communications satellite as we think of it today was TELSTAR 1, which was placed into orbit in 1962 and manufactured by the old Bell System. It was not geosynchronous but instead precessed about the earth each orbit. This track allowed it on one orbit to illuminate simultaneously the eastern United States and part of Europe. Later, on another orbit it would illuminate the western United States and Japan. Each of these occasions were very short-lived; communication was possible for only about 15 minutes.

The first geosynchronous commercial communications satellite was INTELSAT 1, which was launched in 1965. This satellite was the precursor of all commercial satellites in orbit today. A traditional satellite system is best thought of as a relay system. Figure 10-1 shows a simplified model of a satellite system.

The figure also illustrates the operation of a satellite using a single channel or transponder. In each transponder is a signal modulated on a carrier. These carrier frequencies and the bands they operate in are predefined and fixed. Uplink frequencies are those frequency bands where the information is carried up to the satellite. Downlink fre-

Figure 10-1
Simplified Model of a Satellite System

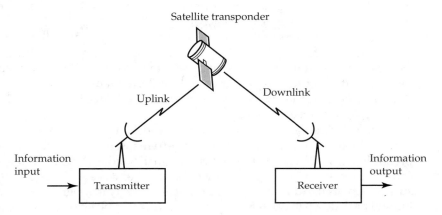

quencies are those frequency bands where the information is carried back down to the earth. Note that the uplink frequency is always lower than the downlink frequency.

Four basic characteristics of communications satellites can be used to classify them. These four characteristics are the orbital altitude at which the satellite orbits, the frequency band the satellite uses to communicate, whether or not the satellite switches the data or simply relays it, and the access technology or air interface that the satellite uses. Orbital altitude has two subclasses: geosynchronous earth orbit (GEO) and low earth orbit (LEO). The next section begins the discussion of the implications of orbital altitude.

■ GEOSYNCHRONOUS ORBITING SATELLITES

Geosynchronous or geostationary orbits are those orbits where the altitude of the satellite is such that it appears stationary to an observer on earth, hence the name *geostationary*. The terms *geosynchronous* and *geostationary* are used interchangeably. At this altitude, approximately 22,270 miles above the earth, the velocity of the satellite exactly matches the rotational, or angular, velocity of the earth.

A more precise definition of a geostationary orbit than "an orbit that is over the equator" would be to say that the angle of the satellite's orbital plane with respect to the equator is 0 degrees. Another name for this is equatorial orbit. In fact there are many orbits possible at 22,270 miles of altitude other than those of commercial importance, such as the equatorial orbits. You could place a satellite in a polar orbit, where the angle of its orbital plane is 90 degrees relative to the equator, or any inclination angle. However, none of these would be called geostationary.

The equatorial orbit is in high demand among various countries and organizations. There is only so much room, however, and this particular orbit, directly above the equator, is tightly packed with satellites around much of the globe. This orbit is the only one where the satellite remains stationary over a particular land mass or region of the earth. Since it is by far the most commercially important geostationary orbit, it is often called just geostationary, not equatorial geostationary.

About 180 GEO satellites orbit the earth. Disregarding those organizations with only a single satellite, about thirty commercial and governmental organizations have satellites in this orbit. The largest of these satellite organizations have about twenty satellites. Satellites in this orbit are used to distribute many different kinds of services. Traditionally, commercial services were restricted to television signal relay and, to a lesser extent, long-distance telephony.

Long-distance telephony, once a significant component of traffic on GEO satellites, is now transferred almost entirely to terrestrial-based optical fiber links. GEO satellites presently cannot be used to link directly to small handheld units like mobile telephones because the receive power required by these small units cannot be generated by satellites in such a high orbit. Additionally, the propagation delay in a signal routed up to a GEO satellite and back down to earth is too great for such an application. As we will see in a later section, such an application is possible using an array of satellites in LEO orbit.

Today, satellites in GEO orbit are still widely used for television signal relay by both networks and many large private organizations. In the latter case, this application is known as business television (BTV). The largest new and fastest-growing application for these satellites is the relay and distribution of internet traffic, both for the public Internet and many privately owned internets.

Geostationary Orbit and Satellite Spacing

The planet earth and a satellite in geostationary orbit, like any two bodies with mass, exhibit a gravitational attraction for each other. This attraction is determined from Newton's second law, the law of universal gravitation. This law states that gravitational interaction between two bodies is attractive and depends directly on the product of their mass and inversely on the square of the distance between them. This relationship can be stated mathematically and appears in Equation 10-1:

$$F = \frac{Gm_1 m_2}{r^2}$$ (10-1)

F = force in Newtons
G = gravitational constant = 6.67×10^{-11} Nm^2/kg^2
m_1 = mass of the earth in kilograms
m_2 = mass of the satellite in kilograms
r = distance of the satellite from the center of the earth in meters

Example 10-1 will illustrate how to apply this famous law.

■ **EXAMPLE 10-1** Find the gravitational force experienced by a geostationary satellite weighing 1 ton.

Figure 10-2
Satellite Spacing

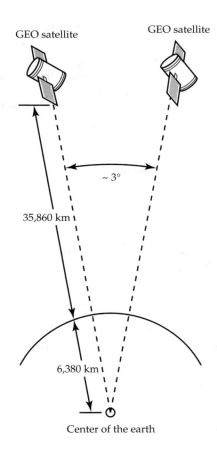

GEO satellite

GEO satellite

~ 3°

35,860 km

6,380 km

Center of the earth

After giving a value for the mass and radius of the earth, this example begins with a unit conversion exercise. We must change the altitude of the satellite, which we know to be 22,270 miles, into meters and the weight on earth to mass in kilograms. (See Figure 10-2 for an explanation of the radius calculation.)

$$m_1 = 5.98 \times 10^{24} \text{ kg}$$

$$r_{\text{earth}} = 6{,}380 \text{ km}$$

$$m_2 = (1 \text{ ton})\left(\frac{2{,}000 \text{ lbs}}{1 \text{ ton}}\right)\left(\frac{2.2 \text{ kg}}{1 \text{ lb}}\right) = 1{,}998 \text{ kg}$$

$$r = r_{\text{earth}} + (22{,}270 \text{ miles})\left(\frac{1.61 \text{ km}}{1 \text{ mile}}\right) = 6{,}380 \text{ km} + 35{,}860 \text{ km} = 42{,}240 \text{ km}$$

$$F = \frac{Gm_1m_2}{r^2} = \frac{(6.67 \times 10^{-11})(5.98 \times 10^{24})(1{,}998)}{(42{,}240 \times 10^3)^2} = 447 \text{ N}$$

In Example 10-1, we found that there is a significant amount of attractive force between this satellite and the earth. Even more force exists between the sun and the earth. An interesting question is, Why does the earth not orbit into, or fall into, the sun? The answer is the same reason why the satellite does not fall into the earth, namely, that the earth in its orbit around the sun and the satellite in its orbit around the earth are moving fast enough so that, by the time the satellite has fallen to where the earth was, it has moved just exactly that distance away. Another way to think of this concept is to imagine the satellite as continuously falling into where the earth was and, just as the satellite reaches that point, the earth has moved in its orbit just exactly that distance away. The only reason any satellite stays in orbit is that it is moving fast enough. If it were possible to slow or to stop the satellite, it would immediately start falling to the earth and would be destroyed quickly. Example 10-2 illustrates just how fast a GEO satellite must be moving to stay in stationary orbit.

■ **EXAMPLE 10-2**   In geostationary orbit the satellite must complete its orbit in just under 24 hours. The earth's rotation on its axis is just slightly less than 24 hours, which is why we have leap years. The actual number is about 23.95 hours, or a sidereal day. The radius of the orbit was calculated in Example 10-1 as 42,240 kilometers. Although all orbits are elliptical, geostationary satellite orbits are very close to a circle, so we will assume a circular orbit for this approximate calculation. Hence, the circumference is found by the following familiar expression:

$$C = 2\pi r = 2\pi(42{,}240 \text{ km}) = 265{,}400 \text{ km}$$

To find the velocity or rate, just apply the familiar relationship of distance = (rate)(time):

$$d = rt \rightarrow r = \frac{d}{t} = \frac{265{,}400 \text{ km}}{(23.95)(3{,}600)} = 3{,}080 \text{ km/s}$$

$$= 3{,}080 \left(\frac{1 \text{ mile}}{1.61 \text{ km}}\right)\left(\frac{3{,}600 \text{ s}}{1 \text{ hr}}\right)\left(\frac{3{,}600}{1}\right) = 6{,}890{,}000 \text{ mph} \approx 7 \times 10^6 \text{ mph}$$

The last step includes a unit conversion from km/s to mph.

Remember, this calculation is for a geostationary orbiting satellite; it appears to be stationary from the perspective of someone standing on the equator. To see how altitude makes an impact (after all, 7 million miles per hour is quite fast), let's find the velocity at which you are moving right now. Because you are on the surface of the earth, and because the earth is rotating, you too are moving at a significant velocity. You are moving at a slower velocity the farther north or south you travel on the earth's surface.

$$r = \frac{d}{t} = \frac{2\pi(6{,}380 \text{ km})}{(23.95)(3{,}600)}\left(\frac{1}{1.61}\right)\left(\frac{3{,}600}{1}\right) = 1{,}040 \text{ mph}$$

Solar Day and Sidereal Day

Example 10-2 introduced the concept of a sidereal day, which is different from what most define as a day, actually referred to as a solar day. In a solar day, the earth makes one complete revolution of its orbit, *plus it moves ⅟₃₆₅ of the way around the sun.* As you can see, a solar day is what is commonly refered to simply as a day. For satellites, however, a solar day would yield the wrong answer for Example 10-2 because we want the time the satellite takes to orbit the earth. A sidereal day is a solar day minus the extra time it takes to move ⅟₃₆₅ of the way around the sun. This difference amounts to just under 4 minutes of each day, hence the approximation of 23.95 hours for one sidereal day.

Geostationary Satellite Spacing

Geostationary satellite spacing is determined by trading off a couple of items. From the point of view of maximizing the number of GEO satellites, it is desirable to have the satellites as close together as possible. There is only one equatorial geostationary orbit around the earth, and many governmental and commercial organizations would like to occupy specific positions, or orbital slots, in that orbit. This topic is a highly charged political issue, and many countries place great prestige on the fact that one of those satellites up in space is "theirs." Another important point is that, to date, GEO satellites used for communications are the only money-making opportunities in space, and they are quite profitable. Billions of dollars of revenue every year is made from these relays in space that were first envisioned by Arthur C. Clarke.

Another consideration must be taken into account. There must be sufficient room between the satellites to address the issues of antenna beamwidths and interference from adjacent satellites using the same frequency allocation (called co-channel interference). If the satellites are so close that the information being relayed up and down interferes with other information, the primary function of the satellite is compromised. This factor and related issues such as the carrier frequency and transmit carrier power limitations have led to regulations that limit the spacing of satellites to approximately 3 degrees of longitude. Figure 10-2 illustrates the concept of satellite spacing.

GEO Satellites Compared to Other Communication Systems

The key advantages of GEO satellites are 1) a single satellite can cover a very large service area and 2) a satellite incurs a low average cost of operation.

Very Large Service Area Covered by a Single Satellite   A single satellite in geostationary orbit can cover an area of the earth as large as one-third of its surface. In fact, three to five satellites can cover the globe, as you can see in Figure 10-3. A single relay in orbit can be used by a very large number of users. Also, no significant terrestrial infrastructure costs prohibit each of these users from communicating with one another. For example, it takes no more infrastructure to speak to someone on a mountaintop halfway across the country than it does to speak to your neighbor across the street. Each situation requires only a single satellite telephone. All of these parties can also see the same real-time sports or news program with no more than a dish antenna mounted on their roofs.

Figure 10-3
Three Satellites Can Cover the Globe

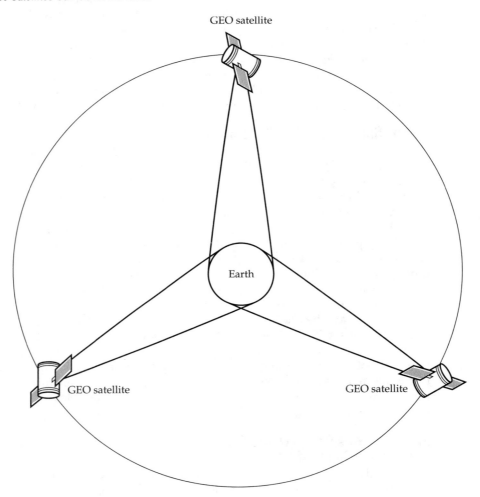

Compare satellite relay to the situation where the connection between two parties must be made without a space-based component. In telephone communication, a land line or microwave link must be established between each pair of individuals who wish to communicate. In the second, a television station must be in each viewing area and both stations must be connected to the original source of the signal.

Figure 10-4 shows the four situations. In each situation, the two subscribers are assumed to be 3,000 miles apart. Part (a) shows a single satellite that allows the two subscribers to communicate via satellite. Here, both signals must be two way and full duplex. In part (b), the use of a satellite as a television relay is shown. Again, a single satellite allows both parties to see the same program simultaneously. In this case, the television studio does

Figure 10-4
(a) A Single Satellite That Allows Two Subscribers to Communicate; (b) Using a Satellite as a Television Relay

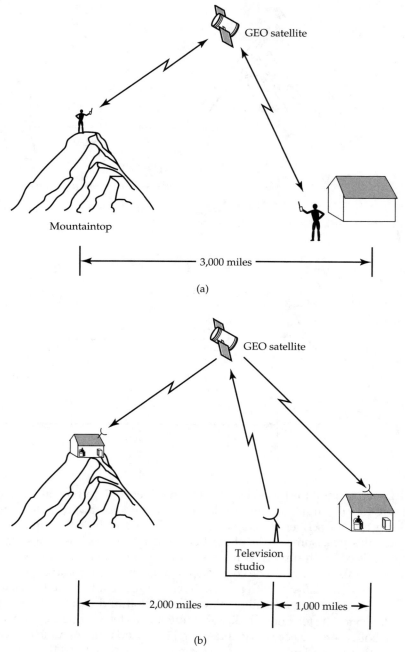

(a)

(b)

(continued)

Figure 10-4 (continued)
(c) Combined Land-Line and Microwave Relay; (d) Two Subscribers Watching Television

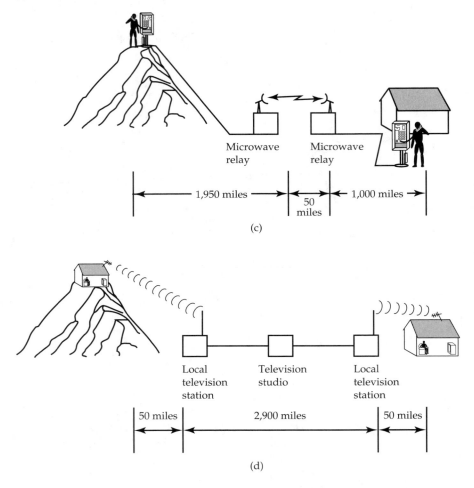

(c)

(d)

the uplink and the two subscribers are receivers only; all signals are simplex. Part (c) of Figure 10-4 illustrates a combined land-line and microwave relay that would be necessary for the two subscribers to converse without a space-based component. The land-line telephones are assumed to be phone booths. Part (d) shows two subscribers at home watching television. Each subscriber receives a broadcast from his or her local station. Both local stations must have land-line connections to the producing studio.

As you can see, with a satellite relay, a single infrastructure element provides the entire link. Without a satellite relay several thousand miles of telephone cable must be installed for the telephone link. For simultaneous television reception, two transmitters close to both users are necessary, and several thousand miles of cable must be laid for each transmitter to simultaneously receive the program material and then broadcast it to each user.

Low Average Cost of Operation   While the initial cost to construct a satellite is high and the cost to place it in orbit is also high, both costs are one-time costs. Over the average lifetime of a typical geostationary satellite, twenty years, there are almost no ongoing costs. When you average the high initial costs over the lifetime of the satellite, the average cost becomes quite reasonable. As satellite technology improves and launch costs drop, the average cost of operation will also drop. If we develop a technology allowing us to refuel and repair satellites in orbit for less than it would cost to replace them, satellites would become very advantageous.

The main disadvantages of a GEO satellite used in communications are 1) high initial cost, 2) large inherent latency, and 3) large inherent signal attenuation.

High Initial Cost   The high initial cost of a GEO satellite is due to five primary factors: the cost to acquire the components and assemble the satellite itself, the cost of the launch, the cost of insuring the launch, the business risk of the launch, and the time value of the investment required.

Satellite components are expensive. They have to be designed for a long lifetime in a hostile environment and must have high reliability. Satellite assembly also requires a clean room environment using highly skilled personnel. System testing prior to launch must be extensive because repair is, at this time, not feasible for the vast majority of commercial satellites. Finally, it takes a significant amount of time to assemble a satellite, not to mention the dedicated space and personnel assignments.

Launching a satellite into a geostationary orbit almost 23,000 miles above the earth and placing it precisely relative to its neighboring satellites is something very few organizations can do reliably. Although costs vary among these providers, all are in the tens of millions of dollars. The risk of a failed launch can be mediated by purchasing an insurance policy that will reimburse the owner of the satellite for its purchase price and, optionally, the cost of a second launch.

No insurance policy can or will address the business risk of a failed launch. If service contracts were in place prior to launch, these services must now be provided in some alternate way. If the launch made sense economically, the alternate methods almost certainly will be more expensive to provide and they may not offer the same value to the customer.

Finally, the time value of money has an impact. The first three costs must be absorbed by the organization that owns the satellite prior to seeing any revenue from that investment. The entire monetary risk is up front. A substantial amount of assets must be committed to a project before any revenue comes in.

These issues explain why only those organizations and nations with large asset reserves can contemplate this kind of investment. It is often impossible for a small organization or nation to fund this investment and have the financial capability to wait for the eventual revenue flow.

Latency   *Latency* is the generic word used to describe any link delay in communications. The term *link delay* is often preferred when speaking of GEO based communications links because it is more descriptive. Both terms are used interchangeably.

Light, like all electromagnetic energy, travels at the speed of light. This speed seems quite fast until you recognize that a satellite acts as a relay and that a geostationary satellite is over 22,000 miles away. A signal traveling from a user to the satellite and back down to the user takes a significant amount of time, about a quarter second, to travel this path. Example 10-3 illustrates the link delay that such a signal would experience.

■ **EXAMPLE 10-3**   Find the total link delay a signal would experience in one round trip from the surface of the earth to a satellite and back again. Assume a standard geostationary orbit altitude of 22,270 miles.

$$d = rt \rightarrow t = \frac{d}{r} = \frac{\left((22{,}270 \text{ miles})\left(\dfrac{1.61 \text{ km}}{1 \text{ mile}}\right)\right)}{3 \times 10^8} = 120 \text{ ms}$$

$$\text{Total delay} = (2)(120) = 240 \text{ ms}$$

As you can see in Example 10-3, the total round-trip delay approaches ¼ of a second. This delay is not significant for television broadcasts or data traffic, but it is unacceptable for real-time conversation or video conferencing. Processing and switching on board the satellite or in the earth station will typically add another 100 to 150 ms to this delay, making the latency number approach a maximum of 400 ms. Low earth-orbiting satellites significantly reduce this time, which is the primary reason that organizations contemplating these applications use satellites in an LEO. See Example 10-4.

■ **EXAMPLE 10-4**   Find the total link delay in the Globalstar LEO system, which orbits the earth at an altitude of 875 miles.

$$d = rt \rightarrow t = \frac{d}{r} = \frac{\left((875 \text{ miles})\left(\dfrac{1610 \text{ m}}{1 \text{ mile}}\right)\right)}{3 \times 10^8} = 4.7 \text{ ms}$$

$$\text{Total delay} = (2)(4.7) \leq 10 \text{ ms}$$

As you can see in Example 10-4, the total round-trip delay is less than 10 ms. Globalstar does not perform any significant on-board switching, so little additional latency is added from the space-based component. Any terrestrial latency must be added to the value calculated in this example.

Signal Attenuation   Since GEO satellites are located at such high orbits, the communications distances involved cause significant attenuations on both the uplink and downlink segments. In each direction, the attenuation is on the order of 200 dB. Put another way, the signal strength is reduced by a factor of about (1,000,000)*(1,000,000)*(1,000,000,000). This reduction has two important consequences. First, the receiver and receive antenna must be high-gain, special-purpose systems to receive and capture this faint signal.

These features make these units expensive, again adding to the up-front costs of implementing a communications system using satellites. Second, because the signal strength is so weak, it is very susceptible to interference from other terrestrial sources, especially microwave links that radiate large amounts of power near the source in a similar frequency band.

■ LOW EARTH-ORBITING SATELLITES

Low earth-orbiting satellites are systems where several satellites orbit in primarily polar orbits at average altitudes of between 300 and 900 miles. A polar orbit is one where the satellite passes over one or both poles of the earth. At these altitudes, the satellites precess about the earth quite rapidly. With the use of spot beam technology, the total array of satellites can reach almost any location in the world.

To understand how these orbits might be defined, imagine a peeled orange and compare it to the earth. The orange has several wedges that meet at the poles. The satellites essentially follow the path of the edge of the wedges in a polar orbit. Each edge of each wedge has several satellites, following one after the other as they orbit the globe. There are other methods whereby LEO satellites might orbit the earth, but our orange-wedge example is representative of the group. Note that all LEO satellites are "hidden" for part of each orbit as compared to a GEO satellite, which is never "hidden."

Because LEO satellites are not geostionary, they are continually precessing about the earth. At any one point in time, several of these satellites will be over the horizon from the perspective of a person viewing from the earth. In other words, because the satellites are in relatively low orbits, many terrestrial locations are over the horizon from the satellites' perspective. Thus, any LEO satellite is hidden for part of each orbit from a stationary observer on the earth. Contrast this situation with that of a GEO satellite. If stationary observers on earth can see a GEO satellite at any point in time, they can always see that satellite as long as it remains in orbit. The orbit is geostationary.

LEO satellites in an orange-wedge orbital arrangement are subject to collisions as they approach either pole. LEO satellites require careful positioning in orbit to avoid this problem. Not all LEO systems use polar orbits, although most do. The Irridium system did; the Globalstar system does not. The LEO Globalstar system uses forty-eight satellites in eight circular orbits distributed around the globe, with six satellites in each orbit. The Globalstar system explicitly sets these orbits so that coverage at the poles is impossible. This arrangement allows the use of fewer satellites and avoids the polar crowding present in a typical polar orbit scheme.

■ GEO AND LEO SATELLITE COMPARISON

The biggest advantages to an LEO system compared to a GEO system are related to the fact that the orbits are much lower than those for GEO satellites. Lower orbits 1) allow lower power terrestrial communication devices (mobile voice/data devices) and 2) they offer a much lower round-trip delay (lower latency).

Because LEO satellites are in such low orbits relative to GEO satellites, the attenuation of signals and the latency due to link delay are also much lower. For many cost-sensitive and latency-sensitive applications, these characteristics are key. For example, satellite telephony requires small antennas on the phones and a strictly limited latency to ensure the perception of full-duplex conversation to the subscriber.

An LEO satellite communications system is sometimes assumed to be less expensive because the satellites themselves are less expensive. Actually, the reverse is usually true. Several LEO satellites are required to accomplish global coverage. Although each individual LEO satellite can be much lower in capability and cost less than a GEO satellite, similar coverage of the globe by a GEO system can be accomplished by as few as three satellites. It is expensive to put such a global system in place, however, even when each satellite itself is comparatively low cost. For example, the Irridium system placed sixty-six satellites in LEO at a system cost of $4.5 billion. The Irridium system orbited at an altitude of just under 500 miles. Globalstar, owned primarily by Loral and Qualcomm, placed forty-eight satellites in LEO at a system cost of $3.5 billion. The Globalstar system orbited at an altitude of just under 900 miles.

Only the Irridium system offers truly global coverage (Globalstar does not provide coverage at the poles), yet both systems are basically global systems, even though their orbits and the number of satellites differ. As you can imagine, the higher the altitude of a particular satellite, the larger the coverage area on the earth. Irridium, orbiting at an altitude of 500 miles, needs sixty-six satellites; Globalstar, orbiting at an altitude of almost twice that, requires only forty-eight satellites. Of course, the penalty of higher orbits is twofold: satellites require higher power transmissions, and propagation delay is increased in round-trip paths.

Other differences between these two systems are the size, weight, and complexity of the satellites used. Irridium satellites, with a full load of fuel, weigh about 300 lbs. The Globalstar satellite weighs about 200 lbs. While both systems are designed to have several satellites launched in a single rocket or launch vehicle, the lower weight and smaller size of the Globalstar reduces launch costs. Both satellites use much less power while in orbit than a typical GEO satellite, around 1 kW instead of 10 kW.

The Irridium satellites are triangular, about 13 feet long and 3 feet across each base. They feature on-board switching with beam-forming methods to distribute traffic across forty-eight spot beams. The switching is capable of handling over 1,000 simultaneous channels. The Globalstar system takes another approach and keeps the satellites simpler. The Globalstar satellite is a straightforward bent-pipe satellite or relay. All call switching must be done terrestrially, but each satellite is less expensive. It also allows switching technology and architectures to be changed easily because they are accessible. Thus, the owner of the satellite can use the satellite bandwidth in a much more flexible way.

When Irridium failed financially, a key barrier to its use in a data rather than voice environment was that the switching was designed for voice calls and voice bandwidths. This design made it almost impossible to market as a data relay for use with Internet traffic. Irridium has resurfaced, however, as a pure voice service. Some governmental agencies had committed to using the Irridium system, and when it came close to financial collapse, the assets were transferred and a financial deal was arranged whereby the new Irridium organization would provide the U.S. government voice service for several years at a fixed price. This price was paid up front and was used to reimburse the original owners of the Irridium assets.

Does this new arrangement mean Irridium will have a long-term future? The answer is probably no. This transfer netted only pennies on the dollar to the original investors, so no new investors are coming forward. Once the satellites reach the end of their service life and have to be taken out of orbit, it is extremely unlikely that new funds will be found to construct and launch replacements. Irridium may have resurfaced, but it will be a short reincarnation. The satellites have a design lifetime of about ten years.

The downside of placing the switching terrestrially is that, instead of being able to route traffic directly between satellites, all traffic must be routed through the terrestrial ground stations. This arrangement adds throughput delay to the traffic. Because the additional hop required has a round-trip delay of only about 10 ms and the switching delay is an order of magnitude greater, however, this additional delay is insignificant. Figure 10-5 shows a Globalstar satellite used in these systems.

Figure 10-5
Globalstar Satellite (*Courtesy of Globalstar*)

■ SATELLITE SUBSYSTEMS

Three satellite subsystems are critical for a satellite's operation. The first is the way a satellite stabilizes itself in orbit, the second is its power supply, and the third is the transponder/antenna arrangement. If any one of these subsystems fails, the satellite is crippled to one degree or another, depending on the magnitude of the failure.

Stabilization Subsystem

Due to variations of gravity, no geostationary satellite remains truly stationary. Specific subsystems must be in place to keep its orientation correct. The inclination angle of the satellite with respect to the ground changes due to the gravitational effects of the sun, moon, and the nonspherical shape of the earth. Although we have assumed the earth to be a perfect sphere, in reality, it is slightly egg-shaped.

These variations cause the satellite to vary from its exact position in orbit. To address this problem, all geostationary satellites carry canisters of Hydrazine gas attached to small thruster rockets. On ground-control commands, the satellites expend this gas to keep their position stable. These small thrusters are either attached to a hinge and the entire thruster assembly is moved, or only the nozzle itself is attached to a swivel. Either approach provides the operator a way to adjust the direction of the applied thrust to control position. As the reserve supply of this gas expires, so too does the life of a satellite.

Another issue is associated with how a satellite keeps its position in orbit relative to the earth constant. Just as the satellite itself may move positionally in orbit due to gravitational effects, it also changes its attitude or orientation to the earth. The thrusters can be used to keep a satellite's position constant, but if they are also used to adjust attitude, the gas will quickly expire. Additionally, gas thrusters turn the satellite rather forcefully and without fine control. Therefore, almost all satellites use a combination of gas thrusters and a more elegant method such as spin stabilization or gyroscopic (sometimes called momentum) stabilization to control attitude.

Spin stabilization is accomplished by actually spinning the satellite when it is placed in orbit. Typically the satellite is spun at a rate of about 200 revolutions per second. Through this approach, the satellite achieves stability in the same way a gyroscope stays upright as long as it is spinning at a rapid enough rate. The use of this method has implications in the design of the satellite. If a satellite will use this approach, it must be cylindrical in shape and cannot use large, extensible solar panels for a power source. Many communications satellites in earth orbit use this approach.

Gyroscopic stabilization uses three metal disks or momentum wheels, one spinning on each dimensional axis. This setup accomplishes stabilization by breaking it into three pieces. Each wheel, of about 10 cm in diameter, is entirely enclosed and provides stabilization in one dimension. By using three, three-dimensional stabilization is achieved. The idea is that they can serve to turn the satellite in different directions through the relative speed each wheel is turning. This type of stabilization is capable of very small adjustments, allowing fine control.

Each metal disk roughly resembles the spinning wheel of a simple gyroscope. Instead of spinning the wheel by hand in a spacecraft, however, spinning is done by an electric motor. Top speed for each wheel is about 3,000 rpm. Occasionally, one or more of these wheels will be spinning at close to its top speed and thus will no longer be effective for orbit correction. With this situation, the momentum inherent in this spinning must be dumped, and a procedure called the angular momentum desaturation (AMD) maneuver is performed.

The AMD maneuver involves spinning the wheels down, which of course changes the orientation of the satellite or spacecraft. Outside earth's orbit, this change can be offset by firing the gas jets. If the satellite is in earth's orbit, sometimes other forces that the satellite experiences or generates can be used to counteract this force and conserve gas. Examples include earth gravity gradient or another device called a magnetic torquer. This simple device consists of a wire loop with a current running through it. It generates a push against the earth's magnetic field.

The gyroscopic stabilization system is set before launch and provides a reference for the alignment of the satellite during its lifetime. Gyroscope systems have a tendancy to drift, so a radio link is maintained for periodic realignment. This link can be used on any satellite shape because the stabilization is achieved internally. Often communication satellites will employ large solar panels for power and will be noncylindrical in shape. Gyroscopic stabilization is used for many earth-orbiting satellites, and almost all NASA exploration satellites use this approach because it is much more suited to the many navigational changes that these probes require. Both of the first-generation space stations, Skylab and Mir, used larger versions of this approach; the international space station will as well.

Figure 10-6 shows two photographs of representative communication satellites. Figure 10-6(a) shows a recent launch from the space shuttle: a large spin-stabilized communications satellite. The surface of the satellite is covered with solar panels. This photo also gives you a good idea of how large some communications satellites have become; this one weighs about 3 tons. Figure 10-6(b) shows a large gyroscopically-stabalized satellite.

Most recently launched satellites use the gyroscopic stabilization technique because it allows more flexibility in the satellite design. Specifically, it means that the solar panel area, and hence power availability, does not have to be linked to the spacecraft surface area. Additionally, and perhaps more important, with gimbled solar panels, the panels can be pointed toward the sun continuously. In a spin-stabilized satellite, a portion of the solar cells mounted on the cylindrical body of the satellite are always facing away from the sun. On the other hand, it is possible to place more electronic processing and functionality in cylindrical bodies, so high-capacity telecommunications satellites tend to be cylindrical in shape and hence spin stabilized.

Power Subsystem

The power subsystem in a satellite can be one of two general types: solar powered and nuclear decay. Only deep space systems, such as the Voyager series, and specialized military satellites use nuclear decay because of earth environmental concerns.

Figure 10-6
(a) Spin-Stabilized Communications Satellite (*Digital Imagery © copyright 2001, PhotoDisk, Inc.*)

Figure 10-6 (continued)
(b) Gyroscopically-Stabilized Communications Satellite (*Photo courtesy of David Ducros/Science Photo Library/Photo Researchers, Inc.*)

All earth-orbiting communication satellites use solar power to generate the electrical power neccessary for operation. As alluded to above, these solar cells can be configured in two ways. For spin-stablized satellites, the solar cells are mounted directly on the cylindrical body of the satellite. A portion of the cells are always pointed toward the sun. As you can imagine, this approach is relatively inefficient because a portion of the cells are always facing away from the sun and contribute no electrical power to the satellite.

Gyroscopically stabilized satellites deploy large arrays of solar cells that are gyroscopically gimbled so that they are always pointed directly at the sun. These cells are folded against the body of the satellite during launch and then deployed in space once orbit is achieved. Small stepper motors adjust the gimbled solar cell arrays every 24 hours to adjust for the earth's orbit around the sun.

The solar panels are rectangular in shape after unfolding, with a total area of several square meters. The solar cells are manufactured from GaAs, and a set of panels suitable for a typical GEO satellite launched today could generate about 10 kW when new. An additional power supply in the form of a battery is always required to power subsystems during launch and eclipses. For a geosynchronous satellite, an eclipse occurs twice a year, near the spring and fall equinoxes, when the earth's shadow passes across the satellite. Today, this battery is typically of a nickel-hydrogen technology capable of supplying between 10 and 25 ampere hours (Ah).

A problem with solar cells is that they gradually lose effectiveness over time. As pointed out earlier, a typical satellite has a lifetime of about ten years. Over this period, the electrical power supplied by the solar cells in either arrangement will decrease approximately 15 to 20 percent. This falloff of power must be considered during the design stage because the power available near the end of the satellite's life will be significantly lower than when it is deployed. This deficiency is usually addressed through a combination of overdesigning the solar cell subsystem and by selectively shutting off communication channel transponders to reduce the power requirements of the communications system as the satellite ages.

Transponder and Antenna Subsystems

The transponder and antenna subsystems are what the satellite uses to implement the relay function. The other subsystems in the satellite support these subsystems. A transponder is nothing more than a repeater that operates at RF frequencies. It amplifies and repeats at its output what was presented at its input. It also frequency converts this signal from the uplink frequency down to the downlink frequency. The uplink frequency is always higher than the downlink frequency because the lower the frequency, the less power required to propagate it any given distance. Because power is limited on a satellite, the lower frequency of the two is always used for satellite-originated transmissions. Each satellite channel requires a transponder. Because this element is critical in the satellite's function, several spares are built into the satellite. They are switched in when any single transponder fails.

Different satellites feature different numbers of transponders or channels. The most common is twenty-four transponders. On any single satellite, each transponder will feature the same bandwidth, typically either 36 or 72 MHz. Most C-band satellites use twenty-four transponders with 36 MHz bandwidths. Each transponder is spaced at 40 MHz increments to allow some guardband between adjacent channels.

The transponder is composed of several discrete elements. Traditionally these elements were implemented using traveling wave tubes and what is called "bent-pipe" engineering, which is another name for analog signal processing using the properties of the physical medium in which the RF signals travel. Today, these elements are constructed using solid-state and MMIC technologies. The eight basic components of a transponder/antenna system in a communications satellite are shown in Figure 10-7(a); Figure 10-7(b) shows the addition of an OBP module.

Figure 10-7(a) illustrates the components of a single channel or transponder of a simple relay satellite, although most modern satellites placed in orbit in the last several years feature

Figure 10-7
(a) Basic Diagram of a Satellite Transponder; (b) Addition of an OBP Module

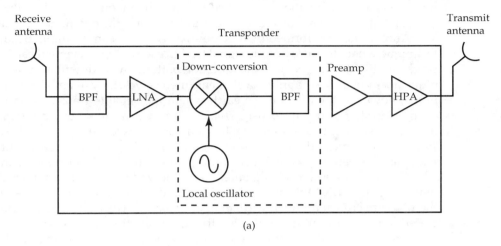

(a)

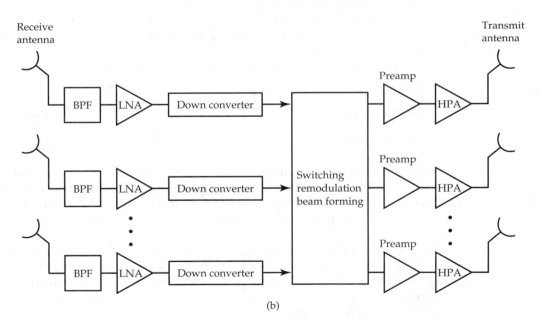

(b)

some kind of on-board processing (OBP). OBP systems allow satellites to switch data streams effectively on the satellite. OBP systems consist of three basic components, and they will be explored in further detail in a section below. OBP systems, by design and necessity, interconnect among several transponders, which you can see in Figure 10-7(b), where several individual transponder antenna pairs deliver independent signals to the OBP module.

This module then performs several potential tasks, including the previously mentioned switching. The first task is remodulation of the information signal to be passed to the downlink antenna. Remodulation is required because the module often does this switching at baseband. It must therefore remodulate it and place it at the downlink frequency.

Another task performed by this module is called beam forming. Beam forming is a technique where the delay of the signal feed to each antenna element can be modified to produce a highly focused antenna radiation pattern on the downlink antenna. This technique does not move the antenna elements themselves; they remain fixed. With this technique, however, the area on the earth where the satellite transmission is sent can be changed in shape quickly, focused sharply, or broadcast in several different locations simultaneously. The antennas for the Irridium used this technique. Examination of Figure 10-5 shows the individual antenna elements (the little squares on the antenna assembly). Each can be steered and combined through use of the beam-forming technique.

As you can see, a receiving antenna is followed by a bandpass filter (BPF) and low noise amplifier (LNA). As we will see in the link calculations, the noise specifications of the LNA are critical for overall system performance. The next three components are enclosed in a dotted outline to indicate that their function is to down-convert the carrier frequency used on the uplink to that used on the downlink. Again note that the uplink frequency is always higher than the downlink frequency. Following that conversion, another two stages of amplification, the RF preamp and the high power amplification (HPA), are performed just prior to passing the signal to the transmit antenna for transmission of the signal to the earth station.

Three basic types of antennas are used on satellites:

Horn antennas

Parabolic reflector antennas

Array antennas

Pure horn antennas are normally employed for global beam applications. They have very wide beamwidths. A reasonable minimum beamwidth for a horn antenna is about 10 degrees. For any application where spot beams are required, horns must be combined with some kind of reflector, usually a parabolic dish. For more information on horn antennas, see Chapter 8.

In satellite applications that require narrower beamwidths, parabolic reflectors are used with multiple horns, often called horn arrays. Figure 10-8 illustrates this configuration for a typical array with a reflector. As you can see from the figure, a horn array is at the focal point of the parabolic reflector. Each horn is fed by a transponder, and the spot beam is designated for a different geographical region on the earth. Here, different cities in northern and central United States are used as an example.

Figure 10-8 shows a line drawing to illustrate the components clearly. Cylindrical horns are combined with a parabolic reflector to increase the gain and narrow the beamwidth. Each horn would be targeted to a specific region on earth. For example, horn 1 would place a spot beam on Chicago, horn 2 would place a spot beam on Atlanta, etc.

Figure 10-8
Diagram of Horn Array with Reflector

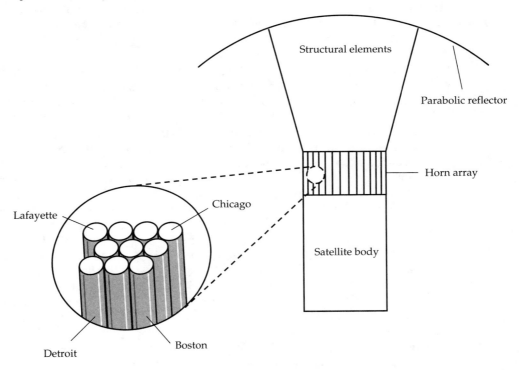

As suggested earlier, narrow antenna beamwidths with high gain require large antenna reflectors. These antennas could not be accommodated in a typical launch vehicle or rocket. Once the satellite is placed in orbit, the parabolic reflector is unfolded and deployed. The reflector is built of a series of "petals" that are folded closed during launch.

Array antennas are widely used in LEO satellites and in many sophisticated applications. Array antennas rely on the beam-forming capability of an OBP module. Again, beam forming can control the beamwidth, power density, and direction of the radiation pattern emitted by the antenna. It is performed by selectively controlling the delay or relative phase of the individual antenna elements in the array. Because beam forming does not require any moving parts, it produces a movable signal that can broadcast in several locations at once.

■ ON-BOARD PROCESSING (OBP)

We discussed OBP briefly earlier in the chapter when it was necessary to introduce the concept as one major classification of satellites and to complement the discussion on antenna types used in satellites. In this section, we will summarize and complement the information presented earlier.

Many modern satellites feature on-board processing or switching, which allows the satellite to operate not only like a relay, but instead to switch, remodulate, and re-multiplex the incoming data streams to particular transponders and shape antenna radiation patterns based on the application requirements of the data stream. OBP systems make use of three major subsystems: switching, beamforming, and multicarrier demodulation.

Figure 10-9(a) shows a block diagram of a satellite with switching capability, but with fixed beams and single carrier demodulation. Note that the switching takes place at IF frequencies and not digitally at baseband. Note also that no demodulation of the individual carriers occurs. This was a first step in performing switching on a satellite. Figure 10-9(b) is a block diagram with two of the components clearly shown; digital switching and multicarrier demodulation allow the information content on the carrier to be switched at baseband in the next module. The third component of OBP systems is the beamforming capability. Beamforming capability is performed in the antenna elements themselves and is implemented in two distinct ways, described below. Modern satellites implement this functionality through the use of phased array antennas.

An example of a relatively simple OBP system is one that would use a series of narrowband SCPC FDM channels in a single transponder on the uplink, and receive and remodulate these channels into a single broadband TDM signal on the downlink. For example, multicarrier demodulation, switching, and remodulation is illustrated by the following example. Suppose the transponder is 72 MHz in bandwidth. Using relatively efficient modulation techniques, such as QPSK, it would be possible to frequency multiplex ninety-six DS-1 signals of 1.544 Mbps into a transponder. The OBP-equipped satellite would be able to receive each DS-1 signal, combine or switch each of them into a single transmission, and remodulate them into a single 150 Mbps downlink.

Digital beam-forming techniques are fundamental to making the best use of OBP systems. Digital beam-forming and another technology called spot beams are closely related. Beam forming is a technique where the delay of the signal feed to each phased array antenna can be modified to produce a highly focused antenna radiation pattern. This technique does not move the antenna elements themselves—they remain fixed—but with this technique the area on the earth where the satellite transmission is being sent can be changed in shape quickly, or the transmission can be focused more sharply or it can be broadcast in several different locations simultaneously. In some cases, spot beams are generated using digital beam-forming techniques; in others, they use narrowly focused antennas.

The switching capability of OBP-equipped satellites is also used to implement a variation of TDMA called satellite switched TDMA (SS-TDMA). In this form of TDMA, a satellite receives a traffic burst from one location, in one time slot. The satellite can buffer the information and retransmit it at some later time in a different location. Spot beam technology can be combined with this approach to relay information to different locations sequentially while transmitting it to the satellite only once.

Spot beam technology can be implemented using OBP beam-forming techniques or with narrowly focused antenna arrays. Figure 10-5(a) illustrates the former; Figure 10-9(b) illustrates the latter. In the latter case, which is often how current satellites in orbit generate spot beams, the number of antennas that can be thus configured is limited. This limitation can be an important factor in defining the satellite's capability to support a specific application.

Figure 10-9
OBP Evolution: (a) Fixed Beams; (b) Scan (Spot) Beams

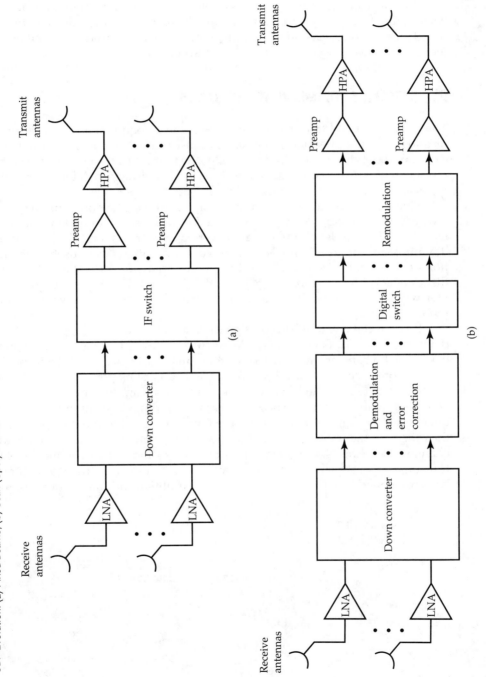

307

A big advantage of spot beam technology in either implementation approach is that because the beam is spread over a relatively small area, the power at the terrestrially based receive antenna is proportionally higher. This increased power allows smaller antennas and lower cost electronics at the destination.

■ SATELLITE COMMUNICATIONS BANDS

The frequency band that the satellite uses is the second major classification of communications satellites. Because many satellites use more than one frequency band, this concept can sometimes be confusing. Usually, the highest frequency band used by the satellite is also used to classify the satellite. This factor is critical in determining what antenna types are used both terrestrially and in the satellite itself.

Five bands are used in communications satellites. The downlink frequency range is always below the uplink frequency range for the reasons pointed out earlier. Table 10-1 lists these five bands with the total band frequency range and the portion used for the downlink and uplink channels. C-band communications were used in the first generation of communications satellites. They used large 3-meter dishes. The newer, smaller dishes of about 1 meter in diameter use the Ku band. The next generation of LEO satellites will use the Ka band. These band names were originally developed for use in the radar industry.

If you receive your television signals on an antenna that is approximately 1 meter in diameter, it is a Ku band antenna. Most current VSAT systems use this communication band. The Ka band antennas will be slightly smaller, approximately 26 inches in diameter, and for home use will be duplex in nature with an uplink power of only about ½ watt.

In any satellite, two broad channels are in use simultaneously: the uplink channel and the downlink channel. These two broad groupings of channels are separated in frequency, with one set of frequencies, the lower, assigned to the downlink channels, and the other set, the higher, assigned to the uplink channels. Each of the broad groups of channels is further subdivided into a number of individual channels used for actual transmission of signals. Each individual channel has its own transponder and antenna. Each transponder contains its own receiver, amplifier, and transmitter designed to work

Table 10-1
Satellite Bands

Band	Frequency Range (MHz)	Downlink (GHz)	Uplink (GHz)
L	1,550–1,650	1.55–1.6	1.6–1.65
C	4,000–6,500	3.7–4.2	5.9–6.4
X	6,500–10,500	7.2–7.8	7.9–8.4
Ku	10,500–17,000	11.7–12.2	14.0–14.5
Ka	17,000–36,000	17.0–21.0	27.0–31.0

Table 10-2
C Band Channel Assignments

Channel Number	Downlink Center Frequency (MHz)	Downlink Polarization	Uplink Center Frequency (MHz)	Uplink Polarization
1	3,720	Horizontal	5,945	Horizontal
2	3,740	Vertical	2,965	Vertical
3	3,760	Horizontal	5,985	Horizontal
4	3,780	Vertical	6,005	Vertical
5	3,800	Horizontal	6,025	Horizontal
6	3,820	Vertical	6,045	Vertical
7	3,840	Horizontal	6,065	Horizontal
8	3,860	Vertical	6,085	Vertical
9	3,880	Horizontal	6,105	Horizontal
10	3,900	Vertical	6,125	Vertical
11	3,920	Horizontal	6,145	Horizontal
12	3,940	Vertical	6,165	Vertical
13	3,960	Horizontal	6,185	Horizontal
14	3,980	Vertical	6,205	Vertical
15	4,000	Horizontal	6,225	Horizontal
16	4,020	Vertical	6,245	Vertical
17	4,040	Horizontal	6,265	Horizontal
18	4,060	Vertical	6,285	Vertical
19	4,080	Horizontal	6,305	Horizontal
20	4,100	Vertical	6,325	Vertical
21	4,120	Horizontal	6,345	Horizontal
22	4,140	Vertical	6,365	Vertical
23	4,160	Horizontal	6,385	Horizontal
24	4,180	Vertical	6,405	Vertical

at the assigned receive and transmit frequencies for that channel. Table 10-2 illustrates the channel assignments for those satellites that use the C-band.

Table 10-2 also shows the polarization of each channel assignment. Polarization allows the satellite to make use of the same frequency space twice. This concept is called frequency reuse or polarization diversity and is responsible for effectively doubling the number of channels that a satellite can use. Examine Table 10-2 closely. The center frequencies of adjacent channels are separated by just 20 MHz. As stated earlier, however, each channel's usable bandwidth is 36 MHz. If this setup seems like an impossibility, it is unless you consider the concept of polarization.

■ POLARIZATION DIVERSITY

This concept is widely used in all types of satellite systems. Recall from Chapter 8 that every antenna has a specific polarization: horizontal or vertical. Also recall that the transmitting antenna and receiving antenna must have the same polarization. A wave with a given polarization will transfer no energy to a receiving antenna that is orthogonally polarized to the wave. Therefore, a vertically polarized receiving antenna will receive nothing from a horiziontally polarized wave, and vice versa. Since each adjacent channel is of a different polarization compared to its nearest neighbors, the effective bandwidth of each channel is really between the adjacent channels of the same polarization. Instead of a 20 MHz channel bandwidth, there is instead a 40 MHz channel bandwidth. Of this bandwidth, 36 MHz is usable, and there is a 2 MHz guardband on each side.

Each antenna for each channel is polarized in a particular orientation, either horizontally or vertically. Hence, one obtains twice the effective bandwidth for each channel. This setup allows signals to overlap in frequency without interference, a technique known as frequency reuse. In Chapter 11, this same term will be used to describe a critical component of terrestrial cellular telephone networks. It needs to be emphasized that, although the term used is the same, the concept is quite different. Satellites use the same frequency range in adjacent transponder assignments by ensuring that the polarizations of adjacent channels are orthogonal to each other. Thus, each pair of transponders can use the same frequency twice, once in a horizontal polarization and once in a vertical polarization. The chief drawback to this technique is that rain or moisture in the atmosphere tends to depolarize the waves and reduce the effectiveness of using orthogonal polarization for the same frequency twice.

In cellular networks, all antennas are polarized in the same direction. The concept there has to do with limiting the transmit power so that the same frequency range can be used nearby because the range of the transmitter is limited. Frequency reuse in cellular networks will be explored in more detail in Chapter 11.

■ RELAYS OR SWITCHERS

Whether the satellite features an OBP module or not is the third major classification of communications satellites. Satellites with no OBP module are called relays; those with an OBP module are called switchers. Traditionally, communications satellites were simple relays; hence the name often applied was bent-pipe satellites. This name referred to the fact that the satellites were constructed using analog microwave engineering practices where much of the signal processing and frequency translation was accomplished by the geometric arrangements of hollow waveguides or bent pipes. These systems did nothing more than amplify and retransmit the signals that were received. Channels were differentiated using FDM techniques. Each channel was uplinked in a single frequency band and downlinked in another. These bands were predefined and could not be changed. Relay satellites do not feature any capability for switching an entire transponder in any way; they are pure FDM.

There are many clever ways to use a single transponder in a bent-pipe satellite to relay several distinct information sources. Four basic techniques will be explored:

1. FDMA techniques
2. TDMA techniques
3. DAMA techniques
4. CDMA techniques

■ FDMA TECHNIQUES

All FDMA techniques use the principle of a single uplink frequency and a single downlink frequency. Each frequency is dedicated to a single information flow. Information flows are distinguished by the frequency used to carry them.

Single Access

Single access was the first and, for some time, the only way to use a transponder in a relay satellite. In this approach, a single baseband signal drives a frequency modulator. This modulated signal is up-converted into a single carrier, uplinked to the satellite transponder, and relayed to another terrestrial location on the downlink. This single modulated carrier uses the entire transponder bandwidth. See the satellite system in Figure 10-10.

As you can see in Figure 10-10, the transponder is dedicated to a single carrier. Thus, the information on that carrier is often composed of several different narrowband sources or a single broadband source. In this arrangement, each signal requires an entire

Figure 10-10
Single Access FDM Approach

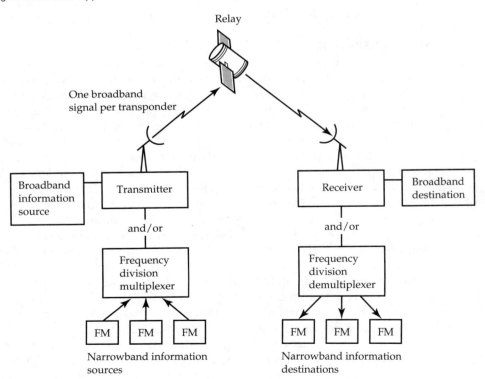

transponder. For this reason, this approach is usually restricted to those signals that can make use of an entire transponder's bandwidth. This approach is used by the vast majority of television signals relayed around the globe. In this application, the television signal, which is really a form of AM, is converted to a frequency modulated signal, up-converted, and passed to a single transponder in a relay satellite.

A television signal is quite broadband and, when frequency modulated, consumes an entire transponder bandwidth, so this application makes sense. In fact, transponders were originally conceived and designed with this application in mind, and the bandwidth specification has become a standard.

Subcarrier Technique   However, many signals are not as wideband as a television signal. In this situation, to approach even remotely a cost-effective solution, several of these signals needed to be combined and relayed simultaneously. A typical example is the combination of approximately 1,000 one-way audio or voice circuits on a single FM carrier. When the single access technique is used in this way, the individual information signals are referred to as subcarriers.

Once the satellite downlinks the signal to a terrestrial earth station and downconverts it, the recovered baseband signal is then frequency demultiplexed back into distinct FM signals. Each narrowband information signal is routed to a satellite receiver and then to an individual subcarrier demodulator, which recovers the original narrowband signal. The subcarrier demodulators are essentially monaural FM radios. If the signal is an audio one and stereo is desired, two separate subcarrier signals must be relayed.

Single Channel per Carrier (SCPC)

The solution to making use of a portion of a transponder without the frequency multiplexing at each end was to subdivide the transponder frequency bandwidth into several smaller preassigned channels. This process is known as the single channel per carrier (SCPC) approach. SCPC implies only a single information signal per carrier, not transponder. In this approach, each transponder may have many hundreds of carriers.

The key difference between a single access and SCPC system is that the single access approach requires that the individual information signals be premultiplexed onto a single carrier, while in the SCPC approach, the individual carriers are uplinked and downlinked through the transponder. Both approaches allow many different sources to be combined into a single transponder, but the SCPC approach is more flexible. Refer to Figure 10-11 and note the contrast to the single access approach.

The main drawback to the SCPC approach compared to the subcarrier approach is that, because the carriers are passed individually through the transponder, the SCPC approach does not use the bandwidth of the transponder as efficiently as the subcarrier approach can. (Refer again to Figure 10-11.) The SCPC approach, with each information signal carried on its own carrier, is required to implement FDM DAMA.

The SCPC approach is often used to implement so-called thin route service. This application is sometimes used to implement voice telephone service in remote areas at a more reasonable cost than stringing a twisted pair cable to a remote or isolated area. It is also used to implement preassigned voice channels for transportable earth stations. In this application, one large earth station is connected to the terrestrial wireline telephone

Figure 10-11
Thin Route Service

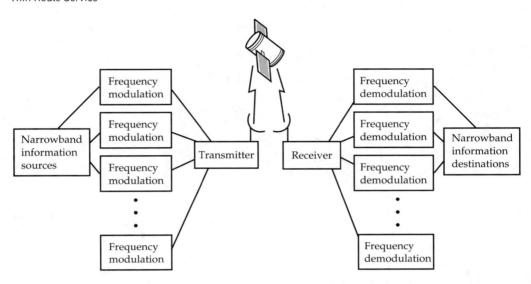

system. An incoming call from a remote location is passed to the remote earth station, uplinked to the satellite in a preassigned carrier location on a transponder, downlinked to the main earth station , and passed to the wireline telephone network. This service is illustrated in Figure 10-12, where each line represents one SCPC simplex link in the transponder.

■ TDMA TECHNIQUES

TDMA is the dominant multiplexing method used in satellite applications. TDMA is the traditional wireline method for taking a single physical channel and placing many distinct information flows in it. These information flows are passed to a central location and combined in specific time slots. Each time slot is a logically distinct channel. TDMA has been used in long-distance trunk lines in the telephone system for years.

In satellite applications, each terrestrial earth station sends a short burst of data at its appropriate time slot in the TDMA frame. Typically, several terrestrial stations use the same transponder at different time slots. The burst of data is modulated onto a single broadband signal on a single carrier frequency, upconverted, and uplinked to an individual transponder in a relay satellite. As in the single access approach, the satellite transponder relays only what was passed to it. The satellite then downlinks this signal to one or more terrestrial sites, where the individually time-multiplexed signals can be broken out and processed.

Each uplinked burst of data must be precisely time-synchronized with every other burst of data so that they do not overlap. Only one burst of data can arrive at the

Figure 10-12
SCPC Approach

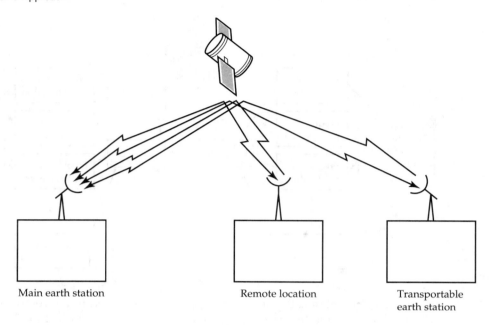

Main earth station   Remote location   Transportable
                                                     earth station

transponder at one time. As must be clear, each station must somehow be synchronized with every other one. Traditionally, this synchronization is done in two steps. First, on every downlink, what is called a reference burst (or, perhaps more familiar, a preamble) is transmitted. Each burst of data is preceded by this sequence of data, which is usually an alternating series of 1s and 0s. This preamble is used to allow all receiving stations to synchronize their carrier recovery systems that demodulate the data contained in the burst. The preamble is also used to derive a local system clock synchronized with the data in the burst.

The second step is to give all local clocks a time reference. This process is done by transmitting a unique bit sequence after the preamble. This bit sequence, usually a series of twenty to thirty 1 bits followed by a single 0 bit, is used by each station to establish an exact time reference for that station. This time reference is required by each station when transmitting its next burst to ensure that the bursts, when arriving at the satellite transponder, do not overlap.

■ TDMA VERSUS FDMA

The primary advantage of TDMA compared to FDMA is that TDMA, by its very nature, is a system designed to transmit digital data. The vast majority of data today is digital, thus giving TDMA a clear edge. The second major advantage is cost. In FDMA, each individual information signal is on a different carrier, so each terrestrial station must be capable of receiving several different carrier frequencies. This feature adds significant cost to the terrestrial station. In TDMA, only a single carrier is used, thus simplifying the terrestrial engineering. On the other hand, TDMA requires each terrestrial station to do precise time synchronization. Precision tends to add cost because it usually comes with a price tag. However, the suitability of the TDMA system for transmitting digital data outweighs this consideration.

■ DEMAND ASSIGNED MULTIPLE ACCESS (DAMA) TECHNIQUES

FDM DAMA

It is possible to place several narrowband signals into a single transponder using FDM or TDM techniques. However, the SCPC approach requires a preassigned frequency slot. It would be much easier to be able to assign some percentage of these slots in an on-demand manner. This on-demand assignment can be done in either frequency or time; both have been used. If a few of these narrowband channels are dedicated for signaling, a terrestrial-based computer, called a network control system (NCS), can simulate switching on a few of the narrowband channels under automatic or remote control.

Figure 10-13 illustrates a transponder that is broken into several narrowband SCPC channels. Each of the control channels and the narrowband information channels occupy the same amount of frequency bandwidth in the transponder. Figure 10-13 also shows two control channels at each end of the transponder bandwidth, with several narrowband information channels located between them. Other arrangements are possible. The number of control channels is directly tied to the number of times that the channels need to be reassigned dynamically using the NCS.

The TDM concept is similar, except that the division would take place in time rather than frequency. Instead of the control channels and information channels occupying frequency slots, they would occupy time slots in a single frequency channel. In principle, the time slots could be of different duration.

Figure 10-13
FDM SCPC DAMA

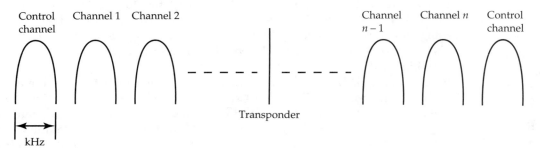

As we already said, Figure 10-13 shows a single transponder being used for several channels using the SCPC approach. Additionally, a couple of control channels are shown. They would be used to implement the DAMA concept. Probably the best known FDM system to use the DAMA concept in this way was the Intelsat SPADE system launched in the early 1980s. This system, along with a VSAT application, will be examined in later sections as an illustration of the DAMA concept. TDM DAMA is described in this chapter in the context of its primary application, VSAT networks.

■ CDMA TECHNIQUES

CDMA techniques are not generally well suited to satellite systems. The fundamental problem is that, with conventional CDMA, each transmission must use a different code word. This requirement implies that each receiver must be unique, clearly a problem given that the chief advantage of satellite communications is that one transmission can be received by large numbers of subscribers. The major advantage of a CDMA approach would be considerable immunity to intersatellite interference, hence the ability to use smaller antennas than might otherwise be possible with any other access approach.

There are two possible solutions to this dilemma. The first is the approach used in the IS-95 CDMA cellular telephony network, described below. The second is the approach used in the IEEE 802.11 WLAN, discussed in the next section. (To review these modifications of classical CDMA, see Chapter 2.) In the IS-95 CDMA cellular system, instead of a large number of different code words, only sixty-four are used. Due to this limited number of code words, only sixty-four different receivers, or a single receiver with sixty-four possible code choices (similar to that done in the IS-95 standard), are needed. These receivers then request access to the transponder on a separate channel, where access is governed by ALOHA techniques. It is not surprising that this solution was similar to the IS-95 cellular approach because it was invented and applied by Qualcomm in its deployments of cellular networks. A satellite was placed into service in 1999 using this technique.

Spread ALOHA Multiple Access (SAMA)

In this technique, CDMA and ALOHA are combined. However, CDMA is applied with a single spreading code, very similar to the way it is done in the IEEE 802.11 WLAN. Since all the transmissions use a single code word, all receivers are identical, a major advantage for any commercial system. The most likely application of this technology is in a VSAT network. A VSAT network generally exhibits large numbers of receivers. If they are all identical, significant cost savings are realized.

Essentially, SAMA works like ALOHA. Each subscriber has random access to the channel or transponder. The CDMA spreading applied yields the major advantage of CDMA, intersatellite interference immunity, hence allowing very small antennas. Typical signal-to-interference ratios are on the order of 6 dB. Since 6 dB amounts to a quadrupling of interference immunity, SAMA techniques can increase the number of subscribers by that same factor. SAMA techniques have been demonstrated in the laboratory, but no systems have been deployed to date.

Coverage Zones   Not all antennas that a satellite uses have the same coverage area. Two broad classifications of satellite antennas are global coverage or spot beam. Modern satellites often feature combinations of both global and spot beam antennas. Global coverage is the traditional way of thinking about a satellite. The satellite orbits the earth and it illuminates a large region of the globe due to its great altitude. This idea was the basis of the relays originally envisioned by Arthur C. Clarke. However, it soon became desirable to refine the audience who could receive the transmission, which resulted in the development of spot beam technology.

The coverage zone is defined to a significant degree by the type of antenna used. Again, there are two classes of coverage zone: global and spot. As discussed above, if the satellite uses spot beam antennas, many different zones may be operating at the same time. If it uses a global antenna, typically there is only one coverage zone at one time. A coverage zone is that part of the earth that is illuminated by the antenna. Figure 10-14 shows the two basic coverage zones in use today, along with typical beamwidths.

Because the number is limited on any satellite, spot beams are now often configured as scanning beams. These beams are spot beams that move from spot to spot in a predefined manner. This feature allows a single spot beam antenna to serve many spots. If you refer to Figure 10-14(b) which shows the NASA ACTS OBP in block diagram form, you will note the presence of scanning spot beams.

Finally, any beam coverage area may be reused more than once and simultaneously. This technique is called frequency reuse and is accomplished by transmitting information in two polarizations, horizontal and vertical. The earth's atmosphere is very unforgiving of this polarization difference and often "twists" the two signals to such a degree that the original polarization can no longer be discerned when the signal reaches the earth. Progress in technology is addressing this issue, and frequency reuse is gaining in acceptance.

■ LAUNCHING A SATELLITE

No discussion of satellite communications would be complete without a word or two about launching, what an earth station is comprised of, and how to point an antenna at an orbiting satellite. When launching a satellite, typically a rocket of some kind is used to move the satellite from the surface of the earth. There are two kinds of launch vehicles: expendable or reusable. A typical rocket is expendable; something like the space shuttle is reusable. Figure 10-15 shows two representative launch vehicles. Part (a) of the figure shows the expendable launch vehicle Arianne 5, the largest and newest in the Arianne family of launch vehicles. Part (b) shows the reusable launch vehicle: the space shuttle.

What is not generally known is that these launch systems do not always place the satellite directly into their ultimate orbits. They generally do for LEO satellites. For GEO satellites, with their much greater altitudes, they do not. Instead, they place the satellite

Figure 10-14
Coverage Zones: (a) Global; (b) Spot Beam

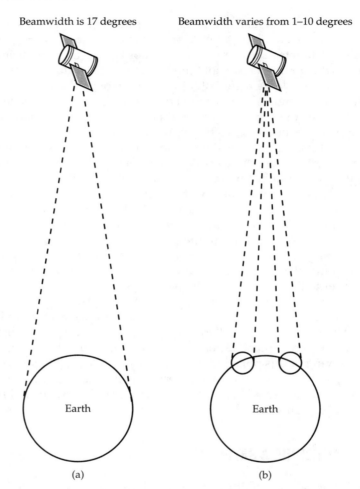

in what is called a transfer orbit. This orbit is very elliptical and is arranged so that the perigee, the point of closest approach to the earth, is at the same altitude as the relatively low altitude initial release altitude. The apogee, the point of farthest distance from the earth, is at the same altitude as the ultimate orbit desired. For a typical geosynchronous satellite, the perigee is about 200 miles in altitude and the apogee is 23,270 miles in altitude. Figure 10-16 shows the relationship among the three orbits.

In the case of an expendable launch vehicle, the rocket itself inserts the satellite into the transfer orbit. In the case of the space shuttle, which must stay in low earth

Figure 10-15
Expendable and Reusable Launch Vehicles. (a) Arianne 5 (*Photo courtesy of David Ducros/Science Photo Library/Photo Researchers, Inc.*)

(continued)

orbit, the satellite is equipped with what is called a perigee kick motor that inserts the satellite into the transfer orbit. In both cases, the satellite is equipped with what is called an apogee kick motor to brake and position the satellite approximately into the geostationary orbit. Fine placement is done by the stabilization motors. It generally takes about thirty to forty-five days for the satellite to insert itself into its final orbital slot and begin operation after it is launched.

Most launch facilities are as close to the equator as possible. The reason for this location is straightforward. The rotational velocity of the earth is at its maximum at the equator because the circumference there is largest. This additional velocity reduces the cost of the fuel used to launch a rocket, thereby increasing

Figure 10-15 (continued)
Expendable and Reusable Launch Vehicles. (b) Space Shuttle (*Photo courtesy of NASA/Science Photo Library/Photo Researchers, Inc.*)

the maximum payload any given rocket can carry. Traditionally, major spacefaring nations of the northern hemisphere placed their launch facilities as far south as possible. Because the earth's surface rotates fastest at the equator, a launch there gives an extra velocity boost. This boost effectively subtracts several hundred miles per hour from the required escape velocity from the earth's gravity. The farther north or south the rocket is from the equator, the more fuel must be used to achieve this velocity. Additional fuel adds significant expense and weight to the launch vehicle. In the United States, proximity to the equator was the reason why Florida won over Texas and California for the primary launch site location of the space program.

However, a new service called Sea Launch is now being offered. In this approach, the rocket is placed on a floating platform, a converted oil drilling platform, and towed to a location as close to the equator as possible. This commercial service is competitive because the same launch vehicle used to launch from the platform can carry additional payload weight than it could in, say, Florida. Because payload

Figure 10-16
Three Orbits for Satellite Launch

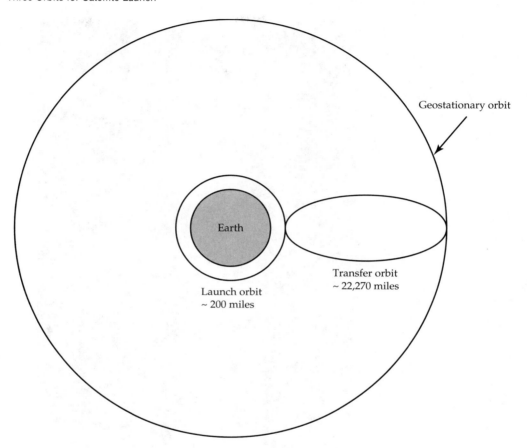

weight often determines the cost of the launch, Sea Launch can undercut competitors' prices for launching the same satellite. The Sea Launch platform with a launch vehicle is shown in Figure 10-17.

■ FINDING A SATELLITE

A satellite in orbit around the earth may appear to be in different locations depending on the position of the earthbound observer. This detail is very important when pointing the ground station antenna to the correct position. The angle and elevation of the antenna is often referred to as the look angles. Two elements or look angles

Figure 10-17
Sea Launch (*Photo courtesy of Sea Launch*)

must be considered. The first is elevation. The second is azimuth, which will be discussed below. The elevation of a satellite is the angle at which the satellite is seen from the observer's location on earth. The larger the elevation, the "higher" in the sky the satellite appears. For an elevation of 90 degrees, the satellite would appear directly above the observer. For an observer on the equator, a geostationary satellite in equatorial orbit could appear at an elevation of 90 degrees. The elevation of a satellite depends on four pieces of data: the longitude and latitude of the observer, and the longitude and latitude of the satellite. Example 10-5 serves as a reminder of these everyday concepts of latitude and longitude.

■ **EXAMPLE 10-5**   Define latitude and longitude and provide a way to estimate both.

Latitude is the angular distance, north or south of the equator, measured in degrees along a meridian drawn on a globe of the earth. Longitude is the angular distance, east or west of the prime meridian at Greenwich, England, to a point on the surface of the earth. Longitude can be measured either in hours, minutes, and seconds or in degrees. Both are equivalent.

It is quite easy to find your latitude if you are located in the northern hemisphere of the earth. First, find the north star. Since it is almost directly over the north pole, just measure your angle from the horizon to the north star. You might use a protractor sighted along your arm. At the north pole, the angle to the north star would be 90 degrees. At the equator, the north star, if visible, would be 0 degrees from the horizon.

Finding your latitude is also easy and does not depend on whether you are north or south of the equator. Imagine the earth as an orange. An orange has eight segments, so each segment corresponds to 360/8 = 45 degrees. Longitude is the measure of the earth in segments. The earth has twenty-four segments as normally drawn on a globe. To find the number of degrees in each segment on a globe, find: 360/24 = 15 degrees. So the distance from any fixed location is in degrees. West from the location is indicated by W, east by E. The longitude segments on earth are referenced to Greenwich, England. If you know the distance from Greenwich, England, you know your longitude. How do you reliably measure your distance from Greenwich? The answer is measure the time. Why? Because there are twenty-four hours in a day, and there are twenty-four segments to the earth. Each hour corresponds to one segment.

■ **EXAMPLE 10-6**   The time zone you are in is 6 hours away from Greenwich, England. What is your rough estimate of your longitude?

$$\frac{6 \text{ hours}}{24 \text{ hours}} = 0.25$$
$$(360 \text{ degrees}) * (0.25) = 90° \text{ longitude}$$

The direction is determined by what direction you are from Greenwich. Let's assume that you are west, somewhere in the United States. Your rough estimate of the longitude is 90 degrees W.

How can you get a more precise estimate? Find how much to the west you are from the closest longitudinal line (a segment). This calculation is illustrated in Example 10-7.

■ **EXAMPLE 10-7**   Suppose you are 200 miles west from the 90 degree line found in Example 10-6. You need to find out how much distance on the surface of the earth corresponds to one segment, or 15 degrees. You know that the diameter of the earth is about 24,900 miles. Approximate this number to 24,000 miles to make the math easy.

$$\frac{24,000 \text{ miles}}{24 \text{ segments}} = 1,000 \text{ miles per segment}$$

This result means that there are 15 degrees per 1,000 miles. So for 200 miles west:

$$\frac{200}{1,000} * 15 \text{ degrees} = 3 \text{ degrees}$$

Your approximation is $90 + 3 = 93°$ W longitude.

The second element that must be considered in finding a satellite is the azimuth of the satellite from the observer's position on the earth. Azimuth is measured in degrees from a point due south of the observer to the satellite, moving in a clockwise manner. Think of the orbit of any geostationary satellite as a great circle in the sky. All the satellites would be placed on this circle. Depending on the location of the observer, these satellites would appear to be placed in orbit in different locations on this circle. For example, an azimuth of 0 degrees would place the satellite on this circle exactly due north of the observer; an azimuth of 90 degrees, due east; etc.

Azimuth is often defined as a horizontal angle referenced from due north. For our example satellite, due north of the observer, this definition would yield an answer of 0 degrees. Prove to yourself that azimuth is an equivalent description to that given for elevation. Either method works quite well in practice.

In summary, the elevation identifies the height of the great circle and the azimuth determines the location on the circle. In other words, your elevation determines what portion of the circle can be seen and the azimuth determines which satellite on that portion of the circle you wish to locate. Knowing how to compute these quantities is critical in determining how to point an antenna toward a satellite. These two elements are often referred to as look angles. Elevation is found by applying Equation 10-2 and azimith by applying Equation 10-3. The use of these equations is demonstrated in the next two examples.

$$E = \tan^{-1}\left(\frac{\cos(S_{long} - A_{long})\cos(A_{lat}) - 0.151}{\sqrt{1 - \cos(2(S_{long} - A_{long}))\cos(2A_{lat})}}\right) \tag{10-2}$$

E = satellite elevation in degrees
S_{long} = satellite longitude in degrees
A_{long} = antenna longitude in degrees

$$A = 180 + \tan^{-1}\left(\frac{\tan(S_{long} - A_{long})}{\sin(A_{lat})}\right) \tag{10-3}$$

A = azimuth of the antenna in degrees

■ **EXAMPLE 10-8**   Find the elevation in degrees of a satellite at longitude 80 degrees from a proposed antenna location at longitude 75 degrees and latitude 40 degrees.

$$E = \tan^{-1}\left(\frac{\cos(80 - 75)\cos(40) - 0.151}{\sqrt{1 - \cos(2(80 - 75))\cos(2(40))}}\right) = \tan^{-1}\left(\frac{(0.996)(0.766) - 0.151}{\sqrt{1 - (0.985)(0.174)}}\right)$$

$$E = \tan^{-1}\left(\frac{0.589}{0.910}\right) = 32.9°$$

The result in Example 10-8 means that, from the proposed antenna location, if you look directly at the horizon and elevate your line of sight exactly 32.9 degrees, you will find the satellite's location or height on the great circle. The minimum angle of elevation that can be used is about 5 degrees. The greater the elevation, the shorter the path length between the antenna location and the satellite. The longer the path length, the greater the attenuation in the channel and the lower the power received at the antenna, all else being equal. While this increase is slight, it makes a real difference for small elevations because of the percentage of the path that lies in the atmosphere. We will explore the difference elevation has on path length in Example 10-10 using Equation 10-4. For now, let's return to the other item necessary to point our antenna: the azimuth calculation.

■ **EXAMPLE 10-9** Determine the azimuth for the satellite and antenna location used in Example 10-8.

$$A = 180 + \tan^{-1}\left(\frac{\tan(80 - 75)}{\sin(40)}\right) = 180 + \tan^{-1}(0.136) = 187.8$$

Therefore, to find our satellite from our antenna position, rotate the antenna to a compass heading of 187.8 degrees and adjust the elevation of the antenna to 32.9 degrees above the horizon. In practice, many technicians set the elevation to an approximate number and adjust the azimuth until the satellite signal is located and is received at approximately maximum strength.

To understand how the elevation effects the path length or distance from the satellite to the antenna, we need to apply Equation 10-4:

$$d = \sqrt{r^2 + r^2_{\text{earth}} - 2rr_{\text{earth}}\cos(A_{lat})\cos(S_{long} - A_{long})} \qquad \textbf{(10-4)}$$

d = distance from the satellite to the antenna in meters
r_{earth} = radius of the earth in meters

■ **EXAMPLE 10-10** Find the elevation and path length from satellite to antenna for the satellite of Examples 10-8 and 10-9, but with antenna latitudes of 5, 10, 20, and 60 degrees.

Recall that the distance of a geostationary satellite from the center of the earth in meters was calculated in Example 10-1.

$r_{\text{earth}} = 6.38 \times 10^6$ m $\qquad r = 3.58 \times 10^7$ m

$$E_5 = \tan^{-1}\left(\frac{\cos(5)\cos(5) - 0.151}{\sqrt{1 - \cos(10)\cos(10)}}\right) = 78.3°$$

$$d_5 = \sqrt{(4.224 \times 10^7)^2 + (6.38 \times 10^6)^2 - 2(4.224 \times 10^7)(6.38 \times 10^6)\cos(5)\cos(5)} = 3.59 \times 10^7 \text{ m}$$

$$E_{10} = \tan^{-1}\left(\frac{\cos(5)\cos(10) - 0.151}{\sqrt{1 - \cos(10)\cos(20)}}\right) = 71.8°$$

$$d_{10} = \sqrt{(4.224 \times 10^7)^2 + (6.38 \times 10^6)^2 - 2(4.224 \times 10^7)(6.38 \times 10^6)\cos(10)\cos(5)} = 3.60 \times 10^7 \text{ m}$$

$$E_{20} = \tan^{-1}\left(\frac{\cos(5)\cos(20) - 0.151}{\sqrt{1 - \cos(10)\cos(40)}}\right) = 57.7°$$

$$d_{20} = \sqrt{(4.224 \times 10^7)^2 + (6.38 \times 10^6)^2 - 2(4.224 \times 10^7)(6.38 \times 10^6)\cos(20)\cos(5)} = 3.63 \times 10^7$$

$$E_{60} = \tan^{-1}\left(\frac{\cos(5)\cos(60) - 0.151}{\sqrt{1 - \cos(10)\cos(120)}}\right) = 15.8°$$

$$d_{60} = \sqrt{(4.224 \times 10^7)^2 + (6.38 \times 10^6)^2 - 2(4.224 \times 10^7)(6.38 \times 10^6)\cos(60)\cos(5)} = 3.95 \times 10^7$$

As you can see from Example 10-10, the path length increases as the angle of elevation decreases. As a percentage value, this increase is small. As pointed out above, however, it is not the total path-length increase that matters but the amount of the path that is found in the atmosphere. For small elevations, almost the entire increase is through the lower atmosphere. The atmosphere contributes noise to the received signal and degrades it significantly. The amount of noise contribution that comes from the atmosphere approximately doubles when the angle of elevation is decreased from 20 degrees to 5 degrees.

■ EARTH STATION

The other half of any satellite link is the earth station. This station can be as simple as a three-piece consumer setup, with an outdoor unit, indoor unit, and antenna, or as complex as a large tracking antenna that serves as a hub for both uplink and downlink purposes. Usually the term *earth station* is reserved for those stations that contain both uplink and downlink capabilities.

An earth station capable of uplinking to a satellite and downlinking from it is composed of several main blocks of functionality. For each satellite transponder pair or channel pair, uplink and downlink, these main blocks of functionality are:

1. Baseband multiplexer and demultiplexer
2. Modulator and demodulator
3. Up-converter and high power amplifier (HPA)
4. Down-converter and low noise amplifier (LNA)
5. Diplexer
6. Antenna
7. Tracking drive and control

These components are illustrated in Figure 10-18. The first unfamiliar component is the diplexer, which separates the two information flows into and out of the antenna. As we discussed previously, satellite channel allocations in a frequency band use orthogonal polarizations on adjacent channels. When using a single earth station antenna both to transmit and to receive, the usual practice is to use one polarization to transmit on the uplink channel and the other to receive on the downlink channel. This practice allows full-duplex communications using a single antenna. The actual choice of polarization for transmitting

Figure 10-18

(a) Earth Station Block Diagram; (b) Earth station located in Gloucester, Ontario, Canada (*Courtesy of Telesat, Canada*)

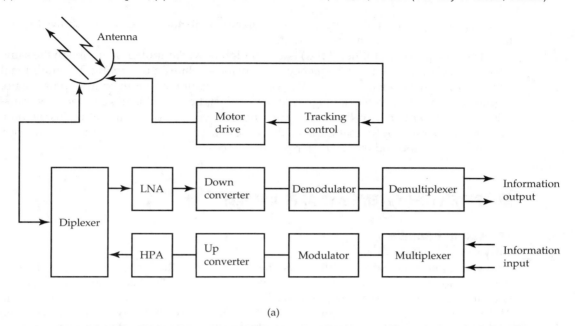

(a)

(b)

or receiving depends on the channel used. Refer again to Table 10-2, which shows the polarization of each channel in the C band.

The high power amplifier (HPA) is a very high power amplifier operating in the same frequency band as the carrier frequency used by the satellite on the uplink. The traditional choice is the traveling wave tube (TWT), operated in saturation for maximum power gain. Figure 10-18(a) shows a block diagram of an earth station capable of simultaneous transmission and reception. Part (b) of the figure shows a typical large earth station installation. This one is operated by Telesat Canada, the same company who built the robotic arm used on the space shuttle and the new international space station.

■ SATELLITE SYSTEMS AND APPLICATIONS

Television Receive Only (TVRO) Systems

Two television receive only (TVRO) systems are in use in the United States: the older C-band sets and the newer Ku-band direct broadcast satellite (DBS) sets. They both work in a similar way, as you can see in Figure 10-19. There are three basic components to this system. The outside unit consists of the antenna, LNB, and feed horn; the indoor unit consists of the tuning and demodulator units; and the display unit consists of the television. The figure is a generic block diagram of a typical home system. The Ku-band DBS system uses a 12.2 to 12.7 GHz frequency band that will accommodate thirty-two television channels, each 24 MHz wide. Obviously some overlap occurs; 36 MHz would be required for no overlap, but this problem is addressed using polarization of adjacent channels, as in C band. The indoor control unit automatically selects the correct polarization.

The indoor unit converts the 950 to 1450 MHz downconverted signal, delivered to it from the LNA and converter combination, to either a 70 or a 140 MHz intermediate frequency. Then conventional FM demodulation occurs. Those interested in modulation techniques will note that DBS frequency modulates the composite signal, which must then be FM demodulated and then remodulated as AM for the conventional television to receive the signal. Conventional television broadcasts are VSB amplitude modulated.

Community Antenna Television (CATV)

The other way many people receive their television broadcasting is through the cable television system, or (as it is officially named) the community antenna television (CATV) system. No special equipment is required at the subscriber's site unless pay-per-view services are desired. At the distribution site, where the television signals are collected prior to being placed on the cable network, there is a set of components similar to those used in the TVRO subscriber equipment. In the CATV system, this reception and signal translation takes place at the distribution site. In a TVRO application, it takes place at the subscriber's site.

The CATV system uses a single outdoor unit with seperate feeds for each polarization so each channel is available at all times at the indoor unit. Instead of having a separate receiver for each channel, all the carriers are demodulated in a common unit consisting of a receiver and filter combination. The channels are then frequency division

Figure 10-19
TVRO Subscriber Equipment

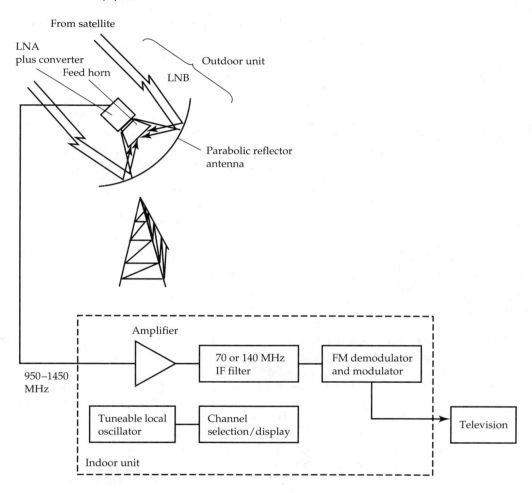

multiplexed for distribution over the cable system. Figure 10-20 shows one method. Locally produced material is inserted as shown in the figure.

Using Satellite Links for Internet Access

Most satellites designed for data transfer are in geosynchronous orbit. The latency involved in a data transfer through a space link will be significant. The protocol used is TCP/IP and TCP requires timely acknowledgments, so this large link delay causes two classes of problems. First, it causes significant retransmission because the acknowledgments do not meet the standard timeout settings appropriate for terrestrial use. Most LAN servers set the

Figure 10-20
CATV Distribution Equipment

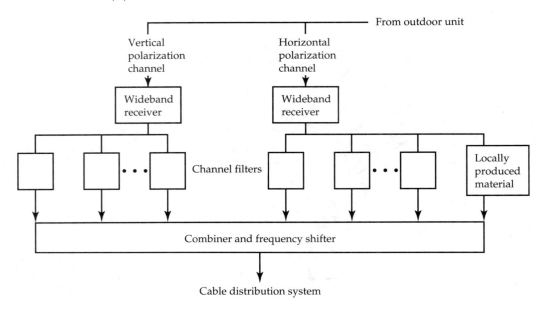

initial retransmission timeout (IRTO) at around 200 milliseconds. Clearly this setting would be a problem when using satellites as relays. Second, TCP will assume that congestion is the cause of the delay and throttle back transmissions.

The international space station also uses a LAN on board and faces the same issues but with a shorter delay. Its orbit is much lower, similar to the orbit of a space shuttle at approximately 200 miles. Because it is not in constant communication with terrestrial stations as it orbits the earth, dropouts of packets will be frequent, again causing TCP to throttle back its transmission rates. E-mail software must be modified to resynchronize itself with the ground periodically. Chapter 16 of this text discusses upper layer protocol issues with wireless links, but an example and a brief discussion at this point is appropriate.

■ **EXAMPLE 10-11**   What effect does round-trip delay have on a TCP session? A GEO satellite has a round-trip delay of 500 ms. The SCPC channel is capable of a data rate of 10 Mbps. If a 100 ms switching delay is present in the network, and the TCP window size is set to 64 Kbytes, can the maximum throughput of a TCP session be calculated? How does this compare to a user datagram protocol (UDP) session?

To solve this problem, take the window size and divide it by the round-trip delay:

$$\text{Throughput} = \frac{\text{window size}}{\text{round-trip delay}} = \frac{(65{,}536)(8)}{600 \times 10^{-3}} \approx 874 \text{ kbps}$$

With UDP, there is no acknowledgment and these restrictions are eliminated. Instead of 874 kbps as a limit, 10 Mbps is the limit for a single user. The limitation is on TCP, not UDP.

The major advantage to using satellite links for Internet traffic is in point-to-multipoint applications to wide ranging sites. Satellites offer distance insensitive pricing. Users pay the same price whether they are dialing from across the globe or across the street from the server. The Internet and satellites share an inherent asymmetric nature. With the Internet, the major portion of the data flow is from provider to user. The satellite channel is also very asymmetric. The data flow is from a single source to very many users. This feature has the potential to offer significant cost savings over traditional circuit-switched wireline links where the data paths are symmetric.

If the number of Internet users is large enough, many of the single requests will be made by many users at the same time and so they can be viewed as one multicast. This arrangement fits the satellite model perfectly. Also, business television and streaming video applications are inherently multicast. In these environments, satellites offer a big advantage over wireline. You pay only for a single transmission to reach many different users. With wireline, a *separate transmission path* would have to be made for each site, even with any sort of multicast protocol. The transmission would still have to travel along many wireline links, each of which generate a charge based on time used and distance traveled. Therefore, satellites are ideally suited to IP multicast.

For these reasons, the latency that a GEO satellite introduces can be of secondary concern. Several satellite systems are designed exclusively for broadband data transfer. The two best known are the Hughes Spaceway system and the Lockheed Astrolink system. The next section will summarize the technology used in the first.

Hughes Spaceway   The Hughes Spaceway system is a constellation of nine GEO satellites capable of switching communications between satellites directly. This capability can minimize the latency in a transmission because it eliminates the double hop that would be required without this capability. Each satellite is capable of an aggregate data rate of over 4 Gbps partitioned in 16 kbps channel widths. The maximum data rate in any one transponder is 6 Mbps. The transponders use the Ku band for both uplink and downlink transmissions. The Hughes Spaceway system provides global coverage, except at the extreme polar regions and in parts of Russia. The typical terminal cost is approximately $1,000.

DirectPC System   The DirectPC system from Hughes offers an alternate approach to Internet access from the home or business. This system still makes use of your Internet service provider (ISP), and the ISP still accesses directly the Web server. However, the Web server information is not routed back to the ISP for transmission to you through the telephone system. Instead, it is passed to the DirectPC network operations center (NOC) at a peak rate of 400 kbps. The NOC passes the data to the GEO DirectPC satellite, and this satellite sends the data at a peak rate of 400 kbps directly to your satellite dish, which then feeds it into your PC through a cable.

The antenna used is about 1.5 feet in diameter. Although this system advertises a download rate of 400 kbps, in fact, this rate can be significantly lower, depending on traffic and usage patterns. Such an architecture gives a high peak download rate because it uses a GEO satellite for the download channel, but there is significant latency in the response. As calculated earlier, the round-trip propagation time between a terrestrial location and a satellite in geostationary orbit is about 400 ms. When switching latency on the terrestrial link is added, typical values approach 500 to 600 ms. For many applications, this range will be acceptable, for an application like interactive gaming or streaming video, the performance may not live up to the expectation implied by the high peak data rate advertised.

SPADE Application

SPADE stands for Single-channel-per-carrier PCM multiple Access Demand assignment Equipment. This system uses 800 channels in each transponder for long-distance telephony. Each FDM voice channel is set at 45 kHz using SCPC techniques. Each of these 800 channels is PCM encoded and digitally modulated onto a separate carrier.

Like the wireline telephone system, each voice channel is allocated 4 kHz of bandwidth. It is then sampled at 8 kHz and converted into an 8-bit PCM code word. This process produces a 64 kbps data rate for each voice channel. SPADE uses QPSK modulators to modulate the 64 kbps signal digitally onto a single carrier. This modulation produces a bandwidth of 32 kHz because QPSK has a bandwidth efficiency of 2. This 32 kHz signal was placed into a 45 kHz SCPC channel, resulting in a 13 kHz guardband between each signal. Since a typical C-band transponder is 36 MHz in bandwidth, 800 channels of capacity exist. In the actual system, only 794 channels are used, the remaining bandwidth being reserved for additional guardbanding and 160 kHz for the common signaling channel (CSC).

Without the CSC, DAMA could not work. The CSC is a TDM signal that occupies 160 kHz of the satellite transponder, just like any other carrier. The CSC is used by the NCS to allocate, on demand, individual SCPC channels in the transponder. All network nodes are permanently connected through the CSC, and each earth station in the network has the facility for generating any one of the 794 carrier frequencies that the SCPC signals use. The individual earth stations in the network assign the voice signals to a particular SCPC carrier location, based on the information carried on the CSC.

Each voice telephone call has three components: call setup, transfer of information, and call termination. The satellite component is used to replace the telephone twisted pair cable that carries the phone call from one central office to another. The control system, a separate network, takes care of the call setup and termination. Like any communications system where the data transmission is carried on a separate channel from the control information, a parallel system must control the flow of data. In a system such as SPADE, that control information is contained in the CSC channel. This channel is used to control the individual SCPC channels that carry the information. Essentially, the CSC is used to establish (call setup) and disconnect (call termination) the individual information channels. When a call is made, the earth station responsible for originating the call reserves an open SCPC channel. This information is passed to all other nodes through the CSC.

As discussed above, the CSC is a TDM signal; therefore, it must have a frame definition. The CSC frame is divided into fifty time slots of 1 ms each. The SPADE system

Figure 10-21
SPADE CSC Frame Definition

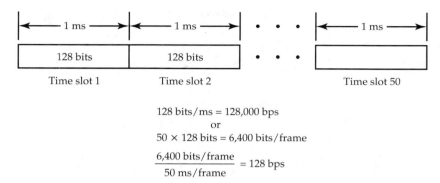

$$128 \text{ bits/ms} = 128{,}000 \text{ bps}$$
$$\text{or}$$
$$50 \times 128 \text{ bits} = 6{,}400 \text{ bits/frame}$$
$$\frac{6{,}400 \text{ bits/frame}}{50 \text{ ms/frame}} = 128 \text{ bps}$$

envisioned a maximum of fifty central offices feeding signals into the satellite channel; therefore, there are fifty time slots in the CSC. Each node transmits control information formatted into 128 bit data patterns during its 1 ms time slot in each frame. The overall data rate of the CSC is 128 kbps. See Figure 10-21.

VSAT Systems

VSAT stands for "very small aperture terminal" and refers to the transmit/receive antennas used to connect to a central hub via a satellite. The systems have entered the marketplace using two distinct frequency bands. VSAT is a term used for antennas that range in diameter from under 0.5 meters to almost 3 meters. Modern systems use antennas smaller than 1.5 meters. VSAT antennas are always parabolic in design.

The term VSAT is somewhat misleading because it implies that any system using small antennas is a VSAT system, which is not the case. The term is reserved for corporate or private networks, and in most cases the system supports duplex communications. Many VSAT systems have more than 500 remote nodes or VSAT stations. Their architecture is very similar to a client server architecture, with the VSAT terminals acting as the clients and the central hub acting as the server. This setup is shown in Figure 10-22.

VSAT-based satellite services are used by several organizations for three primary applications:

1. To exchange information between a central site and remote sites
2. To establish links where wired infrastructure is poor or nonexistent
3. To avoid long provisioning delays

Examples of the first application include oil companies exchanging information and production figures between remote locations and central operations centers. Many oil companies also use this same technology to keep track of sales and inventory data in their retail gasoline stations. They also use it to do pipeline monitoring and to facilitate communication to remote drilling sites. Banks and financial institutions use it to keep track of global money flows and customer accounts. Today almost all ATM machines use

Figure 10-22
VSAT Client/Server Network

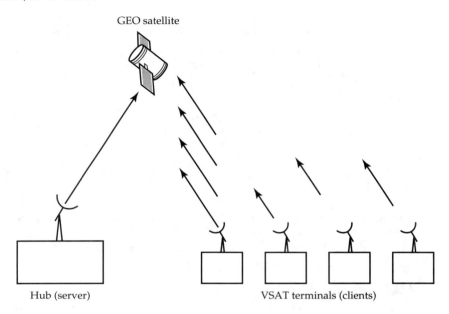

Hub (server)   VSAT terminals (clients)

VSAT networks to link them to the bank. Just a few years ago dial-up land lines were the exclusive solution. Transportation companies use them to aid in inventory control, fleet management, and delivery reconciliation. Hotels use them to aid their reservation systems and support e-mail and frequent traveler programs.

Automobile manufacturers use VSAT-based applications to interconnect dealers, suppliers, and factories. Ford uses its Fordstar network of almost 7,000 sites to deliver training video and enable student response systems throughout the world. In a typical month, Ford broadcasts over 1,000 hours of training to its dealerships in the United States. General Motors has a similar system to implement its distance learning system. Many companies use VSAT systems for this and other business television (BTV) applications.

Because satellites are naturally multicast and broadcast systems, the business pays for only one transmission, even though it may be going to thousands of sites. The fee for this transmission is based on this single transmission. Contrast that fee arrangement to the land line cost, where each site transmission would incur a separate fee.

VSAT systems are easy to set up and install, and they require no wireline infrastructure to operate. The basic equipment requirements are minimal and consist of three components: the indoor unit, outdoor unit, and antenna. The outdoor unit contains all the RF equipment, up-converter, down-converter, and second-stage amplification on the receive side. The outdoor unit must have a clear line of sight to the satellite. The first-stage amplification and frequency conversion on the receive side, or LNB, is typically mounted directly on the antenna. The indoor unit consists of the interface cards and coordination circuitry necessary for TDMA operation. Depending on the date of manufacture, the indoor unit can be bulky and fill an entire rack, or it can be as small and compact as the typical desktop computer.

Usually, the indoor unit and outdoor unit must be no more than 75 to 100 meters apart. This distance limitation between the indoor unit and outdoor unit at the client site is due to the low signal power received. Too much cable length can attenuate the signal and cause poor performance. This distance limitation makes VSAT systems ideally suited for establishing data links with remote sites. Also, rural telephony can be implemented with this technology. For example, Thailand has used the VSAT approach to establish a system of about 4,000 pay phones to link small towns and villages to the rest of the world. This ease of setup and operation underlines the third major advantage of VSAT systems: very little provision delay.

Depending on the service provider, up to six months can be required to provision a typical leased line. This long provisioning delay reflects the time it takes a service provider to coordinate the various organizations that provide the service. Since the breakup of the Bell System in the United States, many choices are available for provisioning a wireline network. LECs, CLECs, IXCs, third-party resellers, etc., are just a few. In addition to these players, a myriad of issues about bringing in the wireline cable to the physical building often arise.

If you own the building and land around it, then the job is made easier because only right of way and city and county regulations must be taken into account. These regulations cover where to dig the hole to place the line, what wall will be penetrated to terminate the line into the building, how the line will be routed inside the building to the data center, etc. Compared to that, placing an antenna on the roof of your own building is simplicity itself.

On the other hand, the vast majority of businesses do not own the building from which they operate. At this point, a whole new set of players come into the game, including building owners, supervisors, leasing agents, other leaseholders, etc. Many of these groups will have input about any antenna placement in the building as well, but many of the "physical plant" objections are diminished with the VSAT approach.

A VSAT terminal can be set up in a few weeks and is much more reliable than many countries' leased lines, particularly international circuits. A typical availability percentage for VSAT network links is 99.5 percent at a BER of 1E10-7 or better. Most of the availability falloff is due to rain fades, and it can be minimized by frequency selection and increased earth station cost. With a VSAT link, there are no local loop issues to contend with and the line rate is distance independent. Finally, the back haul segment can also be implemented over a VSAT link, eliminating any need for a land line.

VSAT networks can take on different architectures but all feature a central hub. This central hub communicates, or broadcasts, to the large number of remote sites where the VSAT terminal and antenna are placed. VSAT networks operate in two frequency bands: the C band and Ku band. Typically, the Ku band systems are deployed in Europe and North America. C band systems (the older and first deployed satellite technology) are often found throughout Africa and in South America. Both technologies are deployed in Asia. Generally, if the antennas are over 2 meters in diameter, the system uses frequencies in the C band. Antennas under 2 meters in diameter use the higher frequency Ku band.

VSAT terminals at remote sites need no on-site staff to operate them. They provide a feed that plugs directly into exiting data terminal equipment (DTE). The central site, often called the master earth station, is always larger than the remote sites and is often located

Figure 10-23
VSAT Star Architecture

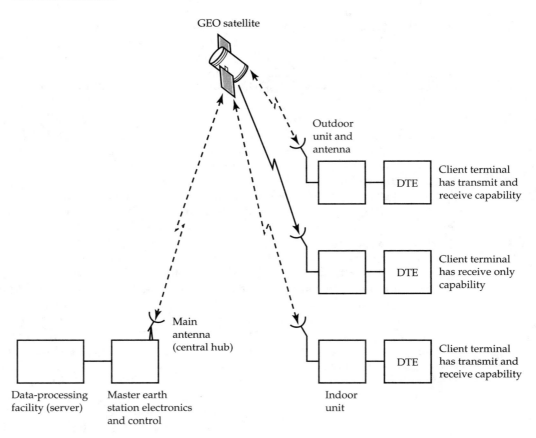

near a main data-processing facility. Many times it is owned by a service provider who will lease portions of the bandwidth to many organizations. However, each organization owns its own VSAT network and has exclusive access to that network and data. Many organizations implement a portion of their virtual private network (VPN) intranet using VSAT techniques. While a VSAT network can support several network configurations, they all feature this central hub concept. Figure 10-23 illustrates a typical VSAT star architecture.

Note in the figure that both receive only and transmit/receive terminals are accommodated. Most VSAT networks feature a star architecture, often called point-to-point. Mesh or point-to-multipoint networks are also possible. The star architecture is called a point-to-point architecture in VSAT terminology because of the point-to-point communication of the terminals to the central hub. Each remote VSAT terminal communicates directly to the hub, and each communication is between the hub and one remote terminal. When two-way, the communication is half-duplex, and it is possible to combine both simplex VSAT stations and half-duplex VSAT station links in the same star topology. Note that communication in both directions uses the same transponder in the satellite.

The mesh or point-to-multipoint architecture allows the master earth station to use a multicast or broadcast mode where several remote VSAT terminals are sent messages in a single time slot. These sytems always use TDMA access technology. Traditional VSAT networks never allow communication directly between remote sites. However, demand assigned multiple access (DAMA) technology allows communication to remote sites on a limited basis. DAMA technology is now available from most suppliers of VSAT systems. As discussed above, in VSAT networks DAMA is an enhancement to TDMA.

In a DAMA VSAT network, direct communication between nodes eliminates the delay of a double hop. Since each traditional VSAT node communicates only with the hub, any communication between remote nodes must follow the path defined by node A to satellite, satellite to hub, hub to satellite, satellite to node B. In a DAMA environment, the path is defined by node A to satellite, satellite to node B. This configuration reduces the delay in the signal. As shown below, the round-trip delay for a signal to a GEO satellite is about ¼ second. Doubling this figure to over ½ second is often an unacceptable latency for certain signals.

You can think of a TDMA VSAT system as similar to conventional packet switched networks. In fact many use the old reliable X.25 protocol. In these networks, many remote sites can communicate with a central site or hub. This setup is the most commonly implemented architecture for VSAT networks. Figure 10-24 illustrates this architecture. In these networks,

Figure 10-24
TDMA VSAT Network

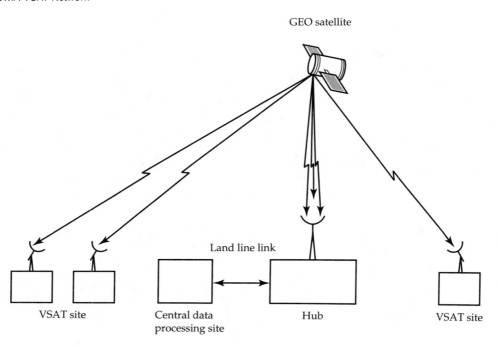

GEO satellite

Land line link

VSAT site   Central data   Hub   VSAT site
                 processing site

the remote sites contend for access to the hub. This contention can result in significantly greater delays for some traffic over others, depending on the traffic conditions. Typical data rates through these channels range from 19.2 kbps to 64 kbps.

VSAT Application   DAMA technology allows VSAT networks to offer direct communication between VSAT nodes using a limited pool of bandwidth in the satellite transponder. The key to DAMA operation in any environment is the NCS, which allocates bandwidth to each site from a pool of frequency or time slot channels on a demand-assigned basis. VSAT DAMA is performed both in FDM, with a pool of bandwidths, and in TDM, with a pool of time slots.

You can think of the NCS as a switchboard. As in the SPADE application, when a station wishes to make use of this service, it sends a request to the NCS over the shared DAMA common signaling channel (DAMA CSC). The NCS then allocates bandwidth from its pool of channels for the communication. Once the communication is finished, the channel is reassigned to the pool for other requests. The NCS can be placed at any node in the network, and all nodes are interconnected through the CSC.

This pool of bandwidth, often referred to as a reserved channel, is both the key advantage and the chief drawback of the DAMA approach. It is difficult to allocate this bandwidth in a fair and equitable way to all the nodes. There are many solutions, and the solution is an important characteristic of a vendor offering this service. As discussed in the previous section, DAMA in a VSAT network offers another key advantage. In a traditional VSAT network, communication between nodes is not possible without going through the central hub. Through the dynamic allocation of time slots, DAMA allows direct communication between VSAT nodes.

This capability was also possible in SCPC systems, but it involved a large cost penalty. SCPC is a preassigned solution, which means that a dedicated channel was defined for just this purpose. Direct communication between nodes is not something used regularly, so this dedicated bandwidth was largely unused. Because the strength of the SCPC technique is efficient use of the space segment, this dedicated, low utilization channel was very undesirable. In any DAMA application, this node-to-node direct communication is assigned in a dynamic way.

The business case for DAMA is a strong one. Costs are reduced, and transponder space is optimized, which can increase revenue by allowing more services to be provided with the same transponder space. DAMA also provides inherent network management capability, thus allowing priority control to certain link requests. This function is widely used in military applications of VSAT networks.

■ GLOBAL POSITIONING SATELLITE (GPS) SYSTEM

The Navstar GPS system is a constellation of twenty-four satellites that, in conjunction with a receiver, can quite precisely identify your location anywhere on or in low altitude around the earth. This system was developed and launched and is maintained by the U.S. Air Force. It provides positioning, timing, and navigation signals free of charge to both military and civilian users. The satellites and receivers use the L band. The current

Figure 10-25
GPS Satellite (*Photo courtesy of Russ Underwood, Lockheed Martin Space Systems*)

satellites use two frequency bands in the L band, designated as L-1, L-2. The next generation of satellites (of which the first was launched in early 2000) will use a third signal in the L-5 band.

All satellites transmit simultaneously on both carriers using spread spectrum techniques with unique chip codes both to provide ranging information and to eliminate interference between the two transmissions. Next-generation satellites will transmit simultaneously on all three carriers. Figure 10-25 shows a current-generation GPS satellite.

The L-1 band, 1,559 to 1,610 MHz, is designated for aircraft navigation and control. An attempt to establish co-use of this band for satellite-to-satellite communications was defeated at the World Radio Communications Conference (WARC-2000). At WARC-2000 the L-2 band was expanded to include the frequency range 1,215 to 1,260 MHz. Also at WARC-2000, the L-5 band was divided into two sections. The frequency range 1,164 to 1,188 MHz is designated for the Navstar GPS system and 1,188 to 1,215 MHz for the envisioned European Galileo GPS system.

The Galileo system is a way for Europeans to reduce their dependency on the Navstar GPS system. They are uncomfortable, as are many commercial enterprises on both sides of the Atlantic, depending on a system controlled by the military. While Galileo is still on the drawing boards, a second GPS system is in orbit today. It is known as the Glonass system and was deployed by the former Soviet Union. The last replacement satellite for this system was launched in 1998, and currently only nine of the twenty-four satellites required are functional. Also, few receivers built today can function with the Glonass system. All these systems work in similar ways but the only existing system in orbit is the Navstar system, so the rest of this discussion will focus on that system. Galileo is slated to be fully operational in 2008.

The Navstar GPS satellites are in six near-circular orbits at an altitude of about 12,000 miles. There is one master control earth station, whose purpose will be defined shortly. While the GPS system is not normally thought of as a communications system, it does perform that function in the process of identifying the location of the GPS receiver. Without GPS timing signals, many communications systems would not function the way we have come to expect. For example, almost all digital cellular telephone systems rely on the timing signals of the GPS system to synchronize between cells, and pagers rely on them as well for similar functions. The long-distance carriers often rely on the GPS system to synchronize long-distance trunk lines.

The idea is that the system depends on the receiver getting simultaneous signals from at least four of the satellites to determine your longitude, latitude, and altitude. Readers familiar with terrestrial surveying might wonder why more than three signals are required for accurate triangulation. The answer is that a fourth value, the time of transmission, is required to determine these quantities accurately because the satellites are in orbit. If you think about it, terrestrial triangulation involves the signals from three *stationary* points. Because the satellites are far from stationary, simultaneous signals from four are needed to establish a time marker that can essentially fix the satellites in space for the other three measurements. This process allows software to determine the location of the receiver with extremely precise values.

The GPS system uses simplex transmission; that is, the only transmissions are from the satellites to the receiver. The transmissions are spread-spectrum-modulated and information dense. The receiver requires no transmitter, and its primary requirement is to measure time differences accurately. Because all the signals travel at the same velocity, accurately measuring the propagation time from each satellite yields the range to each and, from these quantities, the location of the receiver.

Each satellite transmits a table of orbital values, called an ephemeris, which is updated continually by the master control earth station. The ephemeris contains status messages in addition to orbital values. This data, which is broadcast by each satellite to the receiver and includes the timing information, allows the calculation of the receiver's position.

Note that a special time standard, called GPS time, is maintained by the U.S. naval observatory. GPS time differs from the international time standard, Universal Coordinated Time (UTC), due to the addition of leap seconds to UTC time. Each satellite carries its own atomic clock, which is also part of the orbital data monitored by the master control earth station. This data is also subject to correction through the sending of error information to the satellites as part of the update.

The terrestrial component of the GPS system includes the master control station (MCS) and several monitor stations located around the world. The MCS is located at Falcon Air Force Base near Colorado Springs, Colorado. Passive monitor stations are located in Hawaii, California, and Alaska. Other monitor stations, which are not passive, are located on Kwajalein Island in the Pacific Ocean, Diego Garcia in the Indian Ocean, Ascencion Island near Africa, and others.

The MCS monitors the signals from all GPS satellites and periodically sends control information to correct any clock drift on individual satellites, as well as updated ephemeris information, approximately every twelve hours. The nonpassive earth stations use S-band links to update the satellites with this information, which is relayed to the earth stations from the MCS.

It is important to note that, as mentioned above, the GPS system was built and launched for use by the military. The military uses a more sophisticated receiver and two signals from each satellite to get a positional accuracy that is an order of magnitude more accurate than the commercial units widely available now can obtain. The system was seen as a great advantage and has proved so, most recently in the Desert Storm conflict, where the landscape offered few visual markers regarding location. The coordination of ground forces was greatly aided by the GPS system.

The second signal from each GPS satellite that the military receivers use is designed to compensate for variations in the signal as it passes through the earth's atmosphere. The increased accuracy comes from the receiver's ability to measure the time of arrival for two signals that have passed through the same area of the earth's atmosphere at the same time. Corrections for distortion effects of the atmosphere are made by examining any differences in the arrival time of those signals. The second signal also functions as a backup signal should the primary signal from a GPS satellite fail to arrive for any reason.

Until mid-2000 commercial GPS systems offered a positional accuracy of 100 meters in the horizontal dimension and 160 meters in the vertical dimension. These values are combined to give a three-dimensional accuracy to within about 100 meters in any dimension. This accuracy was significantly enhanced at that time by the elimination of what was called selective availability, which was applied purposefully to degrade the signal that was received by civilian receivers. This elimination improved the accuracy to about 20 meters in any dimension. The original idea was that only the U.S. military would have access to such accuracy, and this access would be a battlefield advantage. Current thinking suggests that the commercial advantages of greater accuracy for civilian receivers outweighs restricting use of the GPS system, especially since competing systems, such as Galileo, will be offering military precision at a price. Military precision is on the order of 1 meter in any dimension.

■ NEXT-GENERATION GPS

Given the results of the WARC-2000 meeting, plans are now under way to enhance the civilian GPS system by adding two more frequencies. This addition will require the upgrade of the satellite constellation in orbit. The current generation of GPS satellites are being replaced with model IIR. The IIR system does not implement the additional frequencies, but it is anticipated that the next generation of replacements, IIF, will include this feature. Due to current production and launch schedules of the IIR series, however, the IIF series will not begin launch until after 2005. Once this series begins to replace the IIR series, there will be an additional delay until several of the new series are in orbit. Realistically, wide availability of the second and third civilian signals will not occur for ten years because of the delay in deploying the hardware in space.

Assuming that work goes as planned, a short discussion of the first of the two additional frequencies is in order. Current commercial receivers are classified as L-1 receivers, which means that they operate in the L-1 frequency band and that they receive only a single coded signal from the GPS system. While the satellite upgrade will add two additional frequencies, consumer quality units will likely make use only of the first. Units that implement both bands will offer significantly improved performance. The receivers will require a second RF front end, and cost of manufacture will increase approximately 40 percent for low-cost consumer units. Higher performance units will experience a much smaller percentage increase.

All GPS receivers will then fall into three classes: the current L1 class, future consumer grade commercial units; the L2 class, similar to the current military grade receiver; and the L3 class, which will receive three frequencies for even greater accuracy. In summary, the addition of a second civilian frequency will greatly increase the accuracy and reliability of civilian GPS receivers. The primary benefit will be the large reduction in errors caused by the distorting effects of propagating through the earth's atmosphere. As in the current generation of military receivers, the second signal will compensate for variations in the atmosphere for signals coming from different satellites. Recall that four signals are needed for accurate triangulation and timing synchronization of the triangulation signals. Because all satellites occupy a different region of the sky when being used for position identification, the propagation times from them may suffer from different atmospheric conditions, even though they are passing through the atmosphere simultaneously.

■ MSAT

Existing mobile services do not cover all rural areas well. Mobile satellite (MSAT) addresses this problem by complementing existing cellular and radio networks in the following areas:

Police, ambulance, and fire

Search and rescue

Commercial aircraft

Coast guard operations

Construction projects at remote sites

Oil wells

Wide area vehicle monitoring

Environmental monitoring

This list is only partial. Canada actively promoted MSAT because it is a country of many remote rural areas.

MSAT is really three services combined:

Mobile radio trunking service (MRTS): MRTS is a mobile dispatch service that can be private or shared. If the base station is owned and operated by an organizational user, the service is private. If the base station is owned and operated by a service provider who leases it out (shares it), the service is shared.

Interconnected mobile radio service (IMRS): IMRS connects users to a wireline telephone network where other connectivity is impossible. Calls are forwarded from mobile units to MSAT and then to an earth station, which connects a public switched telephone network (PSTN). IMRS is also used to connect transportable earth stations in remote locations to the public telephone network.

Mobile data service (MDS): MDS provides two-way data links between mobile and fixed (perhaps transportable) terminals. It is used primarily for fleet tracking, cargo monitoring, and data messaging.

MSAT technology uses a GEO satellite in the L band (1.5 GHz downlink, 1.6 GHz uplink) for communication to and from a mobile unit. It uses Ku band for communications to fixed stations. The L-band antenna is a reflector array that generates nine beams covering all of North America as well as Hawaii and the Carribean. Because of the limited bandwidth available in the L band, channels are only 5 kHz in bandwidth. Special compression techniques are used to increase data flow in narrow channels. Low frequency links to mobile units allow omnidirectional antennas for mobile units, eliminating the need for pointing the antenna. Mobile-to-mobile calls are routed in a double hop: mobile to fixed, fixed to satellite, satellite to fixed, fixed to mobile. All other links are single hop.

There are many ways to purchase or rent the earth-end equipment. As an example, briefcase phone service will be described. It is a satellite earth station in a briefcase that can be opened and mounted in a boat, car, plane, etc. It includes the following items:

Base unit

Handset with detachable cradle

Rechargeable battery

AC charger/adapter

Compass

Antenna built into the lid of the briefcase

The RF gear is built into the base unit, and the antenna is usually mounted inside the briefcase cover. These units can be purchased for about $8,000 or rented for about $800 a month.

Most leasing agencies require a one-month security deposit, refundable upon return of the unit. The handset looks almost exactly like a cellular telephone and dialing is pushbutton.

The unit must be aligned to the satellite for best operation. Alignment is accomplished with a signal strength meter on the handset. Rotate the entire briefcase and lift the lid to an appropriate angle for best reception. The units usually feature an RS-232 port for data applications, although the standard modem included is 2,400 bps. Data communications software is identical to what you would purchase off the shelf, for example, Explorer, Outlook.

For remote applications where recharging from an AC power source is not possible, solar panel arrays are available for recharging. In fact, several Internet episodes involved the use of this technology. A typical example is an environmental excursion that folks at home could participate in, along with the crew deep in the jungle. With the addition of a digital camera and a portable PC to interface and view the photos, the data can be posted using the RS-232 port. With the addition of a fax machine or portable PC, facsimilies can be sent and received from anywhere where coverage exists. It is possible to use credit cards to pay for personal calls. This requires the use of an on-site card scanner, which is attached to the RS-232 interface.

■ INMARSAT

The same technology described above is also available internationally, and they are known as Inmarsat terminals. There are three classes: A, B and C. Inmarsat C is a satellite system that allows two-way communications similar to those described above. These terminals are available from over thirty manufacturers and there are perhaps 100 different models and configurations. What was described above is a summary of the most common options among the models.

Inmarsat C transmits all data to and from the satellite at 600 kbps. Anything you can represent as a digital stream can be carried on the channel. In addition to voice calls, other services are supported by just choosing the type of terminal and service subscription. Two-way messaging is supported; however, the maximum message length is 32 kbytes. The mobile unit sends the message on the L band in data packets to the local land earth station (LES). There are about forty LESs around the world. The message is then reassembled and sent to its ultimate address using the international wireline telecommunications systems. (Messages to a mobile unit or group of mobile units can be sent using the ECG concept described below.)

Lots of industries require regular data polling from vehicles, ships, or fixed data-gathering platforms. A data reporting and polling service allows the Inmarsat C terminal to support data polling through a 32-byte message sent either on request or at predefined times. Inmarsat C terminals can also be configured to support position location by linking them to virtually any type (including Loran) of GPS and dead-reckoning equipment on board a ship.

Inmarsat C terminals can be programmed to receive multiple address messages called enhanced group calls (EGC). A special header is applied to the text to indicate the group of mobiles or the geographical area where you wish to send the message. This multicasting is by either region or address of mobile unit. Two main classes of ECGs are offered by the Inmarsat C provider:

SafetyNET: This class is a low-cost means of providing maritime safety information to mobile units. It is used by vessels at sea, search and rescue crews, coast guard agencies, etc. With the ability to target geographically, local conditions can be sent only to those terminals in that area.

FleetNET: This class is intended to allow commercial information to be sent to almost any set of terminals. It is used by news organizations, stock exchanges, departments of public works, etc.

Inmarsat A terminals are the classic analog mobile terminals that you may have seen during the Desert Storm coverage by CNN. They have been a standard for the journalism industry for more than ten years. They provide fax, two-way telephone, e-mail, and data communications at speeds of up to 9.6 kbps. They use the same frequencies and LES stations used by the MSAT and Inmarsat C terminals.

Some units can be configured with what is called the high speed data (HSD) option, which allows arbitrary data communications at speeds up to 64 kbps. This option allows compressed video and high-quality, FM quality (15 kHz) audio links. This option is designed to support video- and audioconferencing. Inmarsat A terminals are composed of two suitcase-size cases with a foldaway parabolic antenna that is about 3 feet in diameter. Each suitcase weighs between 10 and 15 pounds. The unit is designed to work off external batteries or AC power.

Inmarsat A terminals are the workhorse of the shipping industry. Almost every shipping fleet in the world uses them for various commercial, social, safety, and distress applications. At last count there were 18,000 individual ships equipped with these terminals. Inmarsat A terminals have been deployed since the early 1980s.

Inmarsat B terminals are the next evolutionary step compared to the A terminals. The cost to use these terminals is significantly lower than using an A terminal. The B terminals are primarily digital systems. For users with increased bandwidth needs, these terminals are the better choice. They support facsimile, telex, and 64 kbps data ports. The unit itself consists of two components: the antenna and an indoor unit approximately the size of a typical desktop PC. The antenna is slightly smaller than the 1 meter used for the A terminals. The HSD option for the B terminals is designed to interface into the integrated services digital network (ISDN) at the basic rate interface (BRI) rate. BRI service gives the user two 64 kbps bearer channels and one 16 kbps data channel. Generally, only the bearer channels can be used for user data transmission. The data channel is used for call control. Figure 10-26 shows three representative models of Inmarsat B terminals from different manufacturers.

■ SATELLITE LINK CALCULATIONS

Satellite link calculations are very similar to those for terrestrial microwave links, with a few differences. For example, you do not have to worry about Fresnel ellipsoids in a space link because there are no obstacles. On the other hand, there are two links at different carrier

Figure 10-26

Three Inmarsat B Terminals; (a) Nera World Communicator (*Courtesy of Nera, Inc.*); (b) ECI Netlink (*Courtesy of Eurocom Industries*); (c) Thrane & Thrane Capsat Messenger (*Photo reproduced with permission of Thrane & Thrane A/S*)

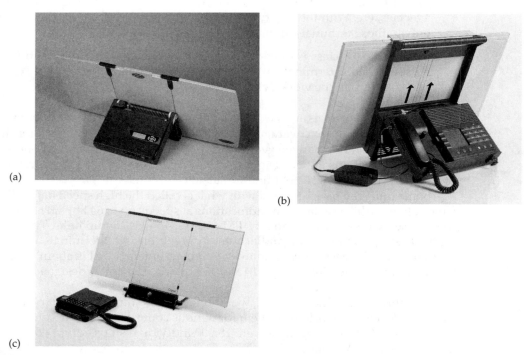

(a)

(b)

(c)

Figure 10-27

Satellite Link Block Diagram

P_i = supplied input power
P_o = power output delivered

frequencies, four antennas, and the satellite transponder gain to worry about. The components we will consider are summarized below and shown in Figure 10-27.

Transmit earth station:	transmitter gain, antenna gain
Uplink:	channel attenuation
Satellite:	antenna gain, transponder gain
Downlink:	channel attenuation
Receive earth station:	antenna gain, low noise amplifier (LNA) effects, receiver gain

■ **EXAMPLE 10-12**   Calculate the power delivered to the customer for a satellite link with the following characteristics. Assume all antennas are parabolic in design.

Input signal power	10 mW
Transmit earth station gain	40 dB
Transmit antenna diameter	15 m
Uplink carrier frequency	6 GHz
Satellite receive antenna diameter	1 m
Satellite transponder gain	100 dB
Satellite noise temperature	200°
Satellite transmit antenna diameter	1 m
Downlink carrier frequency	4 GHz
Receive antenna diameter	6 m
LNA gain	40 dB
LNA noise temperature	120°
Receive earth station gain	40 dB

To begin, we will summarize the variables used to represent each of these quantites:

$$P_i = \text{supplied input power}$$
$$G_{es1} = \text{gain of transmitting earth station}$$
$$G_{txa} = \text{gain of transmitter}$$
$$G_{d,tx} = \text{directive gain of transmitting antenna}$$
$$A_{c,up} = \text{channel attenuation, uplink}$$
$$A_{c,down} = \text{channel attenuation, downlink}$$
$$G_{sat} = \text{gain of satellite}$$
$$G_{d,rx} = \text{directive gain of receiving antenna}$$
$$G_{rxa} = \text{gain of receiver}$$
$$G_{LNA} = \text{gain of LNA amplifier}$$
$$P_o = \text{power of output delivered}$$

First, find the gain of the transmitting earth station. This value consists of two elements: the gain of the transmitter in the earth station and the gain supplied by the uplink antenna, also located at the earth station.

$$G_{es1}(\text{dB}) = G_{txa}(\text{dB}) + G_{d,tx}(\text{dB}) = 40 \text{ dB} + 10 \log\left(6\left(\frac{fD}{c}\right)^2\right)$$

$$= 40 \text{ dB} + 10 \log\left(6\left(\frac{(6 \times 10^9)(15)}{3 \times 10^8}\right)^2\right) = 40 \text{ dB} + 57.3 \text{ dB} = 97.3 \text{ dB}$$

Next, calculate the attenuation due to the uplink segment.

$$A_{c,up}(\text{dB}) = 10 \log\left(\frac{4\pi rf}{c}\right)^2 = 10 \log\left(\frac{4\pi(3.58 \times 10^7)(6 \times 10^9)}{3 \times 10^8}\right)^2 = 199 \text{ dB}$$

Next, calculate the contribution from the satellite.

$$G_{sat} \text{ (dB)} = G_{d,rx} \text{ (dB)} + G_t \text{ (dB)} + G_{d,tx} \text{ (dB)} = 10 \log\left(\frac{6fD}{c}\right) + 90 \text{ dB} + 10 \log\left(\frac{6fD}{c}\right)$$

$$= 100 \text{ dB} + 10 \log\left(\frac{6(6 \times 10^9)1}{3 \times 10^8}\right) + 10 \log\left(\frac{6(4 \times 10^9)1}{3 \times 10^8}\right)$$

$$= 100 \text{ dB} + 20.8 \text{ dB} + 19 \text{ dB} + 139.8 \text{ dB}$$

Find the downlink segment attenuation.

$$A_{c,down} \text{ (dB)} = 10 \log\left(\frac{4\pi rf}{c}\right)^2 = 10 \log\left(\frac{4\pi(3.58 \times 10^7)(4 \times 10^9)}{3 \times 10^8}\right)^2 = 195.5 \text{ dB}$$

Finally, include the earth station antenna, LNA effects, and receive amplifier. Since a C/N value was not requested, ignore the LNA temperature for now.

$$G_{es2} \text{ (dB)} = G_{d,rx} \text{ (dB)} + G_{LNA} \text{ (dB)} + G_{rxa} = 10 \log\left(6\left(\frac{fD}{c}\right)^2\right) + 40 \text{ dB} + 40 \text{ dB}$$

$$= 10 \log\left(6\left(\frac{(4 \times 10^9)(6)}{3 \times 10^8}\right)^2\right) + 80 \text{ dB} = 45.8 \text{ dB} + 80 \text{ dB} = 125.8 \text{ dB}$$

Therefore, the power is:

$$P_o \text{ (dB)} = P \text{ (dB)}_i + G_{es1} \text{ (dB)} - A_{c,up} \text{ (dB)} + G_{sat} \text{ (dB)} - A_{c,down} \text{ (dB)} + G_{es2} \text{ (dB)}$$

$$= 10 \log\left(\frac{10 \times 10^{-3}}{1 \times 10^{-3}}\right) + 97.3 - 199 + 139.8 - 195.5 + 125.8$$

$$= 10 \text{ dBm} - 31.6 \text{ dB} = -21.6 \text{ dBm} = (1 \times 10^{-3})10^{\frac{-21.6}{10}} = 6.92 \times 10^{-6} \text{ W}$$

The power magnitude found in Example 10-12 is a little small. If this were a real system, we would want to get out about the same power we put in, or 10 mW. We can upgrade the earth station segments because the space-based segment, the satellite, is generally not user-configurable. A combination of larger antennas and more amplification could create a system with an overall gain or loss of 0 dB. Since larger antennas tend to cost more than better amplifiers, most designers would seek to increase the amplification of both earth station amplifiers. The LNA would probably be left as is because increasing the amplification while maintaining a low noise figure can be quite expensive. If we add another 30 dB to the uplink earth station gain, we obtain an increase of 3 orders of magnitude. The power transferred to the end equipment would then become approximately what was delivered to the communication link.

$$P_o \text{ (dB)} = -21.6 \text{ dBm} + 30 \text{ dB} + 8.4 \text{ dBm}$$
$$P_o \text{ (W)} = 6.9 \text{ mW}$$

Once the power is computed, finding the C/N ratios for both the satellite and the receiving earth station is straightforward. First find the power delivered to the satellite's radio demodulator by adding together the input power and the earth station's gain. This sum yields the transmitting power from the earth station, the first term. We then subtract the uplink channel attenuation and add the satellite antenna G/T ratio. There are no additional losses in Example 10-12, so subtract Boltzman's constant. This process gives the terms to insert into Equation 9-24, which is reproduced in Example 10-13.

■ **EXAMPLE 10-13**   From Example 10-12, find the C/N values for the satellite and earth station. ERP is given by the sum of the transmit power and the transmitting antenna gain.

$$\frac{C}{N_o} (dB) = ERP (dB) + G_d (dB) - T_N (dB) - A_c (dB) - 10 \log (k) - 10 \log (\text{losses})$$

$$= P_i (dB) + G_{es1} (dB) - A_{c,up} (dB) + G_d (dB) - T_{N,sat} (dB) - 10 \log\left(\frac{1.38 \times 10^{-23}}{1 \times 10^{-3}}\right) - 0$$

$$= 10 \log\left(\frac{10 \times 10^{-3}}{1 \times 10^{-3}}\right) + 97.3 - 199 + G_d (dB) - T_{N,sat} (dB) - 10 \log(1.38 \times 10^{-23})$$

$$= 10 \, dBm + 97.3 \, dB - 199 \, dB + G_d (dB) - T_{N,sat} (dB) + 228.6 \, dB$$

$$= 136.9 \, dB + G_d (dB) - T_{N,sat} (dB)$$

Note how Boltzman's constant is expressed in dB because we chose to express the input power in decibel form. Also note how Boltzman's constant is always 228.6 dB. Just as we did in Chapter 9, to find the G/T ratio, subtract T from G in decibels:

$$G_d (dB) - T_{N,sat} (dB) = 57.3 \, dB - 10 \log(200) = 57.3 - 23.0 = 34.3 \, dB$$

This calculation gives us the value of C/N for the satellite.

$$\frac{C}{N_{o \, sat}} = 136.9 + 34.3 = 142.4 \, dB$$

To find the C/N ratio for the receiving earth station, start with the input power of the satellite:

$$P_{i,sat} (dB) = P_i (dB) + G_{es1} (dB) - A_{c,up} (dB)$$

$$= 10 \, dBm + 97.3 \, dB - 199 \, dB = -91.7 \, dBm$$

Next, find the ERP of the satellite by adding the gain of the satellite:

$$ERP_{sat} (dB) = P_{i,sat} (dB) + G_{sat} (dB) = -91.7 + 139.8 = 48.1 \, dBm$$

Now find the C/N value for the receiving earth station. Begin again with Equation 9-24, where the subscripts reflect the quantities used in Example 10-13.

$$\frac{C}{N_{o \, es}} (dB) = ERP (dB) + G_{LNA} (dB) - T_{N,LNA} (dB) - A_c (dB) - 10 \log (k) - 10 \log (\text{losses})$$

$$= ERP_{sat} (dB) + G_{LNA} (dB) - T_{N, LNA} (dB) - A_{c,down} (dB) - 228.6 \, dB - 0$$

$$= 48.1 \, dBm + G_{LNA} (dB) - T_{N,LNA} (dB) - 195.5 - 10 \log(1.38 \times 10^{-23})$$

$$= 48.1 \, dBm + 40 \, dB - 10 \log(120) - 195.5 + 228.6 \, dB$$

$$= 100.4 \, dB$$

■ SUMMARY

This chapter provided a comprehensive overview of many aspects of satellite communications. Satellite construction and subsystems were explored, the process of launching and finding a satellite in the sky was explained, and the different classifications of communications satellites were described. Access techniques encountered in the satellite industry were identified and several representative satellite systems were summarized. This chapter concluded the three-chapter theme of link calculations with the inclusion of two extended examples that illustrate a round-trip link analysis with a communications relay satellite in geostationary orbit.

REVIEW QUESTIONS

1. Define the angle of inclination of a satellite in orbit around the earth.

2. Calculate the velocity of a satellite that is in circular orbit 8,000 miles above the surface of the earth.

3. Describe the conditions that must be met for a satellite to appear stationary above the equator.

4. Define the angle of elevation of a satellite in geostationary orbit.

5. Write a computer program to calculate the elevation, azimuth, and heading for an antenna when the longitude for the satellite and the longitude and latitude for the receiver are provided.

6. Calculate the round-trip time delay for a satellite in LEO, or 200 miles average distance.

7. What is the typical lifetime of an earth-orbiting satellite? What limits the lifetime of satellites in orbit around the earth?

8. Define apogee and perigee.

9. What is the primary source of power in geostationary satellites?

Questions 10 to 14 refer to Examples 10-12 and 10-13.

10. What is the impact of doubling the diameter of the receiving antenna?

11. What is the impact of doubling the noise temperature of the LNA?

12. What is the impact of dividing the transmitting power of the earth station in half?

13. What is the impact of dividing the gain of the earth station receiver in half?

14. What is the impact of dividing the noise temperature of the satellite transponder in half?

3

Cellular Systems and Protocol Perspectives

11

Cellular Basics

■ INTRODUCTION

This chapter will introduce the basic concepts of systems familiar to all of us as mobile telephones using the cellular technology approach. These systems are also known by the abbreviations PCS (personal communications systems) here in the United States and PCN (personal communications networks) worldwide. The former abbreviation is usually applied to those wireless cellular systems that operate in the 1,800 to 1,900 MHz band.

■ MARKET AND TECHNOLOGY OVERVIEW

Several systems are deployed simultaneously in different parts of the world. Many countries, especially the United States, have several systems deployed, each using a different technology or frequency band, for two reasons. First, the technology was much more rapidly accepted by consumers than any of the providers originally envisioned. This acceptance provided the funding for rapid technological advancement and rapid deployment. Second, because large areas of the United States remain outside cellular coverage zones, it was most profitable for service providers to expand into new territory rather than deploy in areas that already had cellular service. In the few largest markets, of course, deployment occurred when new technology emerged, but this deployment was the exception, not the rule.

Table 11-1
Worldwide Market Share as of 2000

System Name	Market Share
AMPS	10%
D-AMPS and country specific adaptations	20%
GSM	55%
IS-95	15%

Table 11-2
Three Cellular System Approaches

Name	Modulation	Air Interface	Generation
AMPS	Analog FM	FDMA	First
D-AMPS, GSM	Digital PSK	FDM/TDMA	Second
IS-95	Digital PSK	FDM/CDMA	Second

In Europe and to a lesser extent in Asia, the deployment was not so haphazard. In the countries of these regions, cellular deployment was limited until a national standard was agreed upon. Therefore, throughout most of Europe, the cellular systems in each country conform to the same air interface standard, global system for mobile (GSM) communications. In the United States, the majority of installations are advanced mobile phone service (AMPS) and digital AMPS (D-AMPS), with moderate coverage of IS-95 and spotty coverage with GSM systems. Approximate values of the worldwide market share of these systems are shown in Table 11-1. As you can see, the GSM system dominates among currently deployed systems.

While many country-specific adaptations of the technology are used in these systems, three basic approaches are in use today. The first two of these three implementation approaches will be described in some detail in Chapter 12. The CDMA-based systems will be described in Chapter 13. The next-generation phones, the third-generation or 3G phones, will be of several different types of air interface, although most will use some form of wideband CDMA air interface. To summarize, the three system implementation approaches are listed in Table 11-2. Figure 11-1 shows how cellular telephones have evolved from early portables to a product representative of the state of the art in 2001. Each one is a Motorola product. Figure 11-1(a) is representative of the state of the art in 1994; Figure 11-1(b) is representative of the state of the art in 1996; Figure 11-1(c), in 1998; and Figure 11-1(d), in 2001.

After the elements common to every cellular system are presented, several sections on the business and IT staff perspectives of a wireless deployment are provided. These sections are not meant to be comprehensive, but rather an introduction to some of the issues behind the evolution of a traditional wired network to a wireless one. The chapter

Figure 11-1
Cell Phone Evolution: (a) 1994; (b) 1996; (c) 1998; (d) 2001 (*Photo courtesy of Motorola Archives, © 2001 Motorola, Inc.*)

(a)

(b)

(c)

(d)

Motorola Archives, Motorola, Inc., Schaumburg, Ill.

closes with a section on cellular system economics. This section contains the information necessary to form the basis of a running example through the cellular evolution from the service provider's perspective.

■ EVOLUTIONARY BACKGROUND

The three implementation approaches listed in Table 11-2 are evolutionary because there was a major reason for each technological change. The major advantage to evolving from analog modulation to digital modulation was that data channels could be interleaved with the digitized voice signal. The major advantage to moving from a pure FDMA approach to a hybrid FDM/TDMA air interface was greater system capacity. The major advantages in moving from TDMA to CDMA were an increase in the number of channels in the same frequency space, along with much better security protection from radio eavesdropping. The major disadvantage of CDMA compared to TDMA systems is the somewhat higher cost of implementation.

■ BRIEF HISTORY

The basic cellular concept was developed by employees of Bell Labs in the late 1940s. In 1971, AT&T proposed to the Federal Communications Commission (FCC) a cellular-based system for voice transport. The bandwidth allocated to the cellular telephone industry was previously used for UHF TV, channels 70 to 83. This spectrum range from 806 MHz to 890 MHz was seized in 1974, and initial testing was carried out for the rest of that decade. This extensive field testing was used to demonstrate the concept to the FCC.

In 1981, the FCC agreed in principle that the concept worked. Over the next three years, the FCC and other interested parties worked toward an equitable way to assign licenses for this new service. In 1984, AT&T, the parent of Bell Labs, got half the spectrum and a competitor got half. The reasoning was that since Bell had invented the basic idea and carried out the field trials, it deserved the lion's share of the marketplace. At that time, AT&T was still "Ma Bell" and had a virtual monopoly on wireline telephone service.

■ SIX COMPONENTS OF EVERY CELLULAR SYSTEM

Every cellular system has six basic components. Today, all cellular systems follow the same reference model for their architecture, which was developed by the TR-45 standards group. It is not a protocol reference model, like the OSI or TCP/IP model, but an architectural model identifying the names and functions of the subcomponents that every cellular system uses. These subcomponents are grouped into the six main components that every cellular system needs:

1. Gateway into the wireline system
2. Base station (BS)
3. Databases
4. Security mechanism
5. Air interface standard
6. Cellular telephones

The interconnection of these components is shown, in very basic schematic form, in Figure 11-2.

Component 1: Gateway into the Wireline System

There is always a gateway between the base station's switching subsystem and the wireline switching fabric. The basic idea here is that some kind of switching fabric must interconnect the traditional wireline telephone system to the wireless system to allow calls to flow between the two systems. This gateway is composed of two logical components: the mobile switching center (MSC) and the interworking function (IWF), which does protocol translation if required. The MSCs are interconnected with wireline links, and one is designated as the location of the equipment identity register (EIR). Each MSC connects to the wireline network.

Figure 11-2
Six Components of a Cellular System

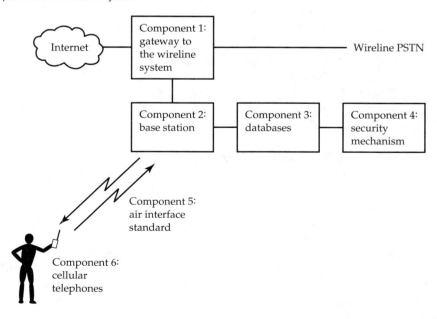

Component 2: Base Station (BS)

The base stations always contain several transmitter/receiver pairs and form one-half of the radio link. These transmitter/receiver pairs are coordinated with a controller unit that connects the transmitter/receiver pairs to the MSC. Therefore, the base station is composed of two logical components: the base transceiver system (BTS) and the base station controller (BSC). There may be, and usually are, several BTSs controlled by each BSC. (Note that the GSM system uses the abbreviation BSS, or base station subsystem, for this component.)

Component 3: Databases

The database is structured to keep track of billing, caller location, etc. Usually each mobile unit has an entry in at least two databases: one to keep track of its "home" in the mobile network, the home location register (HLR), and one to keep track of its current location, the visitor location register (VLR). The latter accommodates roaming. Each cellular telephone has entries in these databases that store the last location, frequency pair or time slot, and other details of the mobile user. The functions of an HLR/VLR pair in a cellular network are quite similar to the function of an SCP in an SS7 wireline network.

Component 4: Security Mechanism

A security mechanism is required to authenticate billing and to provide assurance that the person who is placing the call from the cellular telephone is the actual owner. This function is facilitated by a database, called the equipment location register (EIR), and an authentication center (AC), which manages the actual encryption and verification of each subscriber. There is usually only one EIR and many ACs per system. There are also other security components located in the cellular telephone itself. Note that the GSM system uses the abbreviation AuC for authentication center.

Component 5: Air Interface Standard

Each of the three main implementation approaches to cellular telephony (FDMA, FDM/TDMA, CDMA) uses a different air interface. You can think of the air interface as a way of sharing the bandwidth allocation of the system among all cellular telephones in any particular cell.

Component 6: Cellular Telephone

The cellular telephone itself is called a mobile station (MS). They need not be actual telephones and include various mobile devices that use the cellular system for data transfer as well as voice connections.

A very brief illustration of these physical components was shown in Figure 11-1. Figure 11-3 shows them in more detail and is representative of the TR-45 architectural model. Figure 11-4 illustrates a protocol perspective of the components. Note that Figure 11-4 is the protocol stack for a purely voice service, which will link to the SS7 protocol stack shown on the far right. A data service protocol stack, which would pass through the IWF into the Internet (compared to passing through the MSC to the PSTN), is not shown in this figure.

Figure 11-3
Abridged TR-45 Architectural Model

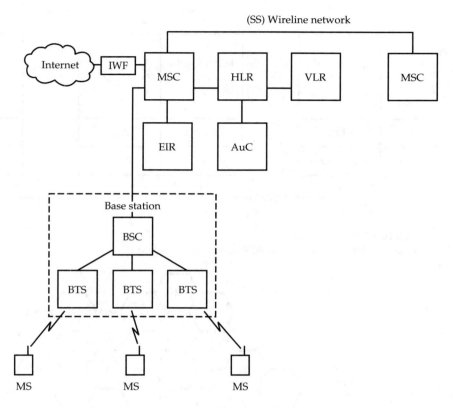

■ BASIC CELLULAR CONCEPTS AND TERMINOLOGY

The basic cellular concept is composed of two fundamental ideas:

1. Reuse the spectrum many times (frequency reuse).
2. Divide the service area into a group of small regions called cells.

Cells can be split to accommodate growth, or frequency groups can be assigned to minimize co-channel interference. A cell is a specific region where a portion of the frequency spectrum is used. Co-channel cells are those cells where the same group of frequencies is used.

Cellular radio systems, more familiar to many as cellular telephone systems, utilize a large number of radio links multiplexed in one of several ways to accommodate a large number of simultaneous calls on the same frequency. This setup is accomplished by splitting up the service region into several hexagonal cells. If you have ever seen a honeycomb, you know how the cells are configured into a grid so that the entire area is covered.

Figure 11-4
Protocol Components of a Cellular System

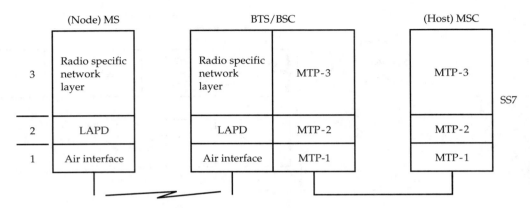

Figure 11-5
Frequency Reuse Concept Showing a Seven-Segment Cluster

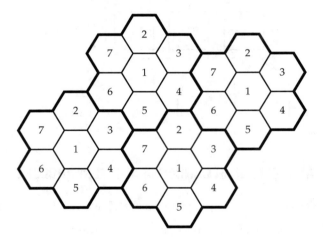

In each cell is one base station with a transceiver capable of spanning the entire cell, but not much beyond. Each radio link is limited to a single cell and a slight overlap with its adjacent partners. This setup might seem like a disadvantage but instead is the key to the wide acceptance of cellular telephones. Since the power output of the base station is low, those frequencies can be used by another cell's base station to communicate with another user just a few miles away in another, nonadjacent cell. Thus, the same frequencies are used many times over in any cellular system. In the honeycomb pattern, as long as no sides of any two cells in the honeycomb touch, a frequency group can be used that has been used before. Figure 11-5 illustrates the concept of frequency reuse, probably the key idea in cellular engineering.

Note that, although the cells in the figure are labeled as 1, 2, etc., each stands for a group of frequencies. A cluster is a group of cells. Figure 11-5 shows a cluster of seven cells. In each cell of the cluster, a different group of transmitting and receiving frequencies is used. Any cellular telephone operating anywhere in a cluster uses a different frequency.

The total spectral allocation for a cellular system is divided by the number of cells in a cluster. Then the base station in each cell in every cluster uses that portion of the total spectral allocation to communicate with any mobile system located in that cell. In any cluster, therefore, the total number of possible communication channels is the same.

Figure 11-5 shows a honeycomb pattern with four clusters of size seven drawn and labeled. Note how the clusters repeat across the pattern. If you imagine the pattern to be the service area, you can see how the frequencies repeat, once per cluster.

The frequency reuse of channels in the same geographic area was necessary because the allocation of bandwidth for cellular services is very limited. As a result, any way to accommodate more users in the same frequency band, shown in Figure 11-5 through reuse, is a big advantage. The number of users who may share the same frequency at the same time, in any small area such as a city, in nonadjacent cells depends on the number of cells, but it is usually at least four and in dense deployments the number is significantly higher. This number is referred to as the frequency reuse factor (FRF). The FRF is found by dividing the number of full-duplex channels defined by the cellular standard by the number of full duplex channels in any particular cell. The calculation is shown in Equation 11-1:

$$FRF = \frac{C}{S} \qquad \textbf{(11-1)}$$

FRF = frequency reuse factor
C = total number of full-duplex channels in a cluster
S = number of full-duplex channels in a cell

Each base station is connected to the wired telephone system through a switching center. In this way, calls from fixed telephone locations can be routed to a mobile subscriber, and vice versa. If too many users try to access the system in any geographical area, congestion will occur and calls will be blocked because there are no free channels.

This expected growth in the system is accomplished by reducing the size of the cell and splitting it into several cells, each with its own base station. This technique effectively allows more calls to be handled by the system as long as the cells do not get too small. The minimum size cell is set at a diameter of about 500 m. With a cell smaller than this, the base stations would interfere with each other and the frequency reuse concept that the system is based on would not work.

A seven-segment cluster was shown in Figure 11-5. If the individual cell size and the spectral allocation per cluster is kept constant, one can vary the number of cells per cluster and thereby vary the coverage area. If you add more cells to the cluster, you end up with a larger overall area coverage for the allocated frequency spectra. This change is desirable, but it comes with penalties. You now have fewer channels per cell, which results in less system capacity in each cell. This approach is taken when the coverage area is large compared to the expected number of subscribers in that area. A semirural environment would be the target application for such an approach.

Figure 11-6
FRF Tradeoffs with Coverage Area

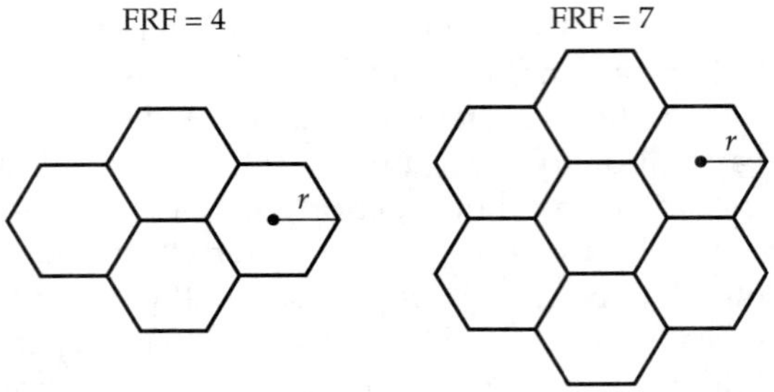

FRF = 4   FRF = 7

Area of hexagon = $3r^2 \sin (60°)$
= $2.6r^2$

Area = $4(2.6)r^2$ Area = $7(2.6)r^2$

On the other hand, it is possible to reduce the number of cells per cluster. Each cluster covers a smaller geographical area, but the number of channels per cell, or system capacity per cell, increases. This solution would be appropriate for situations where the coverage area is relatively small compared to the number of subscribers in that area. A metropolitan center would be the target application for this approach. Figure 11-6 illustrates these two situations. Remember that, in this figure, the cell size is kept constant. In normal circumstances, the cell size would vary as well. Large cells are normally used in areas with low-capacity needs, while small cells are used in areas with high-capacity needs. This proportionality relationship is shown in Equation 11-2:

$$\text{Cell size} \propto \frac{1}{\text{number of subscribers in cell}} \qquad \textbf{(11-2)}$$

The equivalency between frequency reuse and cluster size illustrates how cellular systems can scale as subscriber growth occurs. As the number of cells per cluster decreases, the potential for co-channel interference grows. They are inversely related, as the proportionality in Equation 11-3 shows:

$$\text{Frequency reuse factor} = \text{FRF} \propto \frac{1}{\text{co-channel interference}} \qquad \textbf{(11-3)}$$

As the co-channel cells move closer together, the potential for interference grows. If you think carefully about it, you will see that a small number of same-size cells per cluster means that each unique cell frequency assignment is geographically closer. Alternately, as the number of cells per cluster increases, the mean distance between cell frequency assignments becomes larger.

As already mentioned, this choice of the number of cells per cluster is called the FRF. The number is chosen to maximize revenue per unit area, subject to co-channel interference limitations. Co-channel interference is the interference between cells using the same carrier frequency group. Large FRFs, or a large number of cells per cluster, are very desirable from an interference perspective. On the other hand, small frequency reuse factors are very desirable from an economic perspective. The tradeoff point is always a careful decision that takes into account both technical and economic issues.

■ **EXAMPLE 11-1** How many cells could the District of Columbia (Washington, D.C.) accommodate assuming an FRF first of 4 and then of 7? Compare the number of subscribers that could be supported, assuming a total spectrum allocation that will permit 1,000 simultaneous users and a cell diameter of 2 km.

The District of Columbia is a square 10 miles on a side, so its total area is 100 square miles. To calculate how many cells it will take to provide coverage for this area, we first need to calculate the area of a perfect hexagon. Then using a grid similar to that shown in Figure 11-5, frequency assignments can be made for each FRF. The formula for calculating the area of each hexagon, where r is defined as the distance from the center to one vertex, was shown in Figure 11-6. Using the formula shown in Figure 11-6, and a radius that is half the diameter of the cell (1 km), the area of each cell is:

$$A = 2.6r^2 = 2.6(1)^2 = 2.6 \text{ km}^2$$

It is then a straightforward process to calculate the minimum number of cells required to provide coverage to Washington, D.C., if we take the ratio of the two areas. Note that there are 1.62 kilometers per mile.

$$\text{Number of cells} = \frac{\text{coverage area}}{\text{cell area}} = \frac{(10 \times 1.62 \text{ km/mile})^2}{2.6 \text{ km}^2} = \frac{262.4 \text{ km}^2}{2.6 \text{ km}^2} \approx 100 \text{ cells}$$

This calculation yields only the minimum number because, as you will discover, when you lay out the grid pattern, some cells extend beyond the coverage area and some additional cells are required to cover the edges of the service area. For this example, however, we will assume that this answer is exact.

Now calculate the maximum number of simultaneous users for an FRF of 4 and an FRF of 7 and compare them. First, apply Equation 11-1 and find the number of full-duplex channels per cell:

$$S = \frac{C}{\text{FRF}} \qquad S_4 = \frac{1,000}{4} = 250 \qquad S_7 = \frac{1,000}{7} = 142$$

Now, multiply by the number of cells in the entire service area:

$$\text{Total system capacity}_4 = (250)(100) = 25,000$$
$$\text{Total system capacity}_7 = (142)(100) = 14,200$$

As you can see from Example 11-1, a change in FRF, from 4 to 7 cells per cluster, nearly divided in half the number of simultaneous users. The FRF is indirectly proportional to the number of simultaneous users that a cellular system can support. This relationship is shown below:

$$\text{Total system capacity} \propto \frac{1}{\text{FRF}} \qquad \textbf{(11-4)}$$

Below a minimum value of 3, however, the co-channel interference will dominate, thus limiting the number of simultaneous users. Cellular engineering attempts to find the tradeoff point where these two limiting factors balance. In other words, cellular engineering attempts to answer the question, *"How small can we make the FRF in the interference environment of this deployment?"*

Possible Frequency Reuse Values

The plane geometry of clusters that must fill an entire area is constrained such that it limits the numerical values in a cluster. The only integer values that will work are: 3, 4, 7, 12, 13, and 21. Because these values are all integer values, it is possible to define a ratio, called the *D/r* ratio, that will give us a relative numerical value to gauge co-channel interference. Remember, co-channel interference is wholly determined by the distance between cells using the same frequency group. Because this distance is directly proportional to the FRF, it can be expressed by this value, the *D/r* ratio.

The detailed development of this relationship is outside the scope of this text, but we can define proportionality relationships easily enough. First, the co-channel interference must be indirectly proportional to the distance between the co-channel cells themselves. Second, the distance between co-channel cells is directly proportional to the cluster size. Third, the co-channel interference must be indirectly proportional to the square root of the FRF because the interference is defined by the inverse square law of radiation. Putting these observations into relationships, we find:

$$\text{Co-channel interference} \propto \frac{1}{D}$$
$$\text{Co-channel interference} \propto r$$
$$\text{Co-channel interference} \propto \frac{1}{\sqrt{\text{FRF}}}$$

D = distance between co-channel cells
r = radius of cell

These observations can be combined into what is called the *D/r* ratio. It is expressed in Equation 11-5. The values are illustrated in Figure 11-7.

$$\frac{D}{r} = \sqrt{3\,\text{FRF}} \qquad \textbf{(11-5)}$$

Frequency Assignment Rule

In both Figures 11-5 and 11-7, there seemed to be a relationship between the adjacent clusters shown. For the integer values of FRF, there is a rule specifying the layout for the

Figure 11-7
D/r Ratio Calculations

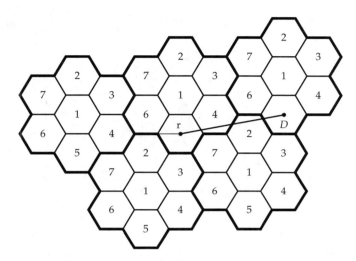

cell frequency plan for any cluster size. This rule is used to assign channel sets to different cells. It is based on two values, i and j, called shift parameters. There are three steps for assigning channel sets:

1. Draw a hexagonal pattern that encloses the coverage area and choose an originating cell somewhere near the geographical center. Label that cell with the number 1.
2. Move i cells along any face of the originating hexagon (along a chain of hexagons that begin adjacent to the originating cell).
3. Move j cells along the face of the current hexagon. The direction of movement is 60 degrees counterclockwise from the direction moved in step 2. Label that cell with the number 1.

These three steps are repeated for each cell in the original cluster, and they are illustrated for a seven-segment cluster in Figure 11-8, where $i = 2$ and $j = 1$. Note that the angle is taken from the direction of the i vector. There is also a relationship among i, j, and FRF. It is shown in Equation 11-6:

$$\text{FRF} = i^2 + ij + j^2 \tag{11-6}$$

■ **EXAMPLE 11-2**   This example is a numerical evaluation of Figure 11-8. Use the values in the figure and verify that Equation 11-6 balances.

$$7 = 2^2 + (2)(1) + 1^2 = 4 + 2 + 1 = 7$$

Figure 11-8
Frequency Assignment Illustration

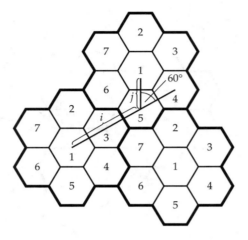

All other integer values of FRF also have integer values of i and j, which is a consequence of the plane geometry used. A three-dimensional system could take on additional values of FRF, and the frequency assignment rules would be modified to take this dimension change into account.

■ CELL SPLITTING

As growth occurs, any cell will reach its capacity limit at some point. This limit occurs when the number of channels available in the cell is equal to the number of subscribers who wish to make a call at any particular point in time. This point is called the maximum traffic load of the cell. The solution is to split the cell into several smaller cells. To illustrate this concept, we will assume that the new cells have exactly half the radius of the old cell and are co-located in the same geographical area. In Figure 11-9 note that when the radius is halved, four new cells fill the original cell area. This process can be demonstrated analytically in the following equations:

$$A_{\text{original cell}} = 2.6r^2$$

$$A_{\text{new cell}} = 2.6\left(\frac{r}{2}\right)^2 = 0.65r^2$$

where r is measured from the center of the hexagon to any apex

Each new cell now occupies one-quarter of the area of the original cell. If each new cell has the same number of channels as the old cell, the maximum traffic load is increased by a factor of 4.

The number of cells in any particular system, the coverage area, cell area, and cluster size (or FRF) are related to each other. This relationship is shown in Equation 11-7:

$$\text{Number of cells} \propto \frac{\text{coverage area}}{(\text{cluster size})(\text{cell area})} \qquad \textbf{(11-7)}$$

In many cases, the proportional relationship shown in Equation 11-7 will closely approach equality.

Evolution of a Cellular Network

As a cellular network matures, more subscribers are added and at some point, cells must be split. After the split, some subset of cells are smaller than the original cell size. This situation complicates the handoff algorithms and co-channel interference effects. In general, the cellular network evolves in three stages. Each of these stages may be repeated several times down to some minimum cell diameter. Different parts of the service zone of any single provider will have clusters in various stages of evolution. These three stages are illustrated in Figure 11-10 for a seven-segment cluster.

Forward and Reverse Channels

All cellular radio systems use two distinct groups of frequencies. The first group is composed of the forward channels, those from the base station to the mobile unit. The second group is composed of the reverse channels, those from the mobile unit to the base station. In other words, the group of frequencies that move from the base station infrastructure, the base station interface to the wireline system, are called the forward channels; those channels that move from the mobile stations back to the existing infrastructure are called the reverse channels. See Figure 11-11.

Practical Issues

The actual practice of cellular deployment is often about two issues. From an economic perspective the question is, How small can the FRF be in this interference environment? From an engineering perspective, this decision is addressed by answering the questions, Where can I put the base station antennas? Do I need to use directional antennas?

You can always build a tower and place the antenna on top of it, but towers are expensive, and acceptance in some communities is not guaranteed. This decision falls under the province of the local city council and zoning laws. Many communities, and often the very communities who use cellular telephony in large numbers, are quite

Figure 11-9
Cell Split

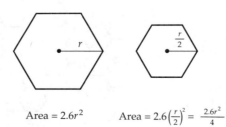

$$\text{Area} = 2.6r^2 \qquad \text{Area} = 2.6\left(\frac{r}{2}\right)^2 = \frac{2.6r^2}{4}$$

Figure 11-10
Cellular Evolution for a Seven-Segment Cluster: (a) Early Stage Deployment; (b) Midstage Deployment; (c) Late Stage Deployment

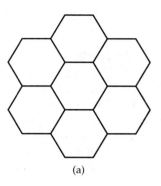

(a)

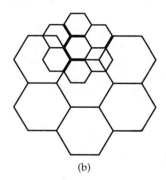

(b)

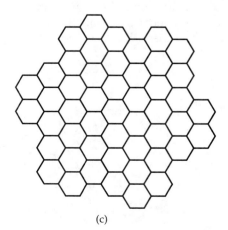

(c)

Figure 11-11
Forward and Reverse Channels

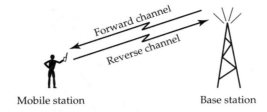

Mobile station   Base station

unreceptive to an antenna tower placed nearby. To meet with each individual city council is a time-consuming labor-intensive process.

As you might expect, placing antennas on the tops of buildings is sometimes a viable option because it reduces the cost of construction (no tower required) and tends to remove them from casual view, thus addressing the main concerns that zoning laws are meant to address. However, there are two problems. First, tall buildings are not found everywhere one might wish to offer cellular service. Second, the rooftops of these buildings are often already filled with antennas for various uses, including television and radio transmission and reception antennas, microwave relay substations, private radio networks, previously deployed cellular systems, etc.

The solution to this problem of antenna placement created a new industry in the last half of the 1990s. A niche now exists for manufacturing antennas that look nothing like an antenna of old. Antennas now resemble shutters on windows, trees, or even some architectural feature on the building, such as a buttress or trim plating. The implementation of an overlay (discussed below) also often requires this type of solution for base station antennas in new systems.

As cellular systems evolve, it is natural that co-channel interference will grow. Thus, antenna deployment practices must also evolve. Early omnidirectional antennas are often used for cost savings. Later, when the cell location features are such that large FRFs are used, co-channel interference will become a major issue. In this case, directional antennas can be used to address the increased co-channel interference. In any cluster, each cell is adjacent to six other cells. The antennas used in this arrangement would then be divided into three groups, each group covering 120 degrees. In extreme situations, six antennas, each with a coverage of 60 degrees, would be employed. The three-antenna configuration is shown in Figure 11-12. Three antennas would be placed in each of these 120-degree sectors, one antenna for transmitting and two antennas for receiving. The two receiving antennas would be used in a diversity configuration to enhance gain on the return path to the base station. The overall antenna configuration would have the parameters shown in Figure 11-13 (a). Figure 11-13 (b) shows an actual cellular antenna deployment like that found on a tower.

The separation of the two receiving antennas depends on the height of the antennas above the ground plane. Usually this height is taken as the height of the tower, but this figure is not always exact. (See Chapter 8 for more detail.) As a general rule of thumb, for an antenna height of 30 meters, a separation of 8λ is required; for an antenna height

Figure 11-12
Directional Antenna Concept (FRF = 7)

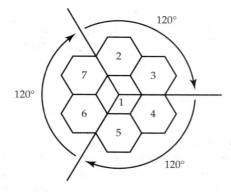

Figure 11-13a
Directional Antenna Configuration (FRF = 7)

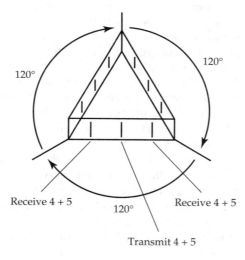

of 50 meters, a separation of 11λ is required. Obviously, this feature has a significant impact on the overall size of the antenna cluster and also affects the esthetic appeal of the tower. To a certain extent, the use of higher frequency bands ameliorates this problem. See Example 11-3.

■ **EXAMPLE 11-3** Compare the antenna cluster size for two cellular systems using a directional antenna approach, as in Figure 11-14. The first system is 900 MHz, the second is 2,400 MHz. Assume a tower height of 50 meters. (The solution to this example is illustrated in Figure 11-14.)

Figure 11-13b
Cellular Antenna (*Courtesy of PhotoDisk. Digital Imagery © copyright 2001 PhotoDisk, Inc.*)

   You should have some idea of the typical performance of the antennas and cellular phones themselves. Typical omnidirectional antenna gains used in the base stations range from 6 to 9 dB. The power supplied to the base station antenna in a typical cell is about 10 W. The signal-to-noise ratio at the cellular telephone for the expected performance is between 15 and 20 dB. A typical design goal when implementing a cellular system is that 100 percent of the users in a cell will get this signal-to-noise ratio 90 percent of the time.

Figure 11-14
Example 11-3

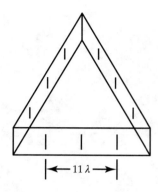

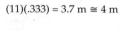

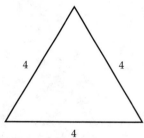

$$900 \text{ MHz}: \lambda = \frac{c}{f} = \frac{3 \times 10^8}{9 \times 10^8} = .333 \text{ m}$$

$$(11)(.333) = 3.7 \text{ m} \cong 4 \text{ m}$$

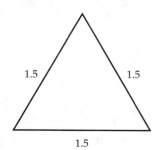

$$2,400 \text{ MHz}: \lambda = \frac{c}{f} = \frac{3 \times 10^8}{2.4 \times 10^9} = 0.125 \text{ m}$$

$$(11)(.125) = 1.375 \cong 1.5$$

Quality Issues

The above paragraphs outlined some of the basic principles behind the cellular concept. When deploying a cellular system, additional concerns must be addressed. The first concerns include various measures of the quality of the service provided:

> Quality of service (QOS): This term is defined in this instance as channel availability. Thus defined, it is a measure of how often a subscriber will attempt to place a call and find all traffic channels busy. It is usually stated in terms of the probability of some amount of time passing before a channel becomes open and available for use.

> Coverage quality: Coverage quality is a measure of the percentage of time an adequate received signal level is received by the subscriber in each location served by the system.

> Transmission quality: Transmission quality is a measure of the fidelity or quality of the signals received by the subscriber. It is measured using the familiar BER measurement for data signals and with a similar metric for voice signals.

These three quality objectives are of great interest to both providers and subscribers. They are usually addressed using the principles described in earlier chapters of this book. For example, coverage computations are used to determine the necessary received signal strength in the area surrounding the base station. The way the power of a signal falls off with distance and the details of a particular antenna's radiation pattern are of importance. The ERP of each base station must be computed and analyzed.

The issue of channel co-location is of great interest when performing coverage quality because each co-channel signal acts as an active interferer. This issue is often associated with the subject of frequency planning, which takes into account these aspects as well as the intermodulation components of the carrier frequencies and the details of the shape of the antennas and various reflecting elements designed to increase the isolation. These concerns have led to specific rules for assigning the individual carriers to cells. Standardized tables for these specifics have been worked out for existing technologies. These tables define the particular channel assignment order in cells.

Often these quality issues boil down to a debate on one philosophical point: what is the threshold where subscribers would rather be blocked than experience a poor quality call? This point is often related to the value of FRF because co-channel interference will often be the dominant factor. However desirable the debate might be, it is usually answered by statistics. The approach often taken is to set the threshold for quality at some compromise level and then adjust it based on subscriber feedback. What is desired is a balance in complaints between too-often-blocked and too-often poor quality. Of course, if either of these problems are allowed to grow too large, subscribers have a tendency to switch vendors. Therefore, quality in cellular networks is usually defined by a single upper limit total of customer complaints, and then this limit is parceled out to different mechanisms that define the fine-tuning of the engineering of the network.

The next several paragraphs define commonly encountered concepts used in the cellular industry.

Blocking   Blocking of a cellular call occurs when the number of full-duplex channels in the cell where the call either starts or ends are insufficient to support another channel. The solution to blocking is either to expand the system capacity by adding more cells per cluster or creating a cellular overlay.

Overlay   Overlaying is widely used in dense metropolitan areas where system capacity is at a maximum using current technology. Overlays require the use of dual-mode or dual-frequency phones, which are cellular telephones that operate in one or more air interfaces or frequency bands. The basic idea is to overlay another cellular system using the same base station locations. This technique can more than double the current capacity in any cell.

Not every type of cellular system can overlay another. As a general rule, those systems that operate in a different frequency band can be overlaid at will. Usually any CDMA based system can overlay a TDMA or FDMA system operating in the same frequency band, although the engineering of this overlay approach is more complex than the first approach.

Locating   A strict definition might be the periodic monitoring of the cellular telephone's signal, another way of saying that it's a way of locating the cellular telephone. Since the phone is mobile, there is no way of knowing what cell it is in prior to the phone being switched on. This is the reason why a cellular telephone must be turned on for you to dial it. If it is not switched on, there is no way of knowing which base station can communicate with the phone. Since the cellular telephone is switched on prior to the call being placed, a relationship already exists between the base station and the individual cellular phone. The gateway consults the appropriate database and instructs the appropriate base station to ring the desired phone.

Handoff   The strict definition of handoff is any change in the assignment of frequency channels for communication to and/or from the cellular telephone. Handoff may occur inside a cell or between cells. When it occurs inside a cell, it is for performance quality or service enhancement. When it occurs between cells, the common usage is to say that the cellular telephone is roaming, but this definition is not precise. In this case the terms *handoff* and *roaming* are being used interchangeably.

Analytically, a handoff due to reaching a cell boundary is straightforward when there is no interference. When the signal strength measured at the base station falls below some threshold, typically in the range of -90 to -100 dBm (1 to 10 μW) the base station coordinates with the next cell base station and performs the handoff. This situation is complicated when significant interference energy is present. Then the received signal at the base station is composed of signal plus noise, and this condition can cause situations when the energy of the combined signal plus noise deceives the base into missing when a handoff should occur, and a dropped call results. This issue is addressed by us-

Figure 11-15
Handoff

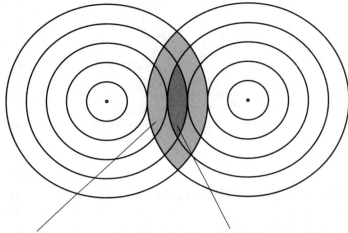

Handoff could occur but hasn't yet.   Handoff has most likely occurred.

ing more than one signal to measure the received power or by using received signal quality and not just raw power as an indicator of when to hand off any particular call. Figure 11-15 illustrates handoff between two adjacent cells.

Roaming   When a cellular telephone moves from one cell into another *and* the new cell is connected to a new MSC, then the telephone is said to be roaming. Roaming is defined as a handoff outside the cell boundary when the MSC also changes. Casual use defines roaming in three possible situations. First, roaming can be from cell to cell under control of the same base station controller. Recall that a base station controller is a subsystem in a cellular system where a single base station controller coordinates several base stations and hence cells. Second, it can be from cell to cell where each cell is under control of a different base station controller, but the MSC stays the same. Third, it can be from cell to cell and require MSC switching. In this last case, the two cells are controlled by base stations connected to different MSCs.

Only the third situation can truly be called roaming. Refer to Figure 11-16. As the subscriber leaves the range of cell C under control by MSC B and enters cell D controlled by MSC A, the two switching systems must coordinate the handoff between two different systems. As the signal strength approaches the handoff threshold in cell C, MSC B searches for a cell under its own control to hand off the user but cannot find one. It then widens the search to other systems and, in this case, finds MSC A, which controls cell D. MSC B sends a handoff request though the wireline PSTN that connects the two systems, and the two MSCs coordinate a handoff to accommodate the roaming user without interrupting the user's conversation on his or her cellular telephone.

Figure 11-16
Roaming

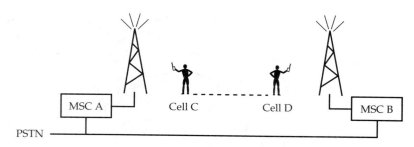

Multiple-Mode and Multiple-Frequency Phones   There are three basic types of cellular telephones. Single-mode phones operate in a single frequency band and use a single air interface. Dual-frequency phones can operate in two frequency bands but with a single air interface. Dual-mode phones can operate in multiple air interface modes, which may or may not be in different frequency bands. In general, the following rules apply to multiple-mode phones. The multiple-mode access rules must have been first standardized and should include at least some of the following items:

1. Spectrum sharing where appropriate.
2. Phone initialization sequence.
3. Handoff between the two overlaid systems in the same cell.
4. Various other call control and billing issues.

Dual-frequency phones for AMPS and DAMPS are available. AMPS and DAMPS dual-mode phones are quite common in the United States. DAMPS and IS-95 phones are increasingly common. Dual-mode DAMPS and GSM phones, as well as DAMPS and GSM phones, will be a growth area in the future, although differences in the MAP specification between the two systems are still being reconciled in the standards bodies.

Near-Far Effect   All wireless systems, including spread spectrum systems, that employ multiple-access techniques share a problem: the near-far effect. If a mobile station is close to the base station, it can effectively capture the channel because its transmission is so much more powerful than other mobile stations farther away. This effect is due only to the reduced channel attenuation of the station close by.

A mobile station far away from the base station may be rendered hidden because the transmitter close by may make it impossible for the receiver to detect a much weaker transmission attempting to communicate with the same receiver. Again, this is due only to the reduced channel attenuation of the mobile station close by compared to the mobile station far away; both mobiles are transmitting at the same absolute power. Effectively, the mobile station that is too far away is hidden. This effect is especially important in CDMA based cellular systems and will be discussed further in Chapter 13.

■ CELLULAR EVOLUTION IN THE UNITED STATES

Many cellular systems are simultaneously deployed in the United States, and the costs to upgrade, refit, and establish a national cellular network with data rates beyond the current standard of 14.4 kbps are very high. Thus, cellular evolution in the United States will be one of increased coverage, not increased data rates, until at least 2003. Cellular service providers will focus their investments on expanding the customer base, which will be done by extending coverage. Extended coverage will yield more revenue and will help finance the very expensive rollout of high data rate services. It will also allow current investments in technology development to be paid for in full. For more information on third-generation (3G) systems, see Chapter 18.

Current data rates from cellular telephony systems are approximately 14.4 kbps. Some service providers offer what is advertised as 56 kbps service, but this service is actually transmitted at 19.2 kbps and uses data compression techniques to simulate 56 kbps. This approach will work reasonably well with text but does little or nothing for graphics, sound, or compressed files. Although the service is transmitted at 19.2 kbps, actual user data rates are about 14.4 kbps because of radio channel issues.

■ METRICOM'S RICOCHET

The combination of these two circumstances (low data rates and service provider bias toward increased coverage over high data rate services in existing coverage areas) offer a window of opportunity. This window of opportunity is the heart of the business plan of a company called Metricom and its high-speed data network called Ricochet. Metricom believes that it can roll out a network offering high data rate mobile services before the cellular providers do.

Metricom was founded in 1985 and has some wealthy investors. Paul Allen (of Microsoft fame), through his Vulcan Ventures organization and WorldCom, invested $300 million in late 1999. In early 2000, Metricom raised close to $700 million through stocks and other financial instruments. Ricochet was first launched in 1995 with data rates of 28.8 kbps in just three markets: San Francisco, Seattle, and Washington, D.C. Coverage growth is the goal of the funding described above.

Ricochet is a service designed for large markets and asymmetric data rates. It will target the twenty-five or so largest metropolitan areas and offer data rates averaging 128 kbps for downloads and 56 kbps for uploads. An enhanced service at data rates of up to 400 kbps will be offered in selected areas. The network operates in the ISM bands 900 MHz and 2.4 GHz, so no licensing is required.

The key to the Ricochet service is that the technology is much less expensive than a cellular solution with the same data rates. Ricochet is a digital packet based service that does not use a circuit-switched connection, something that all cellular networks do.

Because there is no circuit connection, the Ricochet service can be marketed as an "always on" service without occupying a circuit connection. Of course, bandwidth is utilized only during actual packet transfer in both cases, but no circuit connection is required to support that occasional transfer in any true packet data service, of which Ricochet is one example.

The actual hardware is composed of a small shoebox-size white box that sits on top of telephone poles or street lamps. This box is the radio transceiver, and it is connected in a mesh architecture with an average coverage radius of about 300 meters. To increase the number of subscribers in any area, the number of radio transceivers is increased. Depending on the market, the number of transceivers could vary from five to fifty per square mile.

The radio transceivers communicate with each other using the 2.4 GHz band at data rates of up to 1 Mbps. The user modem transmits in the 900 MHz band to the closest radio transceiver. The user modem uses frequency hopping spread spectrum technology, which yields the typical advantages of such a scheme: enhanced security and increased user capacity. The transceiver routes user data packets to Ethernet radios that act as access points into the wireline network. These radios are usually located on the tops of buildings or on radio towers. The wireline network is an engineered Metricom network that connects to Metricom gateways in a company's intranet or other packet based network. As of late 1999, approximately 90,000 transceivers were deployed in the United States. The current generation of user modems are battery-powered devices that attach to notebook computers with Velcro. They are connected using the USB port or a serial port. Later in 2001, a PCMCIA version is expected, and a chip version that can be installed in any portable device is also planned. Figure 11-17(a) shows a Ricochet transceiver. Figure 11-17(b) and 11-17(c) show first-generation user modems suitable for attachment to a notebook computer.

■ APPLYING WIRELESS TO BUSINESS REQUIREMENTS

As in any decision to deploy a new technology, the first question to be addressed is, What part of the business needs to be wireless? In other words, What is the business need that is being addressed by adopting this technology? Often a good way to answer this question is to look at the core technology itself and list the chief differences between the current deployed technology and the new technology under consideration. For wireless adoption, this process is relatively simple: it's all about mobility and untethered access. Therefore, ask yourself, What groups of users are mobile or need to be? This group can include employees who spend much of their time traveling, shipping and stocking staff, etc.

Once business needs have been addressed, it is important to understand that wireless devices themselves have limitations. These very limitations will lead to an analysis of the business process. Like any new technology, wireless offers opportunities to reengineer the enterprise. For example, instead of faxing orders to suppliers, why not link orders directly into the suppliers' databases with an inexpensive hand-

Figure 11-17
Ricochet Components: (a) Sierra Wireless Transceiver (*The Sierra Wireless logo, Heart of the Wireless Machine, and Aircard are registered trademarks of Sierra Wireless. Other registered trademarks that appear on this photo are the property of their respective owners. © 2001 Sierra Wireless Inc.*); (b) Novatel Wireless First-Generation User Modem for Attachment to a Notebook Computer (*Courtesy of Novatel Wireless*); (c) Metricom First-Generation User Modem for Attachment to a Notebook (*Courtesy of Metricom*).

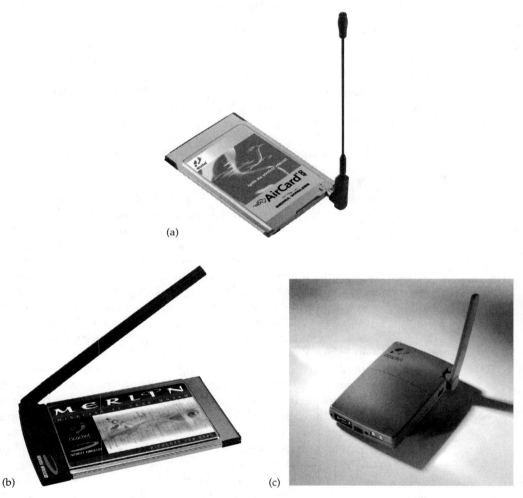

held wireless device? Or instead of a new restaurant advertising the daily special in a newspaper, why not push the advertising content to anyone close by with a mobile telephone?

One key to making sure that the adoption is a success is choosing the right wireless platform. Many solutions are available in the marketplace today, and this diversity will grow as the technology matures. Some will be single-function devices; others will be sophisticated data and voice interfaces. Many will lie somewhere between, and prices will vary widely with functionality. A few important considerations are listed below:

1. Price, processing power, and local storage capability.
2. Form factor, screen size, ruggedness, battery life.
3. Input and output characteristics: keyboard, voice, headset, etc.
4. Wireless network compatibility and operating system (OS) compatibility.

Due to the rapid development of wireless technology, it will likely prove impossible to standardize one device type for everyone. Diversity will rule the day. Many of the critical employees of major organizations have already purchased wireless access devices for themselves in an effort to enhance their productivity. These devices include palm pilots, blackberries, pocket PCs, laptops, cellular telephones, etc. Most information technology (IT) departments will likely have to develop a strategy to accommodate several platforms, operating systems, and device hardware limitations. (Many will probably speak with fond memories of "the good old days," when desktop PCs were the standard, everyone ran Windows, and platform hardware capabilities varied little.)

Push and Pull

From a business perspective, one of the most important issues with wireless devices is the recognition that information has time value: the timeliness of information often determines its relative value. All organizations recognize the need to respond immediately to changes in business information. Wireless access provides a way to reach individuals immediately with breaking information whenever they need it and wherever they may be. In other words, the information goes to the user is instead of the user going to a location where the information can be viewed, as is the case with a wired PC, radio, television, or even a storefront window display. Typically this relaying of information is done through what are called push technologies. E-mail and paging are two primitive examples. The next generation of wireless devices will allow much greater content, rather than a short excerpt or a message to call the office.

Organizations also recognize that getting their message to potential customers in a timely manner can have a dramatic impact on their presence in the marketplace. Push technologies can also be used to place targeted advertising to the consumer based on several methods. By combining GPS and a small Web browser, luncheon specials can be pushed to consumers who are in a several-block radius of a restaurant. Travelers could be attracted by hotel specials based on the hour they arrive in a city and the direction they are traveling, for example, downtown, suburbs, etc.

The flip side of push is pull. Pull technologies are used by the subscriber to request information on demand. The Web browser is probably the best example of this technology. As wireless access technologies improve, the ability of subscribers to a service to pull information will be greatly enhanced. Rather than being tied to a desktop, true broadband wireless access will allow people to access information while stuck in traffic, riding a train, waiting for a plane, etc. The basic idea is that, given an opportunity to pull information at one's convenience, consumers will take advantage of that opportunity and pull a wider and broader array of information. One analogy might be on-line shopping without the visit to the mall.

End Users Are the Drivers for Wireless Access

What are the main drivers for the rapid consumer interest in wireless connectivity? While many issues could be addressed in this section, three ideas seem to capture most consumer interest: teleworking/telecommuting, mobility, and Internet access. The important point to understand about each of these items is that they are being driven by end user demands. In exploring new applications to bring to the untethered user, several considerations must be addressed, including:

Can this application be transported using wireless access?

How many consumers will use this application?

How knowledgeable will the users be?

Where will the users be located geographically?

How often will users access the network?

Will users be grouped in an office or will they access individually?

Will users be fixed or mobile?

Is the application delay-sensitive?

Finally, recall that, for most end users, the four principles of any access technology (usability, flexibility, reliability, and speed) must be addressed to their satisfaction.

Wireless E-mail

Wireless e-mail differs from wireless voice mail because it is a text-based service transported to a wireless device that is a personal digital assistant (PDA) or handheld. These later devices usually feature a miniscreen and a full keyboard. The size of a handheld device is somewhere between a deck of cards and a standard 500-count box of business cards.

One issue with wireless e-mail is that wireless service providers sometimes charge by the number of messages received. In this environment, spam (wireless junk mail) can become very expensive. Usually some baseline number of messages can be received; after that number, each message is individually charged to the consumer. Some estimates of spam content have it accounting for 20 to 30 percent of all e-mail, so the problem of cost becomes obvious. To address this issue, ensure that your provider can:

1. Provide a spam policy in writing.
2. Explain how many e-mail messages are included in the base price.
3. Explain how much each additional message will cost.

It is also useful to take the following steps yourself:

1. Be careful how you forward your desktop e-mail to your wireless device.
2. Understand that attachments that work on your desktop may not work with your wireless device.
3. Be careful how you distribute your wireless e-mail address.

Wireless Data Security

One potential drawback with the transfer of data over a wireless link is the lack of physical security for the path on which the data travels, so the technology is fundamentally less secure. In the vast majority of cases, you will be purchasing bandwidth from a wireless service provider. Thus, how that provider has chosen to implement and specify the level of security offered can have a direct impact on your business.

As we will discuss in detail in later chapters, cellular networks vary a great deal in how they implement security at the most basic level of communication: between the base station and the mobile device. In wireless (unlike wireline), security is not independent of the transport media and access methodology called the air interface. In the United States, the situation is critical because there are several simultaneous deployments of incompatible systems. In most of Europe and a majority of Asia, the standard is GSM.

A critical shortcoming of the mobile devices available today is the lack of processing power required to decrypt transmissions. On the link between the base station and mobile device, the data is in the clear and must rely on the security provisions of the service provider. In other words, there is no end-to-end security unless the end device can both decrypt and encrypt. Hence, the transmission security and air interface offered by the wireless service provider is of critical importance. It is also vital to consider the physical security of the base stations and switching centers themselves. Both are where the switching and billing will be located. It is also the location for the gateways for the emerging wireless application protocol (WAP). Who owns the gateways and who ensures that access to them is secure? The security of the data as it travels from the base station to the enterprise over a wireline link is also an issue. The transmission may be over the Internet or a private line. Whatever level of security you are currently using for data traveling outside the company will probably suffice for this data.

Finally, the biggest threat to mobile devices is the same as it was for laptop computers and cellular telephones. Loss or theft of these essentially portable devices is typically the number 1 threat to security. (Ask to look at a major airport's storage facility for lost mobile telephones; typically there are hundreds tossed into large boxes.) Many believe that this issue will mandate the use of so-called smart cards, which are essentially an identity module for the mobile device. Without the identity module, the wireless device is just so much hardware; it is the card that provides the access to the service for the subscriber. The use of smart cards is the approach taken with the GSM mobile phone system and has worked well there. For more details on this approach, see Chapter 12.

Virus Attacks   Many believe it is only a matter of time when virus attacks are as common in the cellular telephone and handheld environments as they are in the desktop environment. Ironically, the chief barrier to these attacks until now has been the absence of an industrywide standard interface. In the desktop environment, Outlook is virtually a universal standard. A universal standard makes it much easier to propagate viruses widely in the wireline environment. Viruses usually target weaknesses in specific software implementations on specific platforms. Only time will tell if the mobile environment evolves to a universal standard.

Today's wireless clients also have another layer of protection not present in the desktop environment. They are "thin" clients, with limited display, memory, and processing capabilities. Binary attachment content will not generally be accepted and hence cannot be propagated. Related to the "thin" client model is the fact that many mobile devices are functionally hard-wired to a specific application set. Their operating system components are protected in read only memory (ROM). The more function-limited the device is, the less likely candidate it is for a virus attack.

This fact is two-edged, however; because the devices are limited in capability, virus-detection software, which tends to be fairly large and processing intensive, will probably not work on mobile devices. It will be the responsibility of the service provider to address this issue, and it represents a significant business risk for service providers. A few well-publicized virus propagations not blocked by a particular service provider could have a negative effect on that organization's future prospects.

An example is the Phage virus that hit many Palm OS devices in the fall of 2000. The virus had the potential to destroy all programs and files on the mobile device. It traveled on the optical link that many Palm OS devices use to link to fixed PCs. While it was easy to defeat (all you had to do was delete the file Phage.prc), it had the potential to be damaging. Phage had the dubious distinction of being the first widely detected handheld virus. It only targeted devices running the Palm OS, underscoring the point raised above: viruses are designed to attack a specific OS.

The first virus to hit mobile phones was called Timof-nica, and it targeted mobile phones in a region of Spain. It was very similar to an Internet worm virus in structure. Its impact was relatively harmless, at least to the mobile phone subscribers. No data was destroyed because there is little modifiable data in a mobile telephone, but it was damaging to the wireless service provider. The virus sent messages to the mobile phones via e-mail to the short message service (SMS) offered by the provider. The messages were not complimentary to the service provider!

Antivirus vendors are offering new products to combat the emerging problem. As of this writing, most products are targeted at the Palm OS and Windows CE, and most vendors approach the problem from the perspective that the real danger is the handheld device acting as a virus carrier. The handheld is typically infected while the user is reading e-mail with a wireless modem. This class of antivirus product is at least an effective line of defense for the traditional PC being attacked. But it does not protect handheld users from downloading viruses from the Web, e-mail, or even the infrared links of handheld devices that work in tandem with traditional PCs. Products that target viruses intended for the handheld are still in their infancy.

One of the core problems with developing a product for handheld devices is that, traditionally, these virus scanners work by using a database of known virus signatures. That approach is still possible today because the number of viruses is so small, but it looms as a real problem for the future. To get an idea of the scale of the problem, today there are about 50,000 known desktop viruses. If a similar number emerges for handheld devices, the database will become too large to be stored on many of the devices. There are other approaches; one is to look for patterns of behavior. If a program starts to delete a file, for example, the virus protection software would halt the process and alert the user.

In-Building Wireless Telephone Access and Systems
Most in-building private branch exchange (PBX) wireless phone systems or wireless private branch exchange (PBX) modules

Figure 11-18
Nortel Networks Handset Designed for Use with a Companion Wireless Module (*Courtesy of Nortel Networks*)

function in very similar ways to cellular telephony. A base station is connected to the PBX, and channels are assigned to callers on a temporary basis. Handoffs from one base station to another are effectively seamless. Because these handheld phones are mobile, they are susceptible to damage (which seems to occur much more frequently with corporate wireless PBX phones than those owned by individual consumers). As a result, it is almost a requirement that these phones be manufactured to survive a 10-foot fall.

Almost all wireless PBX phones use the TDMA digital access technology and operate at one of three frequency bands: 900 MHz, 1,900 MHz, or 2,400 MHz. Coincidentally wireless in-building phone systems fall into three classes: unlicensed, licensed, and internet protocol (IP) based. The first category, unlicensed wireless PBX modules, extend a wireline PBX to allow wireless access and are by far the most common. Most unlicensed systems operate in the 900 MHz and 1,900 MHz bands. An example would be the Nortel Companion system. A photograph of the wireless handset designed for use with a companion wireless module is shown in Figure 11-18. Few technical issues complicate the installation, and the phones themselves offer most of the advanced features of wireline phones, including the critical corporate requirements of call forwarding and call conferencing. Since these products are proprietary, however, use a wireless module manufactured by the same vendor that manufactures the wireline PBX or some features may not be present. Be diligent when investigating interoperability issues because there are no industry standards. Sometimes wireless PBX modules are used to extend the corporate phone environment into temporary facilities.

The second category, licensed systems, have met with limited market success and they are poor choices for the vast majority of users. These products require participation by a local carrier or service provider who owns the spectrum rights. The advantage of such an approach is no limit to coverage. These systems will work anywhere the carrier has spectrum rights. Typically, the owner of these rights is one of the local cellular providers. Essentially, what these systems offer is package pricing for airtime minutes. While this feature might seem attractive at first, be aware that these systems often bypass the PBX completely. There is no PBX functionality with advanced feature sets, which is often deemed essential for corporate communications. Once these features are fully integrated into cellular telephones, then they will also be available in licensed systems. Today, the product offerings all rely on the AMPS air interface, which does not support these PBX functionality services.

The final category, IP based phones using IP packet technology, represent an emerging choice. For many corporate environments, this option will become cost effective; however, the technology is not yet fully developed. These systems rely on a wireless LAN technology for communications with the IP handsets. (At this time, the LAN technology used consists of the IEEE 802.11 family of networks.) The voice packets share the bandwidth with the data packets. The appeal is natural: one network for two purposes. The drawbacks are fundamental to IP and are not yet adequately addressed. These drawbacks include packet prioritization and control of latency, to mention the two requiring the most attention. Until the throughput and data rates of these wireless LAN systems increase, the capacity to handle both types of traffic with suitable QOS for both classes of users is probably nonexistent. A few companies offer proprietary solutions to these issues, but a prudent consumer will closely question and verify the transition plan, approach, and costs of the supplier when industrywide standards finally emerge.

These systems, like almost every other wireless system in the last few years, have been oversold. Market presence of these systems is not widespread and the core reason is price. Without the large and rapidly growing product volumes predicted by analysts, the economies of scale have not come into play. Quite simply, it costs more to provide in-building wireless PBX access than it does to provide a traditional PBX solution. Typical prices are in excess of $1,000 per line.

■ CELLULAR SYSTEM ECONOMICS

This section identifies briefly the main elements of deploying a cellular system. Prices are, of course, only approximate and are meant to illustrate relative costs. One purpose of this section is to provide the information necessary for the student to write a deployment and financial plan for establishing a cellular system. Depending on the focus of the student, this information can be extended in several ways, some of which are suggested in the Running Example: Cellular System Economics at the end of the chapter.

The information assumes that the deployment of the cellular system is for a first-time entrant to the marketplace, so everything must be purchased new. Because access to the wireline environment is available only through LECs or CLECs, the new organization must rent the switching center access. Administrative costs are also included.

■ SUMMARY

This chapter introduced the basic elements of cellular engineering and described the system components and protocol layers that every cellular telephone system must use. With that introduction to the central concepts of cellular radio, the next chapters will introduce each major air interface approach and provide a system description. These chapters are sorted by the introduction date of the technology. The introduction date closely follows the evolutionary aspects of the air interface. Thus, the first system uses a pure FDMA air interface, the second uses a hybrid FDM/TDMA air interface, and the third uses an FDM/CDMA air interface.

REVIEW QUESTIONS

1. What air interface is most frequently deployed worldwide?

2. List and define the six basic components of every cellular system.

3. Identify the components in the TR-54a architectural model.

4. Discuss frequency reuse and why it is so important to cellular systems.

5. What two factors does FRF rely on and why?

6. What are two key effects of an FRF value?

7. Define overlay and give an example.

8. Compare and contrast handoff and roaming.

9. Discuss the concepts of push and pull from a business perspective.

10. Discuss the concepts of push and pull from a consumer perspective.

11. Discuss wireless e-mail in the context of your experiences.

12. Discuss the possibility of future virus attacks in cellular systems.

13. List the primary categories of in-building wireless telephone access.

14. Investigate and determine the dominant category of wireless telephone access today.

15. Investigate and determine what the market share of the cellular systems listed in Table 11-1 is today in the United States.

In addition to the review questions above, this chapter provides a running example that can be applied to each of the cellular systems reviewed in Chapters 12 and 13. This running example illustrates simplified calculations that cellular providers make prior to establishing cellular coverage in a geographical region.

■ RUNNING EXAMPLE: CELLULAR SYSTEM ECONOMICS

This section identifies briefly the main elements of deploying a cellular system. The prices are meant to illustrate relative costs. One purpose of this section is to provide information necessary for you to deploy and determine the cost of starting a cellular system. Depending on the focus of the student, this running example can be extended in several ways. For example, when new cellular air interfaces are introduced in subsequent chapters, overlays can be evaluated as the subscriber base grows. New technologies may command a higher price per minute from the subscriber, or subscriber demand may require an update in technology. You can study how population density and cell size interact in different ways with the frequency reuse factor.

In this running example, the following assumptions are made:

1. The land on which the antenna towers are to be constructed must be purchased. This figure represents the *land* entry in Tables 11-3 and 11-4.
2. The tower and antenna required in each cell must be purchased. This figure represents the *tower/antenna* entry in Tables 11-3 and 11-4.
3. The service provider must purchase the software to bill and maintain the subscriber base in the switching center. This figure represents the *MSC software* entry in Tables 11-3 and 11-4.
4. A radio is required for each duplex link in each tower (the *radio* entry).
5. The mobile stations (cellular telephones) are offered free to the subscriber by the service provider (the *MS* entry).
6. One base station controller must be constructed per cluster (the *BSC* entry).
7. The service provider must rent switching space from the LEC/CLEC (the *MSC switch* entry).
8. The service provider must hire staff (the *staff* entry).
9. Each subscriber makes three calls a day, 365 days a year. Each call is charged at 5 cents per minute and each call is 5 minutes in duration. This call model is primitive and has no blocking. There are clear opportunities for improvement here.
10. The total coverage area is 30,000 square kilometers.
11. The total number of duplex channels is 1,000.

The abbreviations used are those of the TR-45a architectural model, which is shown in Chapter 12. Tables 11-3 and 11-4 show two possible combinations of parameters that can be used. As the number of users and the hexagonal cell radius are varied, the spreadsheets shown in the tables yield different results. The FRF is limited to 3, and it is suggested that the cell radius be limited to 0.3 km.

In these two tables, the number of users and the cell size are varied. Note that this variation affects the calculated FRF and the revenue. Table 11-3 shows a spreadsheet for a rural area with large cells. Table 11-4 shows a spreadsheet for an urban area with small cells. Each report should be presented in a professional way. The following discussion suggests sections and appropriate content.

Executive Summary

Identify the reason for the analysis. Define the market, location, customer base, and technology utilized. Outline the revenue target and initial costs that are required to implement the system. Summarize the solutions examined, the approach taken, and any special considerations that led to the decision. Identify the principals (authors of the report).

Table 11-3
Rural Area with Large Cell Size

			Year 1	Year 2	Year 3	Year 4	Year 5
Total Area (km²)	30,000						
Total users (thousands)	100						
Duplex channels	1,000						
Cell radius (km)	5						
Calculations:							
Minimum number of clusters needed	100						
Hexagonal cell area (km²)	65						
Number of cells	462						
FRF	3						
One-time expenses:							
Land at $20,000	$20,000	$9,230,769					
Tower/antenna at $50,000	$50,000	$23,076,923					
MSC software at $200,000	$200,000	$200,000					
Radio at $2,000/duplex channel	$2,000	$ 6,000,000					
MS at $100/user	$100	$10,000,000					
BSC 1/cluster at $100,000	$100,000	$10,000,000					
Recurring expenses:							
MSC switch cost at $25/user/yr	$25	$2,500,000					
Staff 2/10,000 users at $60,000/yr	$60,000	$600,000					
TOTAL COST			$61,607,692	$3,100,000	$3,100,000	$3,100,000	$3,100,000
Revenue:							
Calls/day	3						
365 days/year	365						
Charge per call minute	0.05						
Average call duration	5						
TOTAL REVENUE		$27,375,000	$27,375,000	$27,375,000	$27,375,000	$27,375,000
NET			-$34,232,692	-$9,957,692	$14,317,308	$38,592,308	$62,867,308

Table 11-4
Urban Area with Small Cell Size

			Calculations:	
Total Area (km²)		30,000		
Total users (thousands)		1,500	Minimum number of clusters needed	1,500
Duplex channels		1,000	Hexagonal cell area (km²)	2.6
Cell radius (km)		1	Number of cells	11,538
			FRF	12

			Year 1	Year 2	Year 3	Year 4	Year 5
One-time expenses:							
Land at $20,000	$20,000	$230,769,231					
Tower/antenna at $50,000	$50,000	$576,923,077					
MSC software at $200,000	$200,000	$200,000					
Radio at $2,000/duplex channel	$2,000	$24,000,000					
MS at $100/user	$100	$150,000,000					
BSC 1/cluster at $100,000	$100,000	$150,000,000					
Recurring expenses:							
MSC switch cost at $25/user/yr	$25	$37,500,000					
Staff 2/10,000 users at $60,000/yr	$60,000	$9,000,000					
TOTAL COST			$1,178,392,308	$46,500,000	$46,500,000	$46,500,000	$46,500,000
Revenue:							
Calls/day	3						
365 days/year	365						
Charge per call minute	0.05						
Average call duration	5						
TOTAL REVENUE			$410,625,000	$410,625,000	$410,625,000	$410,625,000	$410,625,000
NET			−$767,767,308	−$403,642,308	−$39,517,308	$324,607,692	$688,732,692

Financial Detail

Summarize the market conditions. Identify the pre-revenue infrastructure costs and the yearly administrative costs in detail. Identify any rental versus purchase choices and any regulatory issues that will have a cost impact. If there are competitors in the market, identify how they will affect income and what the risks are. Identify all elements discussed in the chapter section entitled Cellular System Economics as well as any other elements specified by your instructor.

Include chart(s) showing accumulated costs and accumulated revenue, and identify the breakeven point. Do any appropriate comparison studies to support the choices summarized in the executive summary.

Engineering Detail

Identify the FRF, cell size, cluster size, and nearest neighbor co-channel cell. Draw a cell map overlay for the region and identify the frequency group used. Include any further appropriate engineering details.

Appendixes

Place here any supporting documentation, for example, Gant charts for schedules and timelines, spreadsheet calculations supporting tradeoff charts, additional points, etc.

Frequency and Time Division Multiple-Access Cellular Systems

■ INTRODUCTION

This chapter will introduce two of the three basic approaches to implementing a cellular telephone system. The first approach is a pure frequency division multiple access (FDMA) system, while the second combines time division multiple access (TDMA) and FDM to achieve better performance. This hybrid is referred to as an FDM/TDMA system. This text uses one actual deployment to describe each of the different access technologies and to represent the overall system approach.

System approach one, called AMPS, is the traditional analog service deployed widely in the United States, especially in more rural areas. System approach two uses the widely deployed European system, called Global System for Mobile Communications (GSM), as representative of this access approach. Cousins of both systems have been deployed in other countries. A brief summary of these other systems is provided where appropriate.

For this chapter and Chapter 13, the format followed is the same for each system. First, an overview of the number of channels and spectral allocation is provided. The overview is followed by detailed discussion of the six components identified in Chapter 11 and that every cellular system must have. In digital systems, this topic is followed by frame definitions. Then control channel and calling procedures are defined. While the control procedures change from system to system, the calling procedures remain constant, which is natural because the calling procedures must all interface to the wireline environment, where call procedures have been standardized for years. Each section then ends with an overview of similar systems deployed in other countries.

Because digital systems are replacing analog systems, and significant elements of the GSM protocol structure is being used as a prototype for third-generation systems, the examination of GSM continues. Layers 2 and 3 are explored, with a focus on how control protocols are performed above layer 1. Some new frame types are described, and we discuss how control channels are related to traffic channels. Finally, service interface mechanics are briefly summarized.

■ SYSTEM APPROACH ONE: ADVANCED MOBILE PHONE SERVICE (AMPS)

The advanced mobile phone service (AMPS) was originally deployed by the Bell system in 1986. It was the result of the original FCC ruling determining who would provide mobile telephone service and at what frequencies it would be first introduced (as was discussed in Chapter 11). This system was quickly replaced by an upgrade called extended AMPS (EAMPS) to provide additional capacity. The upgrade was done by allocating an additional 10 MHz of spectrum, which allowed the numbers of channels offered to be increased from 666 to 832. This upgraded system is referred to by the original acronym, AMPS. The basic characteristics of this system include:

Analog FM modulation

Worldwide deployment

800 MHz band

50 MHz total bandwidth allocation

    824 MHz–849 MHz: mobile transmit (reverse) channel allocation

    869 MHz–894 MHz: base station transmit (forward) channel allocation

FDMA air interface

30 kHz voice bandwidth

1,664 simplex voice channels

832 duplex voice channels

$$\text{Number of full-duplex voice channels} = \frac{\text{total bandwidth allocation}}{\text{full-duplex channel bandwidth}}$$
$$= \frac{50\,\text{MHz}}{60\,\text{kHz}} = 833.3$$

There are 832 duplex channels in each geographical region or cellular service area. Originally two providers were allocated half of the channels to promote competition. There are approximately seventy-five cellular service areas in the continental United States. First, each of the 832 channels mentioned above is a duplex channel. The actual number of simplex channels is twice this amount, or 1,664. Each simplex channel is 30 kHz wide and is good for only one-way communication (simplex), either from the base station to the mobile subscriber, or vice versa. To carry on a telephone conversation, you need a two-way communication system (duplex). Two-way communication is accomplished by using two of the 30 kHz channels, one for communication to the mobile subscriber from the base station and one for the reverse direction.

Because 80 kHz is held for maintenance, a total of 832 full-duplex traffic (voice) channels are possible in this spectrum allocation. Note also that the spectrum allocation is not continuous; the forward and reverse channels are offset by 45 MHz, with a 20 MHz guardband between the two. The guardband helps to minimize the interference between the traffic channels in the forward and reverse directions. For that reason, any cellular system operating in any particular frequency band will exhibit the same relationships between the forward and reverse channels as any other.

Originally, each region was split among two competitors and, as a result, the 1,664 simplex channels were divided into four groups. Competitor A was to use 416 simplex channels for the base stations to communicate to the mobile subscriber, and competitor B was to use 416 for the same purpose. Therefore, there were 416 duplex FM channels for each competitor, in each cellular service area. As discussed in Chapter 11, competitor A was always the traditional telephone service provider for your region; competitor B was open to competitive bidding among other companies that wanted to participate in this business in any cellular service area.

This situation changed in 1995. A legal decision finalized in that year opened all markets to multiple competitors, not just two. When this decision was made, no new frequency space was allocated, so newcomers to the market were allowed to lease bandwidth from the original two competitors. Again, as mentioned above, each mobile telephone requires two simplex channels so you can both talk and listen at the same time. You will hear people differ on the number of channels. This difference depends on whether they are talking about simplex channels, one competitor's duplex channels, etc. A large number of people are involved in the cellular telephone industry, each with her or his own perspective on the situation.

Out of this total of 832 duplex voice channels are twenty-one channels reserved for management and data traffic between the mobile devices and the base station. These reserved channels are referred to as control channels. One of their primary uses is to identify the voice channel that the cellular telephone will use for voice conversation. The family of control channels has specific frequency assignments. Therefore, there are only 790 duplex channels for customer use. In these control channels, logical paging channels are used by the system to "ring" your portable telephone and provide for other housekeeping and control tasks. The exact frequency allocations showing the twenty-one control channels are illustrated are Figure 12-1. Note that each duplex channel has its forward and reverse channels separated by exactly 45 MHz.

Figure 12-1
AMPS Spectral Allocation

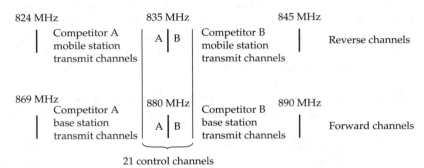

21 control channels

In summary, AMPS control channels are divided into two groups: the dedicated control channels and the paging channels. Every portable telephone uses the dedicated control channels and tunes to the strongest one. The paging channels, primarily used to "ring" a mobile telephone, also transmit the identification number of the subscriber to the entire service area. This function has had a significant impact, as we will discuss below. Sometimes a specific control channel is referred to as an access channel. Access channels are used when a mobile unit (MU) wants to initiate a call. This distinction between an access channel and a control channel will not be made in this text because modern systems do not incorporate this practice. However, AMPS does use special terminology for the mobile unit, which is referred to as the mobile station (MS) in the six components defined in Chapter 11. In this section, MU will be used to distinguish a mobile telephone from a digital MS.

This division of channel types into voice and control channels is a basic characteristic of FDM based systems such as AMPS. While the bandwidth of the control channels may be the same as the voice channels, they are modulated differently and are dedicated to control functions. AMPS uses FSK modulation at a data rate of 10 kbps in these channels. The control channels are used to set up and coordinate access to the voice channels. In this way, the revenue-producing voice channels are available until all call setup functions are complete. For example, no voice channel would be allocated during the dialing and ringing phases of call setup. Once the destination station goes off hook and responds to the ringing, the voice channel is allocated.

This characteristic is retained by the evolutionary cousins of AMPS, identified at the end of the examination of system approach one. In contrast, system approach two—exemplified by GSM and its evolutionary cousins—uses a single channel type composed of an eight-slot TDMA frame, with logical allocations of time slots to voice, control, and data transport. In these cases, voice time slots are allocated based on the logical control channels. In summary, GSM uses a single physical type of channel with the same modulation in each channel, compared to two different physical channels with different modulation in each used by AMPS.

■ THE SIX COMPONENTS

Recall from Chapter 11 that the six components common to every cellular system are the gateway into the wireline system, the base station (BS), databases, the security mechanism, an air interface standard, and, of course, the cellular telephone.

Component 1: Gateway into the Wireline System

The gateway into the wireline system was originally called a mobile telephone switching office (MTSO) in AMPS terminology. In modern usage (and consistent with TR-45), this component is now called a mobile switching center (MSC). It provides the switching fabric between the wireline network and the wireless AMPS network. Functionally, it is a telephone switching central office (CO) and associated software. The best way to think of an MSC is as just another layer or class of switching central offices. The MSCs are interconnected with both the SS7 interoffice network and the voice data network. Its SS7 switching functionality is an SSP. (See Chapter 3.) It is connected directly to the SS7 network and functions as a switching point.

For the purpose of this discussion, we will assume that the interface between the wireline network and the gateway will be wireline itself, for example, either copper-based twisted pairs or fiber-optic links. It is possible for that link to be implemented as a wireless link using a terrestrial line of sight microwave or conceivably an LEO satellite link. These links are rarely used and will not be discussed further. In any case, the logical translation from cellular system switching to switching outside the cellular environment exists.

Component 2: Base Station (BS)

The base station in an AMPS network is simply called a base station. It holds the radio equipment and associated software control for the radio switching. Contrast this setup to switching into and out of the PSTN, which is performed by the MSC. There is one transceiver, transmitter and receiver combined, for each simplex radio channel. In addition, several dedicated FSK based transceivers are used for control or pager communications with the cellular telephone. As discussed earlier, these transceivers are used to set up and tear down calls.

Component 3: Databases

The databases in an AMPS network maintain the time and billing information. They are physically located in the MSC. Figure 12-2 illustrates a block diagram of the MSC, base station, and databases in a typical AMPS installation. Note the interface to both the wireline voice/data network and the SS7 control network.

Component 4: Security Mechanism

There are no real security mechanisms in the AMPS standard. Cloning is a common problem and involves nothing more than a scanner and a PROM burner. Call content security was also absent. The lack of call content security is a consequence of the simplistic FDM approach to channel assignment, which is discussed below.

Figure 12-2
MSC Base Station Block Diagram

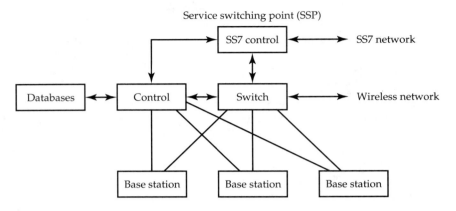

Component 5: Air Interface Standard

The AMPS standard is a pure FDMA system. For each conversation, a single simplex 30 kHz channel is assigned to both the forward and reverse channels. Except in the case of roaming between cells, the entire conversation stays at a fixed frequency location. A simple scanner can find and monitor a phone conversation. (There are several notable examples of this kind of monitoring, including monitoring of key representatives and senators in the United States and of royalty in the United Kingdom.) The simplicity with which a call's content can be monitored was a significant reason why AMPS systems are not widely adopted outside the United States. In the United States, AMPS systems are relegated to rural and semirural areas, where this kind of monitoring tends to be less likely.

Component 6: Cellular Telephone

An AMPS cellular telephone is known by the term mobile unit (MU). MUs are analog devices. AMPS MUs include both single-mode and dual-mode phones, but not dual-frequency phones. Only a single frequency band is implemented for AMPS, so there are no dual-frequency phones. A typical single-mode, single-frequency cellular telephone block diagram is illustrated in Figure 12-3. In pure analog systems such as AMPS, the pulse code modulated (PCM) encoder and decoder blocks shown in the figure would not be present.

As you can see, a cellular telephone includes two basic functional blocks: the user interface and radio transceiver. The user interface includes those components that the cellular telephone subscriber operates: the number or message display, the keypad (numeric or alphanumeric), the microphone to speak into, and the speaker to listen to. The second functional block is the radio transceiver and antenna assembly. The radio transceiver optionally encodes and decodes voice signals and modulates and demodulates those same signals, with amplification provided as necessary. The antenna converts electrical energy into signals that propagate through the radio channel.

Figure 12-3
Block Diagram of a Cellular Telephone

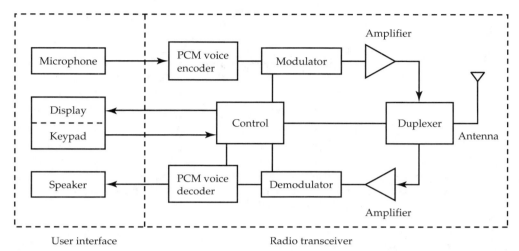

CONTROL CHANNEL SIGNALING

Because the control channels are dedicated in the AMPS approach, this section defines the basic structure and framing aspects of the two classes of control channels, forward and reverse. All control channels use BFSK modulation at a data rate of 10 kbps or a data bit duration of 0.1 ms. Synchronization in the destination device is ensured by the application of Manchester encoding of the data. This encoding causes a bit translation, and hence a frequency shift, for every bit transferred. The frame format is shown in Figure 12-4.

While based on the HDLC structure, the frame has both common and unique elements. The preamble, called dotting in AMPS terminology, is an alternating series of 1s and 0s. The flag, called synchronization words in AMPS terminology, is an 11-bit unique sequence. The unique parts of the frame are the interleaving and repetition of the actual data content, or words, as they are referred to in AMPS terminology.

Each data word on the forward channel is composed of 28 bits of information and 12 bits of error detection and correction using the advanced parity scheme, BCH. On the reverse channel, the data word consists of 36 bits of information and 12 bits of parity. The reverse channel frame format adds more preamble and inserts synchronization flags between each pair of data words. The forward control channel is used to send messages from the MSC to the MU, and vice versa for the reverse control channel.

In each channel, each data word is repeated five times in each frame. This arrangement is shown in Figure 12-4 as A1, A2, . . ., A5 and B1, B2, . . ., B5. This technique addresses the expected fading encountered in any radio channel. Because this fading is typically of short duration, the chance that at least three out of the five data words will be delivered is high. Majority voting ensures that the correct word will be delivered.

Figure 12-4
AMPS Forward Control Channel Frame Format

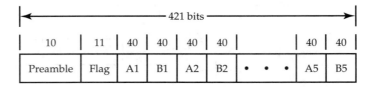

The A and B designations allow two MUs to be communicated with in the same frame. MUs with even telephone numbers use the A channel, while MUs with odd telephone numbers use the B channel. This convention is used, of course, only on the forward channel. On the reverse channel, there are only five data word locations and each is from a particular MU to the MSC.

■ VOICE CHANNEL SIGNALING

This section describes the signaling used on the voice channels. Of course, most of this is speech, but both in-band and out-of-band signaling also occurs in this channel. In-band signaling is nothing more than using the dual-tone multifrequency (DTMF) keypad to access remote electronic devices. You would use the DTMF keypad on a wireline phone exactly the same way. Both keypads are identical in all respects except that signaling is over a wireless interface instead of a wireline interface.

Out-of-band signaling also takes place simultaneously with voice transfer. The primary example is the supervisory audio tone (SAT). Three tones are actually used in the SAT, which is used to verify a reliable round-trip channel link between the base station and MU. The SAT is transmitted at about 6 kHz (5,970, 6,000, or 6,030 Hz) and is used to indicate connectivity even when there is no voice transfer on the link. Each MSC is assigned one of the three tone frequencies, which allows both the MU and the MSC to verify that the link is active. If this signal is interrupted for longer than about 5 seconds, the call is terminated.

The other main out-of-band signaling is called the signaling tone (ST). The ST is a 10 kHz tone used alone to facilitate certain call-processing functions and in conjunction with the SAT to indicate certain actions that the MU may take. When used alone, the ST's primary function is associated with maintenance requests and the termination of calls. (This function is very similar to the in-band signaling used in the plain old telephone system [POTS].) In these cases, the MU sends the ST for a fixed period of time to signal to the MSC that it is waiting for a maintenance response. This process usually indicates that the MU is in a fringe area and is experiencing fading conditions or is preparing to shut down its transmitter. For example, an ST burst of 50 ms is used by the MU after it has received a handoff signal from the base station. Just prior to switching to the new channel, the ST burst is transmitted, confirming to the base station that the MU has now switched. A much longer burst of almost 2 seconds is used by the MU to indicate to the base station that the subscriber has gone "on hook," or hung up. This longer burst signals the termination of a call by the subscriber.

When used with the SAT, these ST tone bursts indicate many more call progress signals than are used in the wireline environment, thus representing the more complex behavior of the wireless environment. These conditions are conveyed to the MSC by using the two tones with several different durations to indicate different conditions.

■ AMPS CALLING PROCEDURE

With the MU switched on and ready to operate, we can list two basic components to any cellular call:

1. Location of the MU and assignment of a duplex channel
2. Security and verification of billing information

Note that AMPS features a few peculiarities in its calling procedure. While there are many similarities, there are some small differences in how this original system uses the control channels and how modern cellular systems do. In the section on GSM, the details of how a modern cellular system processes calls are explored.

Locating the MU and Assigning a Duplex Channel

Base stations in the AMPS system are constantly transmitting on one of the control channels. The MU scans the twenty-one control channels mentioned above and looks for the strongest signal from a base station. The MU has 3 seconds to accomplish this task. When it finds this signal, it assumes that the strongest control channel signal is from the closest base station, and it locks onto that control signal. If for any reason the MU cannot accomplish this task, it has an additional 3 seconds to find and lock onto the next strongest signal, and so on. In an area served by multiple providers, the subscriber typically has identified a preferred system to use, and the MU actually looks for a strong enough signal from the preferred provider. Failing that, it takes the strongest signal from any provider. This arrangement usually results in a higher per-minute cost for the preferred provider, which must pay the nonpreferred provider for this access. This cost is often hidden from the subscriber. The control channels are used for various tasks, including updating the database in the MSC with the current base station association of the MU, ringing the MU in the case of an incoming call, and communicating the channel assignments that will be used for voice conversation, as well as others.

The term *paging channel* is usually reserved for the control channel used to "ring" a cellular telephone when there is an incoming call. The paging channels in the cells broadcast the identifier number of the subscriber. When the MU sees its identifier number on this channel, it knows it has an incoming call, identifies the strongest base station, and activates a ringer. Deciding which cells are used to broadcast this identifier number is based on the geographical coverage of the area code and the exchange of the number dialed.

In AMPS, the channel assignments are not made until an actual call connection is made. In this way, the limited pool of duplex channels can be held open until a revenue opportunity is in place. There is no reason to allocate a duplex channel to an MU when it is not generating revenue for the provider of the cellular service.

If a call is being placed to an MU from the wireline system, the telephone number dialed is used to identify the base station and cell where the MU is located. If the MU is in the on-hook state (it is not being used for a current call), the switch pages the MU, thus starting the process of ringing the MU. The MU then responds to that page, indicating that it is ready for a channel assignment. A maximum of ten attempts are made to locate the MU. The base station assigns a duplex channel to the MU. When this channel assignment is complete, the MU rings and conversation can begin.

In summary, when a mobile unit initiates a call, the AMPS call sequence is as follows:

1. The MU scans the control channels and tunes to the strongest signal from the base station.
2. The MU sends the dialed number and its own identification number.
3. The base station routes these numbers to the MSC.
4. The MSC determines the voice channel assignment and communicates it to the MU.
5. The MSC sends the dialed number to the wireline network.
6. The MU tunes to the indicated voice channel assignment.
7. The MSC completes the circuit connection.

When the call is from a wireline unit to an MU, the procedure is as follows:

1. The wireline network routes the call to the MSC where the MU is registered.
2. The MSC instructs all base stations in that region to broadcast the desired MU's identification number.
3. The MU recognizes its identification number on a paging channel.
4. The MU tunes to the strongest control channel.
5. The MU acknowledges the paging, and the base station relays this acknowledgment to the MSC.
6. The MSC determines the voice channel assignment and communicates it to the appropriate base station and to the MU.
7. An alert tone is sent to the MU on the voice channel and a ring is sent to the wireline caller.
8. Once the MU responds, the circuit connection is complete.

Security and Verification of Billing Information

In the process described above, the first step, after someone tries to call your cellular telephone, is that the base station looks to see if you are in its area. The system can find your individual cellular telephone because each handset is manufactured with a unique code known as an electronic serial number (ESN), which, together with an electronic identification number (EIN), identifies your handset to the base station. The EIN is assigned by your cellular service provider. Both numbers, along with other subscriber data, are stored in the number assignment module (NAM), which is usually a specific memory location in the MU.

The base station uses one of its control channels to obtain these numbers from your telephone when it first turns on and finds the closest base station. Every time you turn on your cellular telephone, you are transmitting your unique information for anyone with the right equipment to hear. Further, the serial number is transmitted in the clear, with no encryption placed on it. This code also determines who gets billed for calls originated from your cellular telephone, so once the ESN and EIN values are known, they can be programmed into another memory in another handset. When calls are made from that modified handset, they are billed by these code numbers, which belong to the owner of the original handset. This telephone fraud uses what are known as cloned phones, and it is a major problem in any location with densely packed cells using the AMPS technology. Since AMPS uses the same frequency assignment for the entire call, monitoring of the call is simple. All that is required is a scanner and a recording device. As long as the MU stays in the same cell, no handoff will occur.

■ AUTHENTICATION

As you might imagine, the billing approach described above needed modification. An authentication approach requires the addition of an authentication algorithm in both the MSC and MU. It is based on and works similarly to the approach used in the GSM system. It uses a random-number generator and solicits an authentication response from the MSC prior to granting access to the voice channels for the subscriber. It allows the service provider to maintain a list of stolen or cloned phones and to deny service to numbers on the list. "Clone-able" phones, once cloned, will be replaced in service with MUs that have implemented this upgrade. Service providers rely on subscriber notification of a cloning situation. Most subscribers pay attention to their bills, so this system works rather well.

■ HANDOFF AND ROAMING

The MU constantly scans the pager channels looking for the strongest signal. Since there is no handoff in AMPS for channel optimization inside a cell, roaming and handoff are often used interchangeably. However, roaming involves much more than a simple frequency handoff.

To begin, recall that every adjacent cell must use a different frequency group to minimize co-channel interference. When an MU moves from one cell to another, the handoff will be to a different duplex channel under the control of a different base station. The original base station will monitor the MU's signal strength. When the signal falls below some threshold, the base station contacts adjacent base stations and asks them to measure the signal strength of that particular MU.

When one of the adjacent base stations indicates the signal strength from that particular MU is above a certain threshold, it will contact the original base station's MSC. That MSC then coordinates the handoff in frequency pairs between the two base stations. The original base station actually sends the commands to the MU, telling it what channels to switch to. Once the new base station hears the MU on the new channels, the MSC switches the landline connection to the new base station. This step completes the handoff.

Completion of the handoff means that the call must be switched at the MSC, the two base stations must coordinate the setup and tearing down of a duplex frequency channel, and the MSC must route the call to the second base station. All must occur in a short enough period of time so there is no noticeable loss of service during switching. The handoff typically takes about 200 ms.

Roaming is defined as a handoff when the two cells that the subscriber moves between are controlled by different MSCs. In any other case, the frequency reassignment resulting from a user moving between two cells is called a handoff. Obviously, the difficulty of accomplishing the frequency handoff increases due to the coordination that must now occur between two MSCs in two different locations.

■ OTHER SYSTEMS THAT USE THE AMPS APPROACH

Total Access Communications System (TACS)

> Analog FM modulation
>
> Deployed in Europe and Asia
>
> 900 MHz band
>
> 30 MHz total bandwidth allocation
>
> > 872 MHz–905 MHz: mobile transmit (reverse) channel allocation
> >
> > 917 MHz–950 MHz: base station transmit (forward) channel allocation
>
> FDMA air interface
>
> 25 kHz voice bandwidth
>
> 1,320 simplex voice channels
>
> 660 duplex voice channels

TACS was first deployed in 1986 in Great Britain and was known as the European TACS (ETACS). A very similar system was deployed in Japan and was known as the Japanese TACS (JTACS). Both systems are based on the AMPS standard, but the frequency allocations are somewhat different and the voice channel bandwidth is a little narrower. The frequency allocations for the systems are country-dependent. Versions of TACS are also deployed in Italy and Spain, as well as most other European countries. Germany, Portugal, Switzerland, and the remaining western European countries use a somewhat different system called NMT, which uses FDMA but employs a different modulation technique, making the systems incompatible.

Examining the characteristics listed above, you can see the similarity to the AMPS system. The channel offsets are the same: 45 MHz between transmit and receive frequencies. Due to a lower total bandwidth allocation (30 MHz rather than 50 MHz), even when combined with a lower voice channel allocation (25 kHz rather than 30 kHz), the TACS system supports a smaller number of simultaneous duplex conversations than does the AMPS system: 600 compared to 832. Voice quality performance is similar for both systems.

■ SYSTEM APPROACH TWO: GLOBAL SYSTEM FOR MOBILE COMMUNICATIONS

The Global System for Mobile Communications (GSM) was originally deployed in Germany in 1992. It was the first digital cellular system deployed. The digital approach pioneered by GSM is based on two key technologies: variable rate pulse code modulation (PCM) encoding of speech and access to the channel through TDMA techniques. The basic characteristics of this system include:

Digital GMSK modulation

Worldwide deployment

900 MHz band

50 MHz total bandwidth allocation

 890 MHz–915 MHz: mobile transmit (reverse) channel allocation

 935 MHz–960 MHz: base station transmit (forward) channel allocation

FDM/TDMA access method

25 kHz modulation bandwidth

200 kHz traffic channels (eight time slots per traffic channel)

992 duplex voice channels

PCM encoding takes the analog voice signal and digitizes it into a series of 1s and 0s. Different PCM systems digitize in different ways. The specific approach used in GSM samples the voice signal at 8 kbps and modulates this signal into a bandwidth of 25 kHz. The 25 kHz voice channel in this system does not exist as a specific frequency allocation but rather as a time slot in a much larger 200 kHz traffic channel. There are eight time slots in each of these traffic channels. Only ⅛ of the time is any individual mobile station

Figure 12-5
GSM Spectral Allocation

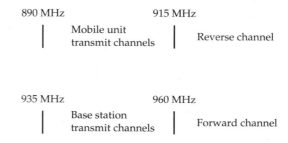

receiving a signal, and only ⅛ of the time is any particular mobile station transmitting a signal to the base station.

The number of duplex channels is found by dividing the total bandwidth by the bandwidth of the traffic channel:

$$\text{Number of duplex channels} = \frac{25\,\text{MHz allocation}}{200\,\text{kHz}} = 125$$

Note that, while there is a total of 125 possible duplex traffic channels, one traffic channel is held back for maintenance and control transmission. This channel is where the bulk of the various logical control channels obtain their bandwidth allocation. The remainder of 124 traffic channels, each with eight time slots, yields a result of 992 voice channels. The frequency assignments for this system are shown in Figure 12-5.

Each base station is allocated one or more of the 124 traffic channels. Each traffic channel is implemented by modulating a carrier. Since there are 124 traffic channels, there are 124 carriers in the GSM system. Stated another way, there are 124 carriers, each spaced 200 kHz apart. Each of these carriers is modulated by a traffic channel composed of eight time slots. Each time slot is used by a cellular telephone to transmit voice or data.

Note that the fundamental unit of time in the GSM standard is the time slot. As we already mentioned, there are eight time slots in each traffic channel. Each physical channel used by a cellular telephone in the GSM scheme is identified by two values. The first is the carrier frequency, and hence the traffic channel associated with it; the second is the time slot associated with the traffic channel.

Note that the 45 MHz offset between transmit and receive channels is the same amount of offset in the AMPS standard. Although the GSM system makes use of a different frequency band and uses a TDMA frame in each traffic channel, those traffic channels are still fundamentally FDM channels. If the system is to operate in the same regulatory environment and use the same frequency channels, the separation of those channels must also be the same.

To summarize, each MS uses one time slot, in each frame, in two distinct traffic channels, to implement full-duplex communication between the MS and base trans-

Figure 12-6
Full-Duplex Communication: (a) Reverse Channel (Uplink); (b) Forward Channel (Downlink)

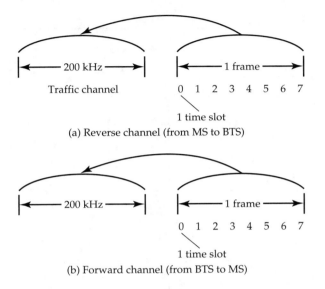

(a) Reverse channel (from MS to BTS)

(b) Forward channel (from BTS to MS)

ceiver station (BTS). The MS uses one time slot, in one frame, in one traffic channel in the forward channel to receive, and one time slot, in one frame, in one traffic channel in the reverse channel to transmit. This setup is illustrated in Figure 12-6. Note that the transmit and receive time slots are offset by three time slots. This offset makes it significantly cheaper to manufacture the mobile station because it does not have to transmit and receive at the same time. The offset is indistinguishable in voice signals.

■ THE SIX COMPONENTS

In Chapter 11, the architectural model that all cellular systems use was briefly referred to and was identified as TR-45a. GSM was the first cellular system to be deployed after that model was developed, so we will introduce an abridged version of TR-45a here as Figure 12-7. Refer to this figure as you go through the six components.

As you can see, there are four protocol stacks in the GSM architecture and three layer 2 protocols in use: LAPDm, LAPD, and MTP-2. LAPDm is the link protocol between the MS and BTS, LAPD is used between the BTS and BSC, and MTP-2 is used between the BSC and MSC. Note that, among other tasks, the BTS must buffer messages between the mobile link protocol and the wireline protocol. The data rate available to LAPDm is a maximum of 22.8 kbps. The LAPD and MTP-2 link layers have a full 64 kbps DS-0 available. Therefore, the BTS and BSC must work together to accomplish rate adoption and buffering and thus mesh these two data rate environments.

Figure 12-7
Abridged TR-45a Architectural Model

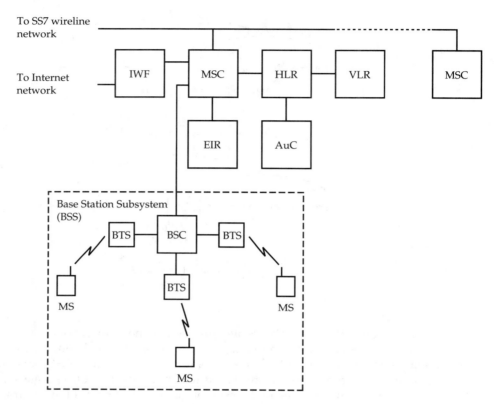

Component 1: Gateway into the Wireline System

There are two subcomponents to this interface: the mobile switching center (MSC) and the interworking function (IWF). Together these subcomponents form the actual point of contact into the wireline network. Each MSC is connected to several other MSCs. Incoming calls are routed to the appropriate MSC based on the number dialed. The IWF located in each MSC acts as the gateway protocol between the GSM network and the internet for IP data traffic only.

The MSC performs all switching functions required for the operation of the mobile stations in the group of cells that it serves. These functions include call control and routing, any protocol conversion required to interface to the wireline network, paging services, location and handoff functions, and authentication. Each MSC is connected to several base station subsystems. In short, the MSC coordinates call setup to and from GSM subscribers.

The MSC is where three key databases are located in the GSM system. The first two databases function together with the switching functions of the MSC to implement roaming. The third is a key component in implementing security. The databases are discussed below.

Component 2: Base Station

The base station in a GSM network is composed of two main components and is known as a base station subsystem (BSS). The BSS groups the radio equipment, controllers, and associated software into two logical components: the base station controller (BSC) and the base transceiver station (BTS). Each BSC is connected to several BTSs. There is one BTS per cell.

The BTS consists of all radio equipment, including transceivers for both the traffic channels and the control channels. It performs all signaling related to the radio interface and is the logical location of the antennas. The BSC is responsible for control and management of the BTS. By instruction of the MSC, it performs the transceiver switching necessary for handoff. Specifically, the GSM system separates the controlling function from the base station transceivers themselves.

The BSC also performs some protocol conversion functions that are necessary for control of the connection to the mobile station. The BSC does protocol conversion at layer 2 of the OSI model. It is the logical location of the CODECs necessary for speech encoding as well as the encryption of the signaling. The BSC also provides service monitoring of the radio channel and controls power transmission levels and frequency hopping of the BTS transceivers. In summary, all control functions of the BTS are provided by the BSC.

Component 3: Databases

Three databases are critical to the operation of the GSM system:

Home location register (HLR)

Visitor location register (VLR)

Equipment identification register (EIR)

The HLR is the database where all important information concerning the mobile subscriber is located. This information includes all billing and subscription data, telephone number, and data related to security as well as the subscriber's international mobile subscriber identity (IMSI). Every subscriber to the GSM system belongs to a "home" network or MSC. The HLR on the MSC contains this information. Information such as the location area (LA) and mobile station roaming number (MSRN) is also kept in this database. Note that, although subscriber billing information is kept in the HLR, the MSC switch performs all actual billing. The MSC switch is where billing is always performed in the wireline environment, and this approach is extended to the wireless environment.

In most cases, the location area (LA) is covered by a single MSC. It consists of several adjacent cells because each MSC controls one or more BSSs, which in turn consist of several BTSs and hence cells. The functions of the IMSI and MSRN will be introduced later. Figure 12-8 illustrates the relationship among LA, MSC, BTS, and BSC. An LA can vary in size, depending on the size of the cells connected to a single MSC. In a metropolitan

Figure 12-8
Location Area Under the Control of a Single MSC

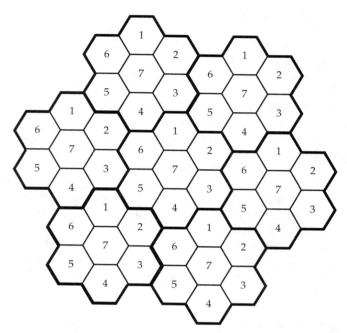

Each cluster is under the control of a single BSC. Each cell contains a single BTS.

area, an LA is typically composed of perhaps a dozen cell clusters, each covering a geographical area of approximately 5 km square. In a more rural setting, the geographical area corresponding to an LA can be much larger.

The HLR also contains the LA where the mobile station last registered. Therefore, when a subscriber roams outside the current LA, the data listing for the LA where the mobile station was last registered is updated. Although the HLR is logically shown to be associated with the MSC, there are several physical MSCs, and often there is only one physical HLR.

The function of the VLR is to address those issues that need to be addressed by the HLR when the subscriber is outside his or her home network or HLR area. The VLR lists all subscribers currently in its LA. VLRs are implemented in each MSC. The VLR data consists of a subset of HLR data that is necessary for the MSC to manage the subscriber. This data includes authentication information, the IMSI, the MSRN, etc. When a subscriber roams into a new LA, the HLR LA information is updated and the appropriate HLR data is copied into the current LA VLR. The VLR subscriber information is always more current than the corresponding information in the HLR when the subscriber is outside his or her home network. The function of an HLR/VLR pair in a cellular network is quite similar to the function of an SCP in an SS7 wireline network.

The EIR is associated closely with the security mechanism and will be discussed in the next section

Component 4: Security Mechanism

GSM features a robust security mechanism, not only for protection of subscriber information and the prevention of cloned phones, but also on the voice and data that passes through the radio channel. Detailed exploration of the security mechanisms are discussed below, but there are three key components:

Authentication center (AuC)

Equipment identity register (EIR)

Subscriber identity module (SIM)

The AuC contains all the information necessary to protect a subscriber's identity and right to use the service, as well as preventing covert monitoring of conversations. The AuC also implements the encryption algorithms on request of the HLR. Logically, there is one AuC per MSC. Physically, the AuC may be distributed as a subcomponent of the HLR.

The EIR is located at one MSC and is the central database for keeping the IMSI for each piece of mobile equipment. Sometimes it is implemented as a subcomponent of the HLR. Each piece of mobile equipment is sorted under one of three lists: white, gray, and black. The color of the EIR listing of the mobile equipment is used to identify valid, broken, or stolen equipment, respectively. It is important to realize that the information in the EIR is never sent over the radio channel. The communication links between an MSC and other MSCs are wireline. The HLR contains the IMSI, which is used by the AuC to authenticate the subscriber.

The SIM is located in the mobile station and stores the subscriber's personal data and billing information, which includes both permanent and temporary data:

Permanent data

Serial number of the SIM

GSM services subscribed to

IMSI

Personal identification number (PIN)

Personal unblocking key (PUK)

Authentication key

Temporary data

LA

Ciphering key

BCCH information (list of carrier frequencies used for call setup and handoff)

The SIM can be either a permanently embedded memory chip or a smart card that can be transferred to another piece of mobile equipment. Note that the SIM card, not the mobile

Figure 12-9
GSM Layer 1 and 2 Protocols

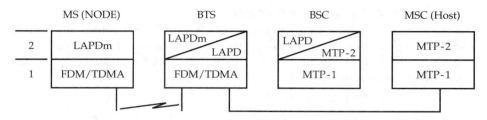

equipment itself, is your identity. For all calls but emergency calls, the SIM card and PIN must be present.

Component 5: Air Interface Standard

The air interface of the GSM system is a series of 200 kHz traffic channels, each divided into eight time slots or burst periods. Technically, this setup is called a hybrid FDM/TDMA air interface because it combines elements of both FDM and TDMA. The frame structure of GSM is hierarchical and includes functions beyond those normally associated with a frame structure. For example, one level of the frame hierarchy has no function other than to enhance the encryption algorithm.

Component 6: Cellular Telephone

The mobile station (MS) is the term used to capture the two functional aspects and subsystems of what is typically a cellular telephone. The two components are:

Mobile equipment (ME)

Subscriber identity card (SIM)

ME is the term for the entire piece of equipment except the SIM card. The ME includes all hardware and software; the SIM is the subscriber's identity. Together, they form the MS. The ME may have several interfaces: the voice interface to the subscriber consisting of a microphone and speaker; some kind of display, usually implemented with LCD technology; and a keyboard and some soft keys. One or more data interfaces to other data equipment may be included, and items such as a scanner may be implemented. Without the SIM card, the ME is just a piece of equipment with no access to the GSM network. Note that the EIR keeps track of the ME and the SIM, and the AuC keeps track of the subscribers. Figures 12-9 and 12-10 illustrate the several components of a GSM network from a protocol architectural viewpoint.

■ FRAME DEFINITIONS

This section on GSM examines the hierarchy of frames used in this system for traffic channels. The most basic component of a frame is the structure of the time slot itself, and

Figure 12-10
GSM Layer 3 to 7 Protocols

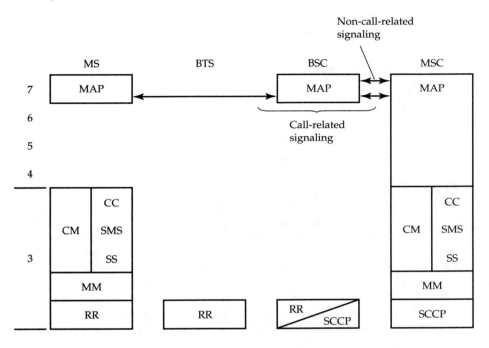

the basic component of the GSM frame structure as a whole is the burst period or time slot. The TDMA scheme that GSM uses divides the channel into traffic channels, which are composed of time slots. The time slot definition is illustrated in Figure 12-11. This text considers only the normal time frame structure. Other frame structures exist in the standard, but they are never used for subscriber traffic, either voice or data. They are the frequency correction, synchronizing, and random access bursts.

Each physical channel in the GSM scheme is identified by two values. The first is the carrier frequency, and hence the traffic channel associated with it. The second is the time slot associated with the traffic channel. Each traffic channel cycles every 4.615 ms. Thus, every MS assigned to that traffic channel has an opportunity to transmit once during this time period.

Because of the guard bits, the time slot has a different time period than the burst period associated with the time slot. Often, the two terms are used interchangeably. To be precise, the burst period represents the transmission time, while the time slot represents the information frame time. In this view, the guard is not considered part of the information frame because it contains no information content. It is considered part of the frame definition, however, and hence is included in the discussion below.

Figure 12-11
GSM Time Slot Structure (Traffic Channel Only)

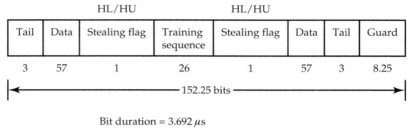

Bit duration = 3.692 μs
Slot time = (156.25) $\times$ (3.692 μs) = 576.875 μs
57 + 57 + 1 + 1 = 116 bits (information bits)

Tail = 3 Bits

Although these bits are specified as tails, they are actually better understood as ramping bits. Note that they are positioned at each end of the active portion of the traffic burst. These bits contain no information but are used to ramp up and down the power out of and into the transceiver, respectively. They always have the same values. This feature has very desirable effects on the overall performance of the system because any transmitter tends to exhibit spectral splash or switching transients when it first turns on. This spectral splash is characterized by emissions beyond its normal channel boundaries and is called adjacent channel interference. It acts like a noise signal to other nearby units. Thus, these bits were placed in the slot definition to allow a smooth transition and avoid this unnecessary and unwanted interference.

Data = 57 Bits

This location is the actual data transport area of the slot definition. Two 57-bit blocks of data are present. They are combined for a total data throughput for each time slot of 114 bits.

HL/HU = 1 Bit

The HL/HU bits, sometimes referred to as stealing flags, are used to signal whether a burst is a normal traffic burst or has been "stolen" by an urgent message. The bits are always used as pairs to establish redundancy in this logical field. When the bits have a value of 0, the data bits contain normal traffic. When the bits have a value of 1, they contain fast associated control channel (FACCH) messages. These time slots are used for urgent messaging. The FACCH grabs an MS time slot, thereby supplementing the voice channel. These channels are designed for bidirectional point-to-point control of the MS and are often used for handoff coordination. They are grouped under the general heading dedicated control channels (DCCH) and will be discussed in more detail later.

Training = 26 Bits

Because training occurs in the center of the time slot, its purpose cannot be some sort of preamble to synchronizing the clock to the time slot. Instead, it is a training sequence for the equalizer contained in the modulator and is sometimes called a midamble. Its es-

Figure 12-12
GSM Frame Definition

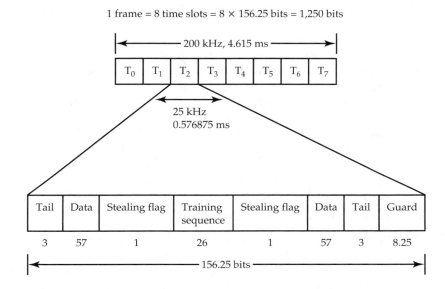

1 frame = 8 time slots = 8 × 156.25 bits = 1,250 bits

sential purpose is to keep the spectral shape sufficiently narrow. It acts to keep the side lobes that are generated during the normal modulation process dampened down to maintain a good main-lobe-to-side-lobe power ratio.

Several sequences will accomplish this task effectively, so this field has been put to another purpose. The eight best sequences have been selected and are used to identify the "color" of the burst. They are referred to as color codes. Because any MS must know which BSS it is communicating with, the neighboring BSSs are assigned different color codes, thus providing an elegant signaling method that consumes no additional bandwidth.

Guard = 8.25 Bits

The guard space prevents individual bursts from overlapping in time due to delay fluctuations. You can think of this space as a pause between words to allow any echo to dissipate. Imagine that you are standing before a large canyon with an echo response. You know that if you speak slowly enough you can hear the echo. If you speak too rapidly, the echo overlaps your next word and your speech sounds garbled. The guard space minimizes this echo effect. Figure 12-12 shows clearly how the bit interval given in Example 12-1 flows naturally from the frame definition.

■ **EXAMPLE 12-1**   Calculate 1 bit duration when the frame duration is 4.615 ms.
Because 1 frame duration is 4.615 ms, then

$$1 \text{ time slot duration} = \frac{4.615 \text{ ms}}{8 \text{ time slots}} = 576.875 \text{ } \mu s$$

Because 1 time slot consists of 156.25 bits, then

$$1 \text{ bit duration} = \frac{576.875 \text{ μs}}{156.25 \text{ bits}} = 3.692 \text{ μs}$$

The TDMA scheme used by GSM divides the channel into time slots, and a group of MSs shares eight of them. As we discussed above, the general term for the transmission generated by an MS is a burst period. GSM defines an eight-time-slot frame, where there are eight logical channels inside each physical radio channel. GSM terminology defines each time slot as a burst period to distinguish it from the traffic burst, which is the aggregate of eight time slots in a traffic channel. To summarize:

Time slot: the period in time when an MS or BTS transmits.

Burst period: the transmission that occurs during a time slot.

Traffic channel: the 200 kHz frequency channel composed of eight traffic bursts modulated onto a single carrier frequency.

Frame: a group of eight time slots in any traffic channel.

As used in GSM, TDMA switches the full 200 kHz radio channel to each MS for a time slot within a traffic burst. Thus, each MS gets 200 kHz of bandwidth for the data portion of the time slot. (See the data rate calculation in Example 12-3.) The MS sends its traffic during this time slot, hence the name *traffic burst*. There are eight logical traffic bursts in each traffic channel.

As you can see from Figure 12-12, there are eight time slots in each frame. These time slots are numbered 0 to 7. Remember, in TDMA, each of these time slots is assigned to a single MS, so eight different MSs use this frame structure. Each MS gets the entire traffic channel bandwidth, 200 kHz, for one-eighth of the frame time.

■ **EXAMPLE 12-2**   Find the time slot duration if the bit interval is 3.69 μs.

Time slot duration = (number of bits)(bit interval) = $(156.25)(3.69 \times 10^{-6})$ = 576.875 μs

■ **EXAMPLE 12-3**   Find the data rate and line rate of the time slot.

$$\text{Data rate} = \frac{\text{number of data bits}}{\text{times slot duration}} = \frac{116}{576.875 \text{ μs}} = 200 \text{ kbps}$$

$$\text{Line rate} = \frac{\text{number of bits}}{\text{times slot duration}} = \frac{156.25}{576.875 \text{ μs}} = 269 \text{ kbps}$$

In each time slot from Example 12-3, an overhead of about 70 kbps is associated with the radio channel characteristics. Having now described the slot structure, it is time to examine how these slots are combined into the frame hierarchy. There are four frame definitions:

Frame (traffic channel)

Multiframe

Superframe

Hyperframe

Of these four, only the first three have real utility for understanding how the frame structure relates to the operation and sharing of the channel. Figure 12-13 illustrates the GSM frame hierarchy for these three frames.

■ **EXAMPLE 12-4**   Find the frame time and frame rate of the GSM frame.

Frame time $=$ (number of time slots)(time slot duration) $= (8)(576.5 \times 10^{-6}) = 4.615\,\mathrm{ms}$

Frame rate $= \dfrac{1}{\text{frame time}} = \dfrac{1}{4.615 \times 10^{-3}} = 216$ frames per second (fps)

Frame rate $= (216\text{ frames/s})(1,250\text{ bits/frame}) = 270$ kbps

---

Examining the next level of the frame hierarchy and identifying specific MS time slots should help clarify how the frames are allocated across the frame hierarchy. The multiframe introduced here is the primary multiframe. GSM actually has several types of multiframes, but only the one described below is used for traffic. (See Figure 12-14.) In GSM terminology, multiframes are used to establish schedules for the use of the time slots contained in the individual frame types that make up a multiframe definition.

Several items should be noted in the figure. First, you can see clearly how each MS gets a time slot to transmit every multiframe (almost every multiframe; we will return to this issue in a moment). Second, the multiframe definition is composed of twenty-six frames, each of 4.615 ms duration. The multiframe defines the multiplexing of the traffic channels.

■ **EXAMPLE 12-5**   Find the multiframe time, which is the period when every MS gets an opportunity to transmit.

Multiframe time $= (26)(4.62 \times 10^{-3}) = 120$ ms

Third, notice in the lower portion of Figure 12-14 that the frames composing the multiframe are labeled T, S, and I. Frames labeled with a T indicate that normal voice or data traffic is passed during that traffic channel. These frames make up the majority of the multiframe, as you would suppose because the point of this frame is to transmit data of one type or another. The S frame is used for synchronization and call-processing functions,

Figure 12-13
GSM Frame Hierarchy

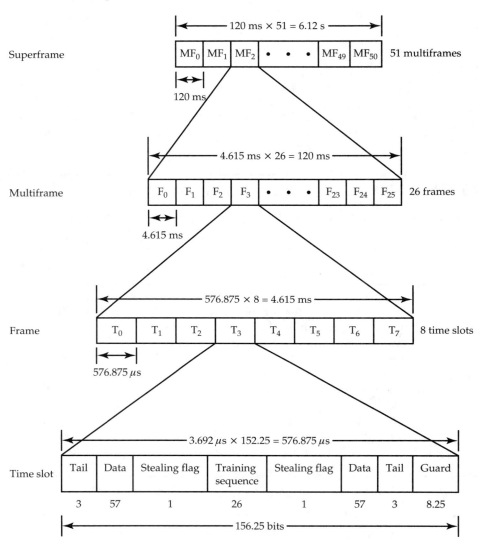

while the I frame is idle. The reasons for these and what signaling they convey will be discussed below. For now, the S frame is used by the slow associated control channel (SACCH).

An interesting concept revealed by examination of Figure 12-14 is that a frame structure is embedded in the multiframe definition but is not stated explicitly. Examine the lower portion of the figure closely. Note that there is a pattern to the frames that do not contain traffic, namely, that there is a signaling burst once every twelve traffic channel bursts. Therefore, the basic structure of the multiframe, from an application point of view, is twelve traffic bursts from any set of MSs, followed by a single signaling burst of one of two types, S or I.

Figure 12-14
GSM Multiframe

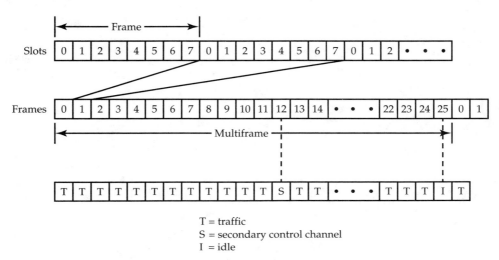

T = traffic
S = secondary control channel
I = idle

To calculate the data rate that results from this multiframe structure, you must also take into account that portion of the time slot itself that actually passes data. On the surface, this data rate would seem to be 114 bits per frame. Example 12-6 illustrates this calculation. Pay careful attention to Figure 12-15 and note that not every frame in a multiframe is used for data.

■ **EXAMPLE 12-6** Calculate the information throughput or data rate of each physical channel used by any one MS. Assume 114 bits of information in every multiframe. The MS uses just one time slot in one traffic channel in every multiframe. See Figure 12-15.

$$\text{MS data rate} = \frac{\text{number of data bits}}{\text{frame time}} = \frac{114}{4.615 \times 10^{-3}} = 24.7 \text{ kbps}$$

As you can see from Figure 12-15, not every frame position is used to transmit data. There is one signaling burst after every twelve traffic channel bursts. The appropriate modification to this calculation is straightforward and is shown in Example 12-7.

■ **EXAMPLE 12-7** Calculate the actual data throughput rate based on the information above. Divide the number of data bits per frame by the frame time and modify the result by the percentage of frames that actually transport data.

$$\text{MS data throughput} = \left(\frac{\text{number of traffic frames}}{\text{total number of frames}}\right)\left(\frac{\text{number of data bits}}{\text{frame time}}\right)$$
$$= \left(\frac{12}{13}\right)\left(\frac{114}{4.62 \times 10^{-3}}\right) = 22.8 \text{ kbps}$$

Figure 12-15
Example 12-15

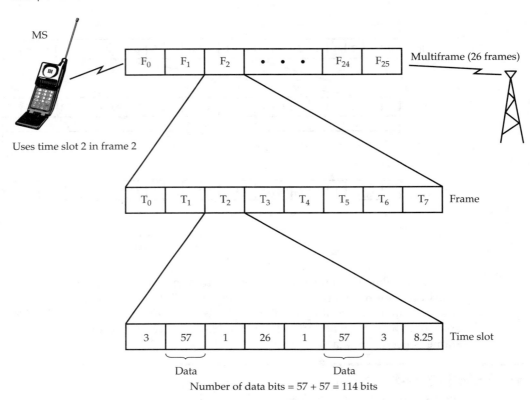

MS

Uses time slot 2 in frame 2

Multiframe (26 frames)

Frame

Time slot

Data Data

Number of data bits = 57 + 57 = 114 bits

Finally, calculate the "reserved" capacity in the multiframe of those frames not explicitly defined as traffic frames. See Example 12-8.

■ **EXAMPLE 12-8**   Calculate the data rate of the nontraffic frames.

Data rate = 24.7 kbps − 22.8 kbps = 1.9 kbps

Before concluding the discussion of the multiframe, we shall touch on the S and I frames from a data rate perspective. These two frame locations are designed to accommodate two separate ideas. The S frame allows a call-processing-related data channel, called SACCH, introduced earlier. This communications channel is designed to transmit low-priority, or slow, control traffic associated with a particular traffic channel. Once each twenty-six frames, bandwidth is allocated for this channel.

■ **EXAMPLE 12-9**   Calculate the data rate supported by the SACCH channel.

$$\text{Data rate} = \left(\frac{1}{26}\right)\left(\frac{114}{4.615 \times 10^{-3}}\right) = 950 \text{ bps}$$

Note that this calculation is not complete. There is no multiplexing of users in the SACCH frame, and the SACCH channel uses all eight time slots. The actual data rate is eight times higher, or 8×950 bps or 7.6 kbps. (The same is true of the I time slot, discussed next.)

The remaining time slot, designated I, is an idle period. It is not really wasted bandwidth because it is intended to offer users a reduced rate service. Since the bandwidth represented is approximately 8 kbps, the intent is to accommodate up to sixteen users, each at a data rate of approximately 500 bps.

The next level to this frame hierarchy is the superframe, which is simply a gathering of fifty-one multiframes into one superframe. The superframe defines the period when all control channels cycle. It is composed of several multiframes, each of which define the period when every MS gets to transmit. It can also be said that the superframe defines the period when both the voice channel and the control channel reset.

■ **EXAMPLE 12-10**   Find the frame time for the superframe.

$$\text{Superframe time} = (51)(120 \times 10^{-3}) = 6.12 \text{ s}$$

Finally, we define the hyperframe. Because the superframe defines the complete period for both voice and control channel recycling, a good question is, What purpose does another frame definition serve? This frame definition exists for one reason only: to enhance the encryption algorithm used. The encryption algorithm requires the frame number as a parameter in its algorithm. If the maximum frame number was the number predicted by the multiframe, this parameter would be limited to $51 \times 26 = 1,326$.

A larger number of permutations is desirable to enhance security. Therefore, an arbitrary structure was imposed to obtain a sufficiently large number. It was decided that a value of $1,326 \times 2,048 = 2,715,648$ was sufficient. Therefore, the hyperframe is a group of 2,048 superframes and has the duration shown in Example 12-11.

■ **EXAMPLE 12-11**   Find the frame time for the hyperframe.

$$\text{Frame time} = (2,048)(6.12) \approx 209 \text{ minutes} \approx 3.5 \text{ hours}$$

■ **HALF-RATE GSM**

Omitted from the above discussion is the concept of half-rate GSM. Half-rate GSM is a scheme where the radio channel's capacity is doubled in terms of the number of subscribers served. In the conventional full-rate GSM, each mobile station uses one time slot in each half of the duplex pair of time slots that form a full-duplex channel. One half is

Figure 12-16
(a) Full-Rate GSM; (b) Half-Rate GSM

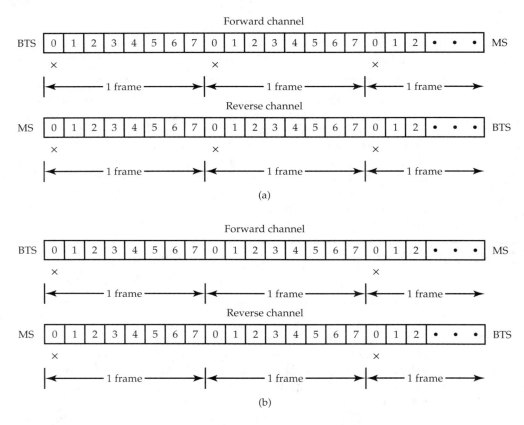

(a)

(b)

used for transmit, one for receive. The half-rate approach dedicates only one time slot every other frame to each subscriber instead of one time slot per frame for transmit and one time slot per frame for receive, as is the conventional approach.

For half-rate GSM, each transmit and receive pair of time slots occurs every two frames, instead of every one, resulting in sixteen simultaneous simplex channels per radio channel. The effective data rate of these half-rate channels is about half that calculated above, or 11.4 kbps. With the appropriate CODECs, these half-rate channels are of acceptable quality in many situations. Figure 12-16 illustrates the two approaches. The time slots marked with an X are associated with any single MS.

■ CONTROL CHANNELS

The frame structure introduced above supports both traffic channels and control channels. The control channels are used for system control, which is not generally passed on to subscribers. The primary purpose of the control channels is to facilitate quick hand-offs and the signaling associated with establishment, maintenance, and release of traffic channels. All but the FACCH and SACCH control channels have their own specific frame structure and multiframe format, which is somewhat different from the traffic frame format introduced above. The detailed description of these frames is unnecessary. They are similar in form to the traffic frame, sporting two tails, a flag or two, and a training sequence in addition to the information field.

Examine the multiframe structure that contains the control channels introduced below. The control channel uses the 200 kHz traffic channel that was held back for these purposes. This channel is sometimes known as the beacon frequency. This control channel is composed of frames used exclusively for control and maintenance of the GSM system. The frames are arranged into a unique multiframe structure that is fifty-one frames in length. Each frame has a duration of 4.615 ms, just as in the traffic frame structure. See Figure 12-17. The letters in the figure correspond to the various control channels discussed below. The channels are summarized Table 12-1.

There are many control channels (CCs) in the GSM standard, and some have already been introduced. The purpose of this section is to place them in some logical order. Three classes of CCs are used in the GSM standard:

Broadcast channel (BCH)

Common control channel (CCCH)

Dedicated control channel (DCCH)

The first class of CC is composed of only one member, the BCCH. As discussed earlier, the BCCH is used to transmit information from the BSS to the MSs in a point-to-multipoint communication. The BCCH is used by the BSS to transmit its LA, and it is used by the MSs to select cell membership based on the power of this broadcast. The BCCH is also used to transmit frequency hopping information to the MSs.

Remember that each cell has a broadcast channel that is used by the various MSs to determine signal quality from the BTS. The BCCH frame also contains information used by the MS to set its transmit power. Earlier we stated that the MS chose the BTS with the highest power. This statement was not quite correct. In a scenario where several BCCH messages are received by the MS and all are of sufficient power to ensure QOS transmissions, the MS actually chooses the one with the lowest recommended transmit power for the MS. This selection extends the battery life of the MS and is a refinement to the algorithm of simply choosing the strongest signal. Usually the strongest signal will also be the closest BTS, but not always. This approach allows the MS to assess the situation and optimize its battery life.

Two subclasses of the BCCH are used to enhance synchronization between the BSS and the MS. The synchronization channel (SCH) keeps track of the current hyperframe. The frequency correction channel (FCCH) addresses frequency drift by the MS. The FCCH provides the MS with a frequency reference. These channels are summarized in Table 12-2.

The second class of control channels is called the CCCH. Three logical channels are grouped under this classification. They are shown in Table 12-3. The first two have already been discussed in some detail. In summary, they are used for notifying the MS of

Figure 12-17
Control Channel Multiframe Structure

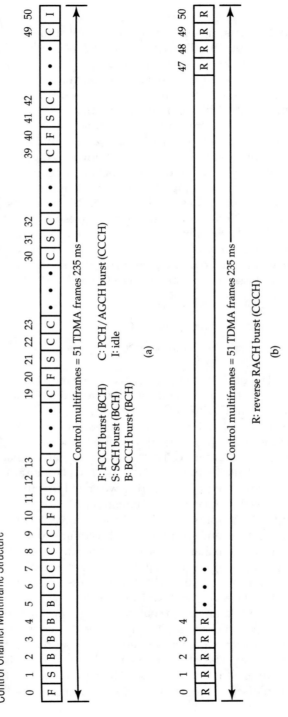

Control multiframes = 51 TDMA frames 235 ms

F: FCCH burst (BCH)
S: SCH burst (BCH)
B: BCCH burst (BCH)

C: PCH/AGCH burst (CCCH)
I: idle

(a)

Control multiframes = 51 TDMA frames 235 ms

R: reverse RACH burst (CCCH)

(b)

Table 12-1
Multiframe Control Channel Frame Assignments

Control Channel	Slot Assignments
F: FCCH	0, 10, 20, 30, 40
S: SCH	1, 11, 21, 31, 41
B: BCCH	2, 3, 4, 5
C: CCCH	6–9, 12–19, 22–29, 32–39, 42–49
I: Idle	50
R: RACH	All slots on forward channel

Table 12-2
BCH Logical Channels

Channel Name	Channel Abbreviation	Direction	Classification
Broadcast control channel	BCCH	BSS to MS	Point to multipoint
Synchronization channel	SCH	BSS to MS	Point to multipoint
Frequency control channel	FCCH	BSS to MS	Point to multipoint

Table 12-3
CCCH Logical Channels

Channel Name	Channel Abbreviation	Direction	Classification
Paging channel	PCH	BSS to MS	Point to point
Random access channel	RACH	MS to BSS	Point to point
Access grant channel	AGCH	BSS to MS	Point to point

an incoming call (PCH) and allowing the MS to request a traffic channel assignment when it has an outgoing call (RACH). The third logical control channel in this group is the AGCH. It is used in most situations to respond to the RACH message from the MS and allocate an SDCCH. The SDCCH is used by an MS after it has obtained access to the GSM network but before it has access to a traffic channel.

The third class of CC is called the DCCH. Three logical channels are also grouped under this classification. They are shown in Table 12-4. The SDCCH is a logical control

Table 12-4
DCCH Control Channels

Channel Name	Channel Abbreviation	Direction	Classification
Stand-alone dedicated control channel	SDCCH	BSS to and from MS	Bidirectional, point to point
Slow associated control channel	SACCH	BSS to and from MS	Bidirectional, point to point
Fast associated control channel	FACCH	BSS to and from MS	Bidirectional, point to point

channel that acts as the default control channel when a traffic channel has not been assigned and a channel is needed to pass control information between the MS and BSS. The SDCCH may also be used to transmit SMS messages.

The SACCH and FACCH control channels are used for urgent messaging. As already mentioned, there are two classes of priority messaging. The SACCH has an assigned time slot, while the FACCH steals an MS time slot, thereby supplementing the voice channel. These channels are designed for bidirectional, point-to-point control of the MS and are often used for handoff-related coordination. An FACCH message can never take over a portion of the SACCH bandwidth because it would greatly complicate the control mechanism.

The SACCH channel is most often used to transmit measurement data from the BSS to the MS on signal strength and receive quality. It is always present and provides a dedicated stream of signaling information between the MS and BSS. The information carried is used to assist in the handoff process.

The FACCH is most often used to pass this same information when timing is critical and the handoff must occur prior to the next available SACCH. It is theoretically possible for up to ⅙ of the traffic frames to be stolen by FACCH messages, which can significantly degrade speech quality. When the system detects this situation, the CODEC repeats the last known speech time slot. So the subscriber hears a slight "stretching out" of the last speech element transmitted.

In Chapter 14, we will see that the general packet radio service (GPRS) data service makes use of three new logical control channels. They are very similar to those defined above. The distinguishing feature is that the control channels described in this chapter are used for control of the circuit-switched voice communications, while the GPRS control channels are used to control the packet-switched data communications provided by GPRS.

Control Channel Layer Interface

Earlier in this chapter, Figures 12-8 and 12-9 illustrated the GSM protocol stack for voice messages. Certain control channels use a somewhat different protocol path in the same stack. The control channels that use this protocol path are those associated with handoff functionality. See Figure 12-18. The first point that should stand out is the unusual layer

Figure 12-18
Control Channel Layer Interfaces

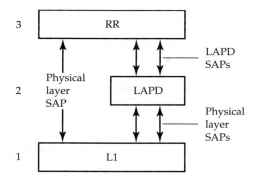

interface between the lowest sublayer of the network layer, the RR sublayer, and the physical layer already described in this chapter. The physical layer SAP locations at this interface are used by the RR network sublayer to provide the fastest possible communications. When the data link layer is bypassed, the latency of response to a request for services is minimized.

The network layer requires access to various physical layer primitives, such as channel frequency and received signal strength, to facilitate functions such as handoff and roaming. This feature is particularly important when the handoff must occur very rapidly. The protocol path is used by the control channels SDCCH and FACCH.

It's also true that, for the control channels, the data link layer adds no value to the service response because there is no CRC checksum calculation in the data link layer used. Therefore, any error detection must take place in the physical layer. Because the data link layer does not contribute added value and because passing through it adds latency, the control channel layer interface bypasses the data link layer.

This channel interleaving that the control channels use is described in the next section. The error-detection coding used at the physical layer is a fire code capable of detecting groups of errors up to 11 bits. Since errors in the physical channel are often bursty in nature, such a coding choice is appropriate because errors will likely occur in groups.

Control Channel Block Interleaving

In the above sections, the basic frame structure has been discussed in some detail. However, the control channel data must be broken up into the 57-bit data segments that the basic time slot structure will support. This section will describe the process that the SDCCH and FACCH control channels experience. The basic difference is that the SDCCH uses both 57-bit blocks of the time slot and the FACCH uses alternate blocks.

In the GSM approach to block interleaving, each 57-bit information field in the traffic frame format represents one block of information. Remember, these control channels steal traffic channel time slots. Note how these 456 data bits representing voice transmissions are actually portioned in the time slot format. Because each time slot has two fields of 57 bits for transfer of information, two methods are used. The first uses four bursts that use, in turn,

Figure 12-19
(a) SDCCH Block Interleaving; (b) FACCH Block Interleaving

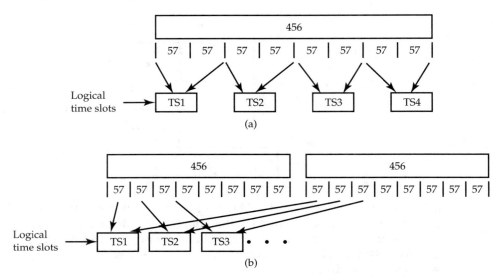

both 57-bit information fields, which is shown in Figure 12-19(a). This block interleaving uses four consecutive time slots and takes 4.615 ms × 4 = 18.46 ms. The other approach, shown in Figure 12-19(b), is to use eight half-bursts. This approach might seem to double the elapsed time for the 456 data bits to transfer, but in practice it does not. The first four whole bursts share the information field with the previous 456 data bits, and the second four whole bursts share the information field with the next 456 data bits, or voice fragment.

We must emphasize that all time slots, including (as discussed above) voice traffic slots, are interleaved this way. Interleaving is designed primarily to convert bursts of errors into random errors that are much easier to recover through coding, but it has a secondary benefit as well. Note that this approach contributes significantly to security for the GSM subscriber. It requires complex systems to decipher the interleaved data once an eavesdropper has broken through any additional coding and security that may be overlaid. For the vast majority of users, this basic functionality, intended to address bursts of errors, provides a secure communication.

The control information is a stream of 184 bits of data. These 184 bits are then appended by 40 fire code bits and 4 tail bits that serve the same function as the 3 tail bits in the traffic channel time slot structure. Additional tail bits provide an extra measure of spectral shaping to the control channels, which may have quite different bit patterns than the essentially random bit patterns of voice. This setup results in a frame length of 228 bits. The origin of these frame lengths is shown in Figure 12-19. As you can see, they are all integer multiples of the core frame bit length of 57 bits. These 57 bits are then convolutionally coded at various rates. The convolutional code used is identical for both traffic and control channels. Depending on the rate used, however, a different frame ex-

Figure 12-20
Bit Interleaving

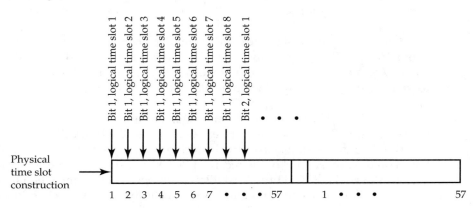

pansion occurs. In the case of these two control channels, the rate used is 2. Thus, the frame (after coding) consists of 456 bits (2 × 228 bits).

■ BIT INTERLEAVING

As briefly mentioned above, the coding done at the physical layer is optimized for small bursts of errors. Understanding the error behavior of radio channels will help you see how the 456 bit frames are inserted into the time slots and how the bits in those time slots are allocated. The first part of this process, fragmentation, was discussed in the section immediately above. The second part is called bit interleaving and is intended to separate components of these frames so that short-term fading or a small burst of errors affects only a single bit for any particular logical channel, be it voice or control.

Bit interleaving places 1 bit from each logical channel into each time slot consecutively. Figure 12-19 showed how the frames were fragmented into logical time slots. The physical time slots are not composed of consecutive bits from a logical time slot. Rather, they are bit interleaved so that each physical time slot is composed of several bits from different logical time slots. Each physical time slot is composed of 8 bits, 1 bit from each of eight sequential logical time slots. This process is known as bit interleaving and is shown in Figure 12-20.

■ GSM CALLING PROCEDURE

With the MS switched on, there are four basic components to any cellular call, as is true with the AMPS calling procedure:

1. Location of the MS and assignment of a duplex channel.
2. Security and verification of billing information.

In the first item, there are actually two situations: the call originates from the MS and the call originates in the wireline environment. A combination of these two situations also captures a third situation: where calls are originated from another cell or another cellular system.

For an incoming call to the GSM system, the MS must first be located. As discussed above, each BTS belonging to an MSC has the same LA, and hence the same location code (LC). A control channel, the broadcast control channel (BCCH) is used by the BSC to identify the LA through the broadcast of an LC. This process tells the MS what LA it is in.

When an MS hears the BCCH message, it knows it is in a valid GSM cell. If it hears more than one BCCH message, it next determines the best cell to associate itself with. The BCCH transmission is used by the MS to determine the closest BTS. This decision is based on the power level of the received BCCH transmission: the strongest signal wins.

This periodic BCCH broadcast allows the MS to determine when it has moved into a new LA. Thus, the concept of an LA significantly reduces the bandwidth that would be required if, each time an MS entered a new cell, it had to register in that cell. Because most users stay within a small group of cells most of the time, this concept has proven quite useful. A periodic registration broadcast from the MS does exist; however, it is infrequent and is used to address the case when an MS has moved out of its coverage area.

If the received BCCH signal is strong enough, the MS synchronizes with it. The BSC then makes use of another control channel, the random access channel (RACH), which is used to exchange registration information with the MS. The RACH is a bidirectional channel with a slotted ALOHA access method. The RACH is also a shared channel used by all MSs in any individual cell. Some kind of shared access method is required to parse out the bandwidth of that channel, hence the use of slotted ALOHA.

The MS transmits a very short registration message to the BSC. The message consists of a random number and a short code indicating the reason for the message, in this case, registration. (The detailed exchange of registration and authentication will be discussed below.) When the BSC receives the message and authentication is complete, the local VLR is updated and the HLR is updated with the new LA. Then the BTS responds by repeating the message on the shared RACH. The MS knows the content of the message it sent, so this repetition allows the MS to identify that the message is intended for it. The MS then sends a second RACH message with a new random number and a code indicating what type of service it requires, along with the telephone number of the call. The BSC then allocates a time slot in a traffic channel, sends a response RACH message to the MS that indicates the frequency information and time slot of the traffic channel, and the MSC routes the call to the destination. If the MS does not hear a response to its message in either of the above situations, it assumes a collision has occurred and will retransmit after a random time, similar to conventional CSMA/CD used in wireline Ethernet networks. CSMA/CD is derived from ALOHA, just as slotted ALOHA is.

Calls to the MS from a wireline number are very similar, except that the location step is not required because the MS automatically registers in an LA when it enters a new one. After the management exchanges required to set up the call and the channel assignment is complete, another control channel, the paging channel (PC), is used to tell the MS to ring, thus alerting the subscriber to an incoming call.

■ HANDOFF

All cellular systems must have some way to hand off calls from one cell to another as an MS moves between adjacent cells. This process is accomplished in a similar way as in the AMPS case. Each MS makes regular measurements of the signal strength it receives from any adjacent cells. In GSM, however, the BSC not only considers signal strength but also factors like congestion in the current cell and overall signal quality.

The MS sends this information to the BSC; the BSC decides when a handoff is necessary. The BSC then communicates with the MS using the RACH in the manner described above and tells it what duplex channel to switch to and when to switch. Note that this process is essentially frequency hopping. The BSC then monitors the new assignment and, when it has confirmed handoff, it tears down the old channel assignment and frees it for a new MS.

■ ROAMING

Roaming is defined as a handoff between cells controlled by different MSCs. The VLR is used to accommodate roaming. When the MS leaves its LA and enters a new LA, the VLR associated with the MSC of that LA obtains the appropriate HLR information and loads it. Simultaneously, the HLR is updated with the new LA of the MS. The local VLR can then work with the local AuC to process any new service requests by the subscriber in the new LA.

■ SECURITY: AUTHENTICATION AND ENCRYPTION

There are two components to the GSM security model: authentication of the subscriber and the ME, along with encryption of messages. Authentication of the subscriber and the ME is intended to prevent the fraud and theft that have been so common among older generation cellular systems. Authentication of the subscriber is done in two steps. First, the subscriber must enter the correct personal identification number (PIN) to activate the SIM. The PIN is a series of four to eight digits. Once this activation is complete, the next threshold is crossed through the use of the international mobile subscriber identity (IMSI) stored in the SIM. As discussed above, the SIM is located in the MS and stores the subscriber's personal data and billing information. This information includes both permanent and temporary data:

Permanent data

Serial number of the SIM

GSM services subscribed to

International mobile subscriber identity (IMSI)

Personal identification number (PIN)

Personal unblocking key (PUK)

Authentication key

Temporary data:

LA

Ciphering key

BCCH information (list of carrier frequencies used for call setup and handoff)

The SIM can be either a permanently embedded memory chip or a smart card that can be transferred to another piece of mobile equipment. The SIM card, not the mobile equipment itself, is your identity. For any call but an emergency call, the SIM card and PIN number must be present. The IMSI is imprinted on the SIM and is not accessible by the subscriber. The whole registration procedure is intended to keep this number secret. However, an exchange of data must occur between the MS and the AuC to authenticate the subscriber and an exchange to authenticate the ME.

As discussed above, the RACH is used to exchange registration information between the MS and BSC. You will recall that the RACH message includes a random number generated by the MS, along with a short flag indicating the type of message. During the registration procedure, this random number is received by the MSC. The MSC then knows the HLR information and can query the AuC for the IMSI associated with that subscriber. The AuC actually generates the IMSI.

Both the SIM and the MSC simultaneously combine the random number with the IMSI and generate a third number. The MSC then forms another RACH message, repeating the original message and appending the third number. The MS receiving this message compares the third number it generated with that sent to it by the MSC. If they agree, then authorization of the subscriber is complete.

In parallel with the above process, the MSC also queries the EIR for the status of the international mobile equipment identity (IMEI) associated with the ME used by the subscriber. The EIR maintains a list of all MEs sorted by three colors. White indicates that the ME is valid, gray indicates that it is broken or out of service, and black indicates that it was stolen. As long as the color returned is not black, the ME is authorized. When the two numbers do not agree, the MS responds to the MSC and the MSC sends the IMEI to the EIR. The EIR then blacklists the ME. The SIM is also blacklisted in the AuC.

Sometimes the user enters her or his PIN incorrectly. If the user does so three times in a row, the SIM is locked. The subscriber must then use the PUK to unlock the SIM. The PUK is a permanently assigned, eight-digit number allocated by the service provider. The subscriber has ten chances to get it right or the SIM will be blocked permanently. It can be unlocked only by the service provider. When an MS is stolen while the unit is active and authorized, the theft will be rewarded only as long as the ME is not turned off or the battery dies. The SIM data is copied into the memory of the MS only for the duration of the active operating period. When power is turned off, the SIM data is automatically deleted.

Note that the IMSI is never sent through the radio channel; thus, no amount of monitoring can reveal its value. Since the random number generated by the MS varies with each new authorization, it is virtually impossible to extract the IMSI from repeated monitoring of messages. This random number generated is from a very large set of numbers. The larger the set of numbers to pick from, the harder it is to guess the next random choice from that set.

■ OTHER SYSTEMS USING THE FDM/TDMA APPROACH

The next four systems are all based on the concepts deployed for the first time in GSM900. The two versions of DAMPS are TDMA based upgrades of AMPS. They use the FDM/TDMA frame structure in a different and simpler version of that used in GSM. The control channel and security aspects are quite primitive compared to the GSM standard discussed above. The GSM1800 and GSM1900 are up-banded versions of GSM900, and the PDC system is a homegrown version of DAMPS deployed exclusively in Japan. It features the same kind of incremental tweaking, namely, slightly narrower frequency bands and slightly more channels, that was done with the Japanese version of AMPS. The major difference in these approaches from GSM is that GSM uses a single type of radio channel, the traffic channel, composed of eight time slots on a single carrier. DAMPS and the other related systems below use two different radio channels, voice and paging control channels, just as AMPS does.

Digital Advanced Mobile Phone System (DAMPS)

DAMPS was first introduced in 1992 and includes the following list of features:

Digital DQPSK modulation

Deployed in the United States

800 MHz (IS-54) band

1900 MHz (IS-136) up-banded system

50 MHz total bandwidth allocation

824 MHz–849 MHz: mobile transmit (reverse) channel allocation

869 MHz–894 MHz: base station transmit (forward) channel allocation

TDMA air interface

10 kHz voice bandwidth

30 kHz modulation bandwidth

832 duplex carriers

Three time slots each result in 2,496 duplex channels. If half-rate CODECs are applied, they result in six time slots for each carrier and a total of 4,992 duplex channels.

The DAMPS system is also widely known as digital AMPS, U.S. digital cellular, or sometimes just U.S. digital. The primary benefit is that its TDMA access method allows many

Figure 12-21
DAMPS TDMA Frame Structure

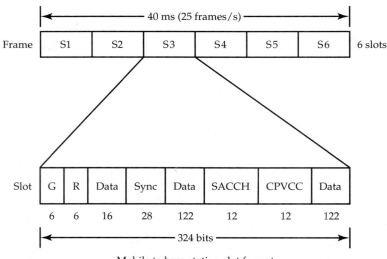

Mobile to base station slot format

more channels to be used simultaneously in a cell. In its original configuration, the TDMA approach allows three simultaneous calls to occupy the same channel. Development and application of half-rate CODECs have allowed system designers to field upgrade these systems to six time slots on each carrier.

The IS-54 standard, approved in 1990, defines the 800 MHz TDMA system and the IS-136 standard as the same air interface, but the IS-136 standard is shifted up in frequency to 1,900 MHz. This process is known as up-banding and describes an existing wireless system being cloned at a higher frequency band. It allows multiple use of the same physical space by two identical, but shifted in frequency, systems. It is the primary way that providers have accommodated an increase in subscribers in certain geographical areas.

The DAMPS system is designed to transmit 48.6 kbps in a 30 kHz voice channel. Each channel accommodates three users at the aggregate rate of 48.6 kbps. The speech encoding is performed at slightly less than 8 kbps, which, after channel coding, becomes 13 kbps. Once control information is added, this figure rises to 16.2 kbps, which yields 48.6 kbps when multiplied by the number of time slots, three. The TDMA reverse channel frame structure is shown in Figure 12-21. As suggested above, half-rate CODECs reduce the speech encoding to 4 kbps, thus allowing twice the channel density of earlier systems.

In its IS-136 implementation, DAMPS also introduced several enhanced features, most important, a sleep mode for battery conservation, the half-rate CODECs discussed above, and dual-mode phones that use either the IS-54 or IS-136 systems. These additional control functions were made possible with the use of DCCH framing. It allows control framing very similar to the GSM standard discussed above.

GSM1800

GSM1800 was introduced in 1994. It is also widely known as digital communication system 1800 (DCS 1800). It includes the following characteristics:

Digital DQPSK modulation

Deployed worldwide

1800 MHz band

TDMA air interface

25 kHz voice bandwidth

GSM1900

This system was introduced in 1996 and is also widely known as personal communications system (PCS1900). It is frequently used as an overlay to the GSM900 system. Here is a list of its features:

Digital DQPSK modulation

Deployed worldwide

1900 MHz band

TDMA air interface

25 kHz voice bandwidth

Personal Digital Cellular (PDC)

This system is the Japanese version of DAMPS and was first deployed in Japan in 1996. The following is a list of its features:

Digital DQPSK modulation

Deployed in Japan

800 MHz band

TDMA air interface

8.33 kHz voice bandwidth

25 kHz modulation bandwidth

■ LAYERS 2 AND 3 IN THE GSM ARCHITECTURE

GSM Layer 2

GSM has an upper layer structure based on the OSI model, but with a few special characteristics all its own. These protocols are what the GSM system uses to manage the wireless

Figure 12-22
GSM Layers 2 and 3

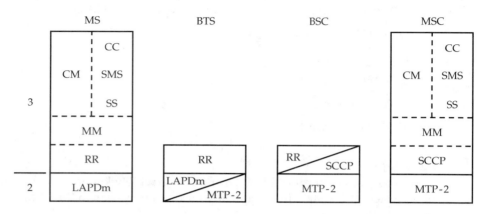

network and interface to the management of the wireline network. The basic protocol stack is based on SS7, the standard control architecture in wireline voice/data networks. Probably the best place to start is with Figure 12-22, which illustrates the detailed structure of the protocol architecture at layers 2 and 3.

As you can see, there are four protocol stacks in the GSM system, and there are two layer 2 protocols, LAPDm and MTP-2, in use. The layer 2 protocol LAPDm is used only between the MS and BTS because this is the only wireless link. Note also that, among other things, the BTS must buffer messages, called rate adoption, between the two layer 2 protocols. The BTS must perform this function because the data rate over the LAPDm link is a maximum of 22.8 kbps; on the wireline link, a full DS-0 is available. Hence it can transmit and receive at 64 kbps.

LAPDm   Like any layer 2 protocol, LAPDm is composed of two logical parts: the frame definition or signaling protocol and the service interfaces. This section will discuss the frame types first, followed by the service aspects of the data link layer.

In GSM, a segment can be thought of as a group of BSSs controlled by a single MSC. The data link layer used in the wireless component of GSM is called LAPDm. LAPDm is very similar to LAPD, with the exception that several wireless specific frame types are defined that are closely coupled with the network layer protocol service requests made to the data link layer. In other words, LAPDm is adopted for mobile wireless systems. See Chapter 3 for additional details about LAPDm.

Frame Types   There are three broad classes of frame types in this protocol: data frames, command frames, and response frames. Data frames transport data imported from the network layer. Command frames include all I-frames used to transport information (hence the name I-frame) from one station to another. A typical example of an I-frame would be one that supports the sequence numbers in an acknowledged communication.

Table 12-5
Sample GSM Control and Information Frames

Function	Name	Application
Unnumbered ACK	UA	Confirmation
Unnumbered	UI	Unconfirmed I frame
Information	I	Confirmed I frame
Receive ready	RR	Okay to transmit
Reject	REJ	Negative confirmation

Another good example would be the frames that transport the window-size negotiation between the two stations. I-frames are always command frames.

S-frames can occur as either command or response frames. Responses to a window negotiation would be examples of S-frames. In general, all signaling over the control channels and paging channels are accomplished with a combination of I-frames and S-frames. In GSM, the signaling by various control channels is accomplished with the use of I-frames and S-frames. Table 12-5 defines a few of the more representative frames used in the GSM standard.

Probably the most interesting feature of the LAPDm frame structure is that there is no error control embedded into the frame. The traditional FCS field in an HDLC frame is not present. All error detection is performed at layer 1 with fire codes. Of course, there are flow control and acknowledgment mechanisms, depending on the service class desired, but no explicit error detection takes place at the data link layer.

Services and Service Interfaces   Any data link layer interfaces to the physical layer and network layer though service access points (SAPs). In standard OSI terminology, the SAP identifies both the service type and its location in memory. In GSM, this feature is broken into two components: the service access point identifiers (SAPIs) and the connection endpoints (CEs). The relationship between SAPI and CE is illustrated in Figure 12-23.

Each SAPI and CE identifies a unique service and mapping to the particular control channel that will be used to implement that service. Together, they identify a data link connection identifier (DLCI). The more familiar DLCI terminology is used to identify particular physical endpoints for a frame in many packet-switched network systems, such as frame relay. In other words, the DLCI identifies the particular service that the data link layer can provide to the network layer. GSM uses the data link layer LAPDm.

LAPDm provides two classes of services, both of which can co-exist. They are the basic unacknowledged CLS and a class of service known as multiple-frame service, which is a COS. Unacknowledged CLS is transported in unnumbered I (UI) frames. As suggested by the name, there are no sequence numbers, acknowledgments, or flow control in this service offering. The COS offering is transported in I frames. Sequence numbers

Figure 12-23
SAPI and CE

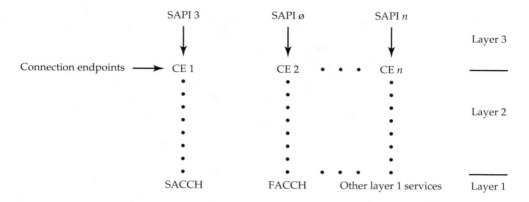

Table 12-6
GSM Control Channel SAPIs

SAPI	CE	Control Channel
0	1	RACH
0	2	BCCH
0	3	AGCH
0	4	PCH
0	5	FACCH
3	1	SACCH

are always used with this service offering and each frame must be individually acknowledged. The flow control used in this service is a simple stop-and-wait protocol with a window size of 1. Each frame must be received and acknowledged prior to the next one being transmitted.

Table 12-6 defines the CE for the two different values (SAPI = 0 and SAPI = 3) of SAPI that have been defined by the GSM standard. The network layer sublayer protocols RR, MM, and CM use SAPI = 0, and SMS uses SAPI = 3. Each of these network layer protocol sublayers will be discussed in detail in a later section of this chapter.

The DLCI, which is determined by the combination of a SAPI and a CE, only defines a service interface to the network layer using some control channel that is identical to the logical channel of the communication. The actual service is defined in an upper layer, and the network layer defines which DLCI, and hence which control channel, will be used by which network layer PDU.

Figure 12-24
GSM Protocols for Layers 1–3

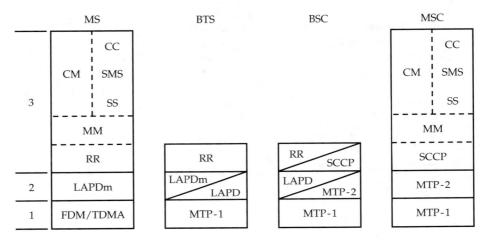

GSM Layer 3

The network layer in GSM is divided into three distinct sublayers:

Radio resource management (RR)

Mobility management (MM)

Call management (CM)

See Figure 12-24. Note that the CM sublayer contains three distinct protocol components. These sublayers will be discussed in that section, which appears below.

RR Sublayer   The RR sublayer is based on the services of LAPDm. It is responsible for setup, maintenance, and termination of each point-to-point radio channel. Maintenance includes the functions of location, handoff, and roaming. RR performs these functions under the direction of the MM sublayer. This management is in addition to the conventional call-handling procedures that occur in a wireline telephone network. The data rate of RR is 22.8 kbps.

RR is responsible for both the physical and logical channels, or layer 2 connections. For example, if a handoff occurs, RR insulates the upper layers from it. Physical channel RR connections may be terminated and reestablished several times during a logical channel connection in which data is to be transferred.

As indicated earlier (see Table 12-6), RR uses SAPI = 0. Note that each MS is allowed only a single RR connection at one time. RR connections are controlled by the MM sublayer, which manages all RR connections and terminations. The RR sublayer is unique because it uses both the services of layer 2 and LAPDm, but also has direct access to layer 1. This setup was illustrated in Figure 12-19.

During connection establishment, RR used two logical channels: SACCH and in most cases FACCH. In some cases, SACCH will be used when FACCH is not available.

MM Sublayer   While the actual channel control involved in location, handoff, and roaming is performed by the RR sublayer, these functions are under the control of the MM sublayer. These MM functions occur only when an RR connection exists.

Location updates occur whenever the MS roams out of reach of its HLR. MM always tries initially to establish a call in the HLR area. If it cannot, then the network searches for a VLR area to establish a call. MM is also responsible for the security management procedures associated with establishing a call. Security management procedures include the authentication and identification steps discussed earlier.

CM Sublayer   The CM sublayer contains three distinct components:

Call control (CC)

Short message service (SMS)

Supplementary services management (SS)

SMS is discussed in Chapter 14, so the service will not be explored here. The SS protocol element provides for specialized call billing. Because of the limited functionality of these two protocol elements as they relate to the overall operation of the basic protocol for voice services, the CM sublayer will often be subsumed into the CC sublayer. CC is responsible for the setup, maintenance, and termination of circuit-switched calls. To some of you, it may seem like each sublayer of the network layer is responsible for these same three tasks. However, the distinct functionality can be seen by recalling exactly what was said earlier. RR is responsible for the radio channel management, which is controlled by MM, which in turn provides the service interface for managing the calls from an upper layer perspective.

The primary function of the CC sublayer is to provide the transport layer with a point-to-point connection between two physical devices. In some cases this connection will be between two MSs; in other cases, it will be between an MS and a wireline telephone or data system. For example, before the RR layer can establish a radio channel to establish a call, the MM sublayer connection must be established. Before the MM layer connection can be established, the CC sublayer service for call establishment must be requested by the transport layer.

One interesting item that the CC sublayer must handle is the coordination between the wireless network and the wireline network. When a number is dialed in the wireline telephone network, the number information is carried over the control network into the MSC and routed to the appropriate BSC. Once the call is established, a traffic channel is used to allow the two subscribers to communicate. Note that between the BTS and the MS are two types of channels: traffic channels and the logical channels intended for data signaling. The traffic channels contain CODECs, so when a number is dialed from a mobile unit, the DTMF tones must be carried over a logical channel. FACCH is used so that the CODEC will not distort the DTMF tones. Once the call is established, the traffic channel is used to allow the two subscribers to communicate. The job of managing these multiple connections falls to the CC sublayer.

■ SUMMARY

This chapter has described the two traditional air interface approaches for cellular systems: pure FDM using an analog system, and a hybrid FDM/TDMA using a digital system. This chapter sets the stage for Chapter 13, which describes the third broad classification of cellular systems: those using an FDM/CDMA air interface approach.

REVIEW QUESTIONS

1. Fill out Table 12-7 with the details of each cellular system technology.

2. Describe the call procedure from mobile unit to PSTN in AMPS.

3. Describe the call procedure from PSTN to mobile unit in AMPS.

4. Define and compare the traffic channel, time slot, and frame as used in GSM.

5. Discuss how the HLR and VLR function and thus accommodate roaming.

6. Describe the three security elements used in GSM.

7. List and describe the control channels used in GSM.

8. List and describe the sublayer protocols of layer 3 in GSM.

9. Discuss the function of the location area in GSM and the process GSM uses to identify the cell in which the MS is currently located.

10. Discuss how the security elements in GSM work to enhance security.

11. Discuss the registration procedure used in GSM.

12. List and describe the function of each field in the GSM frame.

13. List and describe the function of each frame type in the GSM hierarchy.

Table 12-7

	AMPS	DAMPS	ETACS	GSM-900	GSM-1800	IS-95
Frequency band	___	___	___	___	___	___
Modulation type	___	___	___	___	___	___
Bandwidth allocation	___	___	___	___	___	___
Channel bandwidth	___	___	___	___	___	___
Multiple-access method	___	___	___	___	___	___

14. How many time slots per frame does each MS use to transmit information?

15. What relationship exists between the time slot on which the MS transmits and the time slot on which the MS receives?

16. Discuss the color of a traffic burst and its purpose.

17. Describe how the HL/HU bits work in association with the control channels.

18. Describe in detail the handoff mechanism used in GSM. Identify each step of the process.

13

Code Division Multiple-Access Based Cellular

■ INTRODUCTION

This chapter follows the general format of the previous two chapters in its exploration of the third air interface approach, code division multiple access (CDMA). There are two major differences in a CDMA system compared to an FDMA or FDM/TDMA air approach. The first is the air interface. CDMA is quite different in its characteristics than the familiar FDM and TDM techniques. The second is the absence of frequency planning.

Frequency reuse is still present, but every cell uses the same frequency spectrum. In an FDMA or FDM/TDMA based system it was not possible to use the same frequency set in adjacent cells because they would interfere with each other (co-channel interference) and degrade the performance to such an extent that the system would not operate. To illustrate the way CDMA uses frequency reuse but not frequency planning, see Figure 13-1. The naming of the components follows the TR-45 architectural model and hence the six components, with the same functions, remain the same. This chapter assumes that the reader is familiar with CDMA as an access technique and the concepts of spread spectrum. These topics were discussed in detail in Chapter 2.

Figure 13-1
Frequency Reuse in a CDMA Environment

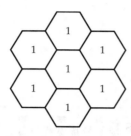

■ IS-95A

The IS-95A standard was first deployed in 1998. Its major characteristics include the following:

Digital OQPSK (uplink) and digital QPSK (downlink)

Deployed in the United States and Asia

800 MHz band (IS-95A)

45 MHz forward and reverse separation

50 MHz spectral allocation

1900 MHz (IS-95B) upbanded system

90 MHz forward and reverse separation

120 MHz spectral allocation

2.46 MHz total bandwidth (many channels share the same bandwidth)

1.23 MHz reverse CDMA channel bandwidth

1.23 MHz forward CDMA channel bandwidth

DS-CDMA access method

8 kHz voice bandwidth

64 total channels per CDMA channel bandwidth

55 voice channels per CDMA channel bandwidth

This system was the first CDMA based system deployed. The IS-95A standard operates at the top end of the 800 MHz band. The IS-95B standard operates in the 1900 MHz band.

The motivation behind the original deployment of this system was to increase the channel capacity of existing cellular systems. It has achieved this objective and offers anywhere from five to thirty times the capacity of the AMPS system. This increase depends on the particulars of the actual deployment and will be explored below. The IS-95 system was pioneered and developed by Qualcomm, Inc.

The unique characteristics of CDMA as an air interface allow the seeming impossibility in the next sentence to be possible. Each full-duplex pair of voice channels in a cell makes use of 2.46 MHz of spectrum, and fifty-five pairs of voice channels in each cell make use of the same 2.46 MHz spectrum. Recall that in CDMA spread spectrum, many channels, distinguished by code, can occupy the same frequency space. The spectrum is split in two: half for the forward channel, half for the reverse channel, each 1.23 MHz in bandwidth. While this number of channels may seem small at first glance, remember that you are obtaining fifty-five channels in the same space where the AMPS accommodated only forty-one. See Example 13-1 for the calculation.

■ **EXAMPLE 13-1**   How many voice channels could be obtained in the 1.23 bandwidth allocation for IS-95A if the AMPS air interface was utilized?

$$\text{Number of channels} = \frac{\text{total bandwidth}}{\text{channel bandwidth}} = \frac{1.23 \times 10^6}{30 \times 10^3} = 41 \text{ channels}$$

Clearly, the result in Example 13-1 is an advantage of this system over the AMPS deployment. A gain of fourteen simplex channels per cell is an increase of over 25 percent in the number of potential subscribers per cell. It is important to note, however, that the IS-95A technique does lend itself well to a direct AMPS channel replacement, as would be the case with an AMPS to DAMPS evolution. In both cases, large groups of channels are replaced at one time. Dual AMPS—IS-95A systems exist because it is possible to share the same frequency space. Very few dual AMPS—DAMPS systems exist.

Since the spectral allocation for cellular systems in the 800 and 1900 MHz bands is independent of the air interface, additional channels can be obtained by applying FDM techniques to CDMA systems. This technique results in several channel sets adjacent to each other in the frequency range defined by the spectral allocation. The forward and reverse channels of the IS-95B system are illustrated in Figure 13-2.

The original IS-95A was designed to replace/complement AMPS systems. Thus, its terminology and the terminology for the IS-95B for the specific frequency assignment of any particular mobile or base station is tied to the AMPS concept of a 30 kHz channel assignment for each simplex communication. This feature has resulted in the retention of some specific terminology, namely, the use of the terms *channel number* and *CDMA frequency assignment*.

The channel number is actually a binary number that corresponds directly to the center of the CDMA frequency assignment. Table 13-1 illustrates this convention for the 1900 MHz system. For convenience, the binary channel number is resolved to decimal form. The channel number has no real significance for understanding how the IS-95 standards operate. It is included for refitting AMPS base stations into IS-95 base stations only.

A good question to ask is, How many 1.23 MHz CDMA channels are in the total spectrum allocation? Example 13-2 provides the answer.

Figure 13-2
IS-95B CDMA Spectral Allocation: (a) Reverse Channel; (b) Forward Channel

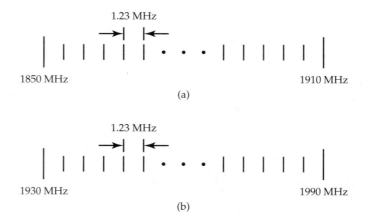

(a)

(b)

Table 13-1
Channel Number and CDMA Frequency Assignment for IS-95B

Source	Channel Number	CDMA Frequency Assignment
Mobile	$1 \leq N \leq 1{,}199$	$0.050N + 1850.000$
Base station	$1 \leq N \leq 1{,}199$	$0.050N + 1930.000$

■ **EXAMPLE 13-2**   How many 1.23 MHz CDMA bandwidths are provided in each direction?

$$\text{Number of CDMA channels} = \frac{\text{spectral allocation}}{\text{channel bandwidth}} = \frac{25 \text{ MHz}}{1.23 \text{ MHz}} = 20$$

Now it is possible to compare directly the number of subscribers this system can support to the system approaches examined earlier in this text. Example 13-3 provides the calculation.

■ **EXAMPLE 13-3**   What is the total number of full-duplex communication channels that the IS-95A system can support?

(Number of CDMA bandwidths)(number of channel numbers) = (20)(55) = 1,100 full-duplex channels

Table 13-2 summarizes the calculations for each major system approach. Note that, to compare apples to apples, so to speak, the IS-95A system is used because its band-

Table 13-2
Potential Channel Comparison

System	Bandwidth Allocation	Simplex Channel BW	Multiplier	Total Duplex Voice Channels
AMPS	25 MHz	30 kHz	1	832
DAMPS	25 MHz	30 kHz	3	2,496
GSM	25 MHz	200 kHz	8	992
IS-95A	25 MHz	1.23 MHz	55	1,100

Table 13-3
True Channel Comparison

System	Duplex Voice Channels per Cluster	Number of Clusters	FRF	Total Duplex Voice Channels
AMPS	832	10	7	8,320
DAMPS	2,496	10	7	24,960
GSM	992	10	7	9,920
IS-95A	1,100	10	1	77,000 (51,300)

width allocation is the same as the other systems in the table. The IS-95B system, with over twice the bandwidth allocation, features many more channels.

The first detail that you may notice in Table 13-2 is that the total number of duplex channels for IS-95A does not seem to match the number inferred in Table 13-1. The reason is that in Table 13-2 (and Example 13-3) only those channels available for voice/data services were shown. Recall that nine channels in each 1.23 MHz channel bandwidth are held back for signaling. Thus, they are shown in Table 13-1 but not in Table 13-2 and Example 13-3.

The next detail that you may notice is that the multiplier column requires some explanation. In the AMPS system, each voice channel occupies the entire simplex channel BW. Evolution to DAMPS meant that each simplex channel was divided into three TDMA channels, thus increasing the revenue channels by a factor of three. In the GSM scheme, each 200 kHz traffic channel is divided into eight TDMA channels, and finally for IS-95, each 1.23 MHz CDMA channel bandwidth is divided into fifty-five CDMA channels. Note that the DAMPS system, widely deployed in the United States, features the highest potential revenue per frequency space. Table 13-2 does not take into account that the CDMA air interface requires no frequency planning. Depending on the cluster size of an FDM or FDM/TDMA deployment, the absence of frequency planning significantly increases the total revenue potential of CDMA and partially explains why it is the air interface choice for 3G system deployment. Table 13-3 addresses this issue.

CDMA is fundamentally different as an air interface than is an FDMA or FDM/TDMA based system. Prior to a discussion of the six components, two sections discussing these differences are provided below. They address two consequences of using CDMA as an air interface: no frequency planning, and a fundamental difference in the mechanism that limits the total number of simultaneous active channels in any cell.

■ NO FREQUENCY PLANNING

It is critical to understand that, although a correspondence with AMPS frequency channels exists, the communication and control channels in IS-95 are differentiated by code and therefore no frequency planning exists. The whole concept of cellular frequency assignment, co-channel interference, etc., does not apply. There is no need to draw a cellular grid or to use the tools introduced in Chapter 11 to assign specific frequency bands to particular physical areas. Thus, CDMA deployment is much easier in some ways than deployment of other air interface approaches. There is a dramatic effect on the number of total voice channels that the system can support. Table 13-2 assumed that the critical element was how many voice channels could be supported in spectral allocation. Of course, this element is affected by how many times the frequency can be reused.

In an AMPS, DAMPS, or GSM deployment, the frequency reuse factor (FRF) varies from a low of 3 to a high of 21. Most deployments are in the range of 4 to 12. Taking these values of FRF and factoring them in yields the results in Table 13-3. This table reflects the fact that the value of the FRF divides the total number of voice channels any particular deployment could support. Recall that in frequency planning, you took the total number of channels available to you and divided them among the number of cells in each cluster, using the FRF value as the number of cells in any cluster. When using CDMA as an air interface, the cluster size becomes ideally 1; hence the FRF = 1. Instead of distributing the channels in any cluster evenly among the number of cells in that cluster, you have all channels in every cell because each cell is, in effect, a cluster itself. For a reasonable comparison, Table 13-3 assumes a total number of cells of 10 and an FRF of 7. Other values would change the calculations significantly, which is why we stated earlier that the channel improvement of IS-95 over AMPS was a variable figure.

While every cell can theoretically be reused by every frequency, in reality a practical value for the FRF in an IS-95 environment is about 1.5. This practical value is a result of interference, as we will discuss in the next section. Substituting a value for FRF of 1.5 in Table 13-3 results in a total number of duplex channels of approximately 51,300.

■ RESOURCE VERSUS INTERFERENCE
FUNDAMENTAL LIMITS

Table 13-3 showed that an increase in channels clearly depends on the value of the FRF. If we assume that the number of cells remains constant, the comparison grows more stark as the FRF factor grows. Since there is no frequency planning required in IS-95, the entire spectral allocation may be used in each cell. This feature has the effect of greatly increasing the total potential revenue from such a system.

The calculations shown in Table 13-3 can be clarified by examining what mechanism is actually responsible for limiting the channel capacity of a cellular system. Digital communications systems perform well in white-noise environments. White noise has no particular characteristic; it is of approximately equal power in any particular portion of the spectrum.

As has been pointed out in earlier chapters, FDM and FDM/TDMA based cellular systems are limited by co-channel interference. There is a minimum distance that can be tolerated from a signal source of the same frequency and power spectra, precisely because this interference is not white noise. Instead, it is characterized as co-channel interference.

Therefore, the limiting factor in FDM systems is the proximity of the next cell that is using the same frequency set. Similarly, the limiting factor in FDM/TDMA systems is the same as the FDM case, in addition to the number of time slots that can fit into each frequency slot. In both cases (FDM and FDM/TDMA), these limitations can be characterized as *resource limited*. The critical resource in an FDM system is how many frequency slots can fit in the spectrum allocated. The critical resource in an FDM/TDMA system is how many time slots can fit into the frequency slots available.

CDMA systems are limited by the number of channels operating. There is no frequency planning; every cell uses the entire spectral allocation. There are no limits on the number of frequency slots, other than the legislated limits on spreading factors. CDMA works by spreading the power of each transmission over the entire spectral allocation. Thus, as more and more transmissions are generated, each transmission adds its own contribution to the background noise. At some point, the interference level becomes unacceptable. CDMA systems are *interference limited*. This level of unacceptability is a system QOS parameter. If the QOS parameters will allow a higher level of interference, more subscribers can be added.

Thus, cellular networks deployed using a CDMA air interface are in a unique situation because the number of subscribers that can be supported in any cell is a business decision. This decision is a tradeoff between an acceptable level of transmission QOS and the increased revenue that would be generated by adding additional channel pairs. As the FRF moves toward 1.0, the transmission quality degrades because the number of transmissions combines to generate a white noise background level that starts interfering with the ability of the receivers to discern the information signal from the background noise.

The key to maximizing the number of transmissions, while minimizing the interference, is transmission power control (discussed below in the appropriate section). Remember that CDMA based systems are also digital communications systems and, just like any other, perform best in a white-noise environment. When every transmission is the same power, they all tend to have the same power signature in each portion of the spectrum.

■ THE SIX COMPONENTS

The six components of the IS-95 are illustrated in Figure 13-3. Each component is discussed in the following sections.

Component 1: Gateway into the Wireline System

There is always a gateway between the base station's switching subsystem and the wireline switching fabric. The basic idea is that there must be some kind of switching fabric

Figure 13-3
Abridged TR-45a Architectural Model

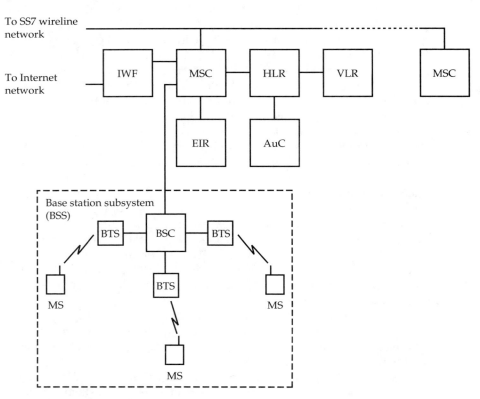

to interconnect the traditional wireline telephone system to the wireless system to allow calls between the two systems. This gateway is composed of two logical components: the mobile switching center (MSC) and the interworking function (IWF), which does protocol translation for most data transfers. The MSCs are interconnected with wireline links and one is designated as the location of the equipment identity register (EIR). Each MSC connects to the wireline network.

Component 2: Base Station (BS)

The base stations always contain several transmitter/receiver pairs and form one half of the radio link. These transmitter/receiver pairs are coordinated with a controller unit that connects the transmitter/receiver pairs to the MSC. Therefore, the base station is composed of two logical components: the base transceiver system (BTS) and the base station controller (BSC). There may be and usually are several BTSs for each BSC.

Component 3: Databases

The databases are structured to keep track of billing, caller location, etc. Each mobile unit has an entry in at least two databases. One, the home location register (HLR), keeps track of the location of the mobile unit's "home" in the mobile network. The other, the visitor location register (VLR), keeps track of where the mobile unit is now and accommodates roaming. Each cellular telephone has entries in these databases that store the last location, frequency pair or time slot, and other details of the subscriber. The functions of an HLR/VLR pair in a cellular network are quite similar to the function of an SCP in an SS7 wireline network.

Component 4: Security Mechanism

A security mechanism is required for authenticating billing and providing assurance that the person who is placing the call from the cellular telephone is the actual owner. These functions are facilitated by a database, called the equipment identity register (EIR), and an authentication center (AC), which manages the actual encryption and verification of each subscriber. Usually there is only one EIR and many ACs per system. There are also other components in the cellular telephone itself.

Component 5: Air Interface Standard

You can think of the air interface as a way of sharing the bandwidth allocation of the system among all MSs in any particular cell. The air interface method used here is CDMA. Refer to Chapter 2 for detailed information on this multiplexing technique.

Component 6: Cellular Telephone

The cellular telephone itself is called a mobile station (MS). These mobile stations need not be actual telephones and in the future will include various mobile devices that use the cellular system for data transfer as well as voice connections.

■ IS-95 LOGICAL CHANNEL TYPES

Five distinct logical channel types are defined and broken into two classes: the control channels and traffic channels. In some cases, certain control channels may carry traffic (thus producing revenue) when the control mechanism does not require that channel.

Control Channels	Traffic Channels
Forward channel pilot channel	Voice/data channels
Forward channel synchronization channel	
Forward channel paging channels	
Reverse channel access channels	

Figure 13-4 shows the physical channel assignments in both the forward and reverse channels for the logical control channels identified. The physical channel assignments are shown relative to the traffic channel assignments. Part (a) of the figure shows the forward channel; part (b) shows the reverse channel for the IS-95B standard.

The pilot channel is a logical control channel that does not carry any information. Its only purpose is to have a channel in each cell where the information content being carried on the channel can have no effect on the channel's power density. For this reason, the pilot channel carries no actual information but rather a specific sequence of bits (all zeros) that is the same for each pilot channel. The pilot channel is always physical channel

Figure 13-4
(a) IS-95B Forward Channel Assignments; (b) IS-95B Reverse Channel Assignments

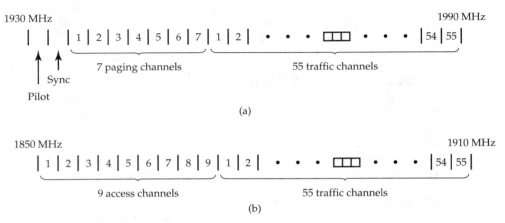

(a)

(b)

number 0 of the forward channel structure and uses Walsh code number 0. The pilot channel is used by the MS to determine the closest BTS. The MS scans all the pilot channels it can access and uses a power measurement to help determine the BTS with which to associate itself. For obvious reasons, there is no data rate specified with the paging channel.

The synchronization channel, referred to as the sync channel, is a logical control channel that does contain a specific information content called the synchronization channel message. It is used to establish time synchronization of the MS with the BTS. This synchronization is critical to the operation of the CDMA approach. The data rate of the sync channel is 1,200 bps, but because of the convolutional encoding and the fact that it is repeated twice, the signaling rate is 4,800 bps.

The synchronization channel message contains a large amount of information needed by the MS. It has a true HDLC type frame structure with a start delimiter, length field, and CRC field. It contains the system identification (SID) number and the network identification (NID) number. Of more importance to this section, it also contains the specific information, the pilot offset index, that the MS needs to identify the pilot channel source.

To understand why this information is needed, recall the fact that frequency planning in CDMA networks does not take place. Therefore, since the pilot channel is always at the same physical channel and always uses the same spreading or Walsh code, how is the MS to identify a unique BTS? The answer lies in the pilot offset index. It contains a number that identifies the chip offset used on the pilot channel. Each BTS in any geographical region offsets its Walsh code by the pilot offset number in the sync channel. A total of 512 possible offsets are available. This convention ensures that any pilot channel with sufficient power for an MS to acquire will have a unique offset value. Thus, the MS differentiates between pilot channels and hence BTSs.

In summary the pilot and sync control channels work together. Therefore, while CDMA systems do not use frequency planning, they still must have a unique way for the MS to identify the best BTS with which to communicate. In the IS-95 standard, this task is done with the pilot and sync control signals.

While each BTS broadcasts only one pilot channel and one sync channel, there are seven additional control channels collectively called the paging channels. Like all the forward control channels, they are used to send information to the MSs. The two previously explained control channels were broadcast in nature; the paging channels are point-to-point. They are used like the paging channels were used in previously examined air interface schemes for cellular telephony. When an MS is about to receive a call, it is notified on one of the paging channels. However, the paging channels, like the sync channel, have a frame structure and carry various types of information to the MS. The paging channels operate at a data rate of either 4,800 or 9,600 bps.

Not only does a paging channel notify the MS of an incoming call, it may provide the calling party's telephone number if that service is part of the subscription package. It also indicates to the MS how it should request a traffic channel to obtain this information, when appropriate. The actual messaging generated by the MS to request a traffic channel is performed on the reverse channel access channel, but the information on which access channel to use is provided to the MS on the paging channel. This channel is also used as a kind of catch-all for any other messages (for example, maintenance and security notifications, frequency handoff instructions, etc.) that may be required to address a particular MS.

The final group of control channels are known as the access channels. They are the only control channels on the reverse channels. All access channels operate at the data rate of 4,800 bps and are used by the MSs in two ways: first, to respond to a signal conveyed to them on the paging channels and second, to inform the BS that the MS requires an assignment of a pair of traffic channels to place a call. As we said earlier, the paging channel message contains the specific physical access channel identification on which the MS responds to the BS.

The access channels are the only channels that the MS uses to communicate with the BS, and the MS can send several different messages to the BS. These messages can convey the telephone number the mobile subscriber wishes to call, coordination in the case of a paging message to the MS instructing it to perform a frequency handoff, registration messages, and authentication responses. The access channel frame format shown in Figure 13-4(b) is representative of the frame format that the sync and paging channels also use.

There are seven paging channels and nine access channels. In fact, these physical channels can be used for voice data traffic when the logical channel requirements permit. Of course, the logical channel requirements have precedence; therefore, the actual number of traffic channels, in each direction, at any one moment in time will vary. In summary, the number of physical reverse channels is 64 (the number of access channels in use), while the number of physical forward channels is 64 − 2 (the number of paging channels in use). In the forward channel, the pilot and sync channels are dedicated and not available for traffic channel use.

■ IS-95 CALLING PROCEDURE

The IS-95 approach to processing cellular calls is exactly like the approach used in all cellular systems. These approaches have been explored in some detail in Chapters 11 and 12, so no further discussion is provided here.

■ SPEECH CODING

The speech coder, or CODEC, used with the IS-95 system generates an output rate of 8,600 bps. A 12-bit CRC code is then added, as well as an 8-bit tail, which increases the total bit rate to 9,600 bps. This rate is the maximum capacity of the channel, but if no speech is being generated, the CODEC automatically detects this situation and progressively lowers the transmitted data rate during times of silence. This process involves four steps at 9,600, 4,800, 2,400, and 1,200 bps. In many cases, the transmitted data rate will jump directly from 9,600 bps to 1,200 bps, the lowest rate.

The 2,400 and 4,800 bps rates are used when control information is sent during the silent periods. This technique is used by the mobile station to reduce its power. Note that this technique also simultaneously reduces the interference to other stations using the same channel. This subject and the exact mechanisms will be discussed in more detail when power control is explored later in this chapter.

The approach taken to reduce the data rate is a duty cycle approach. In other words, the actual bit data rate stays constant, but the number of bits is reduced. The transmission duty cycle varies with the transmitted data rate. When transmitting at 9,600 bps, a 100 percent duty cycle is used; when transmitting at 1,200 bps, a duty cycle of 12.5 percent is used. This technique is called variable rate coding. It is explored in more depth and related to processing gain and framing later in the chapter.

The voice signal is digitized using a variable rate PCM CODEC like in the FDM/TDM based systems, and data can also be sent over any voice channel. The data rates will vary depending on several environmental factors; specifically if the voice channel is being used by the speaker, the parallel data transfer rate will drop significantly. The particular coder implementation also has an impact at the high end of this range.

■ HANDOFF AND ROAMING

While roaming criteria are virtually identical to those discussed in Chapters 11 and 12, handoff is significantly different. Two limiting conditions force a handoff. The first, discussed in Chapters 11 and 12, is based on signal strength. When the signal strength falls below some threshold, handoff occurs. In first-generation systems like AMPS, handoff is solely determined by the base station and administered by the MSC. In second-generation systems, like GSM and DAMPS, the MS also measures the received power from adjacent base stations and signals the results of these measurements to the BSC with which it is currently associated.

The second limiting condition, unique to systems where the limit is interference, applies to CDMA systems in general and IS-95 systems in particular. In such systems, the amount of interference is determined by the number of channels operating, so the handoff criteria depend on the current use of the system and the defined QOS parameters for signal-to-noise ratios in the channel defined by the system operator. The minimum signal-to-noise ratio is set to approximately 18 dB.

■ HARD AND SOFT HANDOFFS

As defined in FDMA and FDM/TDMA systems, handoffs are referred to as hard handoffs because an explicit radio channel reassignment takes place during the channel handoff. In the case of spread spectrum systems like IS-95, the same spectrum is shared by all MS units in the cell (FRF approaches 1). Therefore, no RF channel assignment change takes place. As a result, these handoffs are referred to as soft handoffs because no physical channel reassignment takes place.

In CDMA based systems, a handoff implies another base station performing the radio communication. All IS-95 handoffs are, by definition, soft handoffs. Basically, a call is transferred to the next closest base station as the subscriber approaches the cell boundary. Any base station in a CDMA based system can receive signals from an MS in its cell and an MS near its cell boundary simultaneously because all cells share the same frequency bandwidth. Similarily, an MS will receive signals from multiple base stations simultaneously. Given this information in both the MS and BSC, handoff can be managed from one BSC to another with relative ease and no disruption of service to the subscriber. While both the BSC and the MS receive multiple signals, the actual decision to hand off is performed by the MSC.

Both the MS and the BSC have access to similiar information, so a soft handoff may be initiated by either one. As you might suspect, handoffs initiated by the MS are called mobile-assisted handoffs, while those initiated by the BSC are called base station–assisted handoffs. Two primary conditions trigger soft handoffs. First, the received signal strength can move above or below some threshold. This change is determined by monitoring of the pilot signal. CDMA systems require all transmissions received at a base station to be within a narrow range. This topic is discussed in the next section, but it is important to note here that power control and soft handoff are closely related. In the case of too much power or too little, either the MS or the BSC can initiate a handoff. Second, as a base station approaches its capacity, the MSC looks for nearby base stations with capacity not being utilized. When it finds such base stations, the MSC attempts to balance the load among nearby base stations to maximize efficiency.

Finally, in theory, soft handoffs can guarantee that the MS is always linked to the base station from which it receives the highest quality signal. Hard handoffs cannot guarantee this link. As a practical matter, soft handoffs balance this criterion with other strategies, such as load balancing between base stations.

■ POWER LIMITS AND CONTROL

Power control of the CDMA mobile transmissions is a key aspect of the system's success because, if a mobile transmits at too high a power level, it will interfere with other mobile stations in both its cell and adjacent cells. CDMA systems are very sensitive to differences in received power on the reverse channel.

While the individual transmissions are distinguished by their codes, it is critical that all spread spectrum transmissions received be of a similar power level. The objective is that all signals from the mobile stations are of the same power when they reach the base station. The closer the adherence to this objective, the better the performance of the system, assuming that the power is above some threshold of detectability. IS-95 uses two forms of power control: closed loop and open loop.

It is often asserted that closed loop power control is used on the forward channel and open loop power control is used on the reverse channel. Control implies a round-trip communication. Thus, power control in IS-95 cellular systems is better described by saying that the control loop functions differently on the forward and reverse channels because of the different accuracy required for base station transmissions compared to MS transmissions.

■ FORWARD CHANNEL POWER CONTROL

The base station initially broadcasts with the same power to every mobile station, essentially assigning a default channel attenuation estimate. In general, some mobile stations will be farther away and will thus receive a lower power transmission, while others will be closer and will thus receive a higher power transmission. This dilemma is known as the near-far problem.

Once the MS initiates a request for access to the system, the base station has some information upon which to estimate the forward channel attenuation, using the received power of the access request to adjust its transmissions accordingly. This step is the first in the process of the closed loop control on the forward channel. The second step is when the mobile station, as part of its transmission, sends the base station the desired power it wishes the base station to use in its next transmission. In general, some mobile stations will be farther away and will ask for more power, while others will be closer and will request a lower power transmission. The mobile station measures the receive power and provides a return value based on this measurement, so this process is a closed loop control system on forward channel transmission. It is composed of two steps.

The base station sends power control messages every 1.25 ms to the mobile station, instructing it to increase or decrease its transmitted power in 1 dB increments. The 1.25 ms is derived directly from the fact that IS-95 systems use 20 ms frames in their traffic channels. This power control message is contained in a 1-bit digital gain message, often all that is needed to indicate step up or step down in power because the increment is fixed at 1 dB (±0.5 dB) in either direction. This power control bit is sent on a logical sub-channel in the forward traffic channel frame. Every 1.25 ms, or at a data rate of 800 bps, 2 bits of information content are replaced by the power control bit. A 1-bit decreases the power level, a 0-bit increases the power level. Since the power variation in the forward channel is small, typically 3 dB, a single bit is sufficient.

■ REVERSE CHANNEL POWER CONTROL

On reverse channel transmissions, the control mechanism is different but still consists of two steps. Note that the power variation in transmissions on the reverse channel vary much more widely than those on the forward channel, up to 80 dB. For this reason, power control is much more important and a different control mechanism is required. Fundamentally, power control on the reverse channel is meant to address the near-far effect described in Chapter 11.

Again, there is open loop control of the MS transmissions, and adjustment of the transmission power levels by the closed loop control mechanism is located in the base station. The closed loop feedback loop is entirely contained in the base station. The reverse channel power control begins with each MS estimating the total average power in the CDMA channel. Each MS shares the same frequency channel, so there is always some traffic for the MS to use for this power measurement. This procedure is simple. It requires no timing information and can be made very quickly. Based on this measurement and the control bit status received from the base station, the MS adjusts its transmission power. Note that this measurement must be made quickly to limit the potential interference generated during the feedback cycle. In theory and in practice, all transmissions arrive at the base station at approximately the same power.

The closed loop correction occurs in response to power control messages received from the base station by the MS. This correction is also 1-bit digital power control but is contained in three message frames and allows approximately a ± 24 dB variation around the initial open loop estimate. The combination of the power control bits and the range of the initial open loop estimates allows power control over a range of about 80 dB on the reverse channel. In both forward and reverse channels, the power control bits are inserted into the traffic channels and hence degrade the voice quality.

The voice quality of the conversations in which these messages occur will be temporarily and only slightly degraded. This degradation of the CODEC data bits is spread to all subscribers and so is not noticeable. The location of the control bits is determined by the value of certain bits in a previous frame. This technique also ensures that the output spectrum remains flat.

A summary of the steps involved in power control on the reverse channel is shown below. It should be clear where the feedback occurs and where it does not. (Steps 1 and 2 feature no feedback cycle, while steps 3 to 6 do.)

1. The MS sends an access message to the BS.
2. The BS uses the power in this message to adjust its transmit power.
3. The BS sends frames to the MS.
4. The MS counts bad frames.
5. The MS sends the number of bad frames to the BS.
6. The BS uses this information to adjust power control.

Clearly, step 3 is the power estimate and step 6 is the correction. The closed loop feedback cycle is accomplished in steps 3 to 6. Stated another way, the power level used in step 3 is the open loop estimate, while step 6 is the closed loop revision of that estimate.

There are three power classifications of mobile stations. These power outputs are measured at the antenna connector and are shown in the second column of Table 13-4. Like most wireless systems, there is also a specification on the maximum ERP when an antenna is connected. This maximum prevents the standards bodies from restricting innovation in antenna technology. Both of these classifications are shown in the second and third columns of Table 13-4.

Table 13-4
IS-95 Maximum ERP Values

Mobile Station Classification	Connector Maximum Output	Antenna Maximum Output
I	1 dBW (1.25 W)	8 dBW
II	−3 dBW (0.5 W)	4 dBW
III	−7 dBW (0.2 W)	0 dBW

Finally, note that the average power transmitted by both the mobile stations and base stations is far less than would be experienced with any FDM or FDM/TDMA technology. This feature greatly extends the battery life of the mobile stations compared to other air interfaces.

■ SECURITY: AUTHENTICATION

Each IS-95 MS stores three numbers: the electronic serial number (ESN), the international mobile security identity (IMSI), and the A-key. Recall that the IMSI is also used by the GSM system. The IMSI and the ESN are imprinted on the MS during its manufacture and are not intended to be modified by the subscriber. In the GSM system, however, the IMSI is the critical piece of data used to identify and verify the subscriber. In the IS-95 system, the A-key combined with either the ESN or IMSI are the critical pieces of data. The A-key is 64 bits in length, while the ESN is 32 bits long and the IMSI is 15 bits long.

The BS periodically broadcasts a random number, 56 bits in length. This random number has a special name, the RANDSSD, and is used by both the AuC and MS to generate a 128-bit number called the shared secret data (SSD). The SSD is generated using three pieces of data: the ESN, A-key, and the RANDSSD. Neither the A-key nor the ESN are ever transmitted through the air. Instead, the SSD is transmitted and the AuC uses an algorithm to uncover the A-key. Both the MS and AuC must possess identical SSDs for authentication to be successful.

This procedure has many of the same advantages as does the GSM system approach. The random number changes regularly, so it becomes practically impossible to back out the A-key from the SSD transmissions, and the A-key is never transmitted through the radio channel. On the other hand, the A-key is part of the MS and cannot be carried from one MS to another. There is no SIM module.

IS-95 also incorporates what is known as a unique challenge procedure. Because of the excessive fraud history of the AMPS system, the BS needed an approach where it could, at its own discretion and timing, challenge the MS to authenticate itself. This procedure does not use the SSD approach but a single 24-bit random number. This number is sent to the MS on either a paging channel or an open forward direction traffic channel. In short, the MS then combines this number with a portion of the IMSI number and sends it back to the BS. The BS carries out a similar calculation and compares the value received from the MS. If there is an exact match, the challenge succeeds. If not, the BS will typically terminate the call and deny further access.

■ FRAMING

The fundamental frame in the IS-95 standard is the 20 ms or 96-bit frame. This frame is the fundamental unit of information transfer on the access control channels as well as the traffic channels in both the forward and reverse directions. The pilot and sync channels also have their own frame structures, but these structures will not be explored further than the brief description provided earlier. While the frame length in the access and traffic channels is the same, the frame contents differ slightly.

Figure 13-5 illustrates a high-level view of the frame structure differences used in the two types of channels. Part (a) of the figure shows the traffic frames used in both the forward and reverse channels. Recall that the forward channel transmissions originate from the BTS and the reverse channel transmissions originate from the MS. Figure 13-5(b) shows a representative control channel frame used in the reverse access channel.

There are three fields in Figure 13-5. The information field is where the actual information content is carried. It is a variable length field. The F field, standing for frame quality, is a short CRC. It also is a variable length field. There are three options for this field: 12 bits, 8 bits, or 0 bits. In the first two cases, different generator polynomials are used. Finally, the T field, or tail, is always 8 bits long and is used to flush out the encoder. Table 13-5 summarizes the relationship among the three fields. The information in Table 13-5 relates directly to the variable rate speech coding described earlier. The table shows

Figure 13-5
(a) Traffic Channel Frame; (b) Access Control Channel Frame

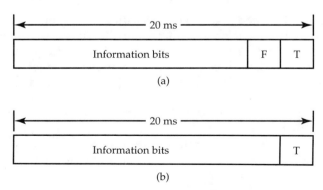

F = frame quality (CRC code)
T = tail bits

Table 13-5
Frame Structure Relationships

Data Rate	Total Bits	Information Bits	Frame Quality Bits	Tail Bits
9,600	192	172	12	8
4,800	96	80	8	8
2,400	48	40	0	8
1,200	24	16	0	8

the four data rates mentioned in the section titled Speech Coding and how that impact is carried over to other frame components.

The approach taken to reduce the data rate is a duty cycle approach, exactly as described in the Speech Coding section. It should be clear why there is no CRC or frame quality field when the data rate is 2,400 or 1,200 bps. In these situations, the information content is so low that to detect errors using a CRC method is of little practical utility.

The frame definitions from Figure 13-5 comprise what is called rate set 1 (RS1). The standard also offers an RS2 option. RS2 increases the standard set of data rates to 14,400, 7,200, 3,600, and 1,800 from the respective RS1 rates. RS2 is an optional implementation. If implemented, it must fall back to RS1 based on the MS characteristics.

As we saw in Figure 13-5, both frame structures had a period of 20 ms. Therefore, the data rates or number of bits per frame will vary according to the traffic rate. However, the actual modulation symbol rate in the radio channel remains the same. Because the modulation symbol rate stays the same and the number of bits per frame varies, it is an example of variable rate coding.

■ VARIABLE RATE CODING

In both forward and reverse channels, the modulation symbol rate remains the same, independent of the information data rate. The data rate values in both directions also remain the same; however, the voice data rates are different. This seeming conundrum is solved by noting the fact that the convolutional encoders in the two channels operate at different rates. In the forward channel, the coder operates at a rate of ½, with $K = 9$. In the reverse channel traffic frames and access control frames, the coder operates at a rate of ⅓, with $K = 9$. (Refer to Chapter 7 for details on what these values mean from a coding perspective.)

Simply put, the lower the rate of the convolutional encoder, the more bits are inserted into the information stream. Therefore, for an information rate of 9,600 bps, a coder operating at a rate of ½ replaces 2 bits for every 1 bit of information. The information transfer rate in the forward channel then becomes $9,600 \times 2 = 19,200$ bps. Similarly, in the reverse channel direction, where the convolutional encoder is operating at a rate of ⅓, the information transfer rate is $9,600 \times 3 = 28,800$ bps.

■ **EXAMPLE 13-4**   Since both forward and reverse channels use the same channel bandwidth and feature different information rates, it should be expected that they would feature different processing gains.
  Forward channel:

$$G_p = \frac{\text{channel BW}}{\text{information rate}} = \frac{1.23 \text{ MHz}}{19{,}200 \text{ bps}} = 64 \rightarrow G_p = 10\log(64) = 18 \text{ dB}$$

  Reverse channel:

$$G_p = \frac{\text{channel BW}}{\text{information rate}} = \frac{1.23 \text{ MHz}}{28{,}800 \text{ bps}} = 43 \rightarrow G_p = 10\log(43) = 16 \text{ dB}$$

Note that the processing gain is greater in the reverse channel than it is in the forward channel. This result should make sense because the reverse channel, originating from the mobile device, will usually be the lower power transmission and hence need all the processing gain it can get.

■ **WALSH CODE USE IN IS-95**

Careful readers of Chapter 2 will recognize that the CDMA systems discussed in this text use one of two basic types of spread spectrum technology, which means there must be a specification for the code length. This specification is shown in Example 13-5. A code set known as Walsh code is used in the IS-95 standard. As identified above, there is a total of 64 channels.
  Each channel uses a different spreading code, called a Walsh code. The codes are all 64 bits long, and the pilot code is composed of all 0s. Walsh code 32, consisting of alternating 1s and 0s, is used for the synchronization channel. Note that the Walsh codes are used to identify the channel; that is, each channel is recognized by the Walsh code used to spread it. Additionally, the Walsh codes are used differently on the forward and reverse links. The 19.2 kbps data stream in the forward channel is multiplied by the 64-bit Walsh code to obtain the 1.2288 Mbps data rate.

■ **EXAMPLE 13-5**   How many bits must the CDMA code in the IS-95 system use?
  To address this problem, you must know two facts: the maximum data rate per channel and the system bandwidth on either the forward or reverse channel. To find the number of bits used to encode the signal, just divide.

$$\text{Number of bits in code} = \frac{\text{total bandwidth}}{\text{maximum data rate per channel}} = \frac{1.23 \times 10^6}{19.2 \times 10^3} = 64 \text{ bits}$$

Security is an important issue and the system addresses it well. The user is protected with the use of a long code of $2^{42} - 1$, which contains the mobile unit's serial number embedded in a code mask. This code mask scrambles the outgoing data stream.

The two main technical problems faced by CDMA were addressed well by the use of Walsh codes that are 64 bits long. (Again, refer to Chapter 7 for details on Walsh codes and spread spectrum CDMA systems.) These limitations are inherent to the CDMA access methods and were cited as the technical reasons why the approach was rejected for earlier mobile phone introduction. These two issues are how to synchronize the mobile unit's code generators to those of the base station, and how to meet the CDMA requirement that all received signals at the base station be nominally the same power.

The synchronization issue was addressed by inserting a synchronization channel and pilot channel in each frame. Each mobile unit looks to the pilot channel to find the strongest signal components, which gives it a way to obtain an estimate of the various multipath components present in the channel, namely, phase and magnitude variance. The chip rate on the pilot channel is locked to a precise timing signal available worldwide, that of the global positioning system (GPS). A comparison of these two signals, along with the knowledge that the synchronization channel is phase locked to its pilot channel, allows the mobile unit to find the information it wants from the base station. This information includes the time of day and a special long synchronization code that allows synchronization of the code generators in both the mobile unit and the base station.

The power control problem discussed earlier is addressed through a similar mechanism. Because the mobile unit has analyzed the path characteristics of the transmission from the base station to itself, it has a good idea of the channel attenuation. Therefore, it uses that information to adjust its output power to meet the requirement that received mobile unit transmissions all reach the base station at similar power levels.

■ BLOCK INTERLEAVING

The IS-95 standard uses block interleaving for exactly the same reasons as were described in the GSM section. Deep fades are a problem in any air interface, and block interleaving is a simple approach that enhances the functionality of the error detection and correction features of the standard. Simply put, interleaving ensures that bursts of errors caused by deep fades or other mechanisms do not affect consecutive bits, thereby rendering the errors more random. Randomness is accomplished by interleaving the data such that logically consecutive bits are not physically consecutive while in the channel. The details of the block interleaving are complex and can be found in the standard. Suffice to say that the interleaver operates on a frame level. Each frame of data is broken into sixteen blocks of 1.25 ms each (16 × 1.25 ms = 20 ms). These sixteen blocks are then placed back into a frame in a specific sequence.

A side benefit of interleaving is that it also serves as a sort of extra layer of security. While this extra layer of security is of less importance in an air interface approach such as CDMA, it still is important given the concerns of many about the privacy of their conversations, be they speech or data, when conducted over radio channels.

■ SUMMARY

In the last three chapters, three basic approaches to cellular communications have been examined. Each of these approaches have been classified by their air interface. Although each system has its own name, and sometimes several, cellular systems can also be classified from a technology evolution point of view rather than by air interface or name. It is possible to view many of these systems as generational. The first-generation systems include systems such as AMPS, EAMPS, TACS, JTACS, and ETACS. These systems use fundamentally analog forms of access, FDMA, and analog forms of modulation. Hence, the phones are referred to as analog phones.

The second generation, or 2G, moved from an analog access scheme to a digital one, TDMA. This group includes standards such as the various GSM standards, DAMPS, and PDC. These phones use both TDMA access methods and digital modulation techniques, and the phones are referred to as digital phones.

Where to place IS-95 systems is problematic. Often they are described as 2.5G. They utilize a digital modulation technique, just as 2G phones do, and have frames, but they also incorporate a fundamentally different access technique, CDMA, and feature soft handoff and sophisticated power control mechanisms. CDMA is the preferred air interface for many 3G systems, and hence a case can be made that IS-95 systems are evolutionary themselves.

REVIEW QUESTIONS

1. Explain why the FRF of the CDMA system is ideally 1.

2. Explain why the number of channels in the CDMA system is more than other systems, such as AMPS and GSM, in the same frequency allocation.

3. How does an IS-95 CDMA system synchronize its base stations?

4. Describe four types of forward channels.

5. Describe two types of reverse channels.

6. Describe the purpose of the pilot channel.

7. Describe the purpose of the sync channel.

8. Describe the purpose of the paging channel.

9. Describe the purpose of the access channel.

10. Give an example of a broadcast control channel and a point-to-point control channel.

11. Explain why power control is so important in CDMA systems.

CHAPTER

14

Cellular Data Services

■ INTRODUCTION

Cellular telephone networks were originally conceived and engineered for voice-only operation, as an extension of the wireline public switched telephone network (PSTN). Mirroring the growth of data over the wireline system, data traffic has grown over the wireless cellular networks as well. At first, growth was managed by using modems attached to analog cellular telephones. Most recently, packet data networks appeared, first as overlays to the cellular voice network and then as integrated logical channels.

This chapter describes several schemes in detail in these last two catagories, from the standard widely deployed in the United States, cellular digital packet data (CDPD), to what were originally developed for GSM systems, the short message service (SMS) and general packet radio service (GPRS). Both SMS and GPRS will be overlaid onto many cellular systems in the United States and around the world as stepping-stones toward 3G data services. While many would not consider SMS to be a data service, in my opinion, this perspective overlooks a potentially inexpensive, admittedly narrowband, alternative.

Both EDGE and GPRS are packet-based data services. Both standards were originally intended to offer additional and upgraded data services to GSM networks. Now, either the EDGE or the GPRS protocol, or both, may be used to enhance FDM/TDMA based cellular systems to offer higher data rate services. The latest trends indicate, however, that EDGE is losing momentum to GPRS. For this reason, GPRS rather than EDGE receives the bulk of the attention in this chapter.

All cellular data services fall into two categories: circuit-switched and packet-switched. The circuit-switched category is further subdivided into two subcategories: analog circuit-switched data and digital circuit-switched data. After dispensing quickly with traditional circuit-switched techniques, the bulk of the chapter discusses packet data techniques.

This chapter introduces many new protocol elements unique to packet data services. As is typical in such treatments, the chapter can become "abbreviation dense." Many new abbreviations are introduced in this chapter, and most of them will not be spelled out for you here. You can refer to the appendix to define this new material.

■ CIRCUIT-SWITCHED DATA SERVICES

Analog Circuit-Switched Data

Analog circuit-switched data transport is implemented by using your cellular telephone as an analog modem and transporting data in those time slots or frequency ranges that are normally used to transport voice. Of course, facsimile data can also be transmitted this way. Subscribers who require only an occasional data transport from a computer or facsimile use the voice time slots to provide data transport instead of voice using digital circuit-switched services. Subscribers who require regular data transport or other specialized services should choose a packet-switched data service.

Analog circuit-switched data services are available only from an analog cellular network, such as AMPS, where an analog channel is available. To use this data service, an attachment (which is nothing more than an analog modem) is inserted into the cellular telephone and dials another computer or facsimile machine. This arrangement is illustrated in Figure 14-1. It is simply an analog modem based data transfer and employs traditional means of error correction and data compression using the MNP family of protocols. Typical throughputs are in the range of 4,800–9,600 bps. This technique works fine as long as the radio channel remains high quality. It is possible to improve the error performance by lowering the bit rate. For each order of 2 that the data rate is reduced, approximately one order of magnitude improvement of the error rate will occur, to a

Figure 14-1
Analog Circuit-Switched Data

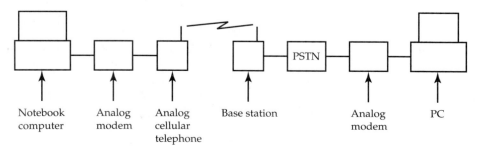

limit of about 1 error in 1 million bits. In practice, well-engineered cellular systems, combined with sophisticated forward error-correction schemes, result in data transfers that are reasonably clean and reliable. On the other hand, some systems that work well for a voice call do not provide the same level of performance for the data connection.

Making use of data compression overlays on the end systems is a potential optimization technique for what are fundamentally private networks. Use of these techniques can result in a fourfold increase, up to approximately 36 kbps, of throughput on analog circuit-switched data links. These practices are well known and make use of such protocols as V.43bis and the widely used Microcom networking protocol (MNP) element 5, MNP-5, long applied successfully in wireline analog modem data links. Optimally, these techniques would be implemented in only that portion of the link between the base station and the modem attached to the cellular telephone. As a practical matter, they are typically installed at the two end devices. Finally, charges for such data transport are independent of the amount of data transported and depend on the duration and distance of the call.

Digital Circuit-Switched Data Services

This section will explore systems where the cellular air interface is FDM/TDMA or CDMA and the voice is digitally encoded prior to modulation. This domain is much larger than the one discussed in the previous section and is capable of being provided by any of the cellular standards in which the voice data is digitally encoded prior to transmission.

The attachment device is used to convert the digital information output by the computer or facsimile machine attached to the mobile phone to the frame format of the wireless standard. This attachment device is commonly mistaken for a modem because it simulates a modem by sending signaling tones that would be expected from a modem connection. It is often integrated into a PCMCIA card and marketed as a "digital modem" or sometimes a digital phone card. The usual configuration is to plug the cellular telephone into a laptop computer via the "digital modem." The cellular link then provides the wireless communication to a base station that connects into the wireline network. This situation is illustrated in Figure 14-2.

Figure 14-2
Digital Circuit-Switched Data

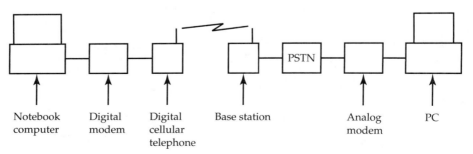

| Notebook computer | Digital modem | Digital cellular telephone | Base station | | Analog modem | PC |

Data transfers in digital circuit-switched cellular networks are performed in two distinct ways: transparent node connections and nontransparent node connections. Transparent connections do not use any additional protocol layer, while nontransparent connections use a protocol called radio link protocol (RLP). Transparent data transmission is the transfer of data without the involvement of the network. In other words, there is no *protocol level* contribution to enhance the performance of the radio channel.

Transparent data transfer is at a fixed theoretical rate of 9,600 bps and an error rate of 1 error in 1,000 bits. Conceptually, this data transfer is analog modem based over a digital air interface and employs traditional means of error correction and data compression using the MNP family of protocols. Again, this method works well as long as the radio channel remains high quality. When the channel error rate becomes too high, one of two events usually occurs: either a handoff to a better performing channel pair, or, if this method is unavailable, circuit breakdown. Again, in practice, well-engineered digital cellular systems combined with sophisticated forward error-correction schemes result in transparent node data transfers that are quite clean and reliable.

If the channel cannot support the desired data rate, an entirely different mechanism is employed to obtain the maximum data rate possible. This second technique is applied to channels that are not of sufficient quality. These channels are called nontransparent channels because sophisticated, nontransparent methods are employed to enhance the data throughput. Nontransparent communication uses protocol layer actions (selective retransmissions and ACK/NAK handshaking) to enhance the performance of the radio channel artificially.

On a per-channel basis, nontransparent data transfer is also capable of a maximum of 9,600 bps throughput. If several time slot channels are combined, much higher aggregate data rates are possible. Nontransparent data transfer requires the use of RLP. The vast majority of digital circuit-switched data services operate in a nontransparent manner and use RLP. The protocol stack used in the CDMA IS-95A standard for asynchronous data transfer is shown in Figure 14-3. This protocol stack assumes TCP as layer 4 and IP/PPP (point-to-point protocol) as layer 3.

Figure 14-3
IS-95A Asynchronous Circuit-Switched Data Protocol Stack

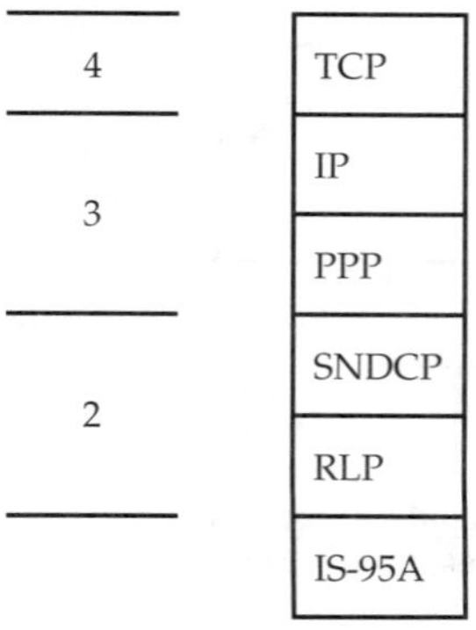

4	TCP
3	IP
	PPP
2	SNDCP
	RLP
	IS-95A

RLP is an error-correcting protocol installed in both the base station and the cellular telephone. The next section describes RLP, which is used to enhance the performance of the radio link and effectively repeats RLP frames that do not make it through. RLP uses the existing logical control channels embedded in the cellular system.

■ RADIO LINK PROTOCOL

Radio link protocol (RLP) makes use of sophisticated error detection and correction and variable data rates to provide an error-free interface to the network layer. RLP is standardized as IS-130. It is used to support CLS nontransparent data transfer. It cannot use external data compression techniques in the same way as transparent data transfers can. This drawback limits the data rate using a single time slot to a maximum of 9,600 bps.

RLP does it own data compression using the V.42bis standard. It also inserts error-detection and error-correction bits into the data packet prior to transmission. Therefore, RLP still does a transformation on the data, but it is a code conversion rather than an analog-to-digital conversion and back again. With RLP, the interface to a PC is a conversion from one form of digital representation to another. RLP is best suited to situations where the mobile unit is in fact mobile, as in a car. Because of the variable data rate protocol approach, RLP copes well with the intermittent connection that occurs during handoff from one MSC to another, which typically takes between 100 and 150 ms.

RLP also works well when the radio channel is noisy or weak because a single bit error in RLP requires retransmission only of the affected frame. The same error occurring with transparent data transfer mode, when making use of data compression techniques, will need several frames retransmitted, depending on the error-correction block sizes. Depending also on the implementation, the number of frames that will need to be retransmitted when using data compression varies from two to fifteen.

RLP is a full-duplex protocol that supports 9,600 bps data service to a mobile unit with the levels of transmission quality that higher layer protocols, especially layer 3 and 4 protocols, have traditionally assumed. In field tests, RLP has delivered typical BER rates of 1 bit in error in 1,000,000 bits. These numbers are very similar to what is experienced in the traditional POTS environment.

RLP is widely used today in most cellular networks to enhance a circuit-switched data service. It is very likely that TCP/IP networks will use RLP to ensure a reliable data stream from layer 2 as well. As you will see below, RLP works by selectively applying retransmissions, using automatic repeat request protocols (ARQ), to achieve a low error rate at the end points. This retransmission is hidden from the upper layer protocols and hence can act to increase significantly the perceived latency of the transmission. This example is part of the larger issue of how to adapt upper layer protocols to the inherent differences of wireless networking, a topic discussed in more detail in Chapter 16.

RLP makes use of rate adaptation. Rate adaptation is a way of adapting the rate that the wireless channel can provide to the rate that the end equipment expects. The MSCs are designed to switch DS-0 digital channels at a rate of 64 kbps. The radio interface to mobile subscribers will not always support this data rate. Functionally, RLP is a layer 2 relay. It breaks the 64 kbps into several smaller time slots of lower data rates. These data

Table 14-1
RA0 Rate Adaptation Speeds

n	Data Rate (bps)
0	300
1	600
2	1,200
3	2,400
4	4,800
5	9,600

rates are all multiples of 300 bps. There are three functional levels of this process: RA0, RA1, and RA2. After a brief description of the three functions, a discussion of how they work together to accomplish the rate adaptation will be presented.

The RA0 function is a simple data conversion utility. It is the lowest level of rate adaptation and is the last step before providing the data to the mobile user. It operates independently of any QOS flags in the header of the data. It converts an asynchronous data flow into a synchronous data flow at a rate of

$$(2^n)(300 \text{ bps})$$

The RA0 function defines the potential rate adaptations that the mobile user will see. The possible data rates are shown in Table 14-1.

The RA1 function takes into account the QOS flags in the header of the data if they are placed there by the LLC portion of the link layer. The RA1 function takes into account the status of the mobile interface, and various control information and synchronization patterns within the frame structure hierarchy. The RA1 function is the intermediate step between the RA0 and RA2 functions. It multiplexes information at an input rate of 64 kbps and outputs data at an intermediate rate of either 8 kbps or 16 kbps.

RA2 exists only in the base stations. Through bit stuffing, it adjusts the bit rate to 64 kbps. As mentioned above, the MSC can switch only 64 kbps segments. If there is insufficient data traffic to provide this amount, the RA2 function bit stuffs until it satisfies this constraint of the wireline system. Figure 14-4 shows the layer 2 RLP components and indicates virtual communication with dotted lines. Note that the BSS acts as a relay for the virtual communication between the MS and MSC because the protocol layer above RLP, such as PPP, does not exist at the BSC. Finally, since RLP is often used for data transfer, it will often be the IWF, which is co-located with the MSC, where the actual protocol communication takes place.

Essentially, RLP breaks up the 64 kbps data slots into smaller slots that can be supported by the radio channel. It then repeats the blocks a sufficient number of times to meet the throughput QOS parameters required by the upper layer protocols. The actual frame structure is somewhat similar to the HDLC template because it features a header, data field, and FCS field. The total length of the frame is 30 bytes or 240 bits. In this

Figure 14-4
RLP Virtual Communication

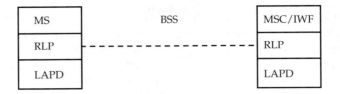

Figure 14-5
RLP Frame Structure

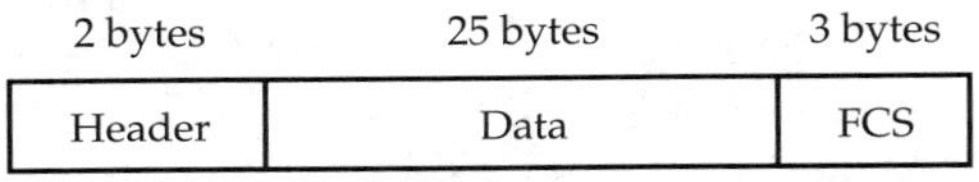

2 bytes	25 bytes	3 bytes
Header	Data	FCS

frame, the data field is 25 bytes, the FCS field is 3 bytes, and the header field is 2 bytes. There are three classes of headers; not coincidentally, there are also three classes of frames in the GSM standard: the I-frame, S-frame, and U-frame. Figure 14-5 shows the RLP frame structure.

Note that the frame structure differs in some significant ways from the HDLC structure. First, it uses no start or end delimiters. The FCS field is stretched to 3 bytes from 2. Perhaps most important, there are no source or destination addresses! The frame structure is used in only one communication at a time. Hence, no addressing is needed.

■ DIGITAL PACKET DATA SERVICES

The fundamental difference in a packet data service compared to a circuit-switched data service is that packet data services allow several subscribers to access multiple locations in the wireless/wireline network. Each packet is individually addressed and routes to its destination in the fastest way possible. In a circuit-switched data service, the data is connection oriented and FIFO is guaranteed. In a packet-switched data service, CLS rules the day and some packets may be lost, arrive out of order, etc. Packet sizes on wireless networks range from a minimum of about 100 bytes to 1,500 bytes of data.

Packet data services are best utilized when the application is bursty in nature. Examples include credit card authorization and database queries, especially in a remote sales environment. Charging for packet based services is based on actual data transported, not call duration. Typically, a user will subscribe for a certain number of kilobytes of data per subscription period. If this level is exceeded, additional data volume is charged on a per-kilobyte basis. A good example of a device that uses packet data is a personal digital assistant (PDA) like the Palm 5, for which several attachments are available with a subscription to CDPD at a bulk purchase discount.

Cellular Digital Packet Data (CDPD)

CDPD was developed for use in the AMPS network. It is essentially a connectionless data network overlay on an analog cellular network. CDPD is a technology that can double the typical data rate traditionally offered by a cellular network, from 9.6 kbps to 19.2 kbps. Additionally, and perhaps more important, it is billed to the subscriber on a usage basis rather than an airtime basis. This feature was very attractive to subscribers because it reduced the cost to transfer data over a mobile network. CDPD was developed to allow subscribers remote access to their e-mail, to provide database queries, and to facilitate mobile point-of-sale terminals.

Therefore, use CDPD for local or regional data applications with no long files to transfer and no need for voice communications, and where communications are from the field to the office. Best application classes are when data is of the short, bursty type, for example, telemetry, point-of-sale terminals, database queries, etc. It is not designed for large user data packets. For reasons that will be explored below, the maximum user data size is 2,048 bytes. When it is used with TCP/IP protocol stacks, as it often is, the maximum IP packet size must be set to 2,048 bytes.

AMPS uses 30 kHz channels for voice traffic; CDPD uses those channels to transport packet data. Recall that AMPS has no time slot structure to accommodate flexible control channels; if it offers a data service, it must use its voice channels. Because they are fixed and predetermined, the control channels are not flexible in application. They are dedicated control channels, whether there is one user in the cell or 100. Contrast this structure with a more flexible mapping of control channels to physical channels, like that used in IS-136 or GSM, and new possibilities emerge.

CDPD makes use of unused voice channels for data transport. It requires a specific PCMCIA card for the mobile device, which doesn't have to be a cellular telephone. CDPD is a packet data service overlaid on the voice service. A typical PCMCIA card that implements the CDPD service is in the range of $300.

Cost for CDPD service varies widely with the type of pricing plan. Typical charges are between 10 and 15 cents per kilobyte of data transferred. As a general rule of thumb, if transferred files grow to an average size of over 5 kbytes, CDPD is not cost effective. At that point, the right choice is often some form of circuit-switched service. Again, CDPD is best suited for small bursty transmissions, not file transfer or browsing on the World Wide Web. E-mail delivery is a potential application fit if the mail does not contain a significant number of attached files.

CDPD is advertised as being capable of data rates of 19.2 kbps; actual data rates will usually lie between 8 and 16 kbps, depending on the number of subscribers in any cell currently in voice conversation. Experience suggests that smaller packet sizes will improve performance. Breaking data frames into packets of between 400 and 500 bytes should be close to optimum.

Each packet is addressed individually so that there are no call setup times. Typical response times are between 1 and 5 seconds. The CDPD response will often be almost instantaneous. On the other hand, if the cellular system has no free channels to use for CDPD, you may wait for some time. In essence, you are riding second class with CDPD. In all cellular voice systems such as AMPS, voice calls always have priority. Having said that, experience suggests that the vast majority of CDPD-equipped cells are underutilized.

Because voice calls have priority, CDPD data transfers usually exhibit a phenomenon known as channel hopping. Channel hopping occurs when a new subscriber requests voice service from the cellular system and the particular channel the CDPD data is using is now preempted for this new subscriber's voice conversation. The packet data flow is then interrupted until a new channel can be found and the CDPD packets transfer to it.

As pointed out above, CDPD uses any open voice channel that appears, so traditional scanning may not be able to eavesdrop on data tranfers. For this reason, CDPD data transfers offer some additional security over voice conversations in the same cellular system. Because CDPD travels second class, the actual frequency channels change occasionally during any data transfer because a voice call takes priority for that particular frequency pair. To an eavesdropper using some type of scanner, the CDPD data transfer appears to be frequency agile, thus frustrating his or her attempt to monitor data transfer illegally. In the AMPS environment, a voice call, because it is connection oriented, makes use of the same channel pair throughout the call duration. In other words, the frequency assignment stays static throughout the conversation, making scanner based eavesdropping easy.

The vast majority of CDPD applications are IP based. In this scenario, each mobile station has a static IP address and is a true internet host. Again, there is no dialing to the cellular system. CDPD uses an IP address to access the net. A static IP address can cause routing issues to arise due to the mobile nature of the mobile end device. This subject is addressed more fully in Chapter 15. An advantage is that CDPD is based on IP, so no protocol translation is required, as is the case for most wireless data services. CDPD can also do encryption based on the data encryption standard (DES) and can provide security using the security management entity (SME) protocol.

From a cost perspective, the IP nature raises another problem because TCP/IP is very ACK oriented. Because CDPD charges by kbyte, all those ACKs can be expensive. ACKs also take bandwidth, which slows down data transfer. The solution is to modify the TCP software to get rid of some of these ACKs.

Finally, although CDPD was developed for use in the analog AMPS network, it is now widely applied as an overlay in other non-GSM cellular systems. GSM has its own GPRS standard to offer packet-switched data services. Specifically, CDPD is available in most DAMPS networks using either the IS-54 or IS-136 standards. For easy comparison, the two protocol stacks for AMPS and DAMPS are shown in Figure 14-6.

The next sections will discuss a CDPD implementation from two perspectives. First, we will examine the physical architecture of a CDPD implementation. Then we will conduct a layer-by-layer protocol analysis.

■ CDPD IMPLEMENTATION

Any CDPD service implementation is in partnership with a cellular provider. A CDPD implementation has five elements, of which three are unique to CDPD. The first unique component is the mobile end station (MES). The MES can be a cellular telephone,

Figure 14-6
CDPD AMPS and DAMPS Protocol Architecture

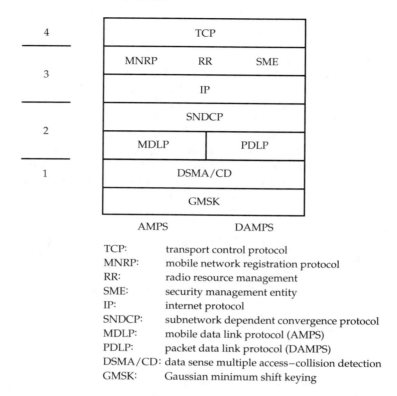

TCP: transport control protocol
MNRP: mobile network registration protocol
RR: radio resource management
SME: security management entity
IP: internet protocol
SNDCP: subnetwork dependent convergence protocol
MDLP: mobile data link protocol (AMPS)
PDLP: packet data link protocol (DAMPS)
DSMA/CD: data sense multiple access–collision detection
GMSK: Gaussian minimum shift keying

computer terminal, or laptop computer. The MES units communicate through the cellular system air interface with the second unique component, a mobile data base station (MDBS). The MDBS is co-located with the cellular base station. It functions as the CDPD radio channel controller and is responsible for the management of the radio interface to the MES.

The MDBS to MES communication makes use of spare voice channels, so it is composed of two simplex channels. These channels are referred to as a CDPD channel pair. As you should expect, the forward channel is from the MDBS to the MES, and the reverse channel is from the MES to the MDBS. Therefore, packets relayed to the base station for transmission to an MES use the forward channel, and packets from the MES to the wireline system use the reverse channel. This forms a full-duplex channel.

The MDBS uses the third unique component to CDPD to interface to the wireline network: a mobile data intermediate system (MDIS). The network architecture is assumed to be TCP/IP, so the MDIS interfaces between the internet and the MDBS. The MDIS is responsible for all mobility management: it provides the necessary routing to the MDBS to which the MES is currently local.

Figure 14-7
CDPD Network Architecture

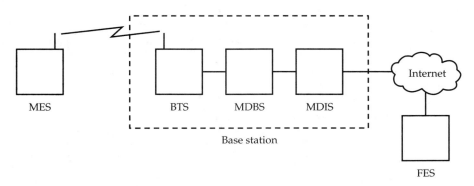

Figure 14-7 shows all five components of CDPD. Three are unique to CDPD: the MES, MDBS, and MDIS. The remaining two components are the cellular base station and the fixed end system (FES). The FES is typically the server to which the MES is exchanging data. A good analogy is to compare the MES to a client and the FES to a server in a client/server architecture.

Mobile End Station (MES) and Fixed End Station (FES)

In general, an MES functions as a client and is mobile. As a result, its IP address does not imply a location. Any MES must support all the functions necessary to perform reliably in a mobile RF environment, including channel registration and acquisition and all other radio resource management tasks (the normal functions of a cellular telephone).

Because an MES IP address is not associated with a physical location, the CDPD network must track all MESs and route IP packets to them. An MES has a true host IP address; it is not associated with the subscriber's network address range. Instead, the MES is associated with the CDPD network's routing domain. This setup has significant protocol implications and will be discussed in more detail later.

In principle, an FES may be owned by either the cellular service provider or the subscriber. When owned by the service provider, it is called an internal FES; when it is owned by the subscriber, it is referred to as an external FES. In the former case, it is inside the security firewall provided by the service provider. In most cases, however, the FES is external to the CDPD network. Typically, it is a database or application server owned by the subscriber and is used to implement the application provided to the client MES.

Mobile Data Base Station (MDBS)

As mentioned above, the mobile database station's primary task is to control the radio interface. It is responsible for radio channel allocation and the media access control

(MAC) interface to the MES. It is always co-located with the cellular voice equipment in a base station. The RF channel pair or CDPD channel pair is composed of two logical simplex links: the forward link from the MDBS to multiple MESs, and the reverse link from multiple MESs to the MDBS. Data packets are modulated and demodulated and relayed appropriately by the MDBS. As you will see in the protocol section, the MDBSs function as bridges and provide a layer 2 relay function.

Mobile Data Intermediate Station (MDIS)

The mobile data intermediate stations provide the logical interface between the wireline and wireless environments. On the wireline side, the MDIS communicates directly with other IP capable devices using standard IP addressing and routing. On the wireless side, the MDIS interfaces to devices that are unique to CDPD, namely, the MDBSs. Since each MDBS is associated with a cell and each cell can be seen as its own routing subdomain, the MDIS also communicates directly to the MESs that happen to be local to the MDBS. This communication is at a higher protocol level than the communication with the MDBS, which simply relays the IP packets between the two entities.

The MDIS is the only network element that has knowledge of mobility, for it must somehow route packets to the MES using the appropriate MDBS. It does so by making use of two functions: the mobile home function (MHF) and the mobile serving function (MSF). Each MES has a listing in one MDIS identifying it as its home MDIS. The MHF contains a list of all MDIS addresses that are currently serving any MES on its home list.

All incoming packets are initially routed to the home MDIS. If the MES is not currently at home, the MDIS consults the MHF and encapsulates the packet, routing it to the MDIS in which the addressed MES is currently. When a subscriber is not in his or her home MDIS, packets take nonsymmetrical paths. In the forward direction, the path is potentially much longer than in the reverse direction because any packets originating from the subscriber will proceed directly to the application server, while those on the forward path (returning the data to the subscriber) will proceed first to the home MDIS and then be forwarded to the current MDIS. See Figure 14-8.

The registration is accomplished using mobile network location protocol (MNLP). It is very similar to other mechanisms described in this text: mobile IP's home/foreign agents. The mobile home function is like the home agent processing in mobile IP and uses encapsulation to forward packets to MDIS in the visited region. The mobile serving function is like the foreign agent processing in mobile IP. It does registration, authentication, authorization, etc.

CDPD roaming between cellular systems is supported only if the two systems have prior roaming agreements, which usually involves both legal and technical issues. For example, if a CDPD subscriber were to roam between two CDPD service providers, the two providers would have to have a prior agreement supporting roaming for a seamless handoff to occur. If one of the networks employed the IS-95 standard and assuming that the MES was capable of operating with both air interfaces, then both providers would also have to implement the shared secret data (SSD) authentication protocol used in the IS-95 standard.

Figure 14-8
Path Differences When Roaming and Using CDPD

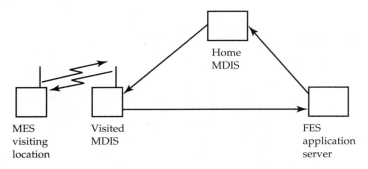

■ CDPD PROTOCOL ARCHITECTURE

The protocol architecture of CDPD is composed of two main segments. Following the TCP/IP model, the network access, internet, and host-to-host layers will be described. Subsequently, the process/application protocol components will be described. We will begin the discussion with layer 1 and move up the stack. Figure 14-9 illustrates the lower four layers of the protocol architecture. In the figure, the MDIS is assumed to be connected to an Ethernet segment running IEEE 802.3. Note the similarity in structure to the GSM protocol architecture except that here a TCP/IP protocol stack is used. In the GSM protocol architecture, an SS7 protocol stack was used.

CDPD Physical Layer
CDPD uses Gaussian minimum shift keying (GMSK) modulation with a raw data rate equal to 19.2 kbps. This data rate explains the basis for advertising CDPD as a 19.2 kbps data service. As stated above, typical throughput values will be significantly less because of occasional dropped packets but also because the Reed-Solomon FEC coding steals data bits for error detection and correction. GMSK is not the modulation technique used in AMPS or DAMPS. Because AMPS occupies its own frequency slot, however, this setup presents no difficulties except that the service provider must provide dual-mode modems for those cells offering CDPD service.

CDPD is restricted to using a pair of analog or digital TDMA cellular channels for each physical CDPD channel. The specific physical layer functions, besides the obvious functions of modulation and demodulation, include the ability to tune to a specified pair of RF channels in a frequency agile manner, set power levels, measure signal strengths, and suspend/resume monitoring of RF channels in the mobile device to conserve battery power.

CDPD Data Link Layer
The data link layer of the CDPD protocol stack is composed of three sublayer protocols: the MAC sublayer, the MDLP sublayer, and the SNDCP sublayer. Each will be discussed in the following sections.

Figure 14-9
Layers 1 to 4 of a CDPD Protocol Stack

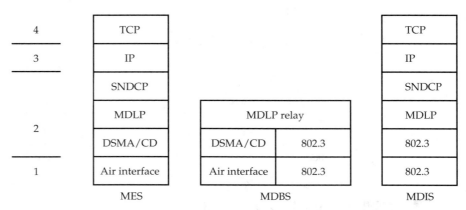

MAC Sublayer   The MAC sublayer in CDPD is based on the protocol data sense multiple access with collision detection (DSMA/CD). DSMA works much like CSMA because it depends on a detection mechanism to determine the status of the channel (to determine if it is busy or idle). However, because there is no wireline interface between the MES and the MDBS, which is the location where this protocol operates, it looks for the status of a bit in the DSMA frame to indicate the status of the channel. This bit is set at the MDBS, and the MESs wait for idle indication prior to beginning transmission.

Of course, as in Ethernet, collisions can occur when two stations begin transmitting almost simultaneously. Again, as in Ethernet, the voice channel currently allocated for reverse direction CDPD transmissions is shared among many MESs. Therefore, some kind of collision detection algorithm must deal with this issue. Note that the forward channel is not shared. No collisions occur in the forward direction because there is only one MDBS in any cell.

So the MDBS listens to the allocated channel for CDPD. If it is idle, MDBS sends out a frame with the "busy bit" set to indicate that the channel is idle. All MESs in the cell hear the idle condition at approximately the same time. As soon as the MDBS senses a transmission on the reverse channel, it indicates on the forward channel that the channel is busy. If only one MES has data packets to send, there is no collision and the packet is received, processed by the MDBS, and passed to the MDIS for delivery. As soon as the MDBS has received the packet and decoded it correctly, it indicates on the forward channel that the packet was received correctly, and the MES knows that its packet is being received correctly. It then continues to transmit packets until it has exhausted its data.

If more than one MES has a packet to transmit, a collision may occur. The only entity that can determine a collision is the MDBS. It does so by examining the packet and trying to decode it. If the bit error rate (BER) is above some predetermined threshold, it assumes that a collision has occurred. The forward channel frame now includes data that indicates the channel is not only busy but the packet was received in error. This data indicates to the MES that a collision occurred. Very similarly to how CSMA/CD works, the MES then goes to its waiting state, using an exponential backoff, until it hears that another idle period has occurred. Then it attempts transmission again.

As must be clear from the above discussion, the frame format of the MAC layer is different in the forward and reverse directions. Both frames use a Reed-Solomon code (63,47) to perform error detection and correction. A 63,47 code indicates that for every forty-seven data symbols, there is actually a total of sixty-three symbols sent, the excess being parity bits used to detect errors in transmission. Both frames arrange data into sixty-three symbol blocks. Each data symbol is represented as 6 bits. This arrangement results in a total of $47 \times 6 = 282$ bits transmitted for every forty-seven symbols of actual data.

In the forward channel, the same frame is transmitted continuously. In the reverse channel, up to three data blocks are transmitted concurrently by each MES. If the MES still has data to transmit, it must then wait for another idle period. This setup underlines why CDPD is best suited for small data packets. Only the forward channel frame will be examined in some detail to illustrate how it signals the MAC protocol and implements the DSMA/CD algorithm.

The forward channel frame consists of a total of seventy data symbols, each of which is 6 bits in length. Every tenth symbol is a control field used for synchronization, the busy/idle indicator, and the successful/unsuccessful decode indicator. Because there are seven control symbol locations, it is clear how the sixty-three symbols from the Reed-Solomon coding combine with the control field symbols to generate seventy symbol frames. Note that the purpose of the control fields is not to represent data but signaling for the protocol. Thus, they are not Reed-Solomon coded and hence do not expand the frame the way additional data symbols would. Because the forward channel is dedicated to MDBS broadcasts, this forward channel frame is sent continuously. Figure 14-10 illustrates the symbol structure of the forward channel frame.

The control fields are broken into three logical segments. One bit from each control field is used to form a 7-bit busy/idle flag. One bit from each control field is used to form a successful/unsuccessful decode flag. The remaining 5 bits are used to form a 35-bit synchronization word that maintains synchronization between the many MESs and the MDBS (and provides the master clock for the system).

The first 8 bits in each block transmitted on the forward channel by the MDBS consist not of data but a color code for identifying the cell frequency set. Each adjacent cell has a different color code corresponding to the frequency set assigned to that cell in the overall frequency reuse strategy. When cells are in the same cluster and controlled by the same MDIS, the first 3 bits of that color code are identical and are referred to as an area color. (*Cluster* and *area color* are synonymous terms.)

MDLP Sublayer The next sublayer up in the data link layer of the CDPD protocol stack is the mobile data link protocol (MDLP). MDLP has one prime function: it compensates, to the degree possible, to create what appears to the sublayer above an error-free

Figure 14-10
CDPD Forward Channel Frame Structure

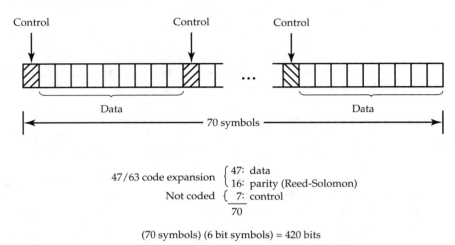

$$47/63 \text{ code expansion} \begin{cases} 47: \text{ data} \\ 16: \text{ parity (Reed-Solomon)} \end{cases}$$
$$\text{Not coded} \begin{cases} 7: \text{ control} \\ \overline{70} \end{cases}$$

(70 symbols) (6 bit symbols) = 420 bits

communications channel. It establishes a connection-oriented service with sequence numbers, retransmissions, and flow control. MDLP also offers an unacknowledged service used only when a response is not expected, such as in broadcast or multicast transmissions. As a secondary function, it also helps to minimize power consumption (always a problem with mobile devices) by providing a sleep function.

If you reexamine Figure 14-9, you will see that the MDLP sublayer is the first protocol layer where the connection exists between the MES and the MDIS directly. The MDBS, located between these two entities, simply relays the MDLP frames between the two endpoints. When reading the previous section on DSMA/CD, some readers may have wondered about the location of the frame definition. They may also have wondered why all the Reed-Solomon coding was necessary. After all, doesn't a link layer frame always include an FCS field? The answer is that the frame definition is provided in the MDLP sublayer. The coding performed in the DSMA/CD layer is substituted for the CRC coding normally done in the classic HDLC FCS field. Figure 14-11 illustrates the MDLP frame format.

This lack of a frame check on the frame as a whole presents an interesting problem when deploying CDPD. Without a check on the address or control field, and the radio channel being a high error rate environment, either field could easily become corrupted during transmission. Because of this architectural choice, it is essential that these errors be addressed by a higher layer in the protocol stack. Many approaches are possible, including the use of additional layer 2 mechanisms such as frame relay or some flavor of LAPB.

The address field in the MDLP frame is a bit complicated. It represents the destination address of either an MES or MDIS. It can facilitate point-to-multipoint connections

Figure 14-11
MDLP Frame Format

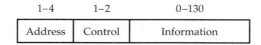

in the forward (MDIS to MESs) direction and point-to-point connections in the reverse (MES to MDIS) direction. The actual addressing format in the field is closely coupled with the concept of a terminal equipment identifier (TEI). The TEI is a variable bit-length number, from 6 to 27 bits, that uniquely identifies each MES.

The TEI is assigned by the MDIS that is serving the MES. This assignment is the first step in establishing a connection between the two devices. The details of how the variable-length TEI assignment is folded into the 4-byte address field are similar to the method used elsewhere. For the purposes of this text, two things are important. First, the last bit (the extension bit) in each byte of the address field is used to indicate whether the next byte is needed. If the value of this bit is a 0, then more addresses follow. If it is a 1, then the next byte begins the control field. Second, the second-to-last bit in the first byte has special meaning. It is used to indicate if the frame is a command or response. This identification can be inferred by the exchange of commands and responses between the MES(s) and the MDIS; however, it is placed here, following the convention used in the LAPD frame in ISDN.

The next field is the control field and has the familiar purpose of defining the type of information contained in the information field. As indicated earlier, two levels of service are offered to the next layer up the stack by the MDLP protocol. These levels of service are a connection-oriented service (COS) and an unacknowledged connectionless service (CLS). The latter is less complex so it will be discussed first.

The CLS service is used by the MDIS only when no response is expected from the MES(s). Typical examples include broadcasts and multicasts. It may be surprising to learn that TEI assignment also uses this service class. The reason is that upper layer protocol elements implement the TEI management and will respond and confirm this assignment. All unacknowledged CLS frames have the same control field value and are called unacknowledged information (UI) frames.

When the communication carries data, it always makes use of the COS. All COS frames are point-to-point and may use either direction. All such frames have sequence numbers in the control field, and the MDLP assumes responsibility for any retransmissions or flow control using these sequence numbers. Seven bits are assigned for sequence numbers, indicating 127 possible values. When this maximum value is reached, the sequence number counter rolls over and begins with 0 again. Of course, the destination system(s) use these sequence numbers to reassemble any out-of-order frames received, as well as to detect missing or lost frames.

One of the most interesting details of how MDLP works is the way it accomplishes recovery from a lost frame. In most protocols designed for wireline application where the medium is reliable and robust, the destination usually requests that the source begin retransmission with the first frame lost, retransmitting from that point on. Depending on the flow control settings, which specify the window size, many frames may have been transmitted from the source prior to receiving the retransmission request from the destination.

MDLP uses an alternate approach called selective retransmission. In this approach, the retransmission request is only for the lost frame. Because the radio channel exhibits a higher error rate and lower bit rate compared to a wireline channel, selective retransmission results in far fewer retransmitted frames and significantly improved end-to-end performance. Selective retransmission is not unique to CDPD. It is also used on some satellite links that feature very large windows and long propagation delays. Here too it can significantly improve overall end-to-end performance.

Approximately a dozen different frames are defined in the MDLP standard. Of these, over half are unnumbered, unacknowledged frames used for link layer control and testing. Three frames are also defined for specific acknowledgments when COS data transfer is being performed. These three frames are known as supervisory frames. MDLP does not use the ACK/NAK approach for these frames. Retransmissions are expected to occur often in the radio channel. If no ACK is received, the source of the frame automatically assumes that the frame was lost in transmission and retransmits. Essentially timeout acts as the mechanism to control retransmissions. Finally, there is one frame for data or information transfer.

Remember that the TEI is valid only between the MES(s) and the MDIS. Thus, a TEI value must correspond to the MAC address of any individual CDPD device. Because of the mobility of MES devices, each MDIS must be able to associate each TEI uniquely with a specific piece of equipment. It is conceivable that an MES with the same TEI as one previously registered with a particular MDIS could roam into its area. For these reasons, part of the TEI registration process is to associate each MES TEI with its 4-byte MAC address. As you will see in the next paragraph, the TEI is assigned on a temporary basis and it maps to a specific MAC address. It can be seen from a higher layer perspective as a dynamic IP address assignment.

Because TEI identification is so important, regular TEI check messages are performed as an audit process. The audit process specifically addresses the issues of what TEIs are in use and if multiple MESs are using the same TEI. Since there are far fewer TEIs than MAC addresses, conceivably they will run out. Each TEI is assigned to a device associated with a timer of four hours. When this timer expires and no frames have been received by the MDIS in that period, the MDIS sends a TEI check message twice to confirm if the device is still in range. If no response is heard, the TEI is made available for another CDPD device.

If two responses to a check message are heard, the assumption is that there are two MESs with the same TEI value. The MDIS confirms this assumption by evaluating the MAC address contained in the response to the check message. One of the MESs is then instructed to removes its TEI, and that MES starts its procedure for obtaining a TEI from the MDIS. Once an MES receives a remove TEI frame from the MDIS, all pending data transfers are purged immediately and this data is lost.

The last topic for this overview of MDLP is the sleep function. The sleep function in MDLP is associated with a data link connection. The TEI assignment is the first step in establishing a connection. During this step, an MES has the option of sleep mode for the duration of this connection. Once it completes the TEI registration period, the MES stays "awake" for the period specified in its sleep request. When this period expires, it enters sleep mode until the connection times out or the MES needs to send a frame. This length of time can be anywhere from a few minutes to a maximum of 4 hours and acts to preserve battery life.

An MDIS can send a message to a sleeping MES because, even when asleep, the MES monitors a regular broadcast that the MDIS sends out about once a minute to synchronize the system. This broadcast also contains a flag and timer indicating what TEIs have messages pending. If an MES sees that it has messages waiting for it to wake up, it does so. A return to sleep mode requires cancellation of the TEI, and the MES must begin the TEI assignment process again.

SNDCP Sublayer   The last of the three sublayers that comprise the data link layer of a CDPD architecture is called the subnetwork dependent convergence protocol (SNDCP). While SNDCP offers both connection-oriented and connectionless service options, it is almost always used as a connectionless service and will be described so in this text. In general, the SNDCP has two primary tasks:

1. Compression/decompression of redundant protocol control information.
2. Segmentation/reassembly of the NPDU.

SNDCP works symmetrically. It performs the first function listed on each line for packets sourced from the network layer above. It performs the second function for frames received from the sublayer below, where the first function has been performed by the ultimate source station. For this description, it will be assumed that a network layer packet has been passed down the stack from the layer above.

Compression and Elimination of Redundant Protocol Control Information   This function takes place in two steps:

1. Compression of the network layer header.
2. V.42bis data compression.

Compression of the network layer header uses well-known algorithms common in wireline networks. They rely on the use of the TCP/IP protocol at layers 3 and 4. These techniques flag header fields that will not change during the TCP connection and transmit them only once during that connection. The premise is that some information is needed only once during a connection and some information is already represented at the network layer.

Probably the two best examples of this approach are either to assign a connection identifier to a source and destination pair or to transmit only the difference between sequence numbers instead of the entire sequence number. In the first example, instead of transmitting the entire IP address in each packet, only the 1 byte connection identifier is transmitted, thus saving bandwidth. Only the first frame transmitted must carry the IP header. At the destination, the state of the M bit indicates the arrival of the last segment, and reassembly can occur. Recall that the MES and the MDIS use the TEI to identify themselves to each other. Only after the packet is reassembled will it be passed to the external IP environment through the destination MDIS system. The second example also saves bits and hence bandwidth in the channel because only the difference between two sequence numbers is transmitted.

While these techniques were developed for wireline implementations, they are also applicable to wireless channels where the bandwidth is an even more valuable commodity. To understand the importance of these techniques, recall that any reduction in payload overhead has a significant impact on subscriber perceptions of transmission rate. Eliminating a few bytes may not make much difference in a 100 Mbps LAN network with a frame size of 1,000 bytes, but it does make a perceivable difference in a 19.2 kbps network with frame sizes of 100 bytes.

After these techniques are applied, the payload will be compressed using V.42bis data compression. V.42bis data compression is also widely used in the wireline environment and is standard equipment on many modems manufactured today. However, it does have some unique aspects when used at the network layer. The fundamental operation of data compression is to substitute a short code word for commonly occurring character sequences. Both source and destination systems maintain a register of all such substitutions. These registrations evolve dynamically during the information exchange. New substitutions are added and those substitutions that have outlived their usefulness are eliminated.

Note that this data compression applies only to the information field of the frame. The frame is exchanged using the COS mode, as specified in the section on MDLP. Thus, the source and destination systems are easily identified, and the registration of code words is easily identified and maintained for the lifetime of the connection.

Data compression is negotiated during the TEI assignment phase. The MES and MDIS agree to use data compression or not, and decide whether it will be applied in both the forward and reverse channels or only in one of them. If both entities agree to use data compression, it will be associated with a particular data connection and the negotiated characteristics will expire when the connection expires.

Because of the length of the data field, CDPD limits the number of code words used by the MES and MDIS. This number must lie in the range of 512 to 8,192. In theory, V.42bis does not limit this value, but most implementations, wireline and wireless, do so as a practical matter by defining the length of the field where this parameter is negotiated.

Segmentation and Reassembly of the NPDU   Segmentation and reassembly of the NPDU is accomplished through signaling in the SNDCP header. The SNDCP header consists of the first 2 bytes of the information field, which is passed to the sublayer protocol, MDLP, below. The header format is illustrated in Figure 14-12. The first bit, called an M (more) bit, is used to implement the segmentation function, the second task identified in the list above. If this bit is a 1, more network layer data segments follow. If it is a 0, this segment is the last segment. A zero also identifies that no segmentation has taken place.

This single-bit approach was first used in X.25. Many modern protocols use more sophisticated approaches that require up to an entire byte of signaling. The single-bit ap-

Figure 14-12
SNDCP Header

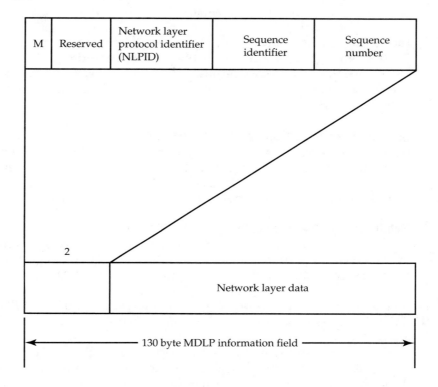

proach is economical. Given the relatively low data rates of CDPD and the maximum packet size of 2,048 bits allowed by CDPD, it is also an elegant and simple approach.

The network layer protocol identifier (NLPID) field is used to identify the particular network layer being serviced by the SNDCP. SNDCP has the capability of multiplexing several network layer protocols to the underlying data link layer subprotocols. This capability allows CDPD to identify what network layer protocol is being used. Two network layer protocol options, IP and the connectionless network protocol (CLNP), are specified in the SNDCP standard. Since this text has assumed a single network layer protocol (namely, IP) as layer 3, this capability is noted but will not be explored further.

There is another use for this field, however; it is used to identify the application/process protocol elements that may use user information fields. The application/process protocols MNRP and SME can both identify their use of the information field. This identification instructs the destination network layer to deliver these packets to the appropriate application/process protocol in the destination system.

The next byte is also used to implement the segmentation of NPDUs into 128-byte segments that can be handled by the lower layer services. The first 4 bits are used to identify the NPDU being segmented. The second 4 bits are used to keep track of the segment number. These two fields, along with the M bit described above, are used to reassemble the NPDU at the destination. Again, it is important to emphasize that the maximum IP data packet size that will work with CDPD is 2,048 bytes.

Encryption   As an option, link layer encryption can be implemented using the SNDCP protocol. Typically, DES is the preferred algorithm. With this algorithm, the SNDCP protocol uses a specific header with fields specifically defined to facilitate the key exchanges required in such an event. SNDCP encryption applies only to the information field.

CDPD Process/Application Protocols

Between the lower layer specification of the CDPD protocol stack and the process/application layers lie the familiar TCP/IP protocols. The protocol elements described in this section are located above TCP. Figure 14-13 illustrates these layers of the CDPD network architecture. As you can see, three protocol elements are defined.

Mobile Network Registration Protocol (MNRP)   The MNRP facilitates MES authorization and registration between the MDIS and MES. Without successful completion of the authorization and registration process, the MES will not have access to services offered by the MDIS. Each MES is individually registered with the MDIS through the use of a network equipment identity (NEI). This 32-bit number, for all intents and purposes, functions as an IP address. It is possible for an MES to have multiple NEIs that map to multiple IP addresses, thus maintaining simultaneous associations with more than one application.

To accomplish the registration, messages are exchanged between the MDIS and MES. These messages are exchanged in a connectionless manner using the services of the layers below, as are all control messages in CDPD. The first step calls for the MES to request the MDIS to activate an IP address for it in the form of an NEI. When doing so, the MES also presents a registration counter, called an authentication sequence number (ASN), and a set of options to the MDIS. The ASN is incremented by 1 every time the MES initializes a data link connection after assignment of a TEI. If it is not the home MDIS, the MDIS relays this counter value to the home MDIS's MHF through the use of the MSF. In either case, the home MDIS checks the value of the ASN to be sure that it is

Figure 14-13
CDPD Process/Application Layer Protocols

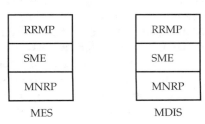

what it should be, namely, one more than it was the last time the MES established a data link connection for the exchange of data.

Simultaneously with the previous step, and in fact in the same message, the authentication occurs, verifying that the MES requesting registration is actually a subscriber. This verification is done through the use of a 64-bit authentication random number (ARN). The MES NEI, ASN, and ARN, taken together, are referred to as the MES credentials. Security management entity (SME) is responsible for the exchange and management of the encryption keys.

The ARN is generated by the home MDIS during the registration request. If the ASN matches, the ARN is generated and is relayed back to the MES. The MES stores these values and must present them for each new data link connection requested. The ASN and ARN values are encrypted in both directions to enhance security.

Security Management Entity (SME)   The SME is responsible for the exchange and management of encryption keys that protect the confidentiality of CDPD user data. Each MDIS and MES in the CDPD system supports SME services. The header in the SNDCP identifies SME messages, and the MNRP is used to provide the necessary services. Essentially, SME runs on top of SNDCP and manages all key exchanges.

In general, four MNRP messages are necessary for initial registration. First, the MES sends what is called an end system hello (ESH) message to the MDIS. In this initial message, the ASN and ARN are set to default values of zero. In the second step, the MDIS increments the ASN by 1 and generates the ARN back to the MES. The MES then confirms the receipt of the ARN by sending another ESH message back to the MDIS, with the ARN included. As indicated in the previous section, these ARN and ASN values are always encrypted. Finally, the MDIS increments the ASN value again, generates a new ARN, and sends this new ARN back to the MES.

Radio Resource Management Protocol (RRMP)   This protocol is responsible for RF channel management in CDPD. There are two reasons why this protocol is important to CDPD. First, recall that data services always ride second class. For example, if a voice circuit is needed by a subscriber, the packet data service provided by the CDPD overlay must move channels. Second, just like a voice connection, when a subscriber moves to another cell, the RF channel pairs must move as well.

This second-class service is a point of pride to CDPD developers. It means that CDPD is undetectable by circuit-switched voice or data subscribers. In other words, the circuit-switched environment is never aware of the packet-switched overlay. Circuit-switched subscribers represent the bulk of the revenue, so this arrangement makes perfect sense. It does mean, however, that the CDPD packet data service must be frequency agile and always take a secondary position when searching for RF channels to deliver its data. (CDPD has no way of determining if the circuit-switched connection is carrying voice or data. Circuit-switched data takes priority over packet-switched data.)

The circuit-switched environment is never aware of the packet-switched environment, but the reverse is never true. CDPD must be able to detect circuit-switched calls to ensure that it never disrupts the service while still contending for the same channels. That functionality is what RRMP provides.

CDPD transmission is frequency agile and mobile across adjacent cells, and performs channel hops or cell transfers under in-band control, unlike AMPS cellular. As a result, RRMP allocates RF channel pairs to CDPD services, controls the hopping to a new RF channel pair due to analog voice demand or CDPD channel load balancing, and is responsible for controlling the MESs transmit power. It also performs configuration information exchange. This exchange is broadcast periodically from the base station at intervals of approximately 10 seconds.

MES channel hopping consists of three logical steps:

1. Determining that conditions for channel hopping are met. Any one of the following conditions can cause channel hop:
   Detection of new color code
   Loss of synchronization
   Weak forward channel signal
   Excessive block errors
   Directed hop message by BS
2. Executing hop: uses information from previous configuration messages to find in-use RF channels and allocated RF channels in current cell, in-use RF channels and allocated RF channels in adjacent cells.
3. Completing hop.

Short Message Service (SMS)

SMS is essentially the inclusion of a textual pager function in a cellular telephone. SMS supports the display of a short alphanumeric message of up to 160 characters on the display screen of an MS. SMS data defaults to coding in 7-bit ASCII, but other mechanisms can be substituted. It is even possible to construct crude and grainy "illustrations" by modifying the character coding. The typical MS is the cellular telephone, but there is no reason why it cannot be a modified pager or other data entry/retrieval device. All that is required is that the MS be capable of interfacing to the cellular system in a standard way.

SMS services are most commonly available on GSM based systems, but many IS-95 based systems also offer it as an enhanced service. The implementation of the protocol stack is the same, with the exception of the underlying air interface. While it is possible also to overlay SMS on AMPS and DAMPS systems, this practice is uncommon because these systems tend to deploy CDPD as the data service.

Typical applications include voice mail, e-mail and facsimile notifications, short e-mail messages themselves, delivery notification, etc. It is probable that on-demand and other alert message services, for example, regular delivery of certain stock prices, stock and commodity price alerts, lottery results, weather alerts, even your up-to-the-minute cellular phone bill, will also evolve. Typical applications will require the message to be typed on the cellular telephone; however, some interfaces allow the text to be typed on a desktop computer, laptop computer, PDA, or similar device; downloaded to the cellular telephone; and exchanged using the SMS service. Gateways between SMS and traditional voice-mail or e-mail systems are widely available.

SMS is a store and forward data service. The messages are not sent directly to the destination from the source but are routed through an SMS message center (SMSMC). In general, there are several SMSMCs in a cellular network that supports SMS. Each SMSMC stores, forwards, and routes the messages when they can be delivered. The SMSMC locates the subscriber's MS for which the message is intended. If the MS is turned on, its location is found in exactly the same way as for a cellular voice call and the message is forwarded at once. If the MS is turned off, the SMSMC retains the message for some period of time. This lifetime of SMS messages is determined by the service provider and is typically a few days.

It is not necessary for a particular service to offer voice and SMS services. SMS exists simultaneously with voice and other data services, and may be offered by providers as an alternative to traditional pagers. Of course, the customer must subscribe to a mobile network that supports SMS. There is no single way that cellular networks provide network access to their SMSMC, X.25, Internet, and dial-up modem interfaces.

SMS offers two classes of service. The first, point-to-point communication, is designed to support messages between two subscribers of the SMS service. It is what is normally described when SMS is defined. The second is a cell broadcast service called SMS cell broadcast (SMSCB). In this service offering, multiple MSs are sent the same message. This service would typically be used to broadcast weather alerts, news reports, and other message subscription options with wide appeal.

It seems very likely that SMS will be used by several application providers as a kind of application gateway into their service offering. An API will be constructed and information, suitably tailored, will be broadcast to subscribers. It is also likely that SMS will extend corporate extranets by allowing message forwarding and selective updating of spreadsheets and databases that reside within the organization. SMS will be used to update the client information stored internally through a series of short messages from the field personnel. Of course, this function would use the point-to-point SMS service.

From the above discussion, it should be clear that the SMS network architecture is based on the GSM network architecture, with the addition of three logical components: the short message entity (SME), SMSMC, and a gateway. The SME is the general name for the component with which the user interacts to enter the message into the SMS system. The SMSMC and the gateway were described above. Figure 14-14 illustrates the network architectural components required by SMS. The underlying cellular service is assumed to be GSM.

In Figure 14-14, the message is assumed to be originating from some SME. It then travels in some manner to the gateway if the originating SME is not a GSM MS. If it is a GSM MS, the message travels directly to the SMSMC. From there it is stored and forwarded to the destination MS. In theory, the message could be transported from the SMSMC to any MS in any GSM network. In reality, agreements between service providers do not exist to support this technique.

The lower three layers of the protocol architecture of SMS depend on the underlying cellular service. For GSM, these layer protocols are identical to the underlying cellular

Figure 14-14
SMS Network Components

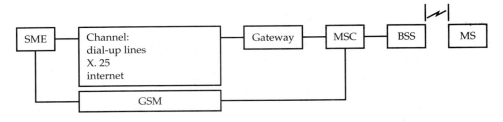

Figure 14-15
GSM SMS Protocol Architecture

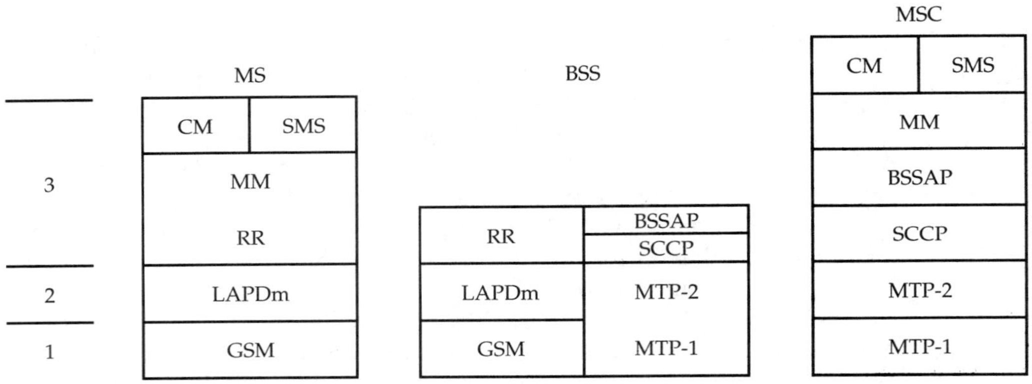

service protocol stack. See Figure 14-15. Layers 4 to 7 are identical to a GSM protocol architecture. Both use LAPDm as a layer 2 protocol, and both use the same layer 3 protocol sublayer set consisting of radio resource management (RR), mobility management (MM), and connection management (CM). These three sublayers provide the same functionality as the GSM architecture. SMS messages are delivered in 140-byte TPDUs at the layer 3 interface.

When the underlying cellular service is IS-95, a somewhat different protocol stack is used. See Figure 14-16. The main difference is that in the IS-95 case, PPP is used as a mechanism for connectivity, and the gateway function is assumed to be IP.

■ FDM/TDMA SPECIFIC PACKET DATA SERVICE

General Packet Radio Service (GPRS)

GPRS is a packet-based data service that uses logical channels in any FDM/TDMA cellular service. It was originally designed to work with GSM, but it has been adapted for

Figure 14-16
IS-95 SMS Protocol Stack

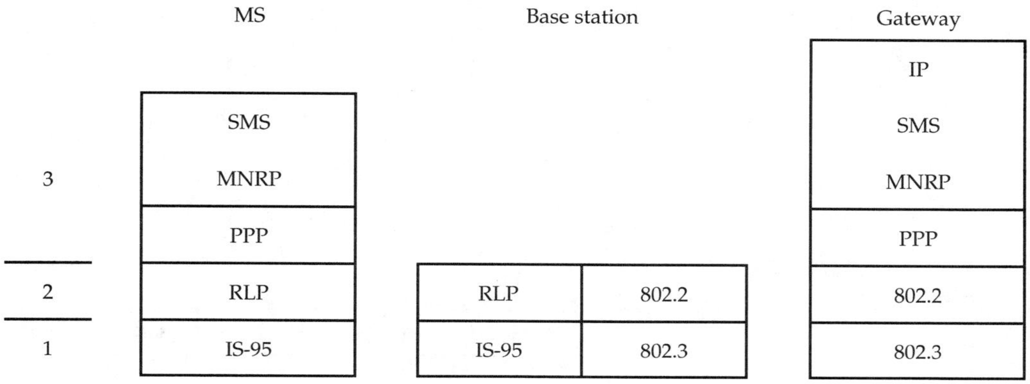

	MS			Base station		Gateway

						IP
	SMS					SMS
3	MNRP					MNRP
	PPP					PPP
2	RLP		RLP	802.2		802.2
1	IS-95		IS-95	802.3		802.3

use with any service with a similar frame structure. Specifically, many DAMPS networks located in the United States apply GPRS to achieve an increase in packet data rates over what is achievable with CDPD. Therefore, while the discussion below describes GPRS in a GSM context, it is also applicable to cellular air interfaces that feature the same fundamental time slot structure within a traffic channel as GSM does.

The basic idea is to integrate packet services into GSM by reserving logical channels for the packet content. These logical channels are made up of unused time slots in any particular traffic channel. GPRS can combine up to all eight of the time slots in a traffic channel. Theoretically, combining these time slots would allow a 160 kbps data channel; however, it is unlikely that this feature will be commonly available to a subscriber. Much more likely will be the situation where four of the slots are available for subscription, resulting in a data service at 80 kbps peak, 64 kbps typical. If nonsymmetrical data services are needed, it is not difficult to imagine that services offered to consumers will be configured as four time slots on the forward channel and one time slot on the reverse channel.

GPRS does not place an upper limit on the size of the packet, but research has shown that as the packet size increases over about 800 bytes, the throughput begins to drop. Therefore, the suggested approach is to transmit several 500-byte packets each minute, with the occasional 1 to 2 kilobyte packet when necessary. As the packet size increases, the average delay in delivering the packet also increases in a wireless cellular environment. For example, as the packet size increases from 1 kilobyte to 8 kilobytes, the average delay increases significantly, from about 500 ms to four times that, or 2 seconds. If lower priority quality of service (QOS) options are selected, the difference in delays decreases as a fractional percentage, but the absolute delays skyrocket to tens and even hundreds of seconds.

Since GPRS makes use of the underlying GSM system, it requires three additional logical components to the existing GSM architecture:

GPRS registers (GR): location of subscriber information and addresses.

Figure 14-17
GPRS Integration

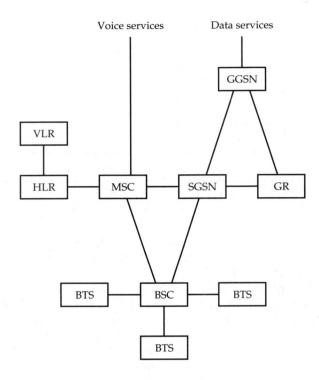

Serving GPRS support node (SGSN): functional support to MS, used to address the GR.

Gateway GPRS support node (GGSN): interface to the packet based wireline network.

The GR is nothing more than a database that can be viewed as a GPRS version of the HLR. It is not often implemented as a separate physical database because much of the information it contains is already present in the HLR. It contains the subscriber information that would normally be associated with the HLR for those subscribers who sign up for GPRS.

The SGSN and GGSN also have counterparts in the voice network. The SGSN is similar to the IWF and the GGSN is similar to the GMSC in their respective operations, but not in their protocol roles, as will be shown later. Unlike the MSC and GSMC, however, these two components, while logically distinct, are often integrated into one unit and deployed at every MSC location where the packet service is to be offered. Figure 14-17 illustrates how these three components interface with each other and the circuit-switched voice service that GSM provides. Note that these components can be implemented as software upgrades to an existing MSC site, with the possible exception of the GGSN, which may require some hardware interface.

From a close examination of Figure 14-17, you might assume that the GPRS data and signaling are all routed through the SGSN. The packet data is routed through the SGSN directly to the BSC. The signaling for the GPRS service, however, is performed over the link maintained by the MSC. There are two reasons for this situation. First, the MSC already contains all the signaling functionality and complexity that the MS can use. There is no reason to duplicate this complexity. Second, all the packet data must travel through the RF links established and maintained by the MSC, so allocation of bandwidth for data packets and the signaling associated with that allocation must take place through the MSC.

GPRS, voice services, and SMS are all intended to run together. However, there is only so much bandwidth available for sharing. The two main classes of service that GSM offers—the circuit-switched voice services and the packet-switched data services—are designed to coexist. For example, it will be possible to initiate a GPRS session during a voice session, and vice versa. Because the GPRS uses the SMS time slots, however, it is not possible to do GPRS and SMS simultaneously.

GPRS will offer both connection-oriented and connectionless service interfaces for point-to-point packet transport. These interfaces are called, respectively, point-to-point connection-oriented network service (PTP-CONS) and point-to-point connectionless network service (PTP-CNS). The error rate on PTP services is quite good and is very compatible with protocol assumptions built into TCP/IP. Typical numbers are in the range of BER values of 1 error in 1 billion bits. Almost all GPRS packet data will be delivered using the PTP-CNS for throughput efficiency reasons.

It is also important to clarify that the PTP-CONS is not actually connection-oriented from an end-to-end perspective because GPRS is, by definition, not end-to-end in the TCP sense. PTP-CONS is connection-oriented-only within the GPRS environment, which usually means MS to MS. I see little practical application for this service offering and discuss PTP-CONS here as an example of an exercise in protocol development. A connection between two MSs would consume several time slots to maintain the connection and would limit the ability to support both circuit-switched voice and packet data services simultaneously.

For point-to-multipoint addressing (which by definition is connectionless in GPRS), there is multipoint addressing within both the mobile network and the wireline network. When the address space where the message is to be sent can be defined by geography, there is no restriction on the addressing. When the address space where the message is to be sent is defined by an address list, then the message is sent only to the areas where the individual addresses are located. Forwarding beyond that location is outside the service definitions of GPRS.

TCP Implications of GPRS   There may be good reasons for not using the point-to-multipoint services that GPRS offers. The reasons lie in the error rate associated with these services. Because any RF system will exhibit a higher error rate than an equivalent wireline system, the packet loss rate is also higher. In a multipoint environment, many packets are sent with one command, so the overall probability of a retransmission is also much higher, as much as 10,000 times higher. The residual BER approximates 1 error in 10,000 bits.

Figure 14-18
GPRS Protocol Stack

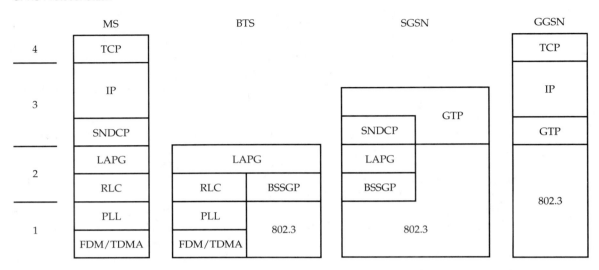

As discussed in Chapter 16, the packet retransmission rate implied at these values of BER are not handled well by the throughput algorithms that TCP/IP uses in a wire-line environment. In this context, it is also important to examine the packet delays outlined earlier in this chapter. Again, the TCP algorithmic responses to large delays can significantly degrade potential GPRS throughput if the protocol is not modified to anticipate this range of packet delays.

In summary, GPRS is a packet service designed to interface to TCP/IP networks as well as traditional X.25 networks. Messages are delivered directly to the MS without any end-to-end communication in the TCP sense. The overhead that normally uses up part of the frame structure can be omitted and, as a result, connectionless data transport can be delivered at 14.4 kbps using a single time slot or as much as 115 kbps using all eight. (A connection-oriented communication would reduce these rates to 9.6 kbps and 76.8 kbps, respectively.) As discussed above, because this service is connectionless, it can offer data transport simultaneously with a voice connection. It is also primarily a software upgrade; like HSCSD discussed below, it will roll out quickly as market demand grows.

GPRS Protocol Stack

A more detailed examination of the protocol stack is now appropriate. Following a more detailed examination of each layer, roaming and mobility management in a GPRS environment are discussed briefly. GPRS is a network layer protocol intended to run by tunneling through IP or some other connectionless network layer. The protocol stack shown in Figure 14-18 is based on the assumptions that the network layer is IP and that the GGSN is running IP over some form of Ethernet.

Compare this protocol stack with that for the GSM circuit-switched voice service, and note that both feature four fundamental architectural locations. In the GSM circuit-switched protocol architecture, those locations were the MS, BTS, BSC, and MSC. In the packet-switched architecture, they are the MS, BTS, SGSN, and GGSN. In the GPRS architecture, the separate roles of the BSC and MSC are not as clearly divided. Both the SGSN and the GGSN provide an element of the MSC functionality. Five new protocols are shown for the first time in Figure 14-18:

PLL: physical link layer

RLC: radio link control

BSSGP: base station subsystem GPRS application protocol

LAPG: link access procedure on the G channel

GTP: GPRS tunnel protocol

While discussed earlier in the CDPD protocol stack, SNDCP will be discussed briefly here, with particular emphasis on its role in the GPRS protocol stack. The next sections will provide a discussion of these layer protocols and focus only on those protocol elements unique to digital packet services.

Physical Layer   In the GPRS protocol architecture, the physical layer is composed of two sublayers, which is the general approach when packet based data services are defined. This same approach, for example, is used in the IEEE 802.11 protocol stack. The two sublayers are the underlying FDM/TDMA air interface and the PLL.

The air interface still performs modulation, demodulation, and all bit level issues associated with the circuit-switched voice and data services offered by GSM. These issues include the error-detection and error-correction through packet retransmission, the actual interleaving of packet data into unused time slots, and some special synchronization issues associated with the variable delay characteristics of the radio channel as they apply to packet transmission. GPRS uses Gaussian minimum shift keying (GMSK) with adaptive coding.

The physical link layer (PLL) is a logical sublayer of the physical layer. It is used to handle the special characteristics of frame based packet relay on the bit level but transmitted in traffic bursts over the GSM network. It performs functions such as framing and coding of data. It is also responsible for the detection and correction of any transmission errors resulting from transmission in the radio channel. In summary, it is used to enhance the standard GSM air interface standard to packet based transmissions.

Data Link Layer   In many protocols, the data link layer is composed of two logical components, the MAC sublayer and the LLC sublayer. The GPRS protocol adds additional functionality to both logical components. The MAC sublayer is expanded into two sublayers, traditional MAC functionality and the RLC component.

The GPRS MAC layer has a unique characteristic that differentiates it from the GSM MAC layer. In the former, multiple time slots in multiple traffic channels can be assigned to an individual MS. The GSM MAC layer does not allow this functional characteristic to ensure QOS parameters associated with circuit-switched voice. GSM circuit-switched

Figure 14-19
PDU Transformation in the GPRS Data Link Layer

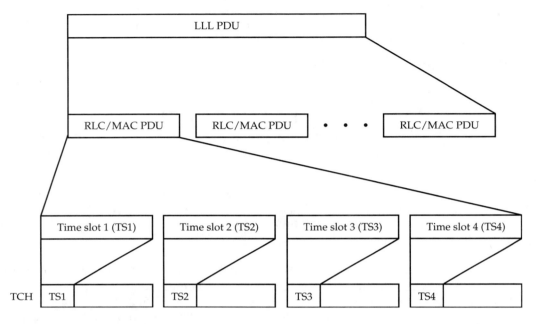

voice traffic allows only a single time slot in a single traffic channel to be used for each simplex channel.

The GPRS MAC allows up to eight time slots in up to four different traffic channels to be addressed to a single MS. Each time slot can support 9.6 kbps in a packet data environment. The actual data rate can vary from 9.6 kbps to 76.8 kbps for a connection-oriented communication and from 14.4 to 115 kbps for a connectionless communication. The RLC protocol allows several logical channels or time slots to be associated with a single MS. RLC uses a slotted ALOHA access scheme to schedule these multiple time slots.

Therefore, RLC controls access to the radio channel by packets. It addresses issues like channel access, collision resolution and retransmission, multiplexing, segmenting, and quality of service assignments. Because an LLC PDU is presented to the RLC service interface, the combined MAC layer protocol segments it into a number of RLC/MAC PDUs based on the size of the LLC PDU. Each RLC/MAC PDU is further segmented into four time slots. These time slots are always transmitted in consecutive TCH or TDMA frames. See Figure 14-19.

When used with GPRS, RLC specifies four different block sizes, known as code schemes (CSs). These four different block sizes each have a different code rate defined. Together with the size difference, they combine to offer four different data rates. As stated above, each RLC block is transmitted interleaved across four GSM physical layer bursts. Block specifics are summarized in Table 14-2.

Note that the time slots may be transmitted simultaneously using four separate traffic channels or consecutively using a single traffic channel. This choice presumably de-

Table 14-2
GPRS Block Summary

Technique	Code Rate	Block Size (Bytes)	Maximum Data Rate
CS-1	0.5	20	8.0 kbps
CS-2	0.67	30	12.0 kbps
CS-3	0.75	36	14.4 kbps
CS-4	1.0	50	20.0 kbps

pends on the QOS parameters associated with the LLC PDU. The LLC protocol controls a standard ARQ FEC error-detection and retransmission based correction scheme. At the destination, an LLC PDU will not be forwarded up the protocol stack until any retransmissions made necessary by one or more time slots being in error are addressed and a completely reassembled LLC PDU exists at the destination.

The BSS controls when and which traffic channels are used by the MS to exchange packet data. As indicated above, this process is controlled in a basic way by the time slot or channel reservations. In another, more sophisticated way this process is controlled by using the HL and HU bit positions as flags. These bit positions in the time slot are renamed uplink state flags (USFs) and are used by the MSC to indicate to the MS that a time slot is available for packet transmission.

Because the MSC controls the time slots, it must have some way of telling an MS when it can send GPRS data and on what logical channel. The state of the USF bits are used to inform the MS of these facts. While the unused time slots are used to communicate GPRS packets, it is important that any particular time slot available for MS transmission be shared among all GPRS MS subscribers under control of any particular MSC. The USF bits tell each individual MS that its time slot is coming up. As you might imagine, these time slots are parceled out to different MSs under control of the SGSN based on subscription level and, within that hierarchy, QOS issues.

The next important issue of this section is how the logical channels discussed above are actually mapped onto physical channels. Each logical packet data channel (PDCH) is associated with a time slot in some traffic channel. Because a PDCH can be bidirectional due to the availability of multiple time slots for a packet transmission, an MS may both receive a packet from the BTS and transmit a packet to another MS at the same time.

As discussed above, this parceling out of time slots is controlled by the MSC in the same way that time slots are parceled out for circuit-switched voice services. Any coordination for this process must be carried out on control channels, again, like the voice and SMS services. Not surprisingly, control channels used by GPRS are similar to those used by GSM for similar purposes. Table 14-3 identifies these six new channels. The last four are sometimes grouped together and classified as four subchannels of the packet common control channel (PCCCH). With this classification, only three new channels are required for GPRS: PBCCH, PTCH, and PCCCH.

The LLC portion of the data link layer in GPRS is the link access procedure on the G channel (LAPG; G stands for GPRS). This protocol element is an enhancement to the

Table 14-3
GPRS Packet Control Channels

Channel Name	Channel Abbreviation	Direction
Packet broadcast control channel	PBCCH	BSS to MS
Packet traffic channel	PTCH	BSS to and from MS
Packet random access channel	PRACH	MS to BSS
Packet paging channel	PPCH	BSS to MS
Packet access grant channel	PAGCH	BSS to MS
Packet associated control channel	PACCH	BSS to and from MS

LAPDm used in GSM. Like any data link layer protocol, the LLC layer is responsible for the service interface to the network layer. As such, it can be seen as transporting the LLC data PDUs from the MS to and from the SGSN. This interface is the virtual communication that exists between any two identical layers in the protocol model. As defined earlier, both point-to-point and point-to-multipoint communications are supported. Logically, flow control for these communication services is also supported at the LLC layer.

The frame structure of the LAPG frame, while based on HDLC, does feature some significant but not unprecedented differences. First, there are no start and end delimiter flags because the frame is encapsulated into a physical layer frame prior to transmission. Because RLC allows a variable frame length, a field is defined to be the data length in bytes, similar to conventional Ethernet. The next major difference is that the address fields are of variable length. These address fields are defined by the packet service subscribed to. There is no one standard for address length. As a result, these fields must accommodate an unknown length. Each data link connection identifier (DLCI) is actually composed of two components. The first is a SAPI, as in the GSM case, and it identifies the connection endpoint at the network layer service interface. The second is called a terminal endpoint identifier (TEI) and it is used to contain the identity of the individual MS. In total, four new SAPIs are defined, one for each proposed QOS class.

The last major difference between LAPG and LAPDm is the enhancement of LAPG to support point-to-multipoint communications. Since LAPDm supports only point-to-point communication, the protocol must be expanded to handle this new addressing technique. Both connection-oriented and connectionless communications are supported with this protocol enhancement.

The functions of the data link layer are related to the mobility and roaming nature of the MS, but only in the sense that as an MS moves between cells, it must securely re-register using LAPG. Hence, when an MS wants to send packet data, it must first "log in" to the packet service. This logging in is called an attachment procedure and takes place between the MS and the SGSN. It is always a connection-oriented data transfer because of the critical nature of this step in using the GPRS. Therefore, prior to a connection being established between the MS and SGSN, the GR is consulted to ensure that the MS has access to the particular packet service offered by the GPRS to which it is making a request. If access is granted, then the SGSN is updated with the current MS location, and packet exchange can begin. The update of the SGSN is very similar to the VLR updating done when an MSC grants access to voice services and transfers subscriber data between the HLR and VLR.

The remaining data link layer protocol to be described is the base station subsystem GPRS application protocol (BSSGP). This protocol is a modified frame relay protocol that provides the same kinds of mobility management functions provided in the GSM protocol element MAP. The three basic elements are support of a GPRS node entering the system and leaving the system, and keeping track of the node while it roams throughout the system (called location updating). The protocol performs these tasks very much like the HLR/VLR performs these operations as described in the GSM section. The main difference is that now the SGSN assumes the role of the MSC.

This protocol also conveys routing and QOS related information between the BSS and the SGSN. Five levels of QOS are supported by GPRS. The appropriate level of QOS for any particular network packet is associated with the appropriate network layer service access point (NSAP). As a result, the network provider examines the NSAP for the service level required by the network user. The default QOS is held in the subscriber's entry in the HLR or it is duplicated in the GR.

Network Layer   The subnetwork dependent convergence protocol (SNDCP) multiplexes several network layer communications onto a single data link communication. This technique allows both circuit-switched voice and packet data communications. Thus, network layer packets coexist with the packet based communications of GPRS on a single data link protocol. SNDCP communicates between the MS and SGSN. SNDCP is responsible for all segmentation and reassembly required at this interface. It also compresses and decompresses both the data and header information for more efficient transmission over the radio channel. Note that this protocol element is common to the two dominant packet data services described in this chapter, namely, CDPD and GPRS. SNDCP supports both acknowledged and unacknowledged data transfer. This protocol and the GTP protocol form the GPRS specific portion of the network layer functionality required to interface seamlessly to IP.

The GPRS tunnel protocol (GTP) encapsulates packets for tunneling under control of IP between the SGSN and the GGSN. It is the key protocol between GSM nodes running GPRS. GTP is the interface not only between SGSN and GGSN in any one service provider's network, but also between GGSN nodes in other providers' networks. Functionally, it uses its signaling to create, modify, and tear down tunnels that the SGSN uses to provide network access for an MS. The MS must be able to communicate with other GPRS nodes both in and outside its provider's network. The MS does not need to be aware of how this communication is done, but it needs the functionality. As a result, only the SGSN and GGSN nodes implement this protocol.

Emerging Protocols for Cellular Data Services

Enhanced Data Rates for GSM Evolution (EDGE)   When it was first introduced, EDGE was viewed as an evolutionary step for GSM packet data. EDGE use 8-PSK as a modulation technique, which allows higher data rates and enhanced data encoding to achieve

Table 14-4
EGPRS MCS Levels

Technique	Modulation	Code Rate	Block Size (Bytes)	Number of Blocks/ Time Slot	Maximum Data Rate/Time Slot
MCS-1	GMSK	0.53	22	1	8.8 kbps
MCS-2	GMSK	0.67	28	1	11.2 kbps
MCS-3	GMSK	0.80	37	1	14.8 kbps
MCS-4	GMSK	1.0	22	2	17.6 kbps
MCS-5	8-PSK	0.37	28	2	22.4 kbps
MCS-6	8-PSK	0.49	37	2	29.6 kbps
MCS-7	8-PSK	0.76	28	2	44.8 kbps
MCS-8	8-PSK	0.92	37	2	54.5 kbps
MCS-9	8-PSK	1.0	37	2	59.2 kbps

superior error recovery. EDGE uses the same architectural components of GPRS: the GGSN and SGSN. EDGE can deliver data rates up to 384 kbps (although in ideal conditions and using all eight time slots, 470 kbps is theoretically possible). EDGE requires that the subscriber not be moving too fast. Thus, the bit rates will be available for stationary or walking users but not those driving down an uncongested roadway.

There are two components to EDGE: enhanced circuit-switched data (ECSD) and enhanced GPRS (EGPRS). The first standard defines enhancements to the circuit mode, while the second does the same for packet mode data. EDGE is usually thought of as a packet data service only, but the enhancements apply to both types of switching. We will examine the EGPRS component briefly.

Enhanced GPRS (EGPRS)   EGPRS is all about using the techniques of rate adaptation to enhance the data rate delivered to subscribers. The basic concept is to take advantage of adaptive modulation, coding, and the combination of multiple time slots to enhance the data throughput. EGPRS accomplishes these tasks through defining nine different modulation and coding schemes, called MCSs. The first four, MCS-1 to MCS-4, use the GMSK modulation technique, and the last five, MCS-5 to MCS-9, use an 8-PSK modulation technique.

Both groups have several levels of code rates, varying from 1 to approximately ⅓. These levels cause each MCS to exhibit different maximum data rates per single time slot. Three different RLC block sizes are defined for EGPRS. Some MCS levels use a single block, while others use two blocks. The block sizes also vary. The variation in block size is how EGPRS implements adaptive data rate. Table 14-4 illustrates the nine MCS levels and the particulars of each.

■ SUMMARY

This chapter provided an overview of both circuit-switched and packet-switched data techniques. Packet data services are typically implemented as a protocol module overlay onto existing voice services in a cellular network. For this reason, the cost to implement these services, from a provider's perspective, is very low. When traffic channels are diverted to data, however, potential revenue from voice services is lost and channels once dedicated to data services cannot be reconverted to voice services in real time. Therefore, a critical number of data users must exist for a cellular service provider to offer packet data services. For this reason, mobile data services in the United States have not yet achieved nationwide coverage. In major markets, CDPD is widely deployed, but this service is not available in rural or semirural areas.

REVIEW QUESTIONS

1. What are the two basic techniques for data transmission?

2. What are transparent data transmissions?

3. What are nontransparent data communications?

4. Describe RLP and explain why it is used.

5. What is data rate adaptation and why is it used?

6. When are packet data services best used?

7. Explain why CDPD must be frequency agile.

8. Describe the functional roles of the MDIS and MDBS.

9. Describe the routing problem in CDPD systems.

10. Explain how DSMA/CD is used to signal when the channel is clear for transmissions from the MES.

11. Describe the function of the TEI as it is used in CDPD.

12. Discuss the role that MDLP has in error recovery of lost frames.

13. What are the main functions that SNDCP provides in CDPD?

14. Discuss why SNDCP is so important for wireless packet networks.

15. Describe the registration procedure used in CDPD.

16. What is the role of RRMP in CDPD?

17. Describe the SMS components and their roles in providing service.

18. What channels does SMS use?

19. What channels does GPRS use?

20. Explain why SMS and GPRS messages cannot be sent to an MS at the same time.

21. Compare and contrast the protocol architectures of GPRS and GSM.

22. Describe the roles that the SGSN and GGSN perform in GPRS.

23. Discuss the role that RLC has in GPRS.

24. Discuss the role that LAPG has in GPRS. Contrast LAPG with LAPDm.

25. What is the role of GTP and why is it important?

26. Discuss how EGRPS offers increased data rates over GPRS.

IEEE 802.11
Wireless LAN

■ INTRODUCTION

The IEEE 802.11 committee is a member of the 802 project of the IEEE. The 802 project has established many LAN standards that have proven themselves (or not) with the test of time and market adoption. The original 802.11 standard was formally approved in late 1997 after many years of work. In 1999, the first amendment to the standard, 802.11b, was approved. This amendment, called the high rate amendment, added two additional data rates, 5.5 and 11 Mbps. In this text, no differentiation will be made between the original standard and approved amendments; all will be referred to as the 802.11 standard. By late 2001, the only members of this family that are commercially available are those products conforming to the 802.11b standard. For this reason, this standard is the focus of this chapter.

In early 2002, products conforming to the 802.11a standard are expected to become available. Products conforming to the 802.11b standard will operate in the 5 GHz band and will have data rates as high as 54 Mbps initially. Vendors are optimistic that those data rates can be extended to as high as 100 Mbps in the near future. A major difference in the 802.11a technology is that it uses many relatively narrowband carriers, each modulated into a 300 kHz bandwidth, that are combined to form a much higher aggregate data rate. Comparing this standard to 802.11b, which uses spread spectrum technology as an air interface, you can see that 802.11a is essentially FDM technology. Both standards will use the same MAC layer.

As data rates increase, quality of service (QOS) issues will become more important to wireless LANs. These issues will be addressed by the 802.11 committee in the 802.11e standard. The IEEE 802.11e standard specifies QOS that will support both 802.11a and 802.11b wireless standards. Basically, the 802.11e standard gives control over when the mobile stations can transmit to the access points. It is very similar to a classic polling scheme where the master, in this case, the access point, sets up a poll consisting of time slots allocated to each mobile station in its range.

The 802 project is restricted to work on layers 1 and 2 of the OSI model. These layers are the physical and data link layers. The 802 committee has also previously defined the 802.2 LLC component of the data link layer, with which all 802 standards must interconnect. Therefore, the 802.11 committee, just like the 802.3 committee, has the task of defining a PHY and MAC layer. See Figure 15-1.

The figure shows a single MAC specification and multiple PHY specifications. This scenario is perfectly permissible because the service definitions of OSI are always from the bottom up; that is, lower layers offer services to layers immediately above them. Because a common LLC layer interface is required, a common MAC layer is also required. The MAC layer definition is totally in the domain of the 802.11 committee, however, and committee members decided to allow multiple PHY definitions. The only requirement is that they all offer a set of services to the MAC layer that can be used to provide an identical set of services to the LLC layer.

If you think about it, this approach was the same one taken in the most successful 802 committee (at least as measured by products deployed): 802.3 or, as it is commonly known, Ethernet. Several 802 physical layer standards use a common 802.3 MAC layer. Two of the more widely deployed are 10BaseT and 100BaseT.

The 802.11b standard defines three alternate physical layers (FHSS, DSSS, and infrared). Because the DSSS physical layer option is the dominant market presence, it will receive the most discussion, followed by the FHSS option. The infrared option will be addressed only briefly because no product releases have appeared. The MAC layer's

Figure 15-1
IEEE 802.11 Protocol Stack

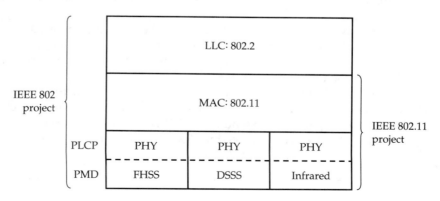

structure, function, and services will be explored. To begin, an overview of the topology and components that make up an 802.11 LAN will be provided.

■ TOPOLOGY AND COMPONENTS

There is really only one required, basic component to any IEEE 802.11 compliant LAN: the wireless station itself. This station is nothing more than a wireless network interface card (NIC) installed in a computing device. Like any NIC, it provides the layer 1 and 2 functionality necessary for communicating on a LAN segment. The standard describes any wireless station as STA. While an STA does not have to be mobile, the vast majority will be. For ease of understanding, from this point forward, this text will refer to the STA as a mobile station.

Of course, for any communication to occur, there must be at least two communicating elements. The first is the mobile station described above. The second can be another wireless station or, more commonly, an access point (AP). The stations may be mobile or fixed. The access point is often referred to as a base station. This terminology is a result of an analogy with the cellular telephony system; however, this text will avoid this analogy and refer to access points directly. The AP and the mobile station communicate using radio technology. The AP provides the link between the radio LAN and a wired LAN to which the AP is typically connected. In this situation, the AP functions like a traditional bridge because it provides a link from one MAC layer to another with a common LLC layer, in this case, IEEE 802.2. Figure 15-2(a) shows a representative 802.11b AP. Figure 15-2(b) shows a representative NIC from the same manufacturer.

Two additional components are part of most 802.11 compliant wireless LANs: the portal and the distribution system. These components will be introduced below. They are related to how the access points are interconnected and how they communicate to other networks. There are three basic wireless networking topological environments that 802.11 is designed to support. The first is ad-hoc networking, which is described in

Figure 15-2
(a) IEEE 802.11 Wireless Access Point; (b) PCMCIA Form Factor NIC Card (*Courtesy of Agere Systems. © Agere Systems*)

(a)   (b)

Figure 15-3
Ad-Hoc Network

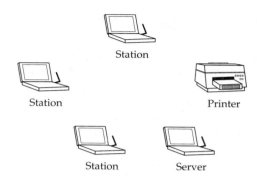

Station

Station

Printer

Station

Server

the next section. In this scenario, no AP is used. The second is intended for stand-alone operation where a single AP exists and is called an independent basic service set (IBSS). The third is intended for enterprise applications and uses a distribution system to tie several APs together to achieve seamless coverage over an extended range. It is called an extended service set (ESS) in 802.11 terminology.

◼ AD-HOC NETWORKING

In ad-hoc networking, no AP is used and the individual stations communicate directly with each other. Stations within range of each other, if they are configured in peer-to-peer mode, automatically form an ad-hoc network. It is important to understand that an ad-hoc network is the only architectural topology where the individual stations communicate directly with each other. Whenever an AP is in range, the mobile stations will automatically direct their communications to the AP.

In the vast majority of ad-hoc networks, there is no method for connecting to networked components outside the range of the individual stations. It is possible, however, to equip a fixed device, such as a switch or router, with a wireless NIC and thus communicate with network devices outside the range of the wireless cards. However, the ad-hoc topology is designed to support the quick and painless networking of units on an ad-hoc basis. It allows temporary networks to be set up and torn down with little effort. Applications include temporary networks set up for meetings and facilitating breakout working groups. Note that all the stations are not required to be mobile in the ad-hoc network shown in Figure 15-3.

◼ NETWORKS USING AN ACCESS POINT

The second and third topological environments for wireless networking require an access point. The range of each access point is very similar to a wired LAN segment. In each segment, several stations are attached using the media appropriate to the LAN standard. In 802.11, of course, this media is the atmosphere around us instead of a metallic or optical conductor.

A wireless LAN segment has a special name in 802.11; it is called the basic service set (BSS). There can be only one AP in a BSS. Each BSS is defined as a separate wireless LAN segment and each has a unique identifier. This identifier is known as a basic service set identifier (BSSID). Both IBSSs and BSSs have BSSIDs. A BSSID is the identifier of the BSS, or wireless LAN segment.

A BSS is defined as the range of the AP. The range is influenced by the features of the surrounding area. For example, a BSS diameter in an open office environment may be as large as 125 meters. In a steel mill, it will be significantly smaller, perhaps as small as 10 meters. A BSS is considered independent (called an IBSS) when there is no distribution system connecting it to the other network resources. When the BSS is combined with another element (the distribution system), the last topological structure for wireless LANs emerges. It is known as extended service set (ESS). ESSs utilize more than one AP and contain multiple BSSs. An ESS appears to the LLC layer as a single network. Thus, mobile stations can move between BSSs in an ESS and remain transparent to the upper layers of the protocol stack. An ESS also has a unique identifier called an extended service set identifier (ESSID). The ESSID is assigned by the network administrator. Typically, the BSSID is the AP MAC address, and the ESSID is the network name.

Both of these topologies, the IBSS and the ESS, use the BSS as a fundamental unit. The distribution system is the component that connects the APs together. The distribution system can be wired or wireless. Figure 15-4 illustrates these two topologies defined by IEEE 802.11. Figure 15-4(a) shows a BSS that is also an IBSS. Note that a single AP is deployed with no distribution system to tie it to other APs or other network resources. Typical applications include networking a small office area or part of a larger floor, a home or apartment, etc. The limiting criterion for this topology to work is that all stations must be within the range of the AP (in other words, be in the same BSS).

Several APs are shown in Figure 15-4(b). Note that the APs are tied together with a distribution system and that individual BSSs in an ESS are not required to overlap. Each group of BSSs linked in such a way is called an ESS. The ESS topology is designed for applications requiring greater range or interconnectivity into an existing enterprise network. Because the ESSID is assigned by the network administrator, Figure 15-4(b) could be configured as a single ESS or two separate ESSs. This decision is made by the network administrator and is not a result of the topology chosen.

When a distribution system is deployed, network resources are accessible by a mobile station in any BSS. Multiple BSSs can be linked to provide coverage over an extended area. In most industrial and business settings, ESS topologies will be used. Enterprise networks usually require access to network resources from any location. Through the use of a distribution system, an authorized user can access network resources from a mobile station. In some home or small office environments, a distribution system is not required when only a single access point is installed. In this case, network resources are not available to the mobile stations unless the network resources are equipped with wireless NIC cards. If they are equipped with NIC cards, access to the network resource, such as a printer, would be identical to the process for communicating to another mobile station in the same IBSS.

Figure 15-4 (a)
IBSS Topology

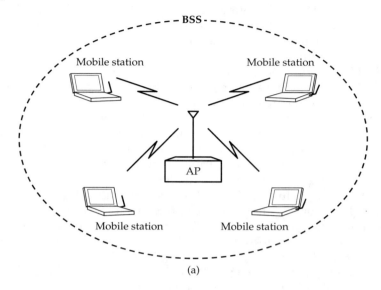

(a)

It might seem that the distribution system must be an 802 network so that the AP, which is only a bridge, can attach directly to it. However, another level of logical connectivity, a portal, is defined by the 802.11 standard. Only one portal is allowed for each ESS. A portal is analogous to the protocol translation aspects of a router or gateway. The portal is the logical connectivity between a distribution system that is not 802 compliant and the AP. Functionally, 802 frames are passed between the portal and the AP for distribution both in and out of the BSS. When the distribution system is 802 compliant, such as an Ethernet network, the portal becomes functionally transparent. It and the AP merge to become an AP.

Therefore, the portal has the function of forwarding frames between wired and wireless stations. When a wireless station sends a frame to a wired station, it arrives first at an AP. Once it has determined that the frame is addressed to a wired station, the AP forwards it to the portal. It is then the portal's responsibility to forward that frame to its destination. Similarly, when frames from wired stations arrive at a portal, the portal forwards them to the appropriate AP, which then transmits the frames to the mobile station.

Again, the BSSs connected in an ESS topology can overlap or not. The stations can move anywhere within the range of an AP and be communicating with other wired and wireless stations. If the BSSs overlap, there will be seamless coverage for roaming stations. If they do not overlap, the station leaves the network when it moves outside any AP's BSS. The ESS topology allows wireless LANs to be constructed to fit any coverage area.

Figure 15-4 (b)
IBSS Topology

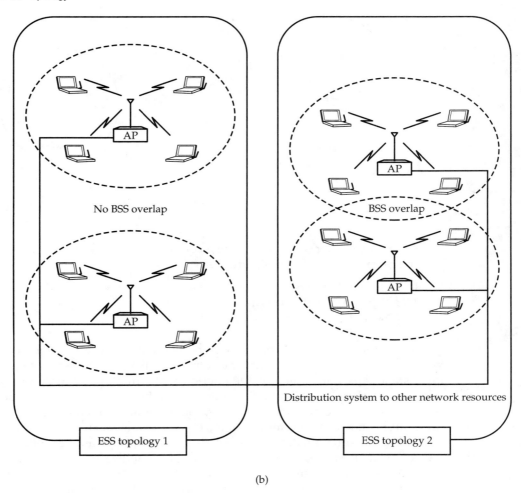

No BSS overlap

BSS overlap

Distribution system to other network resources

ESS topology 1

ESS topology 2

(b)

■ ROAMING WITHIN AN ESS

While the 802.11 standard specifies roaming configurations and the basic messages needed to support roaming between ESSs (and these are discussed below), there is no real standard defined. Into this void have stepped the major vendors, who have adapted a standard called the interaccess point protocol (IAPP). IAPP addresses roaming between ESSs. This standard is discussed in a later section.

Within an ESS, however, 802.11 does provide for roaming between BSSs. The IEEE 802.11 standard defines three configurations salient to this discussion. In this section, we will assume that in each case, the BSSs are part of a single ESS; that is, they are all part

Figure 15-5
BSS Overlap Definitions: (a) BSSs Do Not Overlap; (b) BSSs Partially Overlap; (c) BSSs Co-located

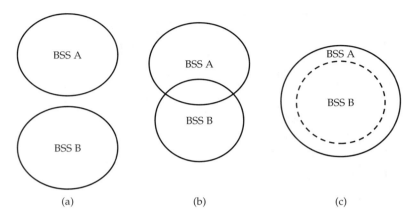

of the same logical topology and share a common ESSID. These configurations are defined below and illustrated in Figure 15-5:

1. BSSs that do not overlap.
2. BSSs that partially overlap.
3. BSSs that are co-located.

As a station moves from one BSS to another in the first case, where the BSSs do not overlap, the station loses its communication to the network. BSSs in a single ESS can be any distance apart; for example, the interior of all buildings on a campus may be covered with overlapping BSSs, but the exterior paths between them may not be. As the station travels from one building to another, communication would be lost. Upon entering any building, however, communication would be resumed. How any application would respond to this situation depends, of course, on the specific application.

As a station moves from BSS to BSS, in the second case, when the BSSs overlap to some degree, it never loses communication with the AP in each BSS and communication is seamless. Applications experience no disruption. To the application, the mobile station appears continuously connected, like a wired station would. This second case is the typical situation. In the overlap area, a station will receive signals from two or more APs. The station has the responsibility for choosing the "best" AP based primarily on signal strength and network utilization. When a station determines that a signal is poor, it begins scanning for another access point. This scanning is usually accomplished through a passive listening approach, but it is possible for the station to probe each channel actively and to solicit a response from any AP on that channel. Once an AP has been selected, association is accomplished. The association process is described in detail in a later section.

The third case, where one BSS co-locates with another, is useful for two application domains and is rarely used. The first reason for adopting this topology choice would be

Figure 15-6
(a) BSS Overlap; (b) Frequency Overlap

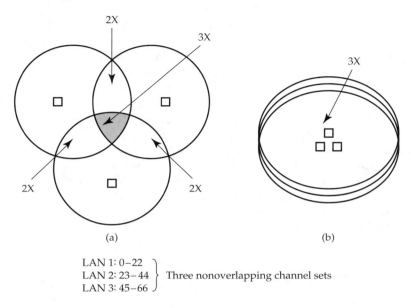

(a)

(b)

LAN 1: 0–22
LAN 2: 23–44 } Three nonoverlapping channel sets
LAN 3: 45–66

redundancy. By overlapping more than one BSS in the same physical area, you obtain redundancy in the single points of failure in BSSs connected to act as an ESS, redundant APs, and redundant distribution systems. The second application domain is where the application running on the wireless unit requires a higher data rate than can be supported running at the advertised data rate of the wireless system. By overlapping up to three BSSs, you can increase the data rate at the mobile unit by a factor of three. However, this application would require significant upper layer protocol modification. Further overlapping is not possible because the frequency space of each overlapping BSS must be unique. Frequency space considerations are important for deployment, and we will discuss it in more depth when the physical layer is examined. Different specifics apply to each physical layer option.

Figure 15-6 clarifies the difference between BSS overlap and the concept of frequency overlap. To realize the increased data rates shown in this figure, a mobile station would require additional NIC cards: 2X would require two NICs and 3X would require three NICs. Again, modifications would have to be made to the protocol stack to combine the individual data flows to the application into a single stream.

■ NO ROAMING BETWEEN ESSs

The 802.11 standard does not address roaming between ESSs because there are many circumstances where it is not desirable for a mobile station moving into another ESS to be granted access rights. To understand this situation, imagine that you could take your laptop computer into any building and immediately be granted full access rights to the LAN through a kind of magic NIC card. Obviously, this access would have significant network security implications. Because the 802.11 committee charter is applicable only to layers 1 and 2, and because a solution to this dilemma required protocol elements above layer 2, roaming outside an ESS could not be addressed.

In another example, a mobile or wireless station intended for use in a warehouse has no need to communicate when the warehouse employee carrying it reports to an administration building for a meeting. Additionally, the warehouse application may require specific hardware or software capabilities in the station that are different than those required in the applications running over the wireless LAN in the administration building.

As mentioned above, however, the major vendors have recognized a need for roaming capability and have specified IAPP to support it. Essentially, what the 802.11 standard has done is to ignore the need for communication between APs of different ESS zones. With this approach, the only way that APs could communicate is if they are all in one ESS. IAPP will allow seamless communications between APs located in different ESSs, and it is discussed briefly in a later section.

■ 802.11 PHYSICAL LAYER

Three classes of physical layers are currently specified in the 802.11 standard. It is clear that higher speed versions will emerge in the near future. This discussion addresses only those aspects that are currently standardized. Before going into the details of what is different between the physical layer options, however, let us discuss what is common across all three.

First, the various physical layer options all communicate to the MAC layer through several service access points (SAPs), as is shown in Figure 15-7. Next, both RF standards use the same frequency band, 2.4 GHz. Both RF standards provide the same services to the MAC layer above, the same layer management, and similar SNMP MIB functionality. Physical layer services focus exclusively on the functions of transmitting data and receiving data from the wireless media. These two functions are combined in a logical sense with a third function used to determine the state of the media prior to transmis-

Figure 15-7
IEEE 802.11 Physical and MAC Layer Interface

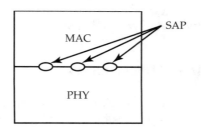

sion. This third function provides the carrier sense capability required for the MAC access methodology, CSMA.

The physical layer has three high-level services that it provides to the layer above: transmission (modulation), reception (demodulation), and carrier sense. All three use the capability of the physical layer to couple to the media through the use of an antenna. These services are best illustrated by discussing examples of how they would operate in two different scenarios. The first illustrates the operations for the transmission of data and the second illustrates the operations for the reception of data.

Transmission

When the physical layer is first powered up, it begins listening to the medium to determine if the channel is being used. This function is the carrier sense capability. There are two cases: either the medium is busy or it is not. If it is busy, the physical layer automatically routes the incoming data to the receive function and tries to determine the message. It also sends a signal to the MAC layer that the medium is busy. In the case where the medium is idle, the physical layer sends a signal to the MAC layer that the medium is idle. The MAC layer always makes the decision about whether to transmit or not. The physical layer provides only the carrier sense service.

Once the MAC layer makes the decision to send a frame, it passes the data frame, along with specific header information, to the physical layer SAP. It also sends to the physical layer the data rate chosen for this message. The data rate must be one standardized by the IEEE 802.11 committee for the implementation to be compliant. After a maximum delay of 20 ms, the physical layer sends a preamble sequence for the required period, and immediately after that sequence is completed, it sends the header and data frame.

Reception

In the case where the channel is busy and an incoming message is detected, the physical layer sends a message to the MAC layer to let it know that there is an incoming data frame. The physical layer uses three steps to differentiate a valid incoming message from noise. First, the RF energy level must be above some minimum threshold; the standard specifies at least 85 dBm. Second, the incoming message must contain a valid preamble. Third, once synchronized with the preamble, the physical layer must detect a valid header. If all three conditions are met, and after the signal that a valid frame has been detected is passed to the MAC layer, the header and frame contents are passed up the protocol stack. Finally, once the entire frame is read in, the physical layer sends a message to the MAC layer that the frame has ended.

The services discussed above are common to all physical layer options in the 802.11 standard. However, not all physical layer options can perform at the same data rate. These options are summarized in Table 15-1. Each physical layer option in the table has relative advantages and disadvantages, which will be addressed in the appropriate section. As long as all systems are deployed with the same physical layer option, data rate interoperability at the lowest common denominator is ensured. The level of that interoperability depends on the vendor mix of implementations chosen. For a unified management interface, for example, it is generally required that all access points be purchased from the same manufacturer.

Table 15-1
IEEE 802.11 Options

DSSS	FHSS	Infrared
1 Mbps	1 Mbps	1 Mbps
2 Mbps	2 Mbps	
5.5 Mbps		
11 Mbps		

■ PHYSICAL LAYER STRUCTURE

The 802.11 committee decided that, because there were three physical layer options, and because these options all used a different modulation technique, the best way (from a logical perspective) to combine them into a single physical layer was to divide the physical layer into two sublayers. The lower of the two, the physical medium dependent (PMD) sublayer, is more tightly coupled to the physical layer option chosen. However, both sublayers have unique characteristics. The upper of the two, the physical layer convergence procedure (PLCP) sublayer, insulates the MAC layer from differences in the PMD. The primary functions of the PLCP are two. The first is to define a physical layer "frame" for each physical layer option, really which PMD option. The second is to function as the logical service location for the three service functions (transmit, receive, and carrier sense) of the entire physical layer. This relationship is shown in Figure 15-8.

In a nutshell, the PMD does the actual transmission (modulation), reception (demodulation), and coupling to the media (through an antenna) under the control of the PLCP. Each PLCP is coupled with a specific PMD. We will discuss the physical layer from a bottom up point of view, taking first the PMD and then the PLCP. Because the direct sequence option is the only standardized high data rate option, 11 Mbps, this physical layer option will be discussed first.

Figure 15-8
IEEE 802.11 Physical Layer Structure

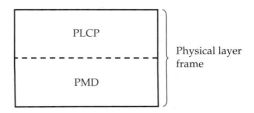

■ DSSS PMD

The direct sequence option has one key advantage over either the frequency hopping or infrared: the important advantage of increased range per access point. This advantage has another important consequence for any large deployment: lower overall costs due to fewer access points required.

A detailed discussion of how DSSS operates has already been provided in Chapter 2, so we will not repeat that here. However, it is appropriate to review certain specifics of the general DSSS approach used in 802.11. The CDMA DSSS technique specified for use in 802.11 networks uses only a *single spreading sequence or code word.*

As it is normally encountered, CDMA uses multiple spreading sequences to allow multiple individual users to share the same frequency channel in the same geographical space. It differentiates users through unique code words, hence, the name code division multiplexing. All mobile telephone systems use the traditional CDMA approach. The 802.11 standard allows multiple sets of users to share a single geographical space by using a single code word on multiple frequency channels. Individual users in a set are distinguished at the MAC layer by a unique MAC address.

Where the mobile telephony standard, IS-95, CDMA DSSS approach is to use the same frequency channel for several users, the 802.11 DSSS approach is to use multiple frequency channels to accommodate several sets of users or LANs. I want to be perfectly clear here. Mobile telephony and wireless LANs depend on the concept of frequency reuse, that is, the use of the same frequency again and again in geographically distinct areas or cells. They just implement the concept a bit differently.

Coverage and System Considerations

The 802.11 DSSS approach uses only a single code word in each BSS. The code word is the same in each BSS. To accommodate multiple sets of users in any single BSS, multiple frequencies with the same code word must be used. This feature is the most significant issue with deploying 802.11 DSSS systems. The important issue is how to lay out the AP so that coverage of mobile stations is assured. There are many choices, and both geographical coverage of the BSSs and frequency assignments to those BSSs must be considered. Figure 15-9 shows several examples of good and bad practices in this area. In the figure, the BSS range is shown by the circles, and the frequency assignments are noted by F1, F2, and F3. Each stands for a nonoverlapping frequency range within the total spectral allocation.

Since the total spectral allocation must be standardized for interoperability, there is a limited number of sets of frequency bands, hence, sets of users, in each BSS. We will calculate this number later. The spectral allocation that the DSSS standard specifies is also different in different parts of the world. The total spectral allocation in the United States is 2.400 GHz to 2.4835 GHz, or 83.5 MHz. In Europe, some countries have standardized on a similar bandwidth; others have standardized on as little as 22 MHz, as Japan has. A 22 MHz spectrum allotment allows only a single BSS or LAN in any area. It also guarantees collisions and poor performance at the boundary of each BSS.

Figure 15-9
Coverage Choices: (a) Good Choice; (b) Bad Choices

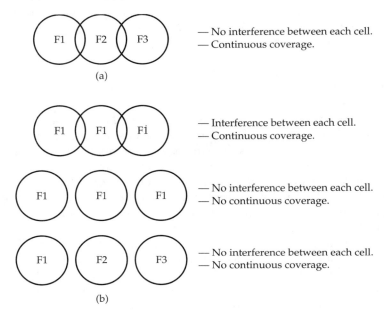

The reasons for the different spectral allocations in different countries are complex and are often intertwined with national pride and previous deployments of systems that made use of the chosen frequency band. As the 802.11 standard becomes commercially dominant, additional channels may open in several countries. If you plan on deploying a system outside the United States and Canada, check national frequency regulations carefully or you may find yourself in violation of the law, not to mention deploying a system that may be only partially or occasionally operational. As a general rule, it is usually best to purchase any wireless system in the country where it is to be deployed. You will thus have a much greater assurance that the implementation is designed with national regulations in mind.

The frequency spectrum is often expressed as a number of fixed bandwidth channels. The reasons why seem to grow from a traditional SCPC/FDM view of the wireless world by regulatory bodies. This conventional approach misleads the newcomer in understanding how the actual system works. Since DSSS uses a spread spectrum technology, each actual channel spans several of the 5 MHz channels specified in the standard. The frequency assignments for the United States, Canada, and portions of Europe are shown in Table 15-2.

These channel identifiers are important when you are configuring an access point in an ESS topology because the channel identifiers you see in Table 15-2 are presented to you. A common mistake is to assume that access points placed in geographical overlap will not overlap in frequency if different channel numbers are used. This assumption is

Table 15-2
802.11 DSSS Frequency Assignments

Channel	Frequency (GHz)
1	2.412
2	2.417
3	2.422
4	2.427
5	2.432
6	2.437
7	2.442
8	2.447
9	2.452
10	2.457
11	2.462

not true because each access point requires 22 MHz of spectrum, and a more careful choice of channel numbers is thus required.

The actual channels used by the standard can be understood by an examination of the processing gain. As we discussed in Chapter 2, the significant figure of merit for a DSSS system is its processing gain. It was given by the relationship expressed in Equation 15-1:

$$\text{Processing gain} = G_p = \frac{\text{channel BW}}{\text{symbol rate}} \tag{15-1}$$

The IEEE 802.11 committee specified a minimum processing gain of 10 dB. The processing gain is again a subject of national standardization. Example 15-1 illustrates this calculation for a 1 Mbps and 2 Mbps DSSS 802.11 system and shows that this minimum is met for both rates. Note that all four data rates standardized in the DSSS physical layer option use the same bandwidth.

■ **EXAMPLE 15-1** Find the processing gain of an IEEE 802.11 wireless LAN at data rates of 1 Mbps (DBPSK) and 2 Mbps (DQPSK) spread spectrum system. Because of the increased bandwidth efficiency of the 2 Mbps system, the symbol rate stays the same in both cases and is 500 Kbps. IEEE 802.11 systems use a channel bandwidth of 22 MHz. Processing gain is always best expressed as a logarithmic quantity.

$$G_p = \frac{11\,\text{MHz}}{500\,\text{kbps}} = 22 \rightarrow G_p = 10\log(22) = 13.4 \text{ dB}$$

If the channel bandwidth is always 22 MHz, how many simultaneous, nonoverlapping channels are possible at this data rate in the United States? The total allocation is

Figure 15-10
An Example of an 11-Bit Barker Sequence

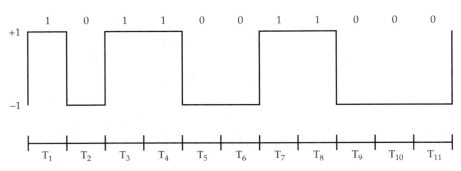

2.4835 − 2.4 = 83.5 MHz. If each channel is 22 MHz, there is room for three nonoverlapping channels, with 83.5 − 66 = 17.5 MHz left over. This leftover is sometimes used to provide for two more overlapping channels. Note that these last two overlapping channels also overlap a bit with one or two of the nonoverlapping channels. With proper geographical considerations, there are actually five nonoverlapping channels available. By making all the channels overlapping, up to six separate channels can be accumulated, albeit with interference effects where they overlap.

To minimize interference and collisions, minimize the frequency overlap. When radios using the same frequency overlap their coverage areas, interference is a result. Imagine that you are in your car and listening to your favorite radio station. As you drive away from the station's transmitter, at some point the signal will grow weaker and you start hearing another station's signal interfere. The analogy is not exact, but it will give you a simple real-world experience for remembering this simple rule.

Don't confuse frequency overlap with geographical overlap. It is necessary to have geographical overlap to maintain continuous coverage. Just make sure that overlapping BSSs do not use the same frequency space. If some geographical overlap of adjacent BSSs did not exist, seamless connectivity would not exist as a mobile station moved between them, and your session would terminate instead of being successfully handed off between BSSs. This feature points out one potential shortcoming of the DSSS option. If you need distinct wireless networks in the same geographical space, for example, a warehouse, DSSS limits you to six overlapping channels or, with proper layout, up to five nonoverlapping channels.

As we explained in Chapter 2 and in Example 15-1, the processing gain and symbol (chip) rate are closely related for DSSS systems. For this reason, the chip code used in 802.11 for DSSS 1 and 2 Mbps data rates is specified at 11 bits. It is a Barker sequence, and it is illustrated in Figure 15-10. The same Barker sequence is used in all APs, and it is always the same code: +1, −1, +1, +1, −1, +1, +1, +1, −1, −1, −1.

Power Limitations

Each country has its own rules and regulations concerning the amount of power that can be transmitted by a radio using the DSSS. This power level is standardized in two ways.

Table 15-3
DSSS PMD Modulation Summary

Data Rate (Mbps)	Transmit Bandwidth (MHz)	Bandwidth Efficiency	Modulation Type
1	22	1	DBPSK
2	22	2	DQPSK
5.5	22	4	CCK/DQPSK
11	22	8	CCK/DQPSK

The first rule concerns the amount of power actually coupled to the atmosphere with no antenna. In the United States, the power is limited to 1 W. Additionally, the 802.11 standard sets a lower limit of 1 mW, which all radios must comply with. Finally, for any radio capable of transmitting greater than 100 mW, a mechanism must be in place to allow the physical layer to be set to produce no more than 100 mW. Power levels in other countries vary but are generally lower than the 1 W rule in the United States. The 100 mW rule is set primarily to accommodate the lower power rules present in much of Europe.

Through the use of either omnidirectional or directed gain antennas, these power levels can be increased significantly. Additional standards are applicable to this direct radiation into the atmosphere, one of which was defined as ERP in Chapter 8. Because operation in this frequency band does not require a license, some maximum power level must be set to allow many different types of communication devices to coexist. The precise measurement specifications are outside the scope of this text. For most users, purchasing an antenna designed to work with the IEEE 802.11 standard and registered for use in the country where the system is being deployed is all that is necessary.

Modulation Techniques

The DSSS PMD at the 1 and 2 Mbps data rates uses a form of digital modulation called differential phase shift keying. Each PMD at a different data rate uses a slightly different form. This modulation technique is very similar to PSK discussed in Chapter 5.

Recall that in differential angle modulation techniques, the change in phase state is used to represent the input data rather than the absolute phase state. As was mentioned earlier in this chapter, each data rate in a PMD option uses the same bandwidth, which is accomplished by increasing the bandwidth efficiency of the modulation technique for higher data rate systems. This process is summarized in Table 15-3. (For a detailed discussion of bandwidth efficiency, refer to Chapter 5.)

Complementary Code Keying

The last two rows in Table 15-3 deserve some additional explanation. When operating at either 5.5 or 11 Mbps, the DSSS PMD uses a form of coding called complementary code keying (CCK). CCK significantly improves the multipath performance of the standard at 11 Mbps. Instead of using the Barker code, a series of complementary code sequences, sixty-four in all, are used. Instead of each bit being represented by a Barker code, up to

6 bits are represented by each code word. This technique increases the bandwidth efficiency of the modulation technique.

The CCK code word is then modulated using the same method as in the 2 Mbps data rate, DQPSK. If all code words were represented by 6 bits, then a data rate of 12 Mbps would result. This data rate is not achievable due to radio channel issues, but 11 Mbps is achievable and was standardized upon. Use of such a complex coding technique at 11 Mbps means that 11 Mbps radios are much more complex than those used for 1 and 2 Mbps.

Finally, because the power of any transmission is regulated by the FCC in the United States, and because additional modulation complexity is required to increase the data rate, the range is reduced as higher data rates are achieved. As a station moves away from an AP, in general the data rate will be reduced because the modulation complexity cannot be supported in the reduced signal level environment. Similarly, as a mobile station moves toward an AP, in general the data rate will increase. This characteristic is called rate shifting. Rate shifting is a physical layer mechanism that is transparent to upper layer protocols.

■ DSSS PLCP

As we discussed earlier, the function of the PLCP is to insulate differences in the PMD flavors from the MAC layer. It does so by encapsulating the MAC frame into a "frame" at the PLCP layer. Historically, the word *frame* has a limited meaning in data communications: it is a group of bits at layer 2 of the OSI model. However, the 802.11 standard uses the word *frame* frequently when describing the function of the PLCP. This word choice will become apparent in the discussion below. (In my view, this word choice was a mistake, even though the approach has a certain functional clarity.)

This book will no longer refer to this grouping of bits at layer 1 as any type of frame but rather as a physical layer convergence procedure PDU (PLCP PDU). I wish to reserve the term physical layer PDU (PPDU) for the group of bits that emerges from the physical layer and is impressed upon the medium. Figure 15-11 illustrates the PLCP PDU, its components, and its relationship to the DLPDU and the PPDU.

As you can see from Figure 15-11, the default or long PLCP PDU consists of seven distinct fields with a header length of 192 bits. The following list summarizes the purpose of each field:

Synchronization (SYNC): This field is the same as the preamble field in other members of the 802 family of protocols, for example, 802.3. It consists of a sequence of alternating 1s and 0s and is used by the receiving PMD to synchronize its bit, or more generally, its symbol clock, to the transmission. (The other form of clock synchronization that must take place involves the chip rate and is called the chip clock. Chip synchronization can be accomplished in various ways, but symbol synchronization is a necessary prerequisite.) This field and the entire PLCP PDU are always sent at 1 Mbps, independent of the data rate specified in the signal field, when the long preamble is used for the backward compatibility reasons stated earlier.

Figure 15-11
DSSS PLCP PDU: Long Preamble Option

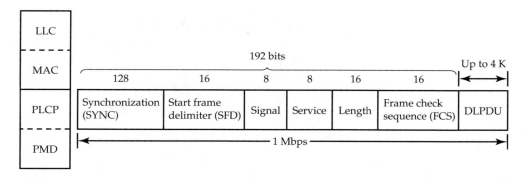

Figure 15-12
DSSS PLCP PDU: Short Preamble Option

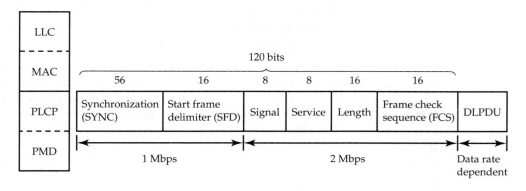

When the short preamble is used, the SYNC field is reduced to 56 bits from the 128 used in the long preamble. This field and the next are always sent at 1 Mbps, independent of the preamble length used. When the short preamble is used, however, the next four fields that follow the start frame delimiter in the header are sent at 2 Mbps instead of 1 Mbps. This technique significantly enhances throughput with the short preamble option. See Figure 15-12.

Start frame delimiter (SFD): This field is a unique series of bits used to define the beginning of the PLCP PDU (or frame). The series of bits never changes and is 1111001110100000.

Signal: This field is a series of bits that identifies the data rate of the DLPDU. Again, it is important to emphasize that the preamble and header of the PLCP are always

Table 15-4
DSSS Signal Code

Data Rate (Mbps)	Signal Code
1	00001010
2	00010100
5.5	00110111
11	01101110

sent at 1 Mbps. The value of the signal field is found by dividing the data rate in kbps by 100 kbps. Table 15-4 defines the values currently defined by the IEEE. As higher data rates are developed and standardized upon, additional entries will be added to this table.

Service: This field is reserved for future use.

Length: This field expresses the length of the DLPDU in microseconds. The format is an unsigned 16-bit integer. This field is used to let the receiving PLCP know when the end of the frame has occurred (a similar function, but expressed differently, as the length field used in the IEEE 802.3 standard).

Frame check sequence (FCS): This field is the CCITT CRC-16 cyclic redundancy code used throughout the IEEE 802 family of protocols. (See Chapter 7 for more detail on how it operates and is used to detect, not correct, errors.) Note that the determination of the code is performed in the data link layer because the circuitry and logic to perform this check and to take action, depending on the result, is already present at that layer.

DLPDU: This frame of data is passed to the physical layer for transmission from the MAC sublayer of the data link layer.

Before concluding this section, we will see the effect of the restriction that the PLCP PDU always transmits at 1 Mbps on actual data throughput of each of the standardized data rates. Example 15-2 will illustrate the performance penalty paid for this backward compatibility.

■ **EXAMPLE 15-2**   What is the maximum actual data throughput for each of the three data rates specified in the DSSS standard? As the size of the DLPDU shrinks, the effect worsens. For a DLPDU of 500 bytes, the data throughput rate at 11 Mbps drops to around 7 Mbps. This example uses the long preamble option and first calculates the data throughput using a 4 kbyte DLPDU length and then a 500 byte DLPDU length.

Actual data throughput can be approximated by taking the DLPDU length and dividing it by the length of the PLCP PDU, then multiplying by the data rate. This approach assumes continuous point-to-point transmission and neglects the header field contributions at all higher layers. These contributions can be neglected in the first order approximation. They will scale well because they will be transmitted at the actual data rate of the transmission, while the contents of the "header" of the PLCP PDU are transmitted at 1 Mbps, independent of the data rate of the rest of the transmission.

$$DT = \frac{(\text{data length})(\text{bit period of data})}{(\text{header length})(\text{bit period of preamble}) + (\text{data length})(\text{bit period of data})}(\text{data rate})$$

DT = data throughput

For a DLPDU length of 4 kbytes:

$$DT_{1 \text{ Mbps}} = \frac{(32,768 \text{ bits})(1 \times 10^{-6})(1 \times 10^{6})}{(192 \text{ bits})(1 \times 10^{-6}) + (32,768 \text{ bits})(1 \times 10^{-6})} = 0.997 \text{ Mbps} \approx 1 \text{ Mbps}$$

$$DT_{2 \text{ Mbps}} = \frac{(32,768 \text{ bits})(0.5 \times 10^{-6})(2 \times 10^{6})}{(192 \text{ bits})(1 \times 10^{-6}) + (32,768 \text{ bits})(0.5 \times 10^{-6})} = 1.98 \text{ Mbps} \approx 2 \text{ Mbps}$$

$$DT_{11 \text{ Mbps}} = \frac{(32,768 \text{ bits})(0.09 \times 10^{-6})(11 \times 10^{6})}{(192 \text{ bits})(1 \times 10^{-6}) + (32,768 \text{ bits})(0.09 \times 10^{-6})} = 10.4 \text{ Mbps} \approx 11 \text{ Mbps}$$

For a DLPDU length of 500 bytes:

$$DT_{1 \text{ Mbps}} = \frac{(4,000 \text{ bits})(1 \times 10^{-6})(1 \times 10^{6})}{(192 \text{ bits})(1 \times 10^{-6}) + (4,000 \text{ bits})(1 \times 10^{-6})} = 0.954 \text{ Mbps} \approx 1 \text{ Mbps}$$

$$DT_{2 \text{ Mbps}} = \frac{(4,000 \text{ bits})(0.5 \times 10^{-6})(2 \times 10^{6})}{(192 \text{ bits})(1 \times 10^{-6}) + (4,000 \text{ bits})(0.5 \times 10^{-6})} = 1.82 \text{ Mbps} \approx 2 \text{ Mbps}$$

$$DT_{11 \text{ Mbps}} = \frac{(4,000 \text{ bits})(0.09 \times 10^{-6})(11 \times 10^{6})}{(192 \text{ bits})(1 \times 10^{-6}) + (4,000 \text{ bits})(0.09 \times 10^{-6})} = 7.17 \text{ Mbps} \approx 7 \text{ Mbps}$$

As you can see from Example 15-2, a penalty must be paid for this backward compatibility when using the long preamble or the short preamble. It is much smaller, however when using the short preamble. Using the same approach as above, note the significantly higher data throughput rates with the short preamble and a DLPDU length of 500 bytes:

$$DT_{1 \text{ Mbps}} = 0.971 \text{ Mbps} \approx 1 \text{ Mbps}$$

$$DT_{2 \text{ Mbps}} = 1.91 \text{ Mbps} \approx 2 \text{ Mbps}$$

$$DT_{11 \text{ Mbps}} = 8.77 \text{ Mbps} \approx 9 \text{ Mbps}$$

The IEEE 802.11 committee felt that organizations that were early adopters and that supported the effort by purchasing 1 and 2 Mbps systems should not now be asked to discard their investment. Actual throughput numbers will only approach these estimates. Realistic data throughput for each of these systems is probably 10 to 20 percent less. While this penalty is demonstrated for the DSSS system only, the concept applies to any 802.11 compliant option when backward compatibility is maintained in this manner. Because FHSS and infrared options are not currently standardized with an 11 Mbps data rate, the effect is not as pronounced.

■ FHSS PMD

The FHSS physical layer is also composed of a PMD and a PLCP. We have already discussed in detail in Chapter 2 how a frequency hopping approach works. Now, we will focus on those aspects of frequency hopping specific to the 802.11 standard. The standard specifies a number of channels for the FHSS PMD to use. These channels are in a similar frequency range as those used in DSSS. Both systems use a portion of the industrial, scientific, and medical (ISM) band. The channel bandwidth, however, is much narrower because frequency hopping moves the location of the transmitted envelope instead of spreading it out. Each channel is 1 MHz wide. For the United States and most of Europe, seventy-nine channel locations are defined. In Japan, twenty-three channels are standardized upon.

In the United States and most of Europe, the frequency band extends from 2.40 GHz to 2.480 GHz, for a total system bandwidth of 80 MHz. In Japan, the frequency span is from 2.473 GHz to 2.495 GHz, for a system bandwidth of 22 MHz. Contrasting the case seen in DSSS, here the idea of a channel as defined in a standard and the channel actually used are the same. The channels are regularly spaced across the spectrum, with specific hopping sequences across this spectrum, to provide the seventy-nine channels.

The speed at which the PMD hops between these channels is regulated by country. In the United States, the minimum hopping rate is set to 2.5 hops per second. Additionally, the minimum distance in frequency from one hop to the next is 6 MHz. As a result of these regulations, in the United States and Canada, three sets of hopping sequences are specified. Each of these sequences has twenty-two different frequency locations to which the PMD hops. These frequency locations are shown below and the pattern is easy to see. The number in the sequence corresponds to the frequency channel number. For the United States:

Set 1: 0, 3, 6, 9, 12, . . . 66, 99, 72, 75

Set 2: 1, 4, 7, 10, 13, . . . 67, 70, 73, 76

Set 3: 2, 5, 8, 11, 14, . . . 68, 71, 74, 77

For Japan, with only 22 MHz allocated, the sequences specified have a fewer number of hops:

Set 1: 6, 9, 12, 15

Set 2: 7, 10, 13, 16

Set 3: 8, 11, 14, 17

One easy way to tell if a manufacturer's implementation meets the 802.11 standard is if the hopping sequences are the same as those listed above. Sometimes a manufacturer will use the upper portion of the allocated spectrum and have hop numbers higher than 77. When an implementation does not conform to the same hopping sequences, it will not interoperate with 802.11 FHSS implementations and should be considered a proprietary system.

The regulations for the minimum hop rate and minimum hop frequency change result from analysis on interference between radios in the same cell. These hopping sequences are theoretically defined to guarantee that any two hopping sequences will only occupy the same frequency space, or guarantee that they will collide with each other only once for every complete sequence. Because of the frequency spillover outside the 1 MHz channel, these collisions occur much more frequently in actual fact. To understand the frequency of these collisions, you must know the actual bandwidth consumed by each transmission. Studies have shown that the actual bandwidth consumption is more like 4 to 5 MHz, not 1 MHz. As a result, typical implementations will feature up to five collisions per sequence.

Power Limitations

Like the DSSS situation, there is a power level specification with no antenna gain. For FHSS, the maximum value is 100 mW, again to comply with European regulatory bodies. The use of antennas may significantly raise the actual ERP into the atmosphere. The general rule is to check with the national body to determine what kind of antennas are permitted. Typically, if the system is purchased in the same country where it is to be deployed, your supplier will already know and comply with these regulations. Like the DSSS situation, the minimum transmit power that must be supported is 10 mW, and there is typically a way to adjust power transmission during initialization of the system.

Modulation Techniques

The FHSS PMD uses a form of digital modulation called Gaussian frequency shift keying (GFSK). Again, like the DSSS situation, higher data rates use the same modulation technique with a higher bandwidth efficiency, essentially packing more bits per symbol. One Mbps systems use two frequencies and the 2 Mbps systems use four frequencies, resulting in a doubling of bandwidth efficiency. For more detail on Gaussian techniques, see Chapter 5. The only real difference is the way the frequencies used are related to each other.

Like any FSK system, the carrier is shifted, or keyed, by some amount to represent a logical 1 or a logical 0. The IEEE 802.11 specification requires that the minimum shift be at least 110 kHz. Refer back to Chapter 5 to learn why an FSK based system, while still featuring the double sideband generation, pays no penalty for it (why its bandwidth efficiency changes with increased modulation density, but its actual bandwidth consumption, *at any one time*, does not).

■ FHSS PLCP

For the same reasons as were stated in the section about DSSS PLCP, we will use the term PLCP PDU for the structure of the protocol unit passed between the two sublayers of the physical layer and reserve the word *frame* for layer 2 PDUs. Figure 15-13 shows the FHSS

Figure 15-13
FHSS PLCP PDU

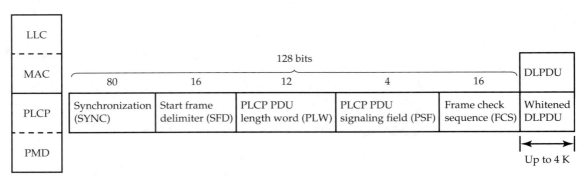

PLCP PDU. As you can see, the PLCP PDU consists of six distinct fields, with a header length of 128 bits. The following list summarizes the purpose of each field:

Synchronization (SYNC): This field is the same as the preamble field in other members of the 802 family of protocols, for example, 802.3. It consists of a sequence of alternating 1s and 0s and is used by the receiving PMD to synchronize its bit, or more generally, symbol clock, to the transmission. There is no chip synchronization concept for FHSS systems. This field and the entire PLCP PDU is always sent at 1 Mbps, independent of the data rate specified in the signal field, as was the case for the DSSS system, for the backward compatibility reasons stated earlier.

Start frame delimiter (SFD): This field is a unique series of bits used to define the beginning of the PLCP PDU (or frame). The series of bits never changes and is 000011001011110. Note that the bit pattern here is almost an inverse of the bit pattern used in DSSS systems.

PLCP PDU length word (PLW): This field expresses the length of the DLPDU in octets. The format is an unsigned 16-bit integer. This field is used to let the receiving PLCP know when the end of the frame has occurred, which is a function similar to the length field used in the IEEE 802.3 standard. Both use length as measured in octets.

PLCP PDU signaling field (PSF): This field is a series of bits that identifies the data rate of the whitened DLPDU. Again, it is important to emphasize that the entire PLCP is always sent at 1 Mbps. Table 15-5 defines the values currently defined by the IEEE. Some entries in this table are not standardized. As higher data rates are developed and standardized upon, nonstandardized entries in this table are likely to change and additional entries added.

Frame check sequence (FCS): This field contains the CCITT CRC-16 cyclic redundancy code used throughout the IEEE 802 family of protocols. (See Chapter 7 for more detail on how it operates and is used to detect, not correct, errors.) Note that the determination of the code is performed in the data link layer because the cir-

Table 15-5
FHSS Signal Codes

Data Rate (Mbps)	Signal Code
1.0	0000
1.5	0001
2.0	0010
2.5	0011
3.0	0100
3.5	0101
4.0	0110
4.5	0111

cuitry and logic to perform this check, and take action depending on the result, is already present at that layer.

Whitened DLPDU: This frame of data is passed to the physical layer for transmission from the MAC sublayer of the DLL. Before transmission, the physical layer whitens (scrambles) the data frame received from the MAC layer by inserting special symbols every four octets. Scrambling minimizes the DC bias of the overall data signal and ensures a more random distribution of energy in the spectrum.

■ INFRARED

The infrared standard is based on the same technology as that used for remote control of your CD player, television, etc. The structure of the standard at the physical layer is very similar to those just discussed. It consists of a PMD and a PLCP. The PMD uses pulses of light in the infrared band (850 to 950 nm) to send data symbols. The modulation method used is a variation of the principle of pulse position modulation, which was outlined briefly in Chapter 6.

Like the remote control standard used by many consumer appliances, each symbol uses a sequence of 16 pulses to represent it. Instead of varying the pulse position within the bit interval, as is done with pure PPM, however, the infrared standard sends a 16 bit sequence where only one bit is a 1 and all others are 0s. In effect, the pulse, the single 1 bit, moves within the 16-bit sequence. Most commercial implementations of PPM use this approach. Table 15-6 defines the 4-bit data word or symbol for each pulse sequence.

Infrared is an attractive technology for short-range wireless links. Many handheld devices use such an infrared link to synchronize their contents with similar content stored on a desktop computer. Early in the standardization process, when many radio ideas seemed too complex for a reasonable commercial implementation of the ad-hoc networking concept, infrared seemed a natural choice. As the radio problems were solved one by one, however, no commercial implementations of the infrared standard ever appeared.

Table 15-6
Infrared Standard PPM Codes

4-Bit Data Word to Be Transmitted	16-Bit PPM Symbol Transmitted
0000	0000000000000001
0001	0000000000000010
0010	0000000000000100
0011	0000000000001000
0100	0000000000010000
0101	0000000000100000
0110	0000000001000000
0111	0000000010000000
1000	0000000100000000
1001	0000001000000000
1010	0000010000000000
1011	0000100000000000
1100	0001000000000000
1101	0010000000000000
1110	0100000000000000
1111	1000000000000000

■ MAC LAYER

Like any layer designed after the OSI model was developed, a layer can be defined by the services it offers to the layer above. The 802.11 MAC layer is no different. The 802 model is constructed so that the MAC sublayer provides services to the LLC sublayer, and the two combine to form the data link layer. In general, the LLC layer defines the service interface to the network layer and is common across all 802.x implementations. The MAC layer defines how bandwidth is shared among members of its address space. Be sure to review the section on CSMA/CA in Chapter 2, where the specifics of how 802.11 access to the medium functions. Many important mechanisms of the 802.11 standard, including the handshake sequences and MAC primitives critical to the effectiveness of the 802.11 standard, are discussed in that chapter.

In a wired environment, the MAC layer address space is defined by the LAN segment to which the stations are attached. The segment is defined by the interconnection points to other LAN segments, usually through a router, bridge, or switch. In a wireless environment, the same concepts hold; however, the definition of an address space is not static because wireless mobile units will roam in and out of any BSS. The wireless LAN segment, or BSS, is dynamic: it changes with time much more rapidly than it would in a wired environment. (I suppose one could argue that any LAN segment, wired or not, is dynamic in the sense that even fixed workstations occasionally are added or dropped. However, this kind of change does not take place during a transmission.)

In a wired environment such as Ethernet, the protocol for sharing bandwidth on the segment is CSMA/CD. The wireless LAN environment is similar. The protocol used for the 802.11 MAC is CSMA/CA. However, the fundamental nature of wireless media requires enhancements to the layer 2 protocol. There are two fundamental differences: wireless links are wireless, and wireless stations are mobile.

Because the transmissions are broadcast into "thin air," and are, in principle, available to anyone with a receiver capable of demodulating them, there is a great deal of interest in encryption. The lower bandwidth of wireless links, as compared to wireline, suggests the use of data compression techniques. Since there is a large difference between the transmit power of the mobile unit and the typical power of a received signal, the transmission of a signal causes the mobile unit to lose detection of any other signal. This characteristic is fundamental to why collision avoidance techniques are used in 802.11 rather than the more reliable collision detection techniques.

Second, because the mobile stations are usually moving, the signal-to-noise/interference level of a received signal changes with change in position. The more rapid the movement, the more rapid and variable this change. The MAC layer defines when the channel is busy or clear and uses this information to facilitate link establishment and maintenance. The physical layer defines how data bits are represented in that channel.

Beyond these two basic concepts, however, there are additional MAC layer functions specific to every protocol stack. These services are specific to the underlying physical layer and, as such, are included in the MAC layer rather than the LLC layer. Recall that the AP was defined as having bridge functionality, which means that it is defined at the DLL layer. Therefore, the 802.11 standard is broken into two logical parts:

1. Additional functionality needed by mobile units.
2. Additional functionality needed by the APs.

There are three broad groups of these services: MAC frame delivery, station services, and distribution services. The first is common to all 802 projects and is nothing more than a mechanism to ensure that all frames are delivered between MAC SAPs; it will not be further explored here. The next two are specific to 802.11 and are discussed next.

■ STATION SERVICES

The two subgroups of station services are driven by one central fact: security on a wireless network is much more problematic than on a wired one. These two subgroups of station services are:

1. Authentication and de-authentication.
2. Privacy.

A wireless network is really a group of wireless stations moving around and transmitting file contents. The bridges or APs are also looking for mobile stations to roam into their BSSs. The APs are eager to establish trust relationships with these mobile stations. It is clear that a decision must be made about who is a member of a particular segment or BSS and who is not. This decision occurs in the authentication step. The 802.11 standard defines two general methods for this process, which are discussed in more detail in the next two sections. One way to think of these two elements is that authentication is a networkwide password, while privacy is offered only if some type of encryption is used.

Open System Authentication

This simple, two-step process is the default authentication procedure. The station wanting to communicate with another station, be it another mobile station or AP, sends an authentication management frame telling the other station its MAC address. The receiving station sends back a frame indicating whether this MAC address is in its authentication table or not. If it is, authentication is accomplished. If it is not, authentication is not granted.

There are some obvious problems with this process, including updating of MAC tables for all mobile stations. Updating would probably be done through broadcasts by the APs. In that case, all a person has to do is passively listen to the MAC table update, and he or she has obtained the information necessary for forging an authentication into that network.

Shared Key Authentication

This approach tries to address the obvious fault identified in the preceding section through the use of a shared key that is never distributed over the atmosphere. Instead, the key is defined for distribution to all stations in some other secure way, independent of the 802.11 network. Stations then authenticate through use of the shared key. This process is defined by the wireless equivalent privacy (WEP) algorithm.

The shared key approach is a four-step process. A station desiring to communicate with another station sends an authentication frame to the other station. Instead of responding as in the open system authentication approach, however, the receiving station sends back 128 octets of "challenge" text. The requesting station then encrypts the text using its shared key and sends this encrypted text back. The receiving station then decrypts using the shared key and compares. Depending on the outcome of that comparison, it will either grant or deny authentication.

This arrangement is obviously superior to the open system approach. The key is never transmitted over the atmosphere for a passive listener to receive. The concern among many adapters of this technology is that the key length is too short, hence the interoperability issue discussed below.

Privacy

The privacy issue with any wireless medium is always a concern. Even if someone does not want to enter the network, he or she can listen and capture data that may be quite confidential. All business enterprises must address the issue of privacy in any wireless communication system. WEP also addresses this issue in a very similar manner to the shared key approach. It is widely considered inadequate, however, and so most manufacturers have upgraded security provisions in their AP product lines.

This situation presents a potential interoperability issue. As mentioned earlier, APs designed by different manufacturers will often not interoperate seamlessly. If you want to upgrade the security, and almost everyone does, you must purchase it from your original AP supplier. This is understandable from two perspectives. First, a better security system will always be available, and second, manufacturers can distinguish themselves in the marketplace by deploying a unique security system. The decision seems like an easy one for the user: go with the best security. However, in 802.11 networks with more than one vendor's AP deployed, this situation will pose interoperability problems because different vendors deploy different security enhancements, and they are not always interoperable.

Since the encryption is placed into the MAC layer, this issue affects both mobile user NIC card selection and AP selection. In some cases, upgraded security is available as a software patch on the AP. In this case, the APs then distribute this operational patch to the mobile stations in the normal manner of communication with mobile units once authentication has been achieved. The shared passwords, of course, are not distributed in this way.

■ DISTRIBUTION SERVICES

As alluded to above, the APs provide the distribution services. Distribution services are needed to provide functionality across several stations instead of to a single station. These services include the necessary functionality to route frames to destinations in a single BSS and to handle mobile stations roaming between BSSs. In general, the distribution of stations is considered to be part of the same BSS, although distribution services also include the functionality to send frames across BSSs and portals. There are five service classes; the first three deal with station mobility.

Association

Association is the name for the service that associates a mobile station to a specific AP within a BSS. Until a mobile station is associated with an AP, no communication between it and any other station on the distribution system can occur. Once completion of the association of a mobile station to a single AP occurs, authentication begins. This association is one-to-one for the mobile stations and one-to-many for the APs. Association is initiated by the mobile station because it senses the signal of the AP. As the signal from the BSS AP that a mobile station is entering grows larger than the signal from the BSS AP where it currently is associated, the mobile station begins an association with the new AP.

Disassociation

This service allows removal of an association. While association is always initiated by the mobile station, disassociation can be initiated by any class of station, either AP or mobile station. Disassociation is a mandatory service; that is, once a station receives a disassociation request, it must grant it. Refusal is not permitted. Generally, mobile stations disassociate when leaving a BSS, and APs disassociate only when removed from service. When mobile stations wish to move to another BSS inside an ESS, the re-association service is used. Mobile stations use this service to *terminate an association*.

Re-association

This service allows mobile stations to maintain communication with other stations in the distribution system when they roam from one BSS to another. A mobile station uses this service to *change its association*. Only mobile stations can initiate a re-association request. Mobile stations use this mechanism to change their association from one AP to another.

Distribution

This service is used by any station to route frames across the distribution system. The distribution service identifies the BSS where the mobile station to which the frame is addressed is communicating. The distribution service provides the BSS to which the frame is addressed to the distribution system. Since a distribution system can be of many different types, no specifics about how the distribution system actually performs the routing are provided. To address a station that is behind a portal, the integration service is used.

Because the distribution system can be of many different technologies, 802.11 cannot specify how frames are routed between APs. Remember, the 802 scope is limited to the lower two layers of the OSI model. The distribution system may very well require the services of layers 3 and 4 to route to a destination, but these services are outside the scope of the project. The committee can specify the layer 2 service primitives that will allow roaming between APs, but there is no guarantee that every vendor will implement them in the same way. This situation raises potential interoperability issues when the APs are not purchased from a single vendor.

Integration

This service is used by any station to route frames through a portal. It gets the frame to the portal. From that point on, how the frame is ultimately routed to its destination is outside the scope of the standard for the same reasons discussed in the paragraph above.

■ MAC FRAME STRUCTURE

No discussion of the MAC layer of a standard is complete without some discussion of the MAC frame structure and its components. As you might imagine, this frame structure is used to hold both data and control frames. Each of the above services has one or more control frames associated with it. The detailed frame exchanges that implement the services are outside the scope of this book.

One of the more interesting aspects of the MAC frame is that it can fragment the data frames at layer 2. The data frames are broken into smaller frames, which tends to lower the susceptibility of the frame to interference. MIB parameters must be set so that they are representative of the actual interference environment in which the station finds itself for maximum information throughput.

Like any HDLC type frame, the 802.11 frame definition falls into two segments: the header fields and data or control information fields. Figure 15-14 shows the 802.11 MAC frame. The following sections give brief descriptions of each field.

Figure 15-14
802.11 MAC Frame Structure

2	2	6	6	6	2	6	0–2,312	4
Frame control (FC)	Duration ID	Address 1	Address 2	Address 3	Sequence control	Address 4	Frame body	Frame check sequence (FCS)

Figure 15-15
802.11 MAC Header

2	2	4	1	1	1	1	1	1	1	1
Protocol version	Type	Subtype	To DS	From DS	More Frag	Retry	Power Mgmt	More Data	WEP	Order

2 bytes

Frame Control (FC)

This field is 2 octets long and specifies the frame type and subtype. Four types of frames are defined: data, control, management, and reserved. Each type has several subtypes. Details of the MAC header are shown in Figure 15-15, where each field is identified briefly.

Protocol version: This field holds the version number of the 802.11 standard. All current version numbers are 0.

Type: This field is one of two that determine the function of the frame. There are three types of frames, each with several subtypes. The types of frames allowed are data, control, and management. A reserved type field is identified for future use.

Subtype: This field is the second of two that determine the function of the frame. All subtypes are one of the types identified in the previous fields.

To DS: This field is set to 1 if the frame's destination is the distribution system.

From DS: This field is set to 1 if the frame must exit the distribution system.

More Frag: This field is set to 1 if the frame is fragmented. A 1 implies that more fragments are to come. If it is set to 0, the fragment is the last one.

Retry: This field is set to 1 if this frame is a retransmission; it is set to 0 for all other transmissions.

Power Mgmt: This field is set to 1 if the station is in power save mode. It is set to 0 otherwise.

More Data: This field is set to 1 if the frames in the sending station are buffered.

WEP: This field is set to 1 if data was encrypted by WEP.

Order: This field is set to 1 if frames must be strictly ordered.

Duration ID

This field is 2 octets long. It is used primarily when packet length is such that packets are fragmented across several frames. In this case, it identifies to the destination station the length of the next fragment. This information is used to enhance the throughput of the network because stations use this data to hold off transmitting. Fewer collisions means higher overall throughput.

Addresses 1, 2, 3, and 4

These fields are 6 octets long and contain addresses. They may contain several different address types, but they always include the source address (SA) and the destination address (DA). Depending on the function of the frame, the transmitting station address (TA), receiving station address (RA), or the basic service set identification (BSSID) may also be included. The BSSID is the address of the access point. It is sometimes known as the network identity because this MAC address is where the distribution network routes any frames directed to a mobile station. The assignment of BSSIDs is an important task for a network administrator. Most APs are shipped with a default BSSID. However, all APs in a single network must have different BSSIDs, which can be changed with the management tools shipped with most APs. There also are provisions for multicast and broadcast addresses. Following traditional practice, the broadcast address is all 1s and it sends the same message to all stations.

Sequence Control

This 2-octet field is where sequence numbers and frame fragment identifiers are kept. Four bits are reserved for fragmentation purposes, and 12 bits are used for sequence numbers. Each fragment of a frame uses the same sequence number and a unique fragmentation number.

The presence and operation of this field is critical to the ability of 802.11 to handle interference. This field allows fragmentation of the MAC frames, which reduces susceptibility to interference. The shorter the frame, the shorter the amount of time it is in the channel and hence the less probable that it will meet with interference. Only point-to-point frames can be fragmented, and the fragmentation size can vary from 256 to 2,048 bytes.

An MIB parameter can be set to define the maximum frame size. If an LLC PDU is received with a data size larger than the one set, fragmentation is performed. As a general rule, this parameter should be set at its maximum in an interference-free environment to ensure the maximum throughput. As interference increases, it should be set progressively smaller.

When interference is high or large numbers of retransmissions are occurring, one way to increase throughput is to decrease the time it takes for retransmissions. Smaller frames can be retransmitted more rapidly than can large frames. Hence, it makes sense to set the frame size small to reduce the probability of a collision due to interference.

When interference occurs, retransmissions are quicker with smaller frames. The retry field in the MAC frame header indicates if the current transmission is a retransmission.

A related MIB parameter allows setting of the retransmission wait time from 1 to 30 seconds. This parameter sets the amount of time that the source station will wait for a response from the destination indicating reception of the frame. It is also possible to set the number of retransmissions that will be attempted by the source station. This parameter can be set from 0 to 64. It is critical that these parameters are set appropriately for the interference environment to achieve maximum information throughput.

Frame Body

This field is used to carry the information in the data frame. It can also be used to carry information associated with either the control or management frame types. The maximum length of this field is 2,312 octets, or just over 2 kbytes. The minimum size is 0 octets, a size used when the frame type has no information to carry. Many control frames have no content other than that expressed in the frame control field. Although the PLCP PDUs allow a data payload of up to 4 kbytes, the MAC frame limits this number to 2,312 bytes, the more important number.

Frame Check Sequence (FCS)

This field is the traditional 32-bit CRC error-detection algorithm used throughout the 802 project.

■ POWER MANAGEMENT

One of the key concerns with mobile units is power consumption. Many of these units are battery powered, so a way to power down the NIC card when it is not being used is of some concern. Power management is implemented in the AP units and will vary depending on the vendor's implementation choices. It should be pointed out, however, that bits in the frame control and duration/ID fields implement this function. This issue is another item to consider carefully when choosing vendors for an AP deployment.

The default mode for these systems is known as constant access mode (CAM). In this mode, stations constantly drain power listening to the network for data directed to them. The power management mode is called polled access mode (PAM). In this mode, stations wake up periodically and listen for a special frame called a traffic indication map (TIM). TIM is a list of all stations having buffered frames at the access point. When this option is implemented by system administrators, it can allow stations to power down the radio for long periods of time and thus conserve power.

All stations on the network must have the same wake-up period so they can all hear the TIM frame at the same time. When a station knows from a TIM frame that it has data frames waiting at the AP, it keeps the radio active until those frames are received. When the frames are received, the station again shuts down the radio and goes to sleep until the next TIM interval. Using the PAM approach can dramatically reduce the power consumption of the stations. Specific ratios depend on the level of traffic in the network, but factors of 10 to 100 are not atypical.

A related frame is the delivery traffic information map (DTIM) frame. This frame also has a timer associated with it, and the time duration is always an integer multiple of the TIM interval. DTIM frames are transmitted when a broadcast message is received at the AP.

■ NETWORK MANAGEMENT

The 802.11 standard has several SNMP MIB commands that allow you to access the MAC layer functions. These functions are accessed using the standard SNMP commands GET-request and SET-request.

■ INTERACCESS POINT PROTOCOL (IAPP)

IAPP is still an emerging standard, so this section is somewhat speculative. However, many vendors have demonstrated interoperability between APs using this protocol. IAPP uses IP/UDP as a network and transport layer protocol and has two basic functions. The first of these functions or subprotocols is the announce protocol; the second is the handover protocol. They work together to allow APs to coordinate and exchange information.

As you might imagine, the announce protocol informs other APs about a new AP and has mechanisms to propagate this information throughout the network address domain. The handover protocol enables the core association and disassociation services to handle inter-AP communications and to provide MAC layer filtering for forwarding frames to those mobile stations roaming between different ESSs.

■ 802.11a EMERGING STANDARD

The current 802.11 standard, which has been the focus of this chapter, has many parts. Commercial interest is concentrated in the first completed extension to that standard: 802.11b, which operates in the frequency band of 2.4 GHz and has a maximum transmission rate of 11 Mbps. The next extension to the standard now expected to emerge is the 802.11a specification. It operates in the 5 GHz band and features a transmission rate of up to 54 Mbps.

The bandwidth available in the unlicensed ISM spectrum at 5 GHz is much greater than that available at 2.4 GHz. In the United States, 300 MHz of spectrum is available; in many parts of Europe, 455 MHz of spectrum is available. In most parts of Europe, however, this bandwidth is already dedicated to a particular system called Hiperlan. As a result, 802.11b will not likely become an international standard. It will be applicable only in the United States, Canada, Mexico, and parts of Asia.

The spectrum in the United States at 5 GHz is broken into two pieces, each 150 MHz in bandwidth. They are known as the unlicensed—national information infrastructure (U-NII) band. The two spectral fragments are found at 5.15 GHz–5.25 GHz and 5.75

GHz–5.825 GHz. Any RF system must generate more transmit power as it moves up in frequency, and 802.11 is no different.

Many similarities exist between the physical layer of Hiperlan and the proposed specification for 802.11a; so many similarities, in fact, that many believe the two can be harmonized to achieve a global specification for 56 Hz wireless LAN technology. This global specification has not yet been realized, but it would be an important milestone toward making 802.11 technology the worldwide standard in wireless LANs, just as it is in wired LANs.

■ SUMMARY

While there was a lot of technical data in this chapter, deploying an 802.11 LAN is really not much different than deploying an 802.3 LAN. The first detail you must ensure is that the data rate of the LAN will support the applications you need. Because the throughput numbers of an 11 Mbps DSSS 802.11 LAN are of the same order of magnitude as a 10BaseT LAN, you can use the 10BaseT as a rough guide, if you err on the conservative side.

Choosing between the DSSS and FHSS or infrared is similar to choosing between using twisted pair or optical fiber in a wired Ethernet environment. Usually, this choice is of little real impact to the applications needs. It should be driven by throughput requirements rather than any particular technology bias. Assignment of AP addresses and the decision about building a large single ESS or multiple zones bring up the same issues as the decisions concerning segmentation in a wired LAN environment. With the appearance of IAPP, roaming between ESS zones seems assured.

Finally, there are some significant issues related to how wireless networks will interact with other stations at layers 3 and 4 of the OSI model. These issues are discussed in Chapter 16 from the perspective of a general wireless network.

REVIEW QUESTIONS

1. What are the three topological choices offered by 802.11?

2. What application areas are most appropriate for each topological choice?

3. Define each of the following: BSS, IBSS, and ESS.

4. What role does a portal take in an 802.11 LAN?

5. Discuss roaming inside an ESS.

6. Discuss roaming outside an ESS.

7. What are the three high-level physical layer services?

8. Describe the steps the physical layer takes to provide transmission service.

9. Describe the steps the physical layer takes to provide reception service.

10. What are the two sublayers of the physical layer? What are their respective roles? Why are the two sublayers defined?

11. Discuss the difference between geographical overlap and frequency overlap.

12. What are the frequency ranges and channel number assignments of:
    a. The three nonoverlapping channels?
    b. The five nonoverlapping channels?
    c. The six overlapping channels?

13. What code is used in 1 and 2 Mbps DSSS systems and why?

14. What code is used in 5.5 and 11 Mbps DSSS systems and why?

15. Discuss the relationship between code word choice in a spread spectrum system and the number of bits represented by each code word. Relate your discussion to the bandwidth efficiency of the modulation technique used.

16. What portion of the long preamble DSSS PLCP PDU is sent at 1 Mbps? Discuss.

17. What portion of the short preamble DSSS PLCP PDU is sent at 1 Mbps? Discuss.

18. What are the four steps in the shared key approach utilized by WEP? Discuss.

19. What are the distribution services? What function does each perform?

20. Discuss the frame fragmentation performed by the MAC layer. Why is it implemented?

21. Discuss the importance of the sequence field in the MAC frame.

22. Discuss the relationship between the MAC frame body size and the PLCP PDU size.

23. Describe the two sublayers of the physical layer in IEEE 802.11.

24. Why does IEEE 802.11 use CSMA/CA (collision avoidance) instead of CSMA/CD (collision detection) in the MAC layer?

25. Describe the access scheme of CSMA/CA.

26. Discuss the two types of authentication in IEEE 802.11.

27. Why is power management so important in IEEE 802.11?

16

Layers 3 and 4 in a Wireless Environment

■ INTRODUCTION

This chapter examines layer 3 and layer 4 protocols in a wireless deployment. Many data oriented wireless networks use the TCP/IP protocol stack, so the bulk of the chapter examines this architecture. However, SS7 is also capable of supporting data flows in a wireless environment and is widely deployed in cellular networks. The two protocol architectures differ in many ways, but the most important element for wireless networks is how the two architectures differ in their control philosophies.

Wireline networks, and the protocol architectures that developed in parallel with them, are built with the implicit assumptions of relatively high data rates, low packet loss rates, and communications throughout a session's lifetime. Wireless networks do not always meet one or more of these characteristics. As a result, when the layer 1 protocol is wireless, some performance enhancement is required to address these issues. Previous chapters have explored several link layer and network specific protocols developed to enhance the performance of the physical layer.

IP and TCP are present in many wireless networks when those networks are intended to provide a data transport service. In many cases, the wireless network is regarded as just another subnet, and no modification of these layer 3 and layer 4 protocols occurs. In this environment, wireless networks do not operate at optimum levels. Therefore, some performance enhancement is needed to realize the full potential of wireless networks. This chapter considers the modifications, specifically the protocols IP and TCP, that can be made at layer 3 and layer 4.

Figure 16-1
TCP/IP and SS7 Protocol Stacks

5–7	Application		TCAP-MAP
4	TCP		SCCP
3	IP		MTP-3
2	Layer 2		MTP-2
1	Layer 1		MTP-1
	TCP/IP		SS7

■ TCP/IP AND SS7 COMPARISON

The TCP/IP and SS7 protocol stacks are presented in Figure 16-1. Note that both protocol stacks indicate no specific layer 1 and layer 2 choices. The layer 1 and layer 2 protocols may be any group of layer protocols examined earlier in this book. In this chapter, they will be regarded as any underlying wireless network and will not be defined further. As we already stated, wireless networks need some performance enhancement to achieve transparency to the upper layer applications. The basic conceptual difference in the TCP/IP architecture and the SS7 architecture is the *location where performance enhancement occurs* because TCP/IP is an end-to-end architecture while SS7 is not.

The TCP/IP protocol stack is the de facto standard for sending packet data over the Internet and many corporate intranets and extranets. It employs a conceptually simple, four-layer protocol architecture. This protocol architecture is characterized by three items. First, the network is philosophically end-to-end. Any complexity that is required due to the nature of the underlying wireless network is pushed up in the protocol stack to the network layer and above. Both IP and TCP are end-to-end protocols, which means that this complexity resulting in performance enhancement must not lie in the network itself, but at the end systems.

Second, the network layer is connectionless and therefore is best effort only. *Best effort* means that IP will make its "best effort" to deliver every packet. When this delivery mechanism fails, no recovery is performed in the network layer. All recovery and performance enhancement mechanisms are placed in the transport layer. Each IP packet contains the source and destination address, but when a packet is lost, recovery is the responsibility of the transport layer. Third, while TCP/IP networks can function with several transport layer protocols (UDP being the most common after TCP), the network layer must be IP for TCP to function properly. We will confine our discussion to those wireless networks where TCP is the transport layer protocol.

TCP is an end-to-end protocol or a true source-to-destination protocol and it executes on the end-system machines exclusively. It is also important to pay attention to the function of TCP. Its name defines its function: it is the transport *control* protocol, which underlines the basic philosophy of this end-to-end approach. The end systems (or, in other words, the ultimate source and destination systems) *control* the protocol. Further, this control occurs in the same network and along the same route that the data traffic uses. In fact, the control parameters encapsulate the data traffic.

SS7, on the other hand, uses the switching elements of the telephone system to perform the appropriate control procedures. SS7 is an entirely separate control system, in parallel with the communications system that carries the voice/data traffic. These two separate networks function as one. Their functions are distinguished using the terminology *data plane* and *control plane*. The data plane refers to the network that actually carries the voice/data traffic. The control plane refers to the network that controls the data plane, and SS7 operates on this network.

This separation of function has several consequences, but the one we will be concerned with is the optimization that can occur when the network path that the data takes is separate from that which the control signaling takes. Since wireless networks are often characterized by relatively low data rates, this separation means that the control signaling is not required to operate at the same transfer rate as the data.

Many intermediate signaling points, all interconnected through a dedicated control network, can anticipate and react to congestion in a much more dynamic way than is possible in the TCP/IP environment. The control signals can propagate faster and independently of the data traffic, which can be a significant advantage. Therefore, in an SS7 controlled network, the control is distributed and composed of a combination of software and hardware. In a TCP/IP network, the control is present only at the end systems and is exclusively software based. This setup has important implications when a network evolution must keep pace with technological progress. Change in a distributed, hardware based network will be slower and more expensive than in a network where evolution can be performed with a software patch on the end systems.

The debate about SS7 and TCP/IP control architectures is lively. It is not the purpose of this chapter to choose sides but to discuss the relative advantages and disadvantages of the two philosophically different approaches in a wireless context.

■ WIRELINE VERSUS WIRELESS PERFORMANCE DIFFERENCES

From an upper layer perspective, there are four main differences between wireless and wireline networks. This section will examine an IP approach to address the following issues:

Lower data rates and lower throughputs in some wireless networks.

Higher packet loss rates in wireless networks.

Intermittent connectivity in wireless networks.

Large delay bandwidth products in some wireless networks.

We will examine each item in some detail and present protocol approaches to address each item. Before doing so, however, it is important to ask a larger question. From a philosophical perspective, where is the best location to adapt to these differences?

The TCP/IP advocate would argue that there is no need for debate. The TCP/IP protocol stack is an end-to-end approach and the layer 4 protocol is the *control* protocol. IP should not be touched. It is the glue that holds together the entire Internet and associated private networks using the same protocol architecture. Others would argue that the whole concept of layering means that, in principle, any layer can address a problem. As long as the service interface to the application does not change, the application riding the protocol stack sees no effect. These points of view are at the core of the two networking philosophies dominant today.

The Internet community would take the first position: complexity is pushed up the stack and into the end system or higher layer protocol. To those who embrace the fundamental nature of the layering concept, complexity can be located anywhere, as long as the service interface is kept pure. Usually, the second community is composed of those who see complexity as an issue to be pushed down the stack and kept as a component of the functionality of the underlying subnet. This approach is fundamental to the traditional PSTN architecture. The idea is to keep the end system simple (the telephone) and put the complexity into the network, thereby isolating the user from that complexity. The choice of a layer 3 protocol for a wireless environment has components of both these approaches.

■ IP IN A WIRELESS ENVIRONMENT

In a TCP/IP environment, layer 3 will always be some version or extension of IP. The layer 3 choice for wireless subnets has created two main camps: those who believe that a new IP layer is required by the roaming nature of mobile stations and those who believe that the existing IP will work well. The case will be made that if the mobile station stays in its own home network, the issues addressed by mobile IP are already addressed by the wireless network at layers 1 and 2. The requirement for mobile IP is driven by the applications and the security those applications require when they are present in mobile, wireless devices.

Background

The primary function of IP is the routing of packets from source to destination. IP accomplishes this function by using routers and routing tables to direct incoming packets to the correct outgoing port of the router and thus direct the packet toward its destination. The routing tables themselves are built using a routing protocol. This protocol finds the shortest path from source to destination and thus picks the correct port for the packet to be directed to by each router that the packet encounters in its journey. You can look at this technique as a connected series of hops. Routing tables are long lists containing successive hop IP addresses that will ultimately lead the packet to the destination system.

Routing tables are large entities, but they would be much larger without a way to organize groups of destinations. The IP addressing system is a hierarchical system that is topology based for fixed or wireline nodes. Because the higher order bits in an IP address identify the network address that the packet is addressed to, and the hosts (end systems) on that network are assumed to be stationary, the network address is a hierarchical structure. Once a host's network address is identified, it remains constant for the entire session lifetime. This concept is also fundamental to how TCP maintains connections. TCP maintains its connection identifiers in a table indexed by IP address at layer 3 and by port address at layer 4. If either of these numbers changes, the table becomes corrupted for that connection and the connection breaks down.

Mobile networking applications almost always run as clients on the mobile station. These clients communicate back to servers attached to wireline networks. The clients and servers exchange packets throughout the session lifetime. This session is usually defined by a TCP connection or a series of connectionless UDP packet exchanges. In the former case, when the connection fails or is torn down, the session between the client and server is lost. The existence of that TCP connection and how it is maintained in a wireless environment with a mobile station is at the heart of the case both for and against the use of mobile IP. The central question is how best to maintain the TCP connection that a sophisticated application will require.

Mobile IP

Mobile IP is intended to provide a solution to mobile stations that roam from one network to another. This situation can be referred to as macromobility, in an analogy to micromobility, which describes the situation of a mobile station moving from cell to cell. This latter mobility is not affected by mobile IP and is handled with the normal cellular mechanisms. Micromobility describes the situation of a mobile station moving from cell to cell, but in both cells the same switching center is maintained. This situation is not defined as roaming because the switching center has not changed. In other words, this mobility is within a single network identity. Macromobility describes the mobile station moving from cell to cell and the switching center also changing. Macromobility is analogous to roaming; minimobility is not. In macromobility, by definition, the network identity changes. It may not always qualify as a roam, but it often will.

The network layer determines paths for packets through multiple interconnected networks and routes packets to a node. These networks consist of various protocols and data rates. It is assumed that layer 4, the transport layer, will be TCP. The advocates of mobile IP would say that it is possible to express the idea behind the function of the network layer from a different perspective in a wireless environment. Instead of imagining a packet routing, or moving, from one network to another, you can think of routing as changing the node attachment from one location to another. Instead of routing the packet, route the node. Once the node is at the correct location, the packet held by the node is also at the correct location. This conceptual idea allows the mobility of packets to be mapped to the mobility of nodes holding packets. In this way, the packets arrive at the correct location. Mobility is performed at the network layer; this idea is central to mobile IP.

A basic requirement of any new IP offering, and mobile IP is no exception, is that layers 4 to 7 are unaffected by this change, which means that the service access points at the layer 3/layer 4 boundary must remain the same. This requirement has the major benefit that the routing infrastructure of the Internet and existing applications will not be affected by the introduction of mobile stations. As seen from the layer above, mobile IP requires that the mobile station uses the same static IP address in a home system and when roaming.

Mobile units in a wireless environment have unique identifiers for themselves, but they often lack a unique identifier for their position. In protocol terms, they have a unique MAC address but a variable IP domain. The MAC address roams across IP sub-networks while requiring active TCP connections and UDP port bindings. If this upper layer connectivity is not required, traditional methods using dynamic host control protocol (DHCP) and domain name service (DNS) updates will suffice and mobile IP is not required. Therefore, the requirement for mobile IP is driven by the applications that are being run on the device and the security those applications require.

Mobile IP Operation

This tight coupling between the construction of the routing table and the structure of the IP address hierarchy is central to the case for mobile IP. To understand this concept, return to the analogy above. Instead of thinking about a packet being routed to a destination, think of a node attachment location being changed. In this scenario, the destination location is no longer static (during the connection). If the destination location is not static, the IP destination address is no longer static and the connection management of TCP will thus fail.

TCP maintains its connection identifiers in a table indexed by IP address at layer 3 and port address at layer 4. If either of these numbers changes, the table becomes corrupted for that connection and the connection breaks down. A mobile station will have mobility; that is, it will move from IP domain to IP domain, thus changing its destination and hence its connection identifiers. To route to a mobile station, the IP address must be dynamic during the connection. Because any connection is usually composed of several packets over time, the dynamic nature of the IP address is of concern. Note that UDP also requires a routing table, and hence either layer 4 protocol as used in the Internet seems to require mobile IP. The approach that mobile IP uses to assign multiple IP addresses to each mobile station is very similar to the concept used in the GSM cellular system called home location and visitor location.

Mobile IP uses two concepts to accommodate mobility: the home address and the care-of address. The home address is used to map a static (during the connection) IP address to the mobile station. The care-of address changes as the mobile station moves from one access point to another. The care-of address captures the current physical location of the mobile station. These concepts will be explained using an example of a mobile station in an 802.11 wireless LAN.

For example, a mobile station begins a TCP connection when the station is in range of one access point. As the mobile station moves through the building and enters the range of another access point, the communication is handed off to the new access point in the normal operation of an 802.11 network. The TCP connection table never sees this handoff when using mobile IP and maintains the connection to the home address. Mobile IP maps the two addresses together and routes the packet to the correct destination—the care-of address—without requiring any modification of TCP. Mobile IP performs this mapping with the mechanism of a home agent. The home agent is associated with the home address at the original access point. When the mobile station moves away from the original access point, two events occur.

First, the mobile station registers its new care-of address with the home agent. Second, it now communicates with another access point. This new network is called a for-

Figure 16-2
Triangle Routing

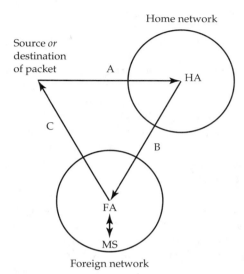

eign network and contains a foreign agent (FA). The home agent gets all the packets for the mobile station and forwards them to the correct access point by modifying the destination IP address. This process is called a redirection and uses the principle of encapsulation. The original packet is tunneled through the IP routing system, as is done in many other cases.

Triangle Routing   Note that the same process is not used in reverse when a mobile station in a foreign address sources a packet during a TCP connection. This situation is called the triangle routing problem and is illustrated in Figure 16-2. In the figure, a triangle is shown with each leg labeled. In the case where a packet is addressed to the mobile station when it is in a foreign network, the packet must route first to the home agent, leg A. Mobile IP then forwards this packet to the foreign agent (shown as leg B). When the mobile station sources a packet while in the foreign network, however, the packet does not travel back to the home network first; instead, it travels directly to the destination (shown in leg C of the figure). In this case, the packet is not encapsulated and is instead sent directly to the destination.

Clearly, on the return path, the mobile station already knows the correct IP address and has no need to pass the packet first to the home agent. This triangle problem of mobile IP is caused by the nonsymmetric delay in the two paths. This problem is well understood, and one way to address it is through the use of route optimization. This technique binds the route path to the packet. In this approach, the home agent functionality disappears and any mobile station packet routes directly to or from the foreign network. This and other solutions are under consideration by the appropriate standards bodies.

Additional problems are basic to the wide adoption of mobile IP. The first is that it is a niche solution. Only those mobile nodes that roam between network identities or IP domains, not within them, need it. Many suppose that this population comprises a small percentage of the transactions of mobile users. It also assumes that all such users practice this macromobility only between networks running TCP/IP. An all-IP network might be in place in the future, but it isn't today nor will it be for at least several years.

Second, the routing table corruption described above is a real problem. Third, current practices of assigning and maintaining dynamic IP addresses are sophisticated and sufficient. It is true that these approaches do not obtain the same level of efficiency with regard to packet exchanges that an assigned static IP address would. In a wireline world, this loss of efficiency is widely accepted as a solution to a noninfinite supply of IP addresses. In a wireless world, where data rates are considerably lower and packet losses are considerably higher, this efficiency gain can be significant.

We return to the scenario described above to illustrate the problem that mobile IP is addressing. The issue was that the mobile station, because of its mobility, would roam out of an IP domain during a TCP connection, but how likely is this possibility? IEEE 802.11 networks can be quite large and seamless when a series of access points are positioned in accordance with a good site survey. A series of 802.11 networks in any one location is often owned by the same provider, and hence all of the networks are in the same IP domain. Of course, it is possible to imagine many situations where this case would not be true, such as airports and various other public areas where many providers may have separate networks. On the other hand, it is not difficult to imagine such a facility having a single designated service provider who sells time on the network to several clients. In this scenario, the IP domain again stays the same during any reasonable transaction duration.

The corruption of the routing table will indeed occur in those situations where the transaction exists across network boundaries. The transaction can be rebuilt, however, and often quite quickly and no data would be lost. Since the mobile station is the client and the fixed unit is the server, the client always initiates the communication (the client sends the first packet when establishing a new connection). Since the server is fixed, its IP address never changes, so the client on the mobile station knows where to send the packet, thus reestablishing the connection. Unless the reestablishment exceeds the timeout provisions of TCP, it may go unnoticed by the user.

The efficiency of dynamic addressing is directly affected by how often it must be performed. For the vast majority of users and applications, mobile stations traverse IP domains infrequently. Modern operating systems implement dynamic addressing as a normal part of their operation because users power up and power down their stations. Therefore, why not make use of these already existing systems that will work for the vast majority of users?

There are advantages to a dynamically assigned address because it tends to be a more optimally routed address for packets than a statically assigned one. In the mobile IP scenario, the static IP address remained in the home system and packets were routed there and then forwarded using an agent to the foreign network. When this unit roams, every packet, including packets that may be sourced close to where the mobile station is currently located, is routed first to the home system and then routed again to the foreign network. This process results in additional routing delay for the mobile IP user. On the other hand, in this scenario, at least the packets arrive and the application is maintained across network boundaries.

To examine how TCP operates, independent of the adoption of mobile IP, the approach taken for the rest of this section will assume that the subnet functionality differences between wireline and wireless networks will be addressed in the end systems through the transport control protocol (TCP). This software resides in the ultimate source and destination systems and embraces the end-to-end, complexity push-up philosophy inherent in the TCP/IP protocol stack.

■ TCP IN A WIRELESS ENVIRONMENT

As stated above, there are four basic differences between wireless and wireline networks:

Lower data rates and lower throughputs of some wireless networks (terrestrial cell-based networks)

Higher packet loss rates in wireless networks

Intermittent connectivity in wireless networks

Large delay bandwidth products in some classes of wireless networks (satellite channels)

In this section, we will look at transport layer solutions to these issues and apply them to the wireless environment.

Lower Data Rates and Lower Throughputs

TCP was developed to accommodate almost any data rate that the underlying network can generate and that the end systems can handle. The source and destination systems each have TCP running and communicating to exchange data. These two systems, once establishing a connection, exchange data in a series of message segments. Each segment has a fixed header of 20 bytes, followed by a data field. In some situations, the two systems will negotiate additional header information that follows the fixed header; however, additional header information is optional.

The data field can vary from less than 100 bytes to just under 64 kB. The header and data field must fit into the IP packet, which has a maximum data field length of 64 kB. The data field size is set by the maximum transferable unit (MTU) parameter. Because of the dominance of IEEE 802.3 networks at layers 1 and 2, the MTU size varies in practice from 500 to 1,500 bytes in the vast majority of systems. However, networks use many different MTU sizes. Table 16-1 illustrates MTU sizes for several network types, which are sorted by MTU size. As this table emphasizes, the MTU is a property of the network, not the end systems. Following the philosophy outlined above, however, network details and performance enhancement take place in the end systems at a higher layer protocol level.

Table 16-1
MTU Sizes

Network Type	MTU Size (Bytes)	Reference
Official maximum MTU	65,535	RFC 791
HYPERchannel	65,535	RFC 1044
SMDS	9,180	RFC 1209
ATM AAL5 (default)	9,180	RFC 1626
IEEE 802.4	8,166	RFC 1042
IEEE 802.5 (4 Mb maximum)	4,464	RFC 1042
FDDI	4,352	RFC 1188
Wideband network	2,048	RFC 907
IEEE 802.5 (4 Mb recommended)	2,002	RFC 1042
Exp. Ethernet nets	1,536	RFC 895
Ethernet networks	1,500	RFC 894
Point-to-point (default)	1,500	RFC 1134
IEEE 802.3	1,492	RFC 1042
SLIP	1,006	RFC 1005
ARPANET	1,006	BBN 1822
X.25 networks	576*	RFC 877
NETBIOS	512	RFC 1088
IEEE 802/source routing bridge	508	RFC 1042
ARCNET	508	RFC 1051
Point-to-point (low delay)	296	RFC 1144
Official minimum MTU	68	RFC 791

* The MTU size can be increased by the negotiation of the two end sites. For example, the MTU size of ARPANET via DDN standard X.25 is 1,007 octets.

If the data passed to the source TCP layer is larger than the MTU, it must be broken into a series of message segments, each with its own header. Again, the maximum size of a message segment is limited by the MTU value and hence is a property of the network. Therefore, if a wireless network is on the route that a particular message must take, and if the MTS of that wireless network is less than that of the wired network, the routers on the edges of the wireless network will break the message into a series of message segments, each with its own header. Thus, TCP handles lower data rate network segments naturally. Of course, a small overhead is associated with reducing the message size because each message must carry its own header. As a rule, lower data rate networks employ smaller MTU message sizes to meet the timeout requirements, discussed next.

Because TCP is a connection-oriented service (COS), each message must be numbered and individually ACKed. The mechanism that TCP uses to discover lost packets is a timer. Every time a message is sent by the source system, the source system performs two tasks. First, it inserts a sequence number into the header of the message and second, it starts a timer. When the message is received by the destination system, it reads the sequence number and writes a response number into a response message. The response number is one more than the sequence number received. This response number is written into the response number field of the TCP header and sent in a message back to the source system. When the source system receives the response message, it examines the response number to verify the ACK from the destination system; then it calculates the time the message took to complete the round trip from source to destination and back to source. This time is known as the round-trip time (RTT).

This RTT is used not only to detect packet losses (which will be discussed in the next section) but also to accommodate changing network conditions. Remember, the TCP/IP method is to push complexity up the stack into the end systems. This method includes any changing conditions in any network segment that the messages pass through on their way between source and destination. Because the RTT estimates the time it takes for a message to travel, it offers an efficient way to accommodate lower throughput values of individual network segments, such as a wireless network. TCP uses a related value retransmission timeout (RTO), based on the RTT estimates, to calculate when too much time has passed for a particular ACK. If too much time has passed, TCP assumes that the packet is lost and retransmits the packet. This process will be examined in more detail in the next section, when packet loss effects are discussed.

Therefore, the insertion of a wireless network segment into a particular wireline message route can be accommodated by TCP using existing methodologies. The protocol adjusts the value of MTU and RTO to address any throughput issues related to the underlying subnets.

Higher Packet Loss Rates

TCP assumes that any packet loss is the result of network congestion. There may be many different reasons for the packet loss, but TCP always assumes it is congestion. Wireless networks often feature lower data rates and thus the MTU size is kept relatively small. If a packet is sourced from a wireline network and is routed through a wireless network, and if this setup is not taken into consideration when the MTU value was originally set, it may be much larger than the MTU size of the wireless network segment. This discrepancy will result in the message being broken into several message segments on the wireless network, as discussed above. Clearly, this situation will add delay to message delivery. If this discrepancy is not addressed quickly by TCP, it is easy to have an RTO value that is too small and thus will expire because of the additional delay. In response, the protocol then adjusts the value of RTO up and tries to send the message again. This process continues until ACKs are received consistently.

The RTO is calculated in the following way. RTO is initially set to approximately 200 ms. This choice has been experimentally set and is a reasonable value for wireline networks. As a history of RTTs for several messages builds and a statistical calculation becomes possible, a mean value and standard deviation are calculated. The RTO is then set to the mean value plus four standard deviations. This calculation is shown in Equation 16-1. It should be noted that, while this formula is a very common method for calculating RTO, it is not the only one; several variations exist. For the purposes of this chapter, we will use the following formula:

$$RTO = RTT_{mean} + 4RTT_{stddev} \qquad \text{(16-1)}$$

The calculation of the RTT is of interest because of what it says about the conceptual bias of the model network used to develop the protocol. This bias is perfectly understandable and reveals itself only with the benefit of hindsight. The RTT calculation assumes a Gaussian, or normal, distribution with a slowly varying mean and standard deviation. The model was chosen using the best data at the time and was a compromise between maximizing throughput when subnets featured packet loss and minimizing unnecessary retransmissions due to slowly varying wireline network congestion.

Additional mathematical refinement was deemed necessary, however, and a smoothing function was overlaid onto this calculation. Any smoothing function alters a calculation in some way to reduce the granularity of the result as the variables change over time. Because the variable in this calculation is the RTT and it is an estimate, the values will be irregular because of the nature of the calculation. Smoothing allows these estimates to be fitted to a theoretical curve more precisely. The mean was shifted slightly by the RTT and the standard deviation was shifted slightly by the difference between the RTT and the mean RTT. Simplified versions of the formulas used are shown below:

$$\text{RTT}_{\text{mean, new}} = \left(\frac{7}{8}\right)\text{RTT}_{\text{mean, old}} + \left(\frac{1}{8}\right)\text{RTT} \qquad \textbf{(16-2)}$$

$$\text{RTT}_{\text{stddev, new}} = \left(\frac{3}{4}\right)\text{RTT}_{\text{stddev, old}} + \left(\frac{1}{4}\right)|\text{RTT} - \text{RTT}_{\text{mean}}| \qquad \textbf{(16-3)}$$

In Equations 16-2 and 16-3, it is explicitly noted which values represent new or updated values and which represent old values. Those variables with no explicit subscript represent the current value of the variable. Again, as in the case of Equation 16-1, Equations 16-2 and 16-3 represent only two commonly employed methods of calculation.

Note that in networks where the RTT is trending upward, the mean value will anticipate this trend slightly. The standard deviation will increase to anticipate this trend, and it will decrease when the RTT is trending downward. This change directly affects the difference between the value of RTT and RTO; in other words, the variance of the measurement affects the RTO calculation. It should be easy to see from the above equations that as the variance of the mean increases, the RTO will be proportionally increased over the RTT. For RTT measurements that vary little, the RTO will be only slightly larger than the RTT.

Historically, wireline networks have traditionally featured bit loss rates of 1 in 100,000 to 1 in 1,000,000. For an MTU of 1,200 bytes, about 1 in 100 packets will have at least 1 bit in error and have to be retransmitted. This packet loss rate is slowly varying and relatively stable. Wireline networks are characterized by guided channels. In guided channels, the physical path traveled by the bits is of fixed length and is insulated, to one degree or another, from interference variations that affect the error rate performance of the channel. Thus, the error rate and hence the packet loss rate is slowly varying about some mean value of performance. This situation fits perfectly the mathematical treatment applied to the RTO calculation.

Wireline networks today feature much lower bit error rates, on the order of 1 in 1,000,000 to 1 in 100,000,000,000. These lower rates result in large measure from the introduction of optical cable, the use of twisted pair balanced circuits, and the transformation of data into digital bit streams. These changes have allowed the extensive application of coding to reduce further the deleterious effects of nature on signal transmission. As a result, the computational model represented by Equations 16-1, 16-2, and 16-3 fits today's more reliable networks even better. The error rates are lower and even more slowly varying due to the initially lower starting value. The insertion of a wireless network thus becomes a real issue for RTO calculations. Remember, the TCP model assumes that packet loss is a function of congestion; most wireline subnets are otherwise highly reliable.

There are two basic problems. First, the much higher packet loss rates in a wireless network segment will introduce an artificially high RTT, which can occur in two ways. First, the normal packet loss effects will lead the algorithm to set artificially high RTTs. High RTTs lead to reduced throughput on any route that includes a wireless network segment. Second, the packet loss rates on a wireless network are much more variable than is the case in wireline networks. Variability in packet loss rates is something for which the model is not well suited. The combination of relatively high packet loss and high variability of RTT is bad for throughput.

The second and unique problem occurs when packet loss rates get very high. When packets are sent multiple times and the source system receives an acknowledgment, how does it determine with which packet the acknowledgment is associated? Since the RTTs of the acknowledgment will vary significantly, this question is important. However, a solution to this problem has been implemented. The rule is to avoid updating RTT for packets that have been retransmitted. Thus, a problem never arises because any packet that needs to be retransmitted automatically disqualifies itself from providing RTT updates. However, this method only solves an internal problem with the original algorithm and does not address the high variability of RTT on wireless networks. It ensures only that RTT does not become artificially high.

Any route that passes through a wireless network, not just the data transfer over the wireless network itself, will be affected. If a route includes a wireless network segment, all connections using that route will see the effects. TCP was designed for networks that feature packet loss rates of 1 percent, which was a common figure at the time TCP was developed. Taking into account the data rates and latencies typical at that time, this percentage translates into no more than 1 packet loss per RTT.

Wireless network subnets feature much higher packet loss rates than 1 percent. This problem is addressed to some extent by coding and power control techniques (which were discussed in earlier chapters), but even with these techniques fully employed, current wireless networks do achieve the 1 percent packet loss rate performance level. Due to the fundamentally unguided nature of wireless channels and the unlicensed operation allowed in many of those frequency bands, it is not clear whether this level of performance will ever be achieved.

What does this situation mean? According to the end-to-end philosophy of the TCP/IP protocol architecture, TCP is where network complexity and performance issues should be addressed. Strictly following the philosophy leads us into the area of modifying TCP in some manner. This subject is known as TCP tuning and we will explore it in some detail. In general, however, this approach requires either that all TCP implementations are modified in some way or that different TCP implementations be in place on systems that potentially route through wireless networks. Further, these approaches all violate in one way or another the end-to-end philosophy inherent in TCP/IP networks.

In the first case, making a change to TCP and applying it universally really means that the algorithms are changed so that exclusively wireline networks see no penalty and mixed routes see a consistent improvement. This solution turns out to be a real problem because the fundamental natures of the network segments are so distinct. There is little enthusiasm to make a change to TCP until wireless networks grow in use and the problem becomes more universal.

In the second case, how does one anticipate, at the end system, that a route will pass through a wireless segment? An organization that uses wireless network segments on its own private network can adjust routing tables and selectively install TCP upgrades to ensure that TCP modifications are made for any route that passes through a wireless network segment. This solution is reasonable for those organizations that are early adopters of wireless technology and want to maximize the overall performance of the enterprise network.

In my opinion, using TCP modifications to address these problems is practical in only selected cases. A better solution would be to make link layer modifications to address the issues outlined above. This solution will work, and TCP/IP has built its reputation on being what works: a de facto standard through adoption. The link optimization approach fits the layering concept, so it may be acceptable to the TCP/IP faithful, even though it does not push subnet complexity and performance issues up the stack into the end systems and thus will violate the end-to-end philosophy. Some, however, advocate modification of TCP to address these issues. These modifications will be discussed next and are collectively called TCP tuning methods. We will examine three: modification of the slow-start algorithm; TCP spoofing as a general technique, along with two variations of it called fast retransmit and the use of multiple acknowledgments; and the adoption of mobile TCP.

TCP implicitly assumes that any packet loss is assumed by the source and that the "connection pipes" do not leak in any appreciable way. TCP tuning is all about hiding any mechanism, other than congestion, that might be responsible for packet loss. It should be pointed out that TCP tuning fundamentally violates the end-to-end philosophy in one way or another.

One way to classify and discuss the methods suggested to address the lower data rate and lower throughput problems is to make an analogy with error-detection and error-correction classification. In that approach (discussed in Chapter 7), error-correction procedures can be classified by the *location* of the information that corrects any errors detected. There were two classifications: either the information is contained in the message itself, allowing correction to take place at the destination, or the information resides in the source and an explicit request must be made for retransmission.

TCP tuning methods can also be classified this way. Either the mechanisms are hidden from the source or they are not. In the first approach, only the TCP code at the destination (assumed to be wireless and a mobile client) is modified. The TCP code in the source (assumed to be wireline and a fixed server) is kept constant. The philosophical approach here is that the problem is local to the wireless node and should be addressed there.

The fundamental problem is that TCP has little to offer with regard to QOS capabilities and rides on the same network as the data it delivers. TCP can address the latency typically found in wireline networks due to occasional dropped packets, but it fails when asked to provide a constant bandwidth stream of data or to deliver data reliably in some specified time period to a mobile user in a relatively high-error-rate environment.

While TCP was described in some detail earlier, some additional concepts must now be addressed to set the context for the discussion below. First, we must answer the question, How does TCP decide how much bandwidth is available for use? Answering this question will lead us into a discussion of the slow-start algorithm used by the vast majority of TCP implementations.

TCP uses a window-based, flow-control mechanism to balance the two conditions that, in a wireline environment, determine whether a packet inserted into a connection can be accessed immediately by the destination. The first condition is when the bandwidth of the connection is smaller than the ability of the source to insert packets. The second is when the bandwidth of the connection is smaller than the ability of the destination to extract packets. In both cases, the bandwidth of the connection is limiting the data rate at which the end systems can communicate. A good analogy is comparing the connection bandwidth to a pipe: the larger the pipe, the more bandwidth is available for the connection to use. Traditional wireline flow control addresses this problem of how to determine if the pipe is becoming congested due to the first of the above conditions. The second is negotiated separately between the source and destination systems and is not a concern in this discussion. It is a negotiation based on the buffer memory size characteristics of the two systems.

The slow-start algorithm assumes that any RTO initiated retransmissions are due to the pipe not being large enough, with congestion resulting. The pipe is assumed implicitly to be a perfect pipe with no leaks. Packets are assumed to make it through the pipe and either be accepted by the destination or not. Extending the analogy we made above, a leaky pipe is apt for a wireless network segment. If a packet is leaked, the destination will never see it and thus will never acknowledge it.

Slow-Start Algorithm   Although it is not slow at all, the slow-start algorithm works in the following way. The source system sends several packets that add up to the initial message size negotiated between the two systems based on their physical characteristics and buffer size. The number of packets sent is determined by how many MTUs will fit into the initial size negotiated by the two systems. This value is used to set the initial value of the congestion window. If an acknowledgment is received by the source system prior to the expiration of RTT, then another series of packets (we will call these a window of packets) of the same size is transmitted and the congestion window size is doubled. Note that the use of a congestion window is flow control by the source and that the advertised window used by the destination in negotiation is flow control by the destination.

As long as each window's worth of packets is acknowledged prior to the expiration of the RTT, the window size is doubled for every nonduplicate acknowledgment received by the source. If you were to plot a curve showing the congestion window increase for any initial message size, you would see that the window growth approaches an exponential curve. Figure 16-3 shows both an exponential curve and a typical "slow-start"

Figure 16-3
Slow-Start Window Growth

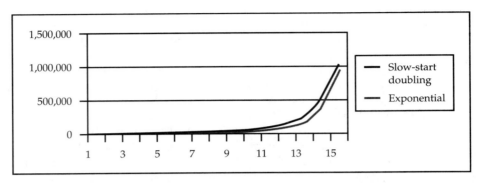

congestion window growth. The exponential growth curve has been scaled to underscore the fact that the growth curve caused by doubling is essentially exponential in character.

At some point, one of three events will occur: the source will receive an acknowledgment from the destination, but not until after the RTT has expired; an intermediate router will start discarding packets; or the destination system will start discarding packets. When packets are discarded, the congestion window is cut in half and the process starts again. This time, however, the window increases only by one additional packet for each window's worth of acknowledgments received prior to timeout. TCP quickly estimates the maximum capacity of the link and ensures that the connection is using all the bandwidth available.

A simple relationship exists between the window size and RTT. That relationship allows a rough calculation of the overall throughput of any connection. The relationship is shown in Equation 16-4 and should help to capture these ideas:

$$\text{Throughput}_{\text{mean}} = \frac{\text{window size}}{\text{RTT}_{\text{mean}}}$$
(16-4)

You can see from the above relationship that, as window size grows, the throughput of a connection also increases. Similarly, as the RTT value increases, it slows the growth of the window and, as such, reduces throughput.

To illustrate the above relationship, we will take two typical values, one for the window size and one for RTT commonly found in personal computers on wireline networks. Many personal computers have their window size defaulted to 8 kbytes. A typical RTT time found in wireline networks is about 50 ms. Placing these values in the relationship above, we find the following throughput maximum value:

$$\text{Throughput} = \frac{65,536 \text{ bits}}{50 \text{ ms}} \approx 1,310 \text{ kbps}$$

Note that this value is approximately a DS-1, framed T-1 line bit rate. Unless your PC has access to a dedicated T-1 line, the default values for the characteristics of the TCP/IP protocol shipped with your operating system will not limit your throughput. The maximum value for window size is 64 kbytes. Placing this value into the above relationship, we find a protocol limited throughput sufficient for a dedicated 10 Mbps link:

$$\text{Throughput} = \frac{524,288 \, \text{bits}}{50 \, \text{ms}} \approx 10,486 \, \text{kbps}$$

It should be pointed out that the behavior of the slow-start algorithm, after sensing congestion once, is combined with the algorithm called congestion avoidance. This modification of the slow-start algorithm provides the linear increase of bandwidth that TCP demonstrates after sensing congestion. Many times these two algorithms are discussed separately, but most TCP implementations combine the two and just refer to them as the slow-start algorithm. Below, when we will explore a couple of modifications of the congestion avoidance algorithm, it will be important to remember that this algorithm is initiated only after congestion is sensed on the segment.

The current value of the congestion window is called the slow-start threshold. The halving of the congestion window after the threshold is reached is called multiplicative decrease. As you can see from the above explanation, the slow-start algorithm responds quickly to both the initial bandwidth available and also continually probes the network for any additional bandwidth that might become available.

While the slow-start algorithm works quite well for many types of networks, it does not perform well in at least two situations. Both will be discussed in the section below titled Large Delay Bandwidth Products in Satellite Networks. This algorithm embraces the end-to-end philosophy on which TCP/IP networks are based. The source is completely responsible for determining the bandwidth of the underlying subnet. All complexity and performance enhancement is based in the end systems.

As we suggested earlier, this algorithm implicitly assumes that any packet loss is the fault of the source and that the "connection pipes" do not leak in any appreciable way. Again, TCP tuning is all about hiding any mechanism, other than congestion, responsible for packet loss.

TCP Spoofing Sometimes spoofing is described as a method of preserving the independence of TCP from the behavior of any individual link. However, it breaks the end-to-end link philosophy on which TCP is based. There are two main approaches to TCP spoofing: fast retransmit and multiple acknowledgments. Both will be explored in the sections below.

Spoofing has occupied a place in data communications for a long time. It is commonly used by protocol converters to bridge between asynchronous bit-oriented networks, like Ethernet, and synchronous, byte-oriented networks, like SNA. The connection is essentially broken into two pieces. The first is composed of the wireline segment(s), and the second is composed of the wireless segment(s). The last node in the wireline segment becomes the virtual endpoint of the TCP connection.

We will assume that the wireline system is the source of the message. When this wireline/wireless gateway node receives messages from the source, it automatically buffers and acknowledges them to the source. In this way, any retransmissions necessary are hidden from the source. The single connection is broken into two separate connections.

The wireline/wireless gateway system performs any retransmissions necessary due to packet loss issues. The wireline/wireless gateway system is the MSC or equivalent in a cellular telephone system and an access point in an IEEE 802.11 system.

Sometimes this idea is extended to eliminating the TCP connection completely on the wireless link. After all, if you are spoofing the end system (the "spoofed"), you can do anything you want on the portion of the network that lies behind the "spoofer" (the wireless/wireline gateway system). Sometimes proprietary systems optimized for the particular wireless environment are employed here, sometimes not. (More will be said about spoofing in the section titled Large Delay Bandwidth Products in Satellite Networks.)

The next two sections describe additional enhancements to the basic spoofing technique. These approaches modify the congestion avoidance algorithm often combined with the slow-start algorithm. Because they employ spoofing techniques, however, they are discussed here. In these approaches, the wireline network is aware of the wireless network segment, and TCP is modified to distinguish packet losses due to congestion from those due to transmission error on the wireless network segment.

Fast Retransmit   Fast retransmit is a TCP spoofing technique that is most useful when only an occasional message or two is lost, such as when a handoff occurs in a cellular system. Of course, these messages may consist of several packets, depending on the MTU and window size negotiation.

To best understand this algorithm, recall that acknowledgments can be received by the source for at least two reasons. Either a message was lost or it was received out of order. TCP tries to distinguish between the two causes by assuming that out-of-order messages will generate only a few duplicate acknowledgments. Therefore, if an end system receives three or more duplicate acknowledgments in a row, it assumes the first condition was the cause and that a packet was lost. When using fast retransmit, this packet is retransmitted prior to RTO expiration. Cellular systems in particular can be enhanced by using this technique, as discussed in the next paragraphs.

When a mobile system moves between cells during the handoff, a few packets may be lost. Fast retransmit requires the mobile system to reinitiate the TCP connection prior to the expiration of the RTO timer for those lost packets. Throughput is not affected while the wireline system waits for the RTO timer to expire before retransmitting the unacknowledged packets. The mobile system, which knows when a handoff has completed, sends several duplicate acknowledgments to the wireline system. This action causes TCP to reduce the window size but also to begin immediate retransmission of the lost packets.

While this approach does reduce the window size and hence reduce throughput, the slow-start algorithm is designed to address any reduction in data transfer rate and will quickly seek out the additional bandwidth. In many cases, this reduction of window size, which would occur in any case when the timer expires, results in higher overall throughputs than waiting for the timer to expire naturally. Note that only in the most general sense could this method actually be distinguishing the packet loss causes. The next form of TCP spoofing explicitly distinguishes packet loss and communicates that through the protocol.

Multiple Acknowledgments   This method distinguishes the losses occurring on the wireline link from those on the wireless link. The acknowledgments are tagged back to the wireline system to indicate if they originate from the wireline/wireless gateway or inside the wireless segment itself. This approach makes use of fast retransmit and breaks the link into two pieces: the wireline segment(s) and the wireless segment(s). Each group has a separate RTT maintained by the wireline system. Basically the approach works by optimizing the window size and timers for each network segment. The name *multiple acknowledgment* comes from the fact that multiple acknowledgments are generated by a single message: one from the wireline portion, one from the wireless portion.

Other methods combine aspects of these approaches in myriad ways. The broad subject is part of active research in many institutions, and new sophisticated approaches are emerging quite regularly. Usually the best are applied to a specific subset of the characteristics of wireless networks that can be addressed by a single modification of the basic TCP algorithm. When these approaches are combined to try to generate a global solution, complexity issues tend to render them into less-than-optimum solutions.

Mobile TCP

Of course, in a cellular network the MSC also knows when a handoff occurs, and this twist has been employed to define mobile TCP. The mobile TCP solution distinguishes the packet loss from handoff in two conditions: a pure handoff and a handoff resulting from a mobile station roaming between two different wireless networks. Mobile TCP is quite reminiscent of mobile IP, and hence its name. Mobile TCP lets the MSC indicate to the wireline source the cause of the packet loss. In both cases, the wireline source then retransmits those packets that might be lost when a handoff completes. Thus, the handoff period does not cause a change in the window size because those retransmissions due to handoff are explicitly tagged. Tagging prevents them from triggering a window size reduction due to retransmission.

In some ways, this approach is similar to the one discussed earlier, where duplicate acknowledgments received at the source due to high packet loss rates do not contribute to the value of RTT. Mobile TCP also has the ability to distinguish when a handoff is a pure handoff and when it results from roaming. These two conditions are distinguished by using two different types of tags. Because the tag type indicates the handoff condition, the resetting of the window size, RTT, RTO, and the slow-start threshold can be performed when appropriate.

In this environment, the performance characteristics of the two wireless networks may be quite different. Superior performance can, in principle, result from starting from scratch when moving to a different network. In practice, the mobile TCP algorithm processes the handoff well, but the enhancement to distinguish between wireless networks does not operate well. This difference is probably due to the fact that most wireless networks between which a mobile station would be roaming during a TCP connection are quite similar.

Intermittent Connectivity

The intermittent connectivity often associated with wireless networks has much in common with intermittent connectivity in wireline networks. First, in some sense, intermittent connectivity is already widely encountered. The best examples may be dial-up sessions from fixed systems and laptop mobility. Dial-up sessions from fixed locations are easily handled with current technology. Dynamic IP address assignment solves this problem well. Mobile laptops are accommodated in the same way. The basic issue with this solution is that each application must address the problem individually. Because the same solutions used in wireline networks will function in wireless networks, no further discussion concerning this problem will be provided.

Large Delay Bandwidth Products in Satellite Networks

Satellites have some special characteristics that must be taken into consideration when passing a TCP connection through them. The two characteristics that directly affect the efficient operation of TCP are the large bandwidth and built-in propagation delay of some satellite channels. A typical GEO satellite channel will exhibit a frequency bandwidth of between 36 and 50 MHz and a round-trip propagation delay of 500 ms. LEO satellites feature lower bandwidths and an order of magnitude lower propagation delays. As a result, they do not require the special attention that a GEO satellite link does.

Typical initial RTO values are in the range of 200 ms. While this value is adequate to address the latency seen in terrestrial based networks, the round-trip time alone when using a GEO satellite is more than twice this value. If you add typical terrestrial based latencies to the orbital-defined propagation, you would have a value of approximately 600 ms. This length of time has several very undesirable effects on TCP, and all are tied to the central fact that TCP uses the same channel for control as it does for data. The most obvious solution is to increase the initial RTO values to accommodate the GEO satellite characteristics. If this solution is applied, the result is the creation of other, more subtle problems. These effects are all a consequence of the long propagation delay of GEO satellite channels. A long propagation delay will have several implications.

The slow-start algorithm for increasing window size depends on receiving acknowledgments from the destination to increase the packet rate. Large propagation delay in a channel reduces the rate at which acknowledgments are received at the source. This delay has the impact of reducing the rate at which the window size increases because increases occur only when a nonduplicate acknowledgment is received. Window size growth is slowed, and so is efficient utilization of the bandwidth of the satellite channel.

The second problem is related to the first. GEO satellites are characterized not only by long propagation delays but also by large bandwidths. For any initial value of window size, larger bandwidth channels need more acknowledgments from the destination to fill the available bandwidth. The acknowledgments take longer and more are required before the end systems are making best use of the available channel bandwidth.

Third, if either end system or some intermediate router decides to discard packets for any reason, the congestion window will throttle back the data rate used by the end systems by half. Half of a thin route, like a DS-0, is 32 kbps; even half a DS-1 is only 750 kbps. Half of a K band satellite transponder is anywhere from 25 Mbps to hundreds of Mbps, depending on the modulation technique. While the percentage is the same, the throughput implications are quite different. As you can see, as the channel bandwidth grows, the sensitivity also grows. The slow-start algorithm clearly has too much granularity for such channels.

Thus, it is not just the large propagation delay that is the issue; the large bandwidth of the channel also has a significant impact. When these two factors are combined, as in GEO satellite links, each individual problem combines to create a larger problem. Therefore, the product of the latency and bandwidth determines when remedies need to be applied.

Note that terrestrial fiber links also demonstrate this problem. The delay is not significantly different than any other terrestrial link, but the bandwidth is much larger. As shown in the previous paragraphs, the critical element is the delay bandwidth product. The larger that product becomes, the more sensitive the channel will be to the effects such as those described. TCP, with its slow-start algorithm, was not designed for these types of channel characteristics.

Some satellite links are only single components in the overall route that a particular TCP connection may inhabit. End systems sharing wireline components with the end systems that have satellite components will not suffer the same penalty when seeking out available bandwidth. The wireline network segments of the link are filled with connections that feature short round-trip times, so the wireline components of the route can become bottlenecks for wireless connections. Essentially, the slow-start algorithm in TCP is biased against large delay links. Short delays mean more rapid acknowledgments, which means more available bandwidth is allocated to closer sessions. End systems linked by a GEO satellite will always appear farther away because their acknowledgments will always arrive later and appear slower due to the large latency in the response.

Three solutions are proposed to address this problem of large delay bandwidth network segments. The first attempts to address the long control feedback loop exhibited by GEO satellites. This large propagation delay causes the slow-start algorithm to wait a long time for the acknowledgments it needs to open the window size and take advantage of the large bandwidth available on most GEO satellite links. One solution is, of course, to start with a larger window size. As was shown in Equation 16-4, a relationship exists between window size and RTT. This relationship is most valid for large window sizes.

Using the typical default value for window or buffer size in most TCP/IP networks and a typical round-trip, two-hop RTT for a geosynchronous satellite, the throughput is:

$$\text{Throughput} = \frac{524,288\,\text{bits}}{520\,\text{ms}} \approx 1,008\,\text{kbps}$$

When this value is compared to the entire transponder capacity, typically several DS-3s, the shortcomings of the approach are clear. Note that the throughput rate is less than what may be supported (1,536 kbps) on a single DS-1 link.

One approach to address this problem quite obviously flows from the above analysis: make the window size bigger to start. Most proposals of this type suggest a doubling or quadrupling of the window size. Doubling gets one past the DS-1 threshold, and quadrupling approaches the effective throughput of most 10 Mbps Ethernet LANs, or about 4 Mbps.

It is possible to refine this technique by substituting byte counting for ACK counting. This alternate method of increasing the window size often works well for GEO satellites because, due to the large propagation time, each ACK represents many bytes. Hence, the window opens up much faster when counting the number of bytes received instead of the number of ACKs received. Difficulties can emerge, however, when the initial window size is not set carefully. If it is set to a large percentage of the maximum throughput that the link can support, the source system can start sending large bursts of data that cannot be supported by the destination system's buffer size.

This approach maintains the TCP/IP philosophical perspective that all systems should be end-to-end both in their control channel and data channel (they are combined into a single channel). The problem, of course, is that the protocol stacks of both source and destination must be customized to the link type.

The next potential solution is an application one that also retains the end-to-end character of both the data and control logical channels and requires no TCP modifications. An application level solution to the slow-start problem is to start several TCP connections in parallel for the same data transfer. The idea is that the application using the protocol stack can be configured to "know" a satellite link is in the route, if by no other means, simply by measuring the RTT measured by the protocol stack. Starting several TCP connections simultaneously addresses the rapidity and throughput limitations imposed by the large RTT exhibited by GEO satellite links. However, it significantly increases the overall aggressiveness of the slow-start algorithm, which is already exponential in nature. This increase will have consequences when the link throughput is reached and multiple TCP connections start receiving ACKs indicating lost or congested packets.

The fundamental problem with this approach is that you are actually increasing the level of TCP activity and at the same time attempting to find ways of reducing its impact. If the problem is that all cars are running out of gas before the finish line, the appropriate solution is not to start more cars, all with the same size gas tanks and same fuel efficiency. Instead, you can start the cars in relays. As one car runs out of gas, start the next one, and so on. This solution would seem to stand a much better chance of solving the fuel problem. Window size and RTT limitations can be compared to gas tank capacity and fuel efficiency. This analogy leads to the last class of approaches, those at the link layer.

Link layer solutions of the general class of spoofing have been around for a long time in data communications. For years, link layer spoofing has been used at the data link layer in asynchronous-synchronous network interfaces. By their very nature, synchronous networks require handshakes with the destination system to maintain the synchronous nature of the transmission. Asynchronous networks, by their very nature, cannot provide handshakes. The solution was to place a protocol converter between the two systems to provide the synchronous network that the ACKs required. Of course, the ACKs were generated by the protocol converter, not the asynchronous end system, but the concept is the same. A good example of this type of device is the IBM 3708, widely used in IT shops to interface between legacy synchronous networks and PC based asynchronous networks.

The fundamental idea is to spoof the TCP protocol by sending back ACKs from an intermediate gateway system rather than the end system. Many GEO satellite links use this approach. TCP spoofing works by essentially eliminating the long round-trip delay from the TCP control function. A gateway device is placed at the start of the satellite link. This gateway device spoofs TCP by providing the ACKs it needs without the long RTT of the satellite link. IP packets are then transferred on the long delay link using an alternate transport algorithm, which is optimized for this link class. This transport algorithm still needs to make best use of the link bandwidth, and so it needs the TCP/IP end system to open its window size accordingly. The gateway device provides the necessary ACKs, and the link bandwidth is quickly optimized.

This section has only touched on the many ways that TCP/IP can be optimized for GEO satellite links and large delay bandwidth links in general. The three solutions outlined above (beginning with a larger window size, byte counting instead of ACK counting, and link layer TCP spoofing) can be complemented by combining them with other protocol solutions discussed elsewhere in this chapter. The above treatment of GEO satellite links assumed that the link bandwidth was symmetric in nature; that is, both the forward and reverse paths operated at the same data rate.

However, asymmetric links are often used to address the much more bandwidth intensive needs of Internet surfers on the forward path compared to the reverse path. Some ISPs make use of GEO satellites on the forward path and dial-up land lines on the reverse path. One such satellite system is the Hughes DirectPC system. This system uses a GEO satellite to provide wide bandwidth forward channels to a large number of terrestrial locations using VSAT technology. The reverse path is a simple dial-up line using a modem. The fundamental problem that TCP has with such an arrangement is also exhibited in asymmetric digital subscriber line (ADSL) technology.

As we said above, TCP assumes that the reverse path has sufficient bandwidth to return the ACKs it requires without congestion. Since the path is not symmetrical, the solutions proposed in the above paragraphs will not work because they address only the forward portion of the link. With a nonsymmetric network, those solutions address only part of the link. A solution must not only address the satellite portion but the combined satellite and slow land line terrestrial link. The ACKs follow the reverse link, and because it is a much lower data rate link than the forward portion, ACKs will be delayed, thus adding to RTT.

Solutions to this problem involve the use of previously described protocol enhancements by limiting the ACKs in some way or smoothing the bursty nature of TCP, thus also reducing the volume of ACKs. One unique approach uses the fact that many of the fields in a TCP header are static during a connection. They can be stripped without risk. TCP ACKs on the reverse path are compressed, thus lessening the bandwidth requirements for successful, timely transport.

■ WIRELESS IN AN SS7 ARCHITECTURE

The SS7 protocol stack is the group of data transport protocols used in the telephone network. It is important to understand that the SS7 protocols perform network control,

Figure 16-4
(a) Typical TCP/IP Protocol Stack; (b) SS7 Protocol Stack

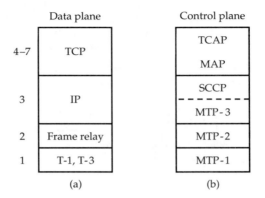

(a)   (b)

Figure 16-5
MSC SS7 Interconnect

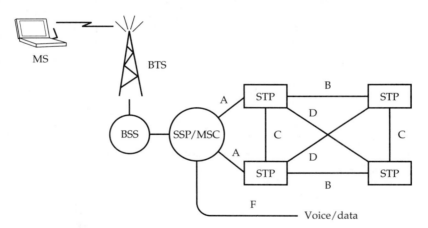

while the digital transmission hierarchy (T-1, T-3, etc.) performs the actual bit transfer. In telephone terminology, there are two planes, the control plane and the data plane. The control plane is used for call control and network management, and the data plane carries binary data signals.

Two protocol architectures are shown in Figure 16-4. Part (a) of the figure shows a typical TCP/IP protocol stack. This protocol stack does the actual data transport over the transmission line system. The SS7 protocol stack is shown in part (b) of the figure. This protocol stack carries no customer data. It is used exclusively for control and management of the transmission line network.

SS7 is used to connect the MSCs to other switching points in the SS7 network. From an SS7 point of view, the MSC is nothing more than a special kind of SSP/SP. The wireless nature of the MS connected to the MSC is visible to the SS7 protocol stack. Figure 16-5 illustrates the situation. Note in the figure that the data travels over the F link, while the control information passes over the redundant A links connected to the MSC. From the control perspective, the MSC is just another type of SSP. The same kinds of issues are present with SS7 as with TCP/IP, with one fundamental difference. The telephone system approach that combines a data plane and a control plane allows you to separate where you do optimization from how you do optimization. In other words, the end-to-end philosophy of the TCP/IP environment requires that these two concepts be tightly coupled. Both are to be performed in the end system. Modern telephone systems do not have this requirement.

Thus, the SS7 protocol stack can be optimized to control and manage data transfer so that quality of service (QOS) parameters can be ensured. This feature doesn't mean that there is any difference in the problems faced by a wireless segment. All the same issues that are present when routing through a wireless segment in a TCP/IP stack are also present when routing through an SS7 stack. In the latter case, there is no philosophical debate about where and how to address the problems. The voice/data network transports the data; the SS7 network routes and performs network management of that data flow.

The critical observation is that the end system in a TCP controlled network must implement both IP and TCP at layers 3 and 4, respectively. An end system in an SS7 architecture does not need layer 3 or layer 4 functionality at all. The control plane will provide the needed layer 3 control using SCCP, and layer 4 is null in the control plane architecture.

The end system in an SS7 controlled network is part of the data plane, not the control plane. Therefore, it needs no functionality at that layer other than what is needed for data flows. The control plane architecture will address all routing and control issues. Since these are two independent networks, this control is independent of the data rate of the data plane. Figure 16-6 illustrates the dual role of many components of a cellular system. The solid lines indicate voice/data transfer using either wireline or wireless links, as appropriate in the data plane. The dotted lines indicate the control plane signaling of SS7.

The fundamentals of wireless data transport are not eliminated by changing the upper layers of the protocol stack. The same four basic differences still exist, but there is no confusion about where and how to address them when using SS7. In an SS7 network, the data flows are separate from the control flows, and this separation can have a major impact in wireless networks simply because now the control data does not have to flow over the same link as the user data. Taking a terrestrial cellular network as an example, the data rates between the mobile station and the base station are strictly limited. Not having to carry the IP and TCP data over this link is a major advantage.

The mobile IP issues discussed above fade in this setting because it is the clear responsibility of the TCAP with MAP extensions to address the issues of roaming outside home switching centers. In fact, MAP was designed precisely to address this issue. Handoff, of course, is handled by the controlling MSC itself and no upper layer protocol contribution is required.

Lower Data Rates and Lower Throughputs

In the previous sections, we discussed in detail how TCP works at a fundamental level to understand the basis of the issue and how TCP handles flow control. As we said, TCP

Figure 16-6
SS7 Links in a GSM Network

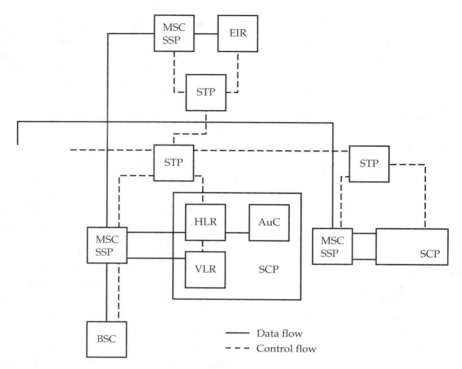

adjusts the values of RTT and RTO to adjust to the data rate and throughput, as measured at the end system. Because the underlying network in SS7 is assumed to be circuit switched, there is no flow control mechanism. Wireless data flow control issues are addressed through a link layer protocol such as RLP.

Higher Packet Loss Rates in Wireless Networks

This issue is addressed in the link layer through the use of the radio link protocol (RLP). As discussed in Chapter 14, RLP is a complex data link protocol enhancement that provides an error-free data channel, but at the cost of a variable data rate.

Because there is no complex calculation for determining the network throughput that is based on packet loss rates, data lost though packet loss is simply buffered and placed in the next available time slot. The data link addresses any packet loss effects. Of course, when excessive packet losses occur, data rates to the mobile station suffer and become variable, but this situation is a consequence of the wireless environment, not a protocol difference.

The central idea here is that because *the control information in an SS7 protocol stack does not pass through the data link, the performance of the data link does not influence the control management.* In the SS7 architecture, the control information is a separate wireline network, and so control issues based on passing control information through the data channel do not arise. The speed limitations and packet loss rates do not affect the rapidity at which the signaling information is exchanged between MSCs. The interoffice network can signal at very high speeds. As a side benefit, this high-speed signaling also means that enhanced services to the mobile station on its slow channel can be configured and provided in parallel with data communication.

The TCP/IP protocol stack will also likely adopt this same approach through the use of RLP, an overlay of the existing layer 2 in mobile end systems. A notable exception to these data link fixes will be satellite segments, which will likely continue to use TCP spoofing because of the unique aspects of satellite links.

Large Delay Bandwidth Products

This issue is not really relevant to the SS7 architecture. Because no separate network is established between the satellite and the terrestrial-based switching center, SS7 is not used in the satellite environment. When data in an SS7 controlled network must flow through a satellite network segment, an alternate control procedure is applied. This alternate control procedure is very similar to the IP/TCP architecture because both control and data flow over the same link. Since the bandwidths are large in most satellite networks, however, this overhead is not an issue.

Unlike the IP/TCP architecture, the flow control in these segments is rendered null because the link is regarded as a fixed bandwidth circuit-switched element. As such, flow control is not required. Instead, the end systems are required to buffer any overflow until they can be transferred on the link.

■ SUMMARY

This chapter focused primarily on the special characteristics that wireless network segments pose for TCP/IP based protocol stacks. Issues and solutions specific to both IP and TCP were described and explored. As a comparison, the SS7 control architecture, when applied to a wireless network, was also explored. The underlying differences between an end system–controlled network philosophy, embodied in TCP/IP, and a network where control is embedded in the network itself were illustrated using this approach.

Many believe that all networks will inevitably move to a single network architecture using IP and TCP. In this chapter, several special characteristics of wireless links were described to illustrate that a universal solution may not be possible without significant changes to these widely deployed layer protocols.

REVIEW QUESTIONS

1. Discuss the basic differences between the TCP/IP protocol stack and the SS7 protocol stack.

2. What are the four main differences between wireless and wireline networks?

3. Describe a high-level protocol mobile IP.

4. Describe the operation of mobile IP and include the triangle routing issue.

5. Why is triangle routing a problem?

6. What is the function of the home agent and foreign agent in mobile IP?

7. What is the relationship between MTU and fragmentation?

8. Discuss the relationship between IP routing and the TCP connection table.

9. Define RTT and RTO. What is the relationship between the two?

10. How would you modify Equations 16-2 and 16-3 for a network that has highly variable latency?

11. Explain the relationship of the slow-start algorithm in TCP to window size.

12. Discuss the effect of lower bit error rates on the TCP algorithm.

13. List the problems incurred by a TCP that uses a slow-start algorithm in a wireless network.

14. Describe the TCP spoofing method in general.

15. Summarize the slow-start algorithm operation.

16. Discuss how the throughput of TCP is controlled by the window size. Use Equation 16-4 in your explanation.

17. Describe the fast retransmission scheme used for congestion control.

18. Describe the multiple acknowledgment scheme used for congestion control.

19. How does fast retransmission help to improve TCP performance in a wireless network?

20. Describe the mobile TCP scheme.

21. Describe the multiple acknowledgments scheme used to improve TCP performance in a wireless network.

22. List the characteristics that make GEO satellite networks unique when TCP is used as the layer 4 protocol.

23. Discuss how large latency and large bandwidth combine to create problems for TCP.

24. What are the problems related to GEO satellites when TCP is used?

25. What are the solutions proposed to address TCP in GEO satellite links?

26. Discuss TCP spoofing in GEO satellite links.

27. How does the SS7 network perform flow control?

28. Summarize the main differences in control philosophy in the TCP/IP environment compared to the SS7 environment.

29. Examine Figure 16-6 and discuss the implications of the different paths shown for data and control flows.

4

Emerging Systems

Emerging Standards: WAP

■ INTRODUCTION

Wireless application protocol (WAP) is a specification for a set of protocols to standardize the way wireless WAP clients can interact with the Internet. Its basic function is to provide wireless subscribers with the ability to view Internet content in a special text format. This content includes items such as e-mail, call management, unified messaging, weather updates, stock and traffic alerts, news, and various information services. It will also include many e-commerce applications as well as banking and proprietary corporate intranet applications.

The language used to interpret this content is called the wireless markup language (WML). WML is a subset of the language called extensible markup language (XML). XML is a stripped-down version of the hypertext markup language (HTML). All WML content is accessed over the Internet using standard HTML requests.

■ OPERATIONAL OVERVIEW

WAP is essentially a specification of a thin-client "microbrowser" and a proxy server that sits between the wireless network and the Internet. Recall that a proxy server is an intermediate program that acts like both a server and a client when the client is making requests for other clients. A proxy server lies between clients and servers when those

Figure 17-1
WAP Overview

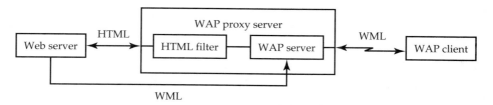

devices have no other means of communicating, typically when a firewall lies between them. Thus, the proxy server intercepts those requests from a client and passes them onto a true server. A proxy server always implements both client and server functionality.

Both the client and server components are specified for any network configuration. The proxy server acts like a gateway providing protocol translation and optimizing data transfer for the WAP client. Recall that a gateway acts like a translator for another server. However, it is distinguished from a proxy server because, from the client's perspective, the gateway is the server. In most cases, the client has no sense that it is not communicating directly with the true server. Figure 17-1 illustrates the relationships among these components and the languages used on each link.

The WAP client communicates through the appropriate air interface to the WAP proxy server. This communication translates client requests into HTML and passes them to the Web server. In the reverse direction, the Web server may provide either HTML content or WML content. As you can see in Figure 17-1, HTML content is first filtered before being passed to the WAP proxy server, and WML content is passed directly to the WAP server. The HTML filter is essentially a translator from HTML to WML.

WAP and WML together comprise an interoperable, global, and open standard that provides tags for organizations that want to offer services to mobile subscribers. There are no restrictions on the type of air access or end device used; WAP is a nonproprietary solution. WAP is applicable to all the cellular telephone standards examined earlier in this book, including, but not limited to, all versions of AMPS, DAMPS, GSM, and IS-95: essentially all 1G and 2G cellular systems. There is no reason to believe that WAP will not be applicable to 3G systems as they are developed. WAP phones based on many of these standards are currently available in the marketplace.

WML is used to create WAP pages in much the same way as HTML is used to create World Wide Web (WWW) pages. Not surprisingly, WML is based on Java, and the WMLScript is based on JavaScript. Just like WWW pages, WAP pages are displayed on a WAP browser, which will be embedded in wireless devices (WAP clients) that are WAP enabled. WAP clients will provide display content to the subscriber and interpret network content using the languages WML and WMLScript. WML is responsible for specifying the presentation and user interaction aspects of WAP clients. WMLScript enhances the browsing experience by adding intelligence to the client and hence minimizing the number of round-trip communications between the client and server. This aspect is very important in the limited bandwidth, high latency environment that will be experienced by the vast majority of WAP clients.

While the above analogy to the WWW is useful for setting the context, the preferred terminology is not WAP pages but WAP decks, which are composed of one or more WAP cards. Unlike JavaScript, which is embedded into WWW pages, WMLScript is not embedded. Rather, WML contains only references to WMLScript URLs. The WMLScript must also be compiled before it can be run on a WAP client. The WAP browser must contain a WMLScript virtual machine to run the compiled code.

WAP also provides for security (although this element of the standard is optional) at the transport layer. The wireless transport layer security (WTLS) features encryption services that many service providers will use in pursuit of e-commerce. Therefore, while this layer is optional, it is likely that the vast majority of protocol stacks will implement it.

■ PROTOCOL OVERVIEW

There are five layers to the WAP architecture. The architecture is compared to the Internet protocol model in Figure 17-2. The abbreviations are listed below:

1. Wireless application environment (WAE): application layer
2. Wireless session protocol (WSP): session layer
3. Wireless transaction protocol (WTP): transaction layer
4. Wireless transport layer security (WTLS): security layer
5. Wireless datagram protocol (WDP): transport layer

The security and transaction layers are new to most readers. These two layers are defined separately because of the low bandwidth of most wireless telephony networks, which is the prime application area for WAP. WAP can be run over three distinct protocol stacks, depending on the security requirements of the application. Because of the low

Figure 17-2
WAP Protocol Architecture

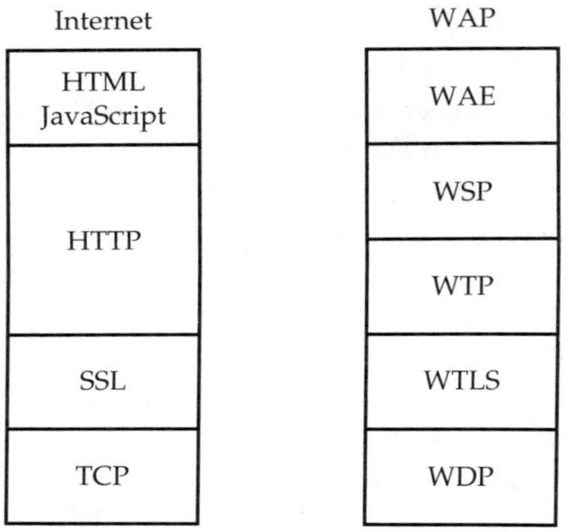

Internet	WAP
HTML JavaScript	WAE
HTTP	WSP
	WTP
SSL	WTLS
TCP	WDP

Figure 17-3
WAP Protocol Stack Options

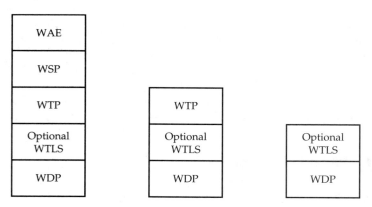

bandwidth of the underlying networks, it was deemed appropriate to granularize this overhead as much as possible. See Figure 17-3, which shows layers 4 to 7 for three applications.

The first stack is an example of the full implementation of WAP. Full security is required, and the data exchanged must traverse through many layer interfaces, each of which require processing. The middle stack does away with the session and application layers and provides a service interface to an application directly at the transaction layer. Finally, the third protocol stack illustrates a bare-bones approach where no transaction services are required and only a direct interface to the transport layer is necessary. In all these protocol stacks, the WTLS layer could be used for those exchanges where security is important.

Note that there is no specification of any layer protocol below layer 4 of the OSI model because WAP is intended to run over IP and whatever air interface the WAP client may use. The next paragraphs outline the function of each of these five layers.

WAE

The WAE is the layer protocol that contains the WML and WMLScript specifications. It also contains the wireless telephony application interface (WTAI) specifications, which detail how WAP applications can access mobile telephone functionality. Separate WTAI addendums have been written for each mobile telephone family. These addendums specify how to perform tasks, such as making a call, using the data services that the air interface offers.

The idea behind WAE is to emulate the success of WWW. Therefore, WAE is intended to establish an interoperable environment that allows service providers to build applications independent of the underlying wireless system. This environment must be capable of functioning despite the limitations of a typical WAP client. These limitations include relatively low bandwidth, relatively high latency, limited battery life, small screen size, and last but not least, limited on-board memory.

It is also important to remember one additional limitation of WAP clients compared to a typical WWW client. In the desktop or notebook world, browsing the WWW is done with a standard keyboard and the form of the display is quite uniform across the client population. In the mobile world, neither of these characteristics will be true. The screen size and input characteristics of mobile WAP clients will vary widely. Second-generation WAP clients will likely be equipped with a flexible voice interface, thereby bypassing the keyboard limitation. This type of interface will also open up the WWW to those for whom English is not their primary language.

One interesting aspect of the WAE model is that it is a so-called push-based content delivery model. Content will be displayed not only based on a WAP client's request, but also at a server's option. This kind of content delivery is accomplished through the use of specific functionality in the WSP and WTA layers that allows servers to push content to the WAP client display without the WAP client's request. These functions will be briefly identified in the appropriate section below.

Because both WML and WMLScript are contained in the WAE layer, a brief discussion of the characteristics of both is appropriate here. WML uses the idea of cards and decks instead of a document. A document consists of a number of pages and the set of pages comprises the document. In WML, a set of cards comprises a deck. As you can see, the concepts of a document and a deck are very similar. As the user interacts with the server, the application running describes this interaction through a set of WML cards.

The user essentially moves through a set of WML cards. As the user encounters each card, he or she may request information, merely examine its contents, or perhaps make some choice. He or she then moves to the next card in the deck, with the possibility that the deck order is modified by the choices made on the previous card. Typically, the deck is stored on the server, and each card is targeted to a specific user interaction.

Because of the variation in client displays, a single card will sometimes be displayed in its entirety. In other cases, two or more screen displays may be required to complete a specific user interaction. Again, voice activation may significantly enhance this interaction and standardize it across numerous WAP clients. The choice of how to display or request the information desired is made by the user agent on the client device. For this reason, WML presents all requests for user information in an abstract way. It falls to the user agent to optimize the interaction for the specific client device. WML has three basic ways of interacting with the user: text entry, option selection, and task invocation.

Text entry is used to support password and basic text entry by the user. The option selection control allows the server to present a list of options to the user. These options may include items such as links to other cards or invoking WMLScripts. Task invocation is intended primarily to activate a history task. It allows the user to jump immediately to any previously encountered card. These controls may be bound to a specific key on the user device or to a region of the display or they may be voice activated.

WMLScript is very similar to JavaScript. It is an extended subset of that language, refining JavaScript for the limitations of WAP clients. For the most part, these extensions focus on reducing the number of round-trip communications with the server, taking into account the limited memory capability of the typical WAP client. Typical applications of WMLScript include checking user input prior to forwarding it to the server and providing the logical hooks to access device peripherals.

WMLScript supports five basic data types: Boolean, integer, floating point, string, and invalid. It has the capability of automatically translating between them as appropriate.

Figure 17-4
WML Architecture

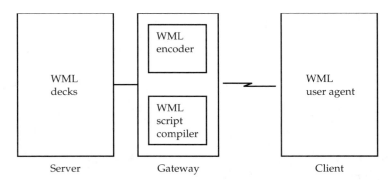

For example, some client devices may not support the display of floating point numbers. WMLScript could automatically display the number in an appropriate data type. WMLScript also supports several operations and functions and defines several libraries.

Figure 17-4 illustrates the primary functional components of the WML architecture. Using this figure, the interaction among a WAP client, gateway, and server can be described. In general, the exchange proceeds as follows:

1. The user submits a request to the server using the WML user agent in the user's WAP client.
2. The server sends back a single deck in text format.
3. The gateway intercepts the deck and translates it, using the script encoder, into a compressed binary form for transmission over the limited bandwidth, high latency radio link.
4. The gateway transmits the deck in binary format to the user. This information is optimized based on user agent negotiation, with the gateway taking into account the WAP client device characteristics.
5. The user submits additional requests based on navigating the cards in the deck and using the WMLScript embedded in them.
6. The gateway translates these requests using the WML script encoder and forwards them to the server. Return to step 2.

WSP

The WSP is the layer protocol that contains services used for browsing applications. WSP is designed with the low bandwidth and large latency aspects of typical wireless telephony networks in mind and features a long session lifetime. It includes two levels of session association: connection-oriented and connectionless. If the connection-oriented session protocol is running, all lower layers of the WAP protocol architecture are used. In this case, WSP runs on top of WTP. If the connectionless session protocol is running, the WSP bypasses both the WTP and WTLS layers and runs directly on top of WDP. Security is passed to the underlying layers.

As the session layer protocol in the WAP architecture, WSP provides the browsing application with three functional tools:

1. Session establishment/release and suspend/resume capabilities.
2. Protocol functionality negotiation.
3. Compressed data content exchange.

Unlike most HTTP sessions, WAP sessions are not tied to the lifetime of the transport connection. A session can be suspended during idle periods. Suspension is intended primarily to optimize battery life. The protocol functionality is negotiated during the session creation process. These functions specify header types, character sets, languages, and device capabilities. Once the headers are negotiated as part of the compression process, they are compressed down to flags to reduce protocol overhead.

As mentioned earlier, both push and pull data services are supported. WSP supports push data in two broad ways:

1. Inside an existing session context.
2. Outside an existing session context.

This push data can run in two ways inside an existing session context: either confirmed or unconfirmed. In both cases, the server has the option of pushing data to the client at any time during the session. In the confirmed case, the server receives a confirmation back from the client that the push was delivered. The confirmation is intended to address reliability of the underlying transport service. If the push data is unconfirmed, the client sends no response to the server upon receiving the push data. Unconfirmed push data is also possible, at the server's discretion, without entering a session with the server. All push data is accomplished asynchronously.

As mentioned above, the session layer can provide two classes of service to the application: connection-oriented and connectionless. The following sections will highlight briefly the characteristics of these two service types.

Connection-Oriented WSP is composed of several facilities. Facilities are implemented through the use of service primitives. Two required facilities and four optional ones are controlled by capability negotiation at session establishment. The facilities are listed in Table 17-1, and the service primitives are listed in Table 17-2.

Session management is the basic functionality required for a client and server to communicate. It is responsible for the negotiation of the capability and protocol options used. It provides for either the client or the server to terminate a session and allows the server to refuse a connection attempt, in which case, it forwards the client request to another server. The client is notified if the connection is terminated due to action by the underlying service provider or the application entity.

Exception reporting allows the service provider, usually a cellular telephone provider, to communicate to the client independent of the session state. Exception reporting is not permitted to change the state of the session. It is intended to be used for cellwide broadcasts and multicasts to which the client may have subscribed. It may also be used for emergency notification to all subscribers.

Table 17-1
WSP Facilities

Facility	Presence	WTP Service Class
Session management	Required	0 and 2
Exception reporting	Required	[Not applicable]
Method invocation	Optional	2
Session resume	Optional	0 and 2
Push	Optional	0
Confirmed push	Optional	1

Table 17-2
WSP Confirmed Service Primitives

Session Management Facility	Method Invocation Facility	Push and Confirmed Push Facilities
S-Connect	S-MethodAbort	S-ConfirmedPush
S-Disconnect	S-MethodInvoke	S-Push
S-Exception	S-MethodResult	S-PushAbort
S-Resume		
S-Suspend		

Method invocation allows the client to request the server to take an action and return the result. Both the client and server are notified about the result of the request and action. Session resume is intended to be used by the service provider when further communication with the client is not possible. Communication breakdown may be due to radio channel congestion or the user roaming out of coverage. The state of the session is preserved until the client can resume communication. Note that session resume is not a required function. Push and confirmed push permit the server to send unsolicited information to the client. In the former case, no confirmation is received. In the latter, the server receives confirmation from the client that the pushed data arrived.

The last column in Table 17-1 shows the transaction class used by each WSP facility. Note that exception reporting has no class because it is intended for use by the underlying service provider, not the WAP protocol stack. A relationship exists between a required service and its WTS class. All connection-oriented clients must support transaction classes 0 and 2. WTP transaction classes define the reliability of the communication between the server and the client. They cannot be negotiated and will be discussed in more detail in the WTP section.

Table 17-2 lists three broad classes of the confirmed service primitives. The first five are used primarily by the session management facility; the next three, by the method invocation facility; and the last three, by the push and confirmed push facilities.

Connectionless The connectionless service is also composed of facilities and service primitives. However, only two classes of service facilities are available. These classes are the optional facilities method invocation and push facilities. The first has two service primitives: S-Unit-MethodInvoke and S-Unit-MethodResult. The push facility makes use of only the primitive S-Unit-Push because it is an unconfirmed service.

WTP

WTP is the layer protocol that contains the transaction protocol options offered by WAP. All these options are low-overhead transaction systems suitable for use by the thin, mobile WAP clients. WTP is message oriented and optimized for transaction services that would be common when browsing. WTP offers three classes of service to the WAP client: unreliable one-way requests, reliable one-way requests, and two-way requests/replies. These WTP transaction classes were identified in Table 17-1; respectively, they are classes 0, 1, and 2. Note that it is possible to have a reliable request without a confirmed reply.

WTP achieves this reliability through the use of explicit transaction identifiers. WTP features no explicit set-up or tear-down capability; it is not a connection-oriented service. Such a service would place too much overhead on the wireless link; eliminating it improves overall performance. WTP improves the reliability of the protocol stack by removing the necessity of retransmissions and acknowledgments by upper layer protocols.

Recall that the entire WAP protocol suite has been constructed with the constraints of wireless links in mind. The WTP layer protocol is designed specifically to minimize upper layer retransmissions in the case of a lost packet and to minimize acknowledgments when using the confirmed service offering. Embedding this layer protocol between the WSP and WDP layers in the protocol stack significantly improves overall throughput performance from the user's perspective. Without the WTP layer, the WSP layer would sit directly on the WDP transport layer. Because WDP is modeled after UDP and the WAP architecture is designed for IP at layer 3, there is no connection-oriented protocol at layer 3 or layer 4, unlike a traditional IP/TCP stack where this control takes place at layer 4.

To understand how WTP accomplishes this task, a more detailed examination of the three transaction classes is appropriate. A transaction is defined as encompassing the request and response pair of an information exchange between a client and a server. For example, a WAP client requests information from a server and the server responds with the information.

Class 0 transactions are unreliable pure datagram services. This class of service is not intended to be a primary way of sending datagrams. If a pure datagram service is desired by an application, the standard recommends the use of the WDP service. A class 0 transaction consists of a single message sent from the source (client) to the destination (server). The destination system provides no acknowledgment and, in the case of a lost message, the source makes no retransmission. This transaction cannot be aborted.

Class 1 transactions are intended to support reliable push operations by the server to the client. A class 1 transaction consists of a single message sent from the source to the destination. The destination system acknowledges the message to the source and holds the transaction in stasis to address the possibility of a lost acknowledgment (signaled by the source retransmitting the initial message). The transaction can be aborted at any time.

Figure 17-5
WTP Transaction Classes

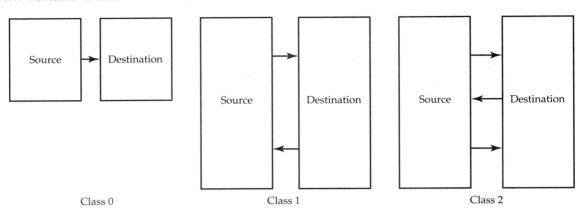

Class 0   Class 1   Class 2

Class 2 transactions are intended to be the basic transaction class used by sessions that require a reliable service. A session will typically consist of several transactions. A class 2 transaction consists of a single message sent from the source to the destination. The destination system replies with a *single response message* that includes, implicitly or explicitly, the acknowledgment. The source then acknowledges the response message to the destination.

Unlike class 1 transactions, the source holds the transaction state information to avoid potential retransmissions of its acknowledgment to the destination system. In the case of the original response being retransmitted due to a lost acknowledgment from the original source to the destination, it would be the responsibility of the original source to retransmit the acknowledgment. A class 2 transaction can be aborted at any time. Figure 17-5 illustrates the three transaction classes.

It should be clear how this layer protocol addresses the connection-oriented capabilities missing from WDP and normally present in TCP. A good question to ask is, Why is a new sublayer protocol necessary when TCP would address the connection reliability issues adequately? The key word is *adequately.* TCP sets up and tears down a connection for every session. This process consumes a large amount of message overhead and, in a wireless, limited bandwidth, large latency environment, it is inappropriate and merely adequate. WTP renders a connectionless WDP into a reliable service from an upper layer perspective without the need for a connection-establishment phase.

WTP class 2 transactions also specifically address the large and variable latency sometimes exhibited by cellular telephone networks. It is possible for a class 2 server to tell a client to wait instead of responding to the client's request. This action prevents the client from retransmitting its request when the server knows that its acknowledgment timer for a response will expire before it can transmit its response. It is another example of how WTP is optimized for its environment. TCP has no such capability and the client would always retransmit its request, using the limited resource of most wireless networks: bandwidth.

WTLS

WTLS is the layer protocol that implements the WAP security option. This protocol is based on and is very similar to the secure sockets layer (SSL) protocol. It provides for encryption, authentication, and data integrity checks to ensure that the data exchanged between the WAP server and WAP client is secure and unchanged. Applications may disable WTLS features selectively based on both the security required for the exchange and those security capabilities featured by the underlying air interface. WTLS is optional; however, it seems clear that implementations using WAP for e-commerce will incorporate it. WTLS uses the same underlying algorithm of SSL: Rivest Shamir Adelman (RSA). Yet it is optimized for use over narrowband channels.

WDP

WDP is the layer protocol that implements the functions of the transport layer. This layer functionally isolates the upper layers of the WAP architecture from any concerns with the underlying air interface and network protocol used. It provides all the traditional functions of a connectionless transport service. In fact, when the underlying network protocol is IP, WDP is identical to the user datagram protocol (UDP).

WDP offers a service interface at its service access point (SAP) interface to either WTP, WTLS, or the application itself, depending on the protocol stack option used. This service interface is independent of the wireless network implementation. WDP implements this service interface through the use of several adaptation layers, each unique to the underlying wireless network. For example, adaptation layers are described for IS-136 packet- and circuit-switched data; GSM, SMS, and circuit-switched data; CDPD and CDMA packet data; as well as many others.

WDP uses a simple datagram consisting of three fields: the source port, destination port, and data field. The length of each of these fields depends on the adaptation protocol. The definition of the data field also varies in its content. For example, it may contain several subfields specifying message type, segment number, and the actual data content of the message.

■ ADOPTION ISSUES

WAP is an effort to standardize Internet and WWW access to mobile devices. It standardizes by optimizing several protocol elements for the current wireless environment in cellular telephone systems. There are two main thrusts to this optimization. The first takes into account the low bandwidth, large latency characteristics of cellular systems, and the second is the limited display and memory capabilities of the clients, which are often modified cellular telephones. Both limitations are now being called into question, and the question is, Is WAP a solution for a problem whose time is almost over?

Third-generation cellular standards are now approved and will be moving forward into product deployment in the next few years. These standards will feature significantly more bandwidth for data links than current systems can provide. Additionally, the pace of technology shows no signs of slowing down, and it is reasonable to expect that cell phones available two years from now will feature much more memory and better battery conservation. These two trends seem to signal that perhaps the transmission and client models on which WAP is based may be phasing out just as the standard is being deployed.

Another more serious concern that WAP faces is the fact that it relies on gateways to translate content into WML. There are problems with gateways from a protocol perspective. Gateways are protocol translators. If the protocol evolves, the gateway must be changed. If it is not changed, the evolutionary aspects of the protocol, usually desirable ones for subscribers, are not available. Hence, gateways effectively wall off sections of the network from new applications that use the new services available from an evolved protocol.

Further complicating the situation is that these gateways are assumed to be provided by the service providers, who will be the cellular providers in the vast majority of cases. Most providers have assumed that their subscribers would use their gateway, hence generating a revenue source from the WAP deployment. They hard-code the gateway address into the phones they sell. Then each time a WAP access is made through the gateway, somebody pays a fee. In the vast majority of cases, this fee is passed along to the subscriber.

My sense is that this situation creates a gaping loophole in the business model. A competitor will figure out very quickly that he or she can offer WAP clients that are not hard-coded to a gateway, thereby allowing the subscriber to shop for services from the entire universe of gateways. If for some reason competitors do not step in with these options, I can imagine lawsuits that seek to change the monopolistic behavior. These lawsuits will be brought by latecomers to the game who are seeking a piece of the business or by consumer groups who want a level playing field.

Another problem with the WAP model using gateways is that if the providers own the gateways, many e-commerce transactions may not be secure enough for the application provider. Banks and financial institutions are well known for their desire to own and secure the network from end to end. Will a WAP gateway, located at a cellular provider's site and being used for all kinds of traffic, be secure enough? My sense is that many financial institutions will say no. Additionally, the whole reason why providers want to own the gateway is to derive revenue from it. Banks in particular have already claimed that territory for financial transactions, for example, ATM fees, etc. Will they agree to pay a percentage to service providers?

WAP has another problem from the content developer's point of view. Every WWW page must be rewritten for WML. Although WML has some mechanisms for addressing different capabilities in WAP clients, they are limited in scope. (See the discussion above on WML.) For example, if the display size is different in two WAP clients, the WML must be rewritten for both clients. Without standardization, many different display types will emerge into the marketplace.

Content developers will have to address these issues and, in the long term, products that require yet another rewrite will probably not appear. In the meantime, however, this situation seems ripe for poor functionality in many WAP clients because the WML code is not optimized for their client machine. How likely is it that all these translations are going to be 100 percent accurate? If WAP earns a reputation for working on only some machines or for a lack of uniformity among applications, it won't help its rate of adoption in the marketplace.

I am also not convinced that a user population accustomed to full content, multicolor screens will hurry to embrace a WWW experience that consists of a monochrome, four-line, text-only display. To me, it is a risk to "promise" WWW access through a WAP client. It seems obvious that once subscribers have tried it, the experience will pale. After all, their only experience with WWW browsing is on full-color, relatively high bandwidth displays. Will they even recognize a four-line text interface as WWW browsing?

If the WAP interface is text only and features only four or so lines, why do we need an entire protocol to display something that the SMS service already does? SMS provides a 160-character display and is not presented as something it is not. Its presentation seems a much more consumer-friendly approach to marketing a new service rather than the risky one of selling more than can be provided, as it seems WAP is set up to do. (SMS is described in Chapter 14. Please refer to it for more information.)

SMS is attractive for another reason, especially here in the United States. The United States, unlike many parts of Europe and Asia, features many different cellular telephone technologies, as described earlier in this text. SMS can work with any of them with only minor modifications to the protocol and to on-site hardware/software. SMS already works directly with HTML, so there is no need for content developers to modify their pages for each and every WAP client. SMS works with the display on almost any cell phone currently manufactured. SMS also avoids the gateway problems discussed above because it does not require a gateway.

SMS is not an interactive environment like WAP because there is no provision for user responses. An SMS message is a one-shot message of limited size. SMS users can have a dialog by exchanging sequences of SMS messages. As indicated above, however, such messages can be instantaneous or delayed. In the latter case, true interactivity is not present.

How many content applications really require interactivity on a cell phone? It seems that the content most cell phone users want is one-way, for example, stock quotes, weather reports, news headlines, disaster warnings, advertising content, etc. Users will e-mail through a voice interface very soon now, and that technology seems to capture the features that most users want. I have a hard time imagining shopping with my cell phone, selecting items by their text description, rather than looking at graphics, which WAP cannot provide. Is WAP really suited for e-commerce?

It is possible that the growth figures of cellular systems are so high that much of the WAP hype is about maintaining market momentum. Until 3G systems, with their much improved data rates, are deployed, providers are hard-pressed to offer consumers any new capabilities. This marketing is an attempt to live up to previous successes by applying a new and trendy feature to cellular phones. I have already indicated that this solution is at best short term.

■ SUMMARY

This chapter explored the emerging standard WAP. While the technology is impressive, several issues still cloud its future adoption. Only time will tell if these issues will be overcome and if they will be addressed before wireless technology advances render the issue mute.

There are no review questions for this chapter or any of the chapters in the fourth part of this book. The technology is not yet mature, so technological questions are inappropriate at the time of this writing. Business and adoption issues can best be addressed after the technology appears in the marketplace; again, review questions at this point are premature.

The best approach for more learning about the emerging technologies described in the fourth part of this book is to read trade press and journal publications as current sources of material and compare what has emerged into the marketplace with what was projected in this chapter. Analysis of how issues raised in this part were addressed by the manufacturers and the marketplace over time can provide a useful tool for additional exploration of the topics presented here.

Emerging Standards: 3G Cellular Systems

■ INTRODUCTION

Third-generation wireless networks seem to be the answer to everything, from a new business paradigm that will transform how marketing is done to becoming the holy grail of corporate data communications. The day is (supposedly) only around the corner when everyone under thirty will abandon their wireline telephones and desktop computers, for they will have this functionality in their mobile telephones. Now they can talk to their friends and surf the Web, all while walking down the street to a coffee shop. The level of hype around this technological evolution is extreme. The fundamental problem is that third-generation cellular systems are discussed as revolutionary when, in fact, they are evolutionary. If one were to believe all that one reads, it would appear that "ultimate" product rollout is "just around the corner." One goal of this chapter is to educate the reader so that he or she will be able to distinguish the fluff from the fact when choosing the right mobile networking solution for a particular business goal.

This chapter will begin with an examination of 3G networks. First, we will define what is meant by the phrase *third-generation* (3G). The four air interface methods capable of achieving 3G goals are listed. This list is followed by a short discussion of how 3G networks will evolve in Europe and the United States. Because 3G networks are still evolving, the standards organizations and participating bodies are introduced to allow the reader to track the evolution as it emerges.

The next section will illustrate one approach to migrating to an all-IP network. The discussion will include the services and the protocol architectures that will be used to support these services. Finally, a vision of the cellular networks of 2010 and beyond is provided.

■ WHAT IS 3G?

The phrase *third-generation networks* encompasses two concepts: worldwide roaming and high data rate services. Both goals are enabled by a spectral assignment that has a wider bandwidth than those of cellular systems deployed today. The roaming goal can be achieved only with current technology in one of two ways. Either unification of the air interface occurs so that a single mode phone can work everywhere, or multimode phones are designed to work with all 3G air interfaces. The first choice relies on a unified spectral allocation across all countries. It is now clear that a unified spectral allocation will not occur in the foreseeable future.

The promise of software-defined radios (SDRs) does offer a solution, but this technology is still in its early stages of development. Current technology requires a mobile telephone (radio) to be designed so that the hardware present in the mobile telephone is closely linked to the air interface technique and the frequency band of operation. SDR designs will allow the mobile device to identify the air interface and frequency band of operation being used and to adjust its operation automatically to accommodate that environment.

The original and still dominant path for 3G cellular systems will be with a wideband CDMA (WCDMA) air interface. Most systems will deploy it; however, there are actually three types of CDMA. While all are wideband, only one type is properly called WCDMA. There is one more approved method of access media: an FDM/TDMA based approach. While it is a wider band than GSM, it is fundamentally quite a similar air interface. Since four different air interfaces have been approved, true global roaming, one of the two key goals of the 3G effort, appears to be compromised. In sum, there are four fundamentally different approaches for the air interface and all are 3G.

The second major goal of the 3G effort was to provide a high data rate service to mobile subscribers. High data rate services require wide bandwidth spectral allocations. With this basic concept in mind, 3G systems can be defined as cellular systems that use the new spectral allocations approved at WARC-2000.

With the wideband spectral allocations, 3G networks will offer higher data rates than those achievable with 2G mobile networks. However, the wild claims of 2 Mbps data links are just that: wild claims. It is true that if each user had access to the entire cell, a single 2 Mbps data link could be achieved in ideal environmental conditions. Typical data rates experienced in 3G networks will move from the order of 10 kbps to the order of 100 kbps, depending on the level of subscription. All-IP networks are also part of the promise of 3G networks, but they are seen as an evolutionary step. Standards bodies are developing the protocols necessary to move this idea forward, but it is clearly a secondary goal.

In summary, the original primary goal of worldwide roaming now seems complicated by several related issues, many of which will be explored in this chapter. Until

Table 18-1
3G Air Interface Options

Air Interface	IMT Class	Technology	Standard	Name
CDMA	IMT-DS	Direct spread	UTRA-FDD	WCDMA
CDMA	IMT-MC	Multicarrier	IS-95C/CDMA 2000	Multicarrier
CDMA	IMT-TC	Time code	UTRA-TDD	TD-SCDMA
TDMA	IMT-SC	Single carrier	UWC-136	[Not applicable]

SDRs appear, the primary goal of the 3G effort is to offer higher data rates to mobile subscribers. Once that goal is achieved, serious planning and architectural efforts will allow a transition to an all-IP network.

A separate issue involves the so-called 3G services. Note that services are introduced at different times in different markets, and at least a few 2G services available in Europe and Asia will be introduced in the United States and elsewhere and marketed as 3G services. For the most part, these services are protocol overlays that enhance data rates and offer limited IP data connectivity in the exiting air interface and switch and signaling environment.

What 3G specifically does not mean is a common air interface. Instead, a family of air interface standards will be allowed. These air interface standards are listed in Table 18-1. Additionally, the ITU has decided not to define the protocols used in the existing wireline core network. There are two widely deployed approaches (ANSI-41 and GSM-MAP), and both will be allowed to evolve independently. The protocols used to interface between the two environments will be specified by the ITU, however, to facilitate worldwide roaming.

■ 3G NETWORKS

As discussed above, 3G networks are distinguished by high data rate capabilities and operation in new spectral allocations. All 3G networks will feature one of four types of air interface. The three variations of a wideband CDMA air interface should dominate the choices, but at this time all four are moving forward.

The entire set of standards required to implement 3G networks fully was originally referred to as IMT-2000. The required standards included the air interfaces and all circuit and packet data protocols. The 2000 in the name was not meant to mean that 3G networks would be standardized by the year 2000. It is the name for the ITU project. The air interfaces are referred to as classes of IMTs. Because the data protocols are now being unlinked from the air interfaces, the air interfaces are referred to by their original abbreviations. Table 18-1 summarizes the air interfaces.

Table 18-1 does not show the digital enhanced cordless telecommunications (DECT) system. It is a 2 Mbps system designed for cordless telephones and closely resembles a standard for a wireless PBX. This system is formally part of the ITU-2000 framework; however, it is not relevant to 3G cellular systems and so it is omitted.

Universal mobile telephone service (UMTS) is the path that many European and Asian countries will take to 3G. The air interface selected by this group is the first listed in Table 18-1 and is called UTRA-FDD. UTRA stands for UMTS terrestrial radio access. UTRA and WCDMA are often used interchangeably. However, UTRA has two forms: UTRA-FDD and UTRA-TDD. Frequency division duplex (FDD) indicates two separate frequency bands (separated by a guard band), one for the forward channel and one for the reverse channel. Time division duplex (TDD) indicates a single frequency channel with forward and reverse channels operating in different time slots. UTRA-FDD is the dominant choice of the two, but both will move forward in different regions of the world.

The other CDMA choice, CDMA-2000, or multicarrier CDMA, will be the path that the currently deployed CDMA systems will take to 3G. These current systems include the IS-95A and IS-95B systems. Service providers (namely, Sprint and Verizon) that have deployed these CDMA based systems in the United States are planning to move ultimately to CDMA-2000. From an air interface point of view, multicarrier CDMA and UTRA-FDD have many similarities. CMDA 2000 will probably evolve in two distinct steps. First deployed will be CDMA 20001x using the same 1.23 MHz channels as does the existing IS-95 systems. Later, CDMA 20003x will replace the 1x systems. CDMA 20003x uses three times the bandwidth, or about 3.69 MHz.

Service providers in the United States who do not have currently deployed, CDMA based networks will instead continue to use 2G air interfaces and employ a protocol overlay. As the need for higher data rate services emerges, EDGE will be overlaid, thus allowing data rates in the 100 kbps range. AT&T has announced that it will follow this approach. Other service providers with similar networks will also likely adopt this approach, if for no other reason, to offer increased coverage using the same standard. Note that this approach is not a 3G air interface but will likely be marketed as 3G services.

In the 2006–2008 time frame, these TDMA based networks will likely move to either UWC-136 or UTRA-TDD. (UWC stands for universal wireless communication and has its roots in IS-136.) In the United States, portions of the 3G spectral allocation are already allocated to existing cellular systems, and the single-carrier TDMA approach requires much less bandwidth. Because there are revenue advantages in moving to some kind of CDMA air interface, the UTRA-TDD approach has appeal for most service providers. Until the spectral issues are resolved, however, this transition must wait. At this point, it is not clear which will come first, 4G or a more unified 3G environment.

The WARC-2000 meeting specified a worldwide standard for the spectral allocation for this air interface. This spectrum consisted of two spectral components: 1,885 MHz to 2,025 MHz and 2,110 MHz to 2,200 MHz. The goal of that allocation was a worldwide commonality for all cellular systems. This allocation was critical to realizing the goal of a truly worldwide roaming capability because all mobiles would operate in the same spectral location.

The service aspects of this interface were left to service providers and third parties, with the intention of fostering innovation. To understand the evolution of the 3G network and how it will proceed, we will examine the standards committee (and their relationships) that produced the specifications. After a brief examination of how 3G networks are expected to roll out in Europe, Asia, and the United States, we will return to an examination of the standards organizations.

3G in Europe and Asia

In Europe and most of Asia, there is only one 3G specification, WCDMA. Third-generation evolution in these parts of the world will thus be relatively rapid compared to the situation in the United States. Where GSM systems are deployed, which is in Europe and many parts of Asia, 3G evolution will move forward in three distinct steps. In the first step, already in progress and quite advanced in some areas, GPRS will overlay existing GSM networks and thus offer a packet radio service to those with compatible cellular phones. This setup gives users a packet network interface to the Internet.

The second step will involve building UTRAN-compatible equipment to replace the base stations and controllers with UTMS-compatible equipment. During this step, the radio interface will remain GSM. Simultaneously, it is highly likely that EDGE will be used to enhance data rates. The third and final step will be the transition to a WCDMA air interface, which will require replacement of the cellular phones with UTMS-compatible equipment.

Note that CDMA-2000 and WCDMA are very similar but also display some critical differences. The system designers of WCDMA chose several parameters to make it incompatible with IS-95 CDMA based systems for two reasons. First, they wanted to speed the adoption of WCDMA; second, they wanted to make sure that WCDMA did not have to pay royalties to Qualcomm. This simple fact of backward incompatibility will accelerate 3G adoption in Europe and most of Asia.

3G in the United States

Third-generation evolution in the United States will not be so simple. Here, there is no universal standard for wireless cellular access. Three 2G standards are deployed in the United States: IS-136, GSM, and IS-95A/IS-95B. Unlike Europe, there is no single standard for 2G or 3G. Carriers will be slow to adapt 3G because it is not in the interests of these service providers to move directly to 3G. Each provider will have its own transition strategy from its existing system. Clearly, each service provider will extract as much revenue as possible from existing deployments before investing in 3G. Some estimates have placed the cost of transition to a universal 3G system in the United States at over $100 billion.

The situation in the United States is very different than that in most of the rest of the world for some basic reasons. At the time of this comparison, the population of the United States was approximately 270 million people, while there were about 100 million more than that in Europe. So the pool of subscribers is greater in Europe than in the United States. The United States is also much bigger, almost three times larger in area than Europe. Therefore, cell site density in the United States is about 2 sites per 100 square miles; in Europe, it is much higher. For example, in the United Kingdom it is about 20, and in Germany it is almost 35. As mentioned earlier, many different systems are used in the United States, and large areas have no cellular coverage at all.

Because 3G really means high data rate transfers of packet data, and because service providers in the United States will choose to increase coverage rather than abandon current systems, software overlays will be the approach for offering packet data and access to the Internet. This approach is somewhat similar to the first phase of the GSM evolution to 3G. While CDPD, a packet service, is currently available in AMPs deployments,

GPRS or EDGE will be used for GSM and IS-136 deployments. CDPD is a relatively low data rate service, and GPRS is only incrementally better. Since EDGE offers significantly higher data rates than GPRS, providers may move directly to EDGE.

AT&T originally announced a two-step process for offering packet data services. Currently, it is reconsidering and will likely move directly to EDGE, although this remains to be seen. Whatever AT&T decides to do, it is likely that SBC/BellSouth will follow the same approach for the simple reason that data services are much more valuable to business subscribers as the coverage becomes broader. In each cell cluster, a few voice channels will be replaced with either GPRS or EDGE data services. As demand for increased data services occurs, additional voice channels will be replaced.

CDMA networks in the United States and elsewhere now running IS-95A or IS-95B will most likely transition to IS-95C (also known as CDMA 2000) using the multicarrier form of CDMA. CDMA 2000 is the evolutionary form of the existing IS-95 systems and will be required to pay royalties under the Qualcomm patent. For this reason, it is likely that only those providers already paying royalties will follow this evolutionary path. Because this scenario is undesirable from a business perspective, a transition to WCDMA is being defined. This transition strategy will allow current IS-95 providers to move to WCDMA in two steps, thus spreading the cost over a longer period of time. The high data rate (HDR) protocol overlay will offer packet data services during the transition. HDR is very similiar in performance to EDGE. Both EDGE and GPRS were originally developed for GSM based systems, but they have been modified to be independent of the underlying access method. The typical subscriber will see GPRS or EDGE data rates in the 20 kbps to 40 kbps range, although EDGE will allow premier subscribers to achieve double that rate (with appropriate charges).

The only long-term solutions to CDMA based 3G networks in the United States will be a reallocation of spectrum to these services and a refit of existing cellular networks. These solutions will not be implemented in the forseeable future because many regulatory and business issues remain to be resolved. In the United States, 3G will mean 2G networks with protocol overlays that will offer packet based data services and increased data rate access to the Internet. A transition to a CDMA based air interface 3G network will have to wait for the reallocation of spectrum and the realization of financial goals.

To illustrate the spectrum issue in the United States, we will review the FCC policy. The FCC will allocate an additional 90 MHz for 3G services, bringing the total allocation to 180 MHz. Most observers recognize that this allocation is between one-third and one-fourth of what is needed. There are really only two options for obtaining the needed spectrum. First, it is possible to move existing users from exclusive use of the spectrum in question. The two key groups are the military, at 1,755 MHz to 1,850 MHz, and fixed wireless carriers, at 2,500 MHz. Both groups resist this option. The military estimates a thirty-year time frame and a cost of $300 million to complete the move. The fixed wireless carriers, currently in a booming market, have no interest in even estimating the job.

Second, it is possible that the entire UHF television spectrum, channels 14 to 69, would be vacated. The success of this approach depends on the entire country moving to cable or satellite access. This process is expected to start soon because the spectrum channels 52 to 59 are going up for sale in an auction currently scheduled for the last quarter of 2002. Whether this sale is a harbinger of the future, only time will tell.

◼ INTERNATIONAL TELECOMMUNICATIONS UNION

The ITU can trace its history back to 1865, when it was known as the International Telegraph Union and was established to address rules for enabling the telegraph systems of different countries to interoperate. Since that time, various bodies have been created and aggregated under its organizational structure. These subdivisions address many issues, including radio and spectrum assignment issues along with telephone issues. The ITU's name was changed to reflect the broad reach of technology it was being called upon to regulate, and ITU now stands for International Telecommunications Union. Figure 18-1 illustrates the organizational structure of this group that is so vital to the future of all forms of telecommunications. As you can see, three main organizational elements have responsibility for technical issues: the ITU-R (radio), the ITU-T (telecommunications), and the ITU-D (development). All 3G activity, and the partnership projects that provide input to them, are organized under the ITU-R.

A few items in Figure 18-1 are not obvious. A double abbreviation indicates where a name change has occurred but the committee role remained the same. For example, the WRC is now known as the WARC, the CCIR is now known as the RA, and the CCITT is now known as the ITU-T. In the alphabet soup of standards organizations, an understanding of such relationships is valuable in tracking progress. A careful examination of the figure may also uncover the fact that some of the abbreviations do not match the full name of the committee exactly. With an international organization, some of the committee names are English translations from other languages, especially French. Two examples are the radio communications bureau (BR) and the telecommunications development bureau (BDT).

◼ PARTNERSHIP PROJECTS

Two partnership projects are of interest in the specification of 3G networks. One goal of 3G networks was global roaming, but there were naturally many national standards bodies also working in this area. These standards bodies were responsible for much of the definition of the protocols that would be combined into the IMT-2000 family of standards. To use this valuable resource and to harmonize the work in a global sense, the concept of partnership projects was conceived.

A third body, the Operators Harmonization Group (OHG), actively participates in these deliberations. It is composed almost exclusively of representatives from the service providers themselves. Its goal is to unify the two partnership projects and arrive at a single architectural, air interface, and intelligent network specification. The operator perspective is that a single standard in these areas means lower cost deployment through higher volume pricing.

We should emphasize at the outset that these bodies, once combined into partnership projects, have no authority to issue standards. Instead they prepare, approve, and maintain a set of technical specifications and technical reports. These reports and specifications are released periodically, as work is finished. They are used by organizations to develop products, often far in advance of formal international approval. The

Figure 18-1
ITU Organizations

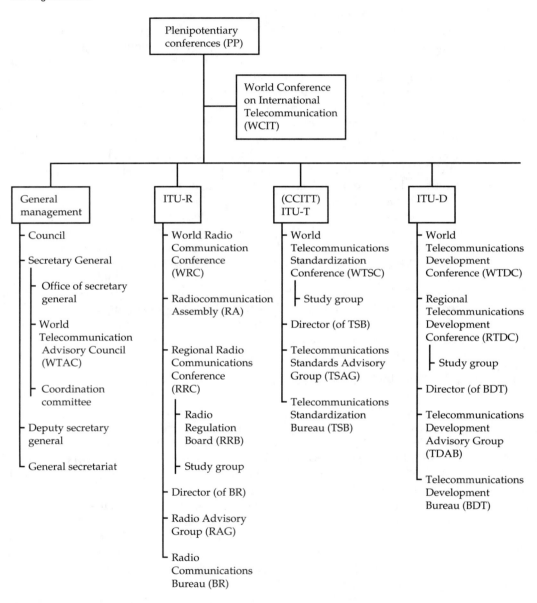

reports and specifications are submitted to the ITU standards bodies, which actually issue ITU standards.

As you might expect, many of the national bodies had similar agendas, but there were some fundamental differences in approach. These differences finally separated the bodies into two groups: Third-Generation Partnership Project (3GPP) and Third-Generation Partnership Project 2 (3GPP2). Each is composed of several technical study groups (TSGs). They differ primarily in their architectural view of the future network and the specification of the intelligent network environment that will work with it.

The group called 3GPP is comprised primarily of those bodies that were responsible for GSM standards. Naturally, they believed that the signaling components of 3G networks should derive from the protocols defined and in use by GSM. In the network sense, signaling should follow the GSM-MAP approach and transition to CAMEL. This group also believes that the air interface should adhere to the WCDMA approach.

The group called 3GPP2 is composed primarily of those bodies that were responsible for the IS-136 and IS-95 standards. Naturally, they believed that the signaling components of 3G networks should derive from the protocols defined and in use by these networks. In the network sense, signaling should follow the ANSI-41 standard and transition to WIN. This group also believes that the air interface should adhere to the multicarrier approach.

Therefore, the two groups differ on two main points. They are close to agreement on the basic access method (CDMA), but they differ on how it is to be accomplished. Second, they differ in how the services provided to subscribers through the signaling network are to be developed. The 3GPP group wants a new approach based on the CAMEL standard; the 3GPP2 group believes that the existing AIN with wireless extensions can do the job.

The ITU has decided to let both groups move forward. In other words, all air interfaces will be allowed, and both core network signaling approaches will go forward. The presence of the OHG has forced both groups to realize that a common interface must exist between the two core network approaches, and this interface will be standardized by the ITU. Without this harmonization, roaming across these core network protocols would be difficult. The next two sections will briefly describe the membership, protocol efforts, and timelines of these two bodies: 3GPP and 3GPP2.

■ 3GPP

The 3GPP includes partners drawn primarily from Europe and Asia and is most concerned with the UTRA-FDD and UTRA-TDD standards. The 3GPP body is composed of five TSGs and one organizational partner (OP) under one project coordination group (PCG). These groups are illustrated in Figure 18-2, and the abbreviations are defined below.

Figure 18-2
3GPP Organizational Chart

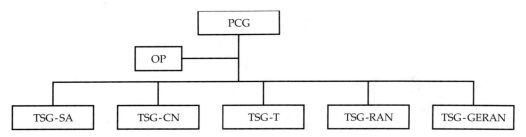

Organizational partners (OP) are other standards bodies that have representation to enhance and encourage international harmonization. They include the following:

Association of Radio Industries and Business (ARIB); Japan

China Wireless Telecommunication Standard Group (CWTS); China

European Telecommunications Standard Institute (ETSI); Europe

ANSI Committee One (T1); the United States and Canada

Telecommunications Technology Association (TTA); Korea

Telecommunications Technology Committee (TTC); Japan

The technical study groups include the following:

TSG-SA (services and architecture): This group is responsible for defining the services and architecture of the network, including areas such as security, speech coding, operation, and maintainence.

TSG-CN (core network): This group is responsible for defining the protocols (including CC, SM, and MM at layer 3 and GSM-MAP) used to facilitate the interaction between the core network and the mobile terminal.

TSG-T (terminals): This group is responsible for coordinating the issues around the mobile terminals' architecture, complexity, and feature set. It is primarily a forum for ensuring conformity across mobile terminals and thus ensuring global roaming capability. This conformity will emerge as a critical issue because terminals are becoming more open and their complexity and capabilities are increasing. At the same time, they are becoming more vulnerable to various security and reliability related threats, including viruses, etc.

TSG-RAN (radio access network): This group is responsible for all radio aspects of the network. It developed WCDMA, it is responsible for the RR protocol at layers 2 and 3, and it addresses issues surrounding the performance of the radio link.

TSG-GERAN (GSM/EDGE): This group is primarily responsible for issues surrounding the base station architecture and feature set. It developed EDGE and it is a forum for ensuring conformity across base stations and thus ensuring global roaming capability.

As discussed above, this last group issues reports periodically on work completed. We will summarize the last report released prior to publication of this text and the probable results of the next two reports, to be issued in 2001 and 2002.

Report R3 (issued in 1999): The key result in this report was the UMTS terrestrial radio access network (UTRAN), the key 3G technology. It also defined the handoff between GSM and universal mobile telecommunications service (UMTS) based networks and the common packet channel (CPCH) specification. Multimedia messaging was defined using the SMS protocol module and extending it to allow attachments. Location services, called LCS, were developed to meet FCC requirements for E911 services in the United States. A set of APIs was developed to allow access to the network for third-party developers. This specification is called the open service architecture (OSA).

Report R4 (scheduled to be issued in the second quarter of 2001): The key issue in this report will likely be the logical separation of the MSC into two components: a media gateway and a server. This step is an important one toward the all-IP network of the future. Multimedia will see additional enhancements in new CODEC specifications for high-quality audio and a streaming audio and video specification. Enhancement of the R3 version of LCS is also scheduled to extend this approach to all 3G networks. The addition of network security, probably using IPsec, will probably also be finalized.

Report R5 (scheduled to be issued in the first quarter of 2002): This report will present the definition of an all-IP multimedia subsystem. It will include the use of SIP with support for VOIP services and functions for call control. It will also include a QOS specification, probably MPLS, and will specify the use of IPv6 addressing for all end-to-end signaling. It is likely that the wireline interface to the Internet will be IPv4, and any vendor-engineered networks designed to support IP services in a mobile environment will be expected to use IPv6.

The 3GPP group has an architectural model that identifies the network elements developed in the committee. A brief introduction to the model is helpful when pursuing current developments in this area. Figure 18-3 illustrates the 3GPP model, often referred to as the UMTS model. As you can see, there are four fundamental components in this model: the mobile station, base station transceiver (BTS), the radio network controller (RNC), and the core network interface (CNI). The mobile station (MS) is the familiar cellular telephone, PDA, or some evolutionary derivative. The base station provides gateway services between the MS and the RNC. The RNC is essentially an evolved BSC and may connect to many BTSs. The CNI, or core network, refers to the other terrestrial core network infrastructure connected to the RAN. It includes gateway functions to intranets,

Figure 18-3
3GPP UMTS Model

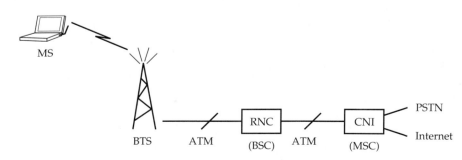

the public Internet, and the PSTN. The 3GPP group specifies ATM as the protocol used between the RNC in both directions.

■ 3GPP2

The 3GPP includes partners drawn primarily from the United States and is most concerned with the CDMA 2000 and UWC-136 standards. The 3GPP2 group is composed of five TSGs, one organizational partner (OP), and one ad-hoc group under one steering committee (SC). These groups are illustrated in Figure 18-4, and the abbreviations are defined below.

Organizational partners (OP) are other standards bodies that have representation to enhance and encourage international harmonization. Note the large overlap with OP organizations in 3GPP. Note also that ETSI is conspicuously absent (they are only an official observer of the group).

Association of Radio Industries and Business (ARIB); Japan

China Wireless Telecommunication Standard Group (CWTS); China

Telecommunication Industry Association (TIA); NAFTA signatories (the United States, Canada, and Mexico)

Telecommunications Technology Association (TTA); Korea

Telecommunications Technology Committee (TTC); Japan

The technical study groups and ad-hoc group are very similar to those in 3GPP, as you might expect because each partnership project has similiar goals.

TSG-A (access network interfaces): This group is responsible for the interface specifications located within the radio access network, as well as those interface specifications between the radio access network and the core network.

TSG-C (CDMA 2000): This group is responsible for the radio access part, radio specifications at layers 1–3, and the internal structure of systems based on the 3GPP2 specification.

Figure 18-4
3GPP2 Organizational Chart

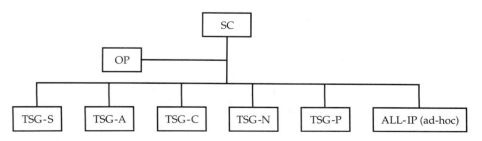

TSG-N (ANSI-41/WIN): This group is responsible for the core network part, including call interfaces and call and noncall signaling.

TSG-P (wireless packet data internetworking): This group is responsible for the specifications of the internet and IP multimedia core network part.

TSG-S (service and system aspects): This group is responsible for the development of service capability requirements and the coordination of high-level architectural issues across TSGs.

ALL-IP (all-IP requirements): This group is responsible for developing high-level requirements and architecture for an all-IP network. The group will transition to a TSG once its initial deliverables are completed.

■ AN ALL-IP MOBILE NETWORK

Note that the achievement of an all-IP network is still several years away. Initial technical specifications and reports will not be released for several years. Product development will follow that, and deployment will follow that. Many speak of an all-IP network as 3G, and the standards committees will issue all-IP network specifications, but the all-IP network is several years in the future. A realistic date for the earliest deployment is 2006, with 2010 a target date for mature deployment.

■ PACKET PROTOCOL DIFFERENCES

The partnership projects' all-IP network reports in both 3GPP and 3GPP2 have fundamental differences about what packet protocol will be supported. This issue must be resolved before true universal IP mobile networks can be deployed. At one level, this difference over packet protocols is about whether to use the HLR/VLR model or the mobile IP specification for roaming mobility. Mobile IP was explored in Chapter 16 and will not be discussed further here, except to point out a few issues of interest.

Mobile IP uses the concepts of a home agent (HA) and foreign agent (FA) in a simliar way that GSM uses the home location register (HLR) and visitor location register (VLR). As envisioned by 3GPP2, mobile IP would function as the data session mobility overlay in transition to an all-IP network. These networks would still use ANSI-41 techniques for voice calls, like the wireline network. Once transition to an all-IP network is complete and VOIP is used for voice calls, then mobile IP would be the standard for mobility.

Mobile IP does have some limitations, however, in a fast handoff environment. As a 3GPP study shows, the latency involved in a handoff using mobile IP is significantly more than that using the HLR/VLR approach coupled with the GPRS tunneling protocol (GTP). The position of 3GPP is that mobile IP is useful only at the edge of a network and that the primary handoff protocol will be the coupling of GTP and the HLR/VLR model. It really boils down to the fact that 3GPP already has a packet protocol developed called GPRS. By combining GPRS with GTP, handoff is solved. On the other hand, 3GPP2 must develop a packet protocol, the function of TSG-P, so it is using the mobile IP module as part of that protocol development effort as a way to speed time to market.

■ ANSI-41 VERSUS CAMEL

At a higher level, the issue is the structure and architecture of an intelligent network (IN). Generally, any IN should offer a signaling interface to service providers who want to control the operation of the network. Any IN is independent of the physical switching infrastructure.

Currently, all 2G systems use the SS7 network to support the signaling that the IN uses for service offerings. One goal of the 3G effort is to move to some form of an advanced IN (AIN). The 3G GSM AIN is a developing standard called customized applications for mobile network enhanced logic (CAMEL). CAMEL is being developed by 3GPP, and an early version was released as release 3 in 1999. Final versions of this standard are due in release 5, which is expected in 2002. CAMEL offers significant functionality in addressing the all-IP network. CAMEL is an optional network feature. From a standards perspective, it is an optional protocol overlay. As a practical matter, all 3G networks that evolve from a 2G GSM environment are expected to use CAMEL. For the most part, these networks are located in Europe and Asia.

When it is extended to include wireless features, the ANSI-41 standard is also referred to as the wireless intelligent network (WIN). Fundamentally, it is a wireless version of the envisioned SS7 AIN features, with support for roaming. This standard is being developed by 3GPP2; however, progress is much slower than that of CAMEL. The United States has very few deployments of GSM and has many IS-136 and IS-95 networks, all now running under signaling control of SS7. These networks are expected to make use of WIN extensions to SS7.

Any IN is intended to be service independent; thus, it lacks any feature specific interaction. However, SS7 was not developed in this way. For reasons of signaling efficiency, the AIN interface to services is closely coupled with the service type. Thus, WIN standards are particularly useful if they detail a service you want to use. On the other hand, the WIN interface will tend to become skewed toward a particular service. Thus, WIN will trade off a generic service interface for tight coupling with the particular service offering. WIN also focuses on those service offerings already present in SS7. Because SS7 has no capability to route IP packets, it seems a poor choice for an all-IP environment.

CAMEL takes the opposite approach. CAMEL is feature independent. While it is truer to the ITU model of what an IN is supposed to be, it has the drawback of being less intuitive in its interface. As a result of this higher level of abstraction, it may prove more difficult to develop features that work across networks than would be the case with WIN. A core issue of the whole 3G effort is to develop features that work across networks, for roaming support. Roaming may become an issue once the standard is completely defined and can be tested and evaluated.

It is now clear that both approaches will move forward and a translation protocol will be defined to interface between them. Each has some significant shortcomings, and the target date for an all-IP network deployment is still several years in the future. If a new alternative emerges, perhaps it would capture the best features of each approach. The following section discusses the capabilities supported by WIN.

◼ WIRELESS INTELLIGENT NETWORK (WIN)

The basic difference between WIN and AIN is the capability to support roaming. The fundamental difference in supporting wireless roaming from the AIN perspective is that a wireless subscriber has a choice of service providers, while a wireline subscriber has none. Any WIN must support some method of cellular service provider selection. The primary purpose of the ANSI-41 standard is to support roaming.

Another aspect to the choice of service providers is that service providers themselves have agreements with each other regarding how they will hand off calls and what billing levels will be applied. The handling of these issues will vary widely among service providers and in many cases will be different for subscribers roaming in the exact same geographical area.

Packet based messaging services, such as SMS, are very complex from an AIN perspective. WIN must provide a method for obtaining the message from the appropriate database and providing the appropriate headers for routing the message to its destination, and because the network is not IP based, encapsulation and de-encapsulation must take place to tunnel these packets to destination. Therefore, the WIN network must support several classes of messages that have nothing to do with the traditional three steps in a call. Instead, mobilty, multiple vendor billing arrangements, and packet transmissions must all be accommodated.

The usual approach is to remove services from the switching matrix. In the wireless environment, the MSCs are simplified and the higher level services are inserted into the SCPs. Because mobility is the big new requirement, the trend is to merge the functionality of the HLR with the SCP. Currently, the MSC must interact with the HLR to obtain the information that supports roaming. Through moving this functionality into the SCP, the MSC becomes both less complex and better at performing its primary job, that of switching telephone calls.

More important than this efficency improvement is the recognition that MSCs are localized devices. A database structure, particularly a networked database structure like an SCP unified with an HLR, means that subscriber modifications and additions need to be done at one location only and will propagate through the WIN naturally. If the MSC and HLR remain as they are today, linked together, then subscriber information must be provided at each switch.

◼ TRANSITION TO AN ALL-IP NETWORK

Transition to a future all-IP network will probably take place in two or three steps. It should be pointed out that these steps are speculative at this time. The first step will be an evolution from using an IWF gateway to allow IP data flows into and out of the MSC,

Figure 18-5
2G IP Connectivity

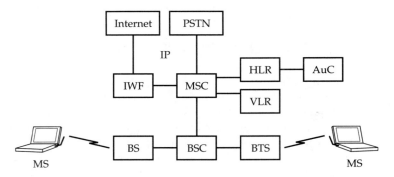

Figure 18-6
2G Packet Service Connectivity

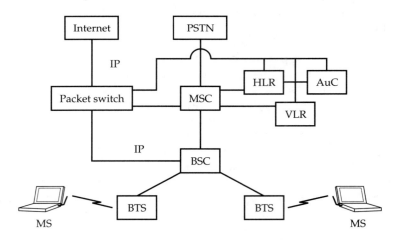

as shown in Figure 18-5. As you can see in this figure, IP packet delivery occurs via the MSC directly. Protocol conversion done by the IWF is required and, in this evolutionary step, has been combined into the MSC.

In parallel with the approach shown above are networks that have taken the next step and introduced true packet service directly to the BSC without requiring the IP data to pass through the IWF and MSC. These networks use the protocol GPRS (discussed in Chapter 14). This arrangement is shown in Figure 18-6. The existence of the GPRS protocol allows direct packet delivery to the BSC. The wireline connection between the BSC and the BTSs are still circuit-switched using TDM technology.

In a 3G all-IP network, the basic architectural difference is that the IP path now extends directly to the BTS. Now the connection between the BSC and the BTSs are packet switched, probably using some flavor of high-speed Ethernet to link them and thus

Figure 18-7
3G IP Connectivity

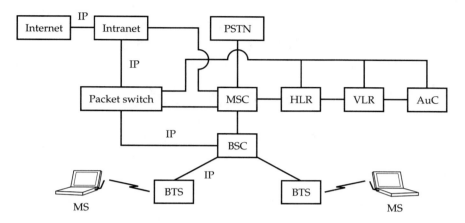

avoid the efficiency problems with a TDM approach. The connectivity for 3G IP is shown in Figure 18-7.

Note that, with this stage, the connectivity between MSCs is now through the use of an engineered IP network. While connectivity to the PSTN is retained for legacy reasons, IP is the technology used to link the cellular network together. This engineered IP network will have the capability to support the QOS requirements of such an arrangement. Access to the public Internet will be through this engineered IP intranet.

■ SUMMARY

This chapter has presented the most recent information on how current cellular systems will evolve to third-generation systems. Several paths are available, and which ones are chosen depends largely on current deployments and cost. The main point is that cellular service providers have to make this transition smooth and find some way to harmonize the worldwide efforts to achieve true global roaming capability with a single wireless device or cellular telephone.

There are no review questions for this chapter or any of the chapters in the fourth part of this book. The technology is not yet mature, so technological questions are inappropriate at the time of this writing. Business and adoption issues can best be addressed after the technology appears in the marketplace; again, review questions at this point are premature.

The best approach for learning more about the emerging technologies described in the fourth part of this book is to read trade press and journal publications as current sources of material and compare what has emerged into the marketplace with what was projected in this chapter. Analysis of how issues raised in this part were addressed by the manufacturers and the marketplace over time can provide a useful tool for additional exploration of the topics presented here.

Emerging Specification: Bluetooth

■ INTRODUCTION

This chapter introduces the Bluetooth specification (not standard). Note that Bluetooth is not standardized by any national or international body, which may have implications for its stability and the degree to which Bluetooth devices will interoperate. As of this writing, conformance test systems required to verify interoperability among products from different manufacturers were just starting to appear. Until these test systems are widely used and tested, interoperable products cannot appear. As mentioned below, wide adoption of Bluetooth in a relatively short time is critical to its success.

Bluetooth was originally developed as a cable replacement technology. The idea was that all kinds of cables could be replaced with short-range, point-to-point radio links. The maximum range was 10 meters and cost was the driving factor because the consumer market was the intended target. Cables in this market are usually very inexpensive, which underscores the fact that the cost of a Bluetooth link must be held very low. The Infrared Data Association (IrDA) standard, widely used for short-range wireless communications in PDAs, is also a target market for Bluetooth.

It was also recognized that small ad-hoc networks were also possible with this technology. The Bluetooth specification is named after a tenth-century Danish king who consolidated parts of modern-day Norway and Denmark into a single kingdom. His last name translates roughly to "blue tooth." The name Bluetooth is meant to convey the idea of consolidating small groups of individuals into ad-hoc networks, like the way the Danish king consolidated small groups into a kingdom.

Today, the idea is for both application domains (cable replacement and ad-hoc networking) to move forward, but the developers and the media are clearly more excited about the possibilities of what they call "unconscious" networking than they are about cable replacement. Unconscious networking, hidden networking, and ad-hoc networking are all synonyms. Technically, the Bluetooth specification calls such networks piconets. For the latter application, achieving a critical mass of users is absolutely necessary. Without large numbers of users, the ability to network unconsciously and with almost anyone will be disappointing.

For this reason, the target platforms for Bluetooth technology are numerous. Advocates predict that wireless devices of every sort, from cellular telephones to PDAs, to wireless notebooks, to as yet unavailable wireless appliances, will use Bluetooth to communicate with a similarly wide range of end devices. These end devices include printers, PCs, vending machines, shopping registers, etc. To make your purchase, point your Bluetooth-equipped device and charge it to your linked bank account. To get your e-mail from your desktop or notebook PC, point your Bluetooth-equipped device and download the mail (of course, the desktop or notebook must also be Bluetooth-equipped). The concept of a wearable computer relies on Bluetooth to connect the various subsystems worn on your body. The forecasts of Bluetooth applicability are essentially universal. Only time will tell if these forecasts reflect reality.

Organizations that support the idea of Bluetooth, which depends on wide marketplace acceptance to achieve its most ambitious goals, have formed a special interest group (SIG) to complete the specification, move adaptation forward, and coordinate conformance testing of products. The founding organizations of this SIG are all big players in the computing and networking marketplace: Ericsson, Intel, Nokia, IBM, and Toshiba. Today, thousands of organizations participate in one way or another in the Bluetooth SIG. In 1999, the first Bluetooth specification was published by this organization. An updated and expanded version 2 is expected in late 2001.

Some aspects of the Bluetooth specification will not be examined in this chapter. Specifically, infrared transmission and audio capabilities are given short shift. Only a small sampling of the application profiles are examined. This chapter is not meant to offer a detailed treatment, only an overview of an emerging wireless specification.

■ OPERATIONAL OVERVIEW

All Bluetooth devices take on the role of either a master or a slave. Any device can assume either role, but there can be only one master device in range. The master device role is limited to synchronizing communications among all devices in range. The device that initiates communications takes on the master role initially, but this role is subject to change if certain conditions change.

Recall that Bluetooth has two operational domains, cable replacement (or point-to-point communications) and piconet communications. In the first case, the master-slave roles have little meaning. In the second case, these roles can be application dependent, hence the requirement for the role to change, depending on conditions.

Traditional master-slave communications required the master to perform one level or another of control over the communications of the slaves. In some manner, the master gave permission for the slave to communicate. The granting of permission occurs here as well in the sense that the master sets the time slot for communications. But the role of master is limited. All high-level communications occur on a peer-to-peer basis.

Slaves can communicate with the master under specific guidelines. From a logical perspective, slave communications can be in either direction but always terminate or originate with the master. At a high level, two classes of slaves are defined by the Bluetooth specification: active slaves and parked slaves. In any piconet, there can be a maximum of seven active slaves and up to 255 parked slaves.

Basically, active slaves are powered up and are listening for transmissions from the master. Because they are actively receiving packets, they remain synchronized with the master and retain an immediate capability to communicate. Parked slaves, of which there are three subclasses, are in some state of low-power operation. This operational state can vary depending on the subclass to which the slave belongs. These operational states and subclasses will be explored in more depth later in this chapter.

A piconet is established automatically. Whenever two Bluetooth devices come into range of each other, they automatically establish a link. Suppose two sets of Bluetooth devices have established separate and distinct piconets. As they move toward each other, devices in one piconet will enter the range of devices belonging to another piconet. This arrangement forms what the specification calls a scatternet, in which individual slaves may take on dual roles in both piconets or they may remain in a single piconet, as long as they are not in range of the other master.

For example, some slave devices in piconet 1 may become parked slaves in piconet 2, and vice versa. By making use of one or more of the substates of parked mode, slaves may communicate with masters from both piconets. In this case, they would cycle through various modes, as determined by the specification. There can never be more than one master in any piconet. With that operational overview, this chapter now turns to an examination of the protocol specification and provides more detail on the functions.

■ PROTOCOL OVERVIEW

The Bluetooth protocol architecture is composed of three groups, the lower two of which are composed of one or more layer protocols. These three groups are illustrated in Figure 19-1. The upper layer group, or application group, is the location where the applications reside that make use of the services provided by the lower two groups. As such, this group is not a true protocol group at all, and applications that lie here have all the traditional advantages of a layer protocol environment, including no awareness of the functionality of the lower layer protocol functionality. To applications here, the wireless nature of Bluetooth is transparent. Software applications that reside here are called profiles. There are many profiles and if adoption projections reflect reality, many more are still to come. Some of the more important profiles will be examined later in the chapter.

The middleware group corresponds reasonably well to layers 3 and 4 in the OSI model. This group is the location for IP and TCP. WAP-enabled Bluetooth devices would

Figure 19-1
Bluetooth Groups

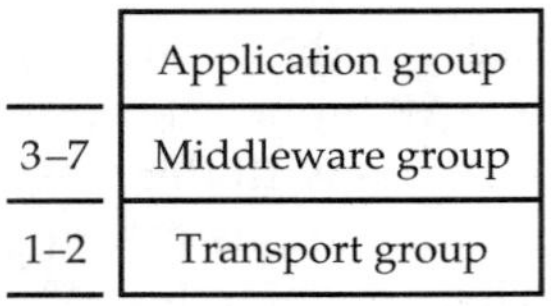

	Application group
	Middleware group
3–7	Middleware group
1–2	Transport group

have WAP in this group. The intent is that any layer protocol or third-party software that exists above layer 2 and is not a true profile application would reside here. It may be confusing to many that the Internet transport layer protocol, TCP, lies in the Bluetooth middleware layer, as does the WAP transport layer, while the Bluetooth transport group holds no traditionally identified transport layer protocol.

The transport group of the Bluetooth standard is a specification of physical and data link layer protocols required to locate and communicate with other devices. All physical and logical links required by any upper layer protocols or applications are performed in this group.

Each group will be examined in more detail. The following sections are organized in a bottom-up fashion, and the Bluetooth groups will be examined beginning with the transport group.

■ TRANSPORT GROUP

The transport group is shown in Figure 19-2, where four layer protocols are shown. From a protocol architecture perspective, however, only the physical and data link functionalities reside here. Comparison to the OSI model is difficult. Roughly, the three lower components of the transport group combine to provide the functionality of the MAC and physical layers of a traditional IEEE 802 specification. LLC operations in that same 802 architecture are split between the LMP and L2CAP components.

Radio

The Bluetooth radio component is a spread spectrum FHSS radio. This type of physical layer has been described earlier, so this section will focus primarily on the similarities

Figure 19-2
Transport Group Layer Protocols

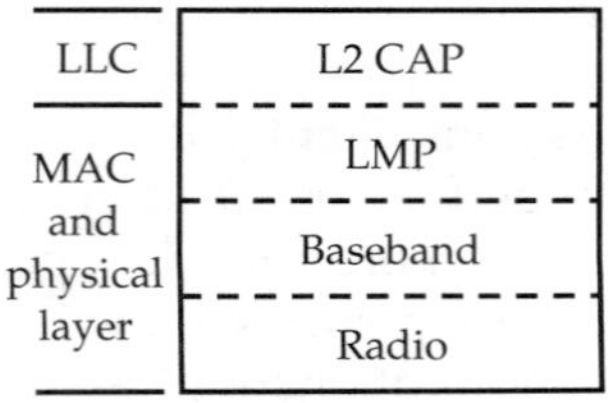

LLC	L2 CAP
MAC and physical layer	LMP
	Baseband
	Radio

Table 19-1
Bluetooth Transmit Power Classes

Class	Transmit Power
1	20 dBm or 100 mW
2	4 dBm or 2.5 mW
3	0 dBm or 1 mW

and differences between the Bluetooth radio component and the IEEE 802.11 radio, which operates very similarly. Bluetooth and IEEE 802.11 operate in the same band but use different air interface methods. The dominant air interface used in 802.11 wireless LANs is 11 Mbps DSSS. Bluetooth uses spread spectrum technology as well but has adopted an FHSS approach. The Bluetooth FHSS is very similar to the 1 Mbps FHSS specified in the 802.11 standard, but it operates at a lower power level at all output power classes.

Like the 1 Mbps FHSS 802.11 option, the radio subsystem used in Bluetooth uses 1 MHz wide channels, with seventy-nine unique channels in the ISM band. This implementation approach is not suprising because, until very recently, it was the only way to get an FHSS radio licensed to operate by the FCC. The radio operates in the ISM band, so it must conform to the spectral allocations of each country. As in the 802.11 case, there are seventy-nine channels in most countries, but not all. While there are many similarities between these two specifications, it is important to note that the radios are not interchangeable. The Bluetooth specification seems to allow for a DSSS radio. To date, however, no implementations of this approach have appeared. A DSSS radio in general costs more than an FHSS implementation, and Bluetooth has a very tight cost budget. Thus, the lack of implementation should not be surprising.

The Bluetooth modulation technique is GFSK, with a symbol rate of 1 Mbps. The maximum transmit power originally was limited to 1 mW, primarily due to cost considerations. The standard specifies three classes of devices, however, sorted by the transmit power. These classes are shown in Table 19-1. Power control is required only on devices designed to conform to power class 1. For comparison, 802.11 devices normally transmit at a power level approaching 1 W.

Because the two standards operate in the same frequency band, it is appropriate to consider potential interference aspects. If we assume that the 802.11 LAN is operating in an 11 Mbps DSSS mode and the Bluetooth device is operating in a 1 Mbps FHSS, class 3 mode, the likely result will be that as Bluetooth devices enter the BSS of an 802.11 AP, the range and throughput performance of the Bluetooth device will be reduced. In this case, most 802.11 devices will see little performance degradation. As the power level of the Bluetooth device increases, the two systems will interfere with each other more and more. This interference will be especially evident near the edge of a BSS. In these cases, both devices will experience reduced data rates in the interference environment. The receive performance of a Bluetooth radio is significantly worse than the equivalent FHSS

802.11 radio. This performance is in line with the low cost focus of the specification. The stated performance of a Bluetooth radio is 1 bit error in 1,000 bits at a receive power level of at least −70 dBm.

Baseband

The baseband sublayer has several functions, including some that would normally be associated with a physical layer protocol. The physical layer functions will be described first, followed by those associated with the data link layer.

The frequency hopping sequence is specified in the baseband component. The reason seems to be that the same clock is used to drive both the frequency hopping sequence and the channel hopping sequence. The master clock runs at 3.2 kHz, exactly twice the frequency hopping rate, which occurs at a rate of 1,600 hops per second. The master clock, divided by two and inverted, yields the time slot clock period of 625 μs.

To make the distinction clear, the frequency hopping sequence determines the 1 MHz spectral band slice on which the carrier is centered. This spectral slice is either 79 or 23, depending on the country where the device is located. The channel hopping sequence has nothing to do with spread spectrum operation. Rather, it is a way of allocating time slots for communications between slave devices and the master device. Each time slot is 625 μs. During this channel slot, a designated Bluetooth device can transmit a single baseband packet data unit (BBPDU) of data.

This terminology choice is unfortunate. The term *packet of data* is normally reserved for layer 3 transmissions. The Bluetooth specification uses this term independent of the layer protocol. For this reason, all informational transfers that occur in the transport group will be referred to by the technical terms used in the specification in the hope of minimizing confusion. The term *packet* will be used in the normal sense; for example, a packet of data is exchanged at the IP protocol layer.

Piconet Operational States   The operation of the piconet is controlled by the baseband component. A Bluetooth device in a piconet can find itself in four operational states: standby, inquiry, page, and connected. Normal operation makes use of the latter three; standby is a passive state.

As mentioned above, the master Bluetooth device is the one to communicate first because it holds the clock synchronization. In most cases, the first communication the master makes is an inquiry. The inquiry mode is how a device searches for other devices in its range. The inquiry mode is primarily designed to allow a device to synchronize with the master.

There are actually several substates in the inquiry mode as the various Bluetooth devices sort out amongst themselves who is the master and who are the slaves. In this mode, the potential for communications collisions is high because there is no defined master. Another substate of the inquiry mode reduces this potential through a backoff mechanism similar to that used in CSMA/CD.

The next mode is the page mode. It is implicit that a defined master of the piconet is established for a device to enter this mode. This method is used by the master to invite another device to join its piconet. Pages are sent at every clock period, or 312.5 μs. The frequency channel upon which the page is made depends on a page hopping sequence. It is defined to ensure that about every second a new frequency is used. This process ensures that any Bluetooth devices entering the range of an already established piconet will encounter an invitation to join in a relatively short time, typically 2 minutes or less.

Once located and invited to join, the devices are now in the connected state. Only in this state can devices exchange information with the master. All BBPDUs exchanged between the various slaves and the master in any piconet have a data payload capacity of 2,745 bits. Note that this value is not an integer number of bytes. The BBPDU, of which there are sixteen different types, also contains a header field that identifies the device address, payload type, an 8-bit CRC on the header bit contents, and several 1-bit fields for sequence and flow control reasons. The CRC generator polynomial is the same fifth-order polynomial used to illustrate CRC operation in Chapter 7.

The communications between a master and a slave are of two priority levels: asynchronous connectionless (ACL) BBPDUs and synchronous connection-oriented (SCO) BBPDUs. The SCO BBPDUs are considered high priority because they are exchanged in reserved channel slots. All ACL BBPDUs are exchanged in those time slots not already reserved for SCH communications. A total of three SCO time slots are in a piconet, while each slave in a piconet has a single dedicated ACL time slot. SCO BBPDUs can be used to send data or digitized audio with reasonable quality. Voice transmissions and other analog audio source transmissions are also possible with SCO BBPDUs using a 64 kbps CODEC. ACL BBPDUs are used for data only. Both types of BBPDUs can be multiplexed on the same RF link.

Link Manager Protocol (LMP)

The LMP component is responsible for all link management. Link management includes the classic two elements of link security, authentication and encryption, as well as flow control issues related to BBPDU data unit size. LMP is also responsible for power management. LMP is accomplished with several specialized LMPPDUs, which have a different frame structure than that of the BBPDUs. This setup is natural because no data, only management or control information, is exchanged at this component.

The security aspects of LMP are accomplished in a two-phase process. First, all Bluetooth devices must verify their identity during the link establishment phase. This process consists of multiple steps in which the two devices must first have a link key established. Once both devices have a link key or have agreed to use temporarily a generated key known as an initialization key, they assume either a verifier role or a claimant role.

The device that has assumed the verifier role sends a 128-bit random number to the claimant. An operation is performed on the random number, and the result of this operation is sent back to the verifier. If the verifier is able to generate the same response internally, authentication is established. Like all modern wireless authentication schemes, the keys are never transmitted over the air. Once authentication has been established, the second phase of security takes place through an encryption key, which is derived from the link key. The encryption key changes with each transmission, and its length is negotiated between the two devices.

Power management is accomplished through the use of the three parked submodes: hold, sniff, and park. Hold mode can be forced by the master for any slave, or a slave may request to enter hold mode. In hold mode, a device suspends all ACL communications with the master. Any SCO transmissions are maintained. The time a slave device stays in hold mode is negotiated with the master. A slave in hold mode is not considered an active device in a piconet.

Sniff mode can be forced by the master for any slave, or a slave may request to enter sniff mode. In sniff mode, the slave listens on every other channel slot for an ACL transmission from the master, as in the hold mode. SCO transmissions, if any, are maintained by the slave. The time a slave stays in sniff mode is negotiated with the master. A slave in sniff mode is not considered an active device in a piconet.

Park mode, like the previous two modes, can be forced by the master for any slave, or a slave may request to enter park mode. This power state is the lowest that a device can be in because the slave device removes itself logically from the piconet but remains in timing synchronization with the master. This state allows a quicker return to membership than if the sense of timing had been lost. The length of time a slave stays in park mode is determined by special transmissions by the master called beacons. Like the previous two cases, a slave in park mode is not considered an active device in a piconet.

Logical Link Control and Adaptation Protocol (L2CAP)

The L2CAP component of the transport group corresponds quite well to the traditional LLC sublayer functionality. Its primary purpose is to provide the service interface of the data link layer functions to the upper layer protocols in a way that isolates those layers from concerns about how the lower layers are implemented.

Both connection-oriented and connectionless mechanisms are provided for in the lower layers of the transport group, but no service interface for connection-oriented services is provided at the L2CAP component. Only ACL BBPDUs can be accessed by the upper layer protocols. L2CAP does have a defined connection-oriented PDU, which for this version is dedicated to supporting the DLCI connections required by the middleware group.

L2CAP also performs the encapsulation and de-encapsulation neccesary to adapt the data field size limitations of the BBPDUs to that of the upper layer IP packets. To an upper layer protocol, L2CAP appears to be a full-duplex communications channel between Bluetooth devices.

L2CAP virtual communications to its peer L2CAP components in other devices are called channels. Each channel has two endpoints, which are identified by a unique channel identifier (CID). There may be several logically distinct CIDs in any single Bluetooth device.

Another component, the host controller interface (HCI), logically resides in the same relationship to the upper layer protocols as does L2CAP. HCI is an optional component of the transport group. The intent of this component is to allow nonintegrated Bluetooth devices to use the Bluetooth transport group. HCI supports a serial physical interface if the Bluetooth-equipped device features one.

■ MIDDLEWARE GROUP

The middleware group of the Bluetooth specification contains two components unique to Bluetooth: the RFCOMM component and the service discovery protocol (SDP) com-

Figure 19-3
Middleware Group

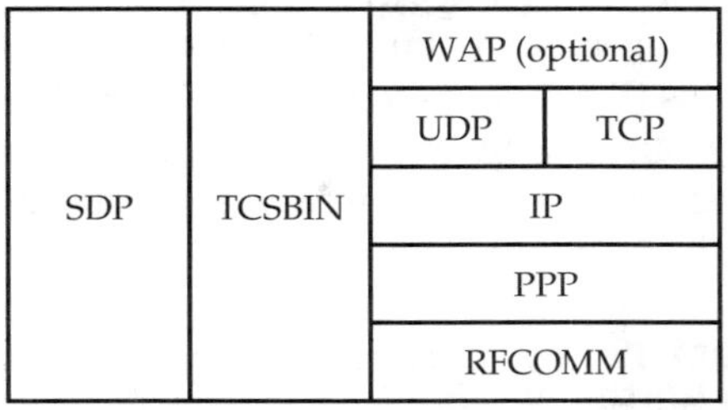

SDP	TCSBIN	WAP (optional)	
		UDP	TCP
		IP	
		PPP	
		RFCOMM	

ponent. RFCOMM is an abbreviation for RF communications. Above these components lie the familiar protocols: IP at layer 3 and UDP/TCP at layer 4. See Figure 19-3.

RFCOMM

RFCOMM interfaces directly to the application group and effectively emulates data transport over a serial cable. Think of RFCOMM as a serial cable replacement protocol to any upper layer applications that want to use the Bluetooth transport group. RFCOMM is used to support serial communications when the application is integrated onto the same platform as the Bluetooth transport group. It is accessible through the protocol stack rather than an external serial interface.

RFCOMM provides all the control signaling normally associated with a serial port, including RTS/CTS, DTR/DSR, XON/XOFF, and remote status queries on the serial interface. In the Bluetooth sense, each link is identified by a data link connection identifier (DLCI) through the connection-oriented mechanisms of L2CAP. The specification supports up to sixty multiplexed serial links over a single RFCOMM connection. Therefore, up to sixty logically distinct, multiplexed serial links, each identified by a unique DLCI, can be supported by a single RFCOMM Bluetooth link using the SCO BBPDUs provided by L2CAP. When multiple logical links are multiplexed over a single serial physical link, the throughput performance of each individual logical link will be reduced.

SDP is intended to support discovery services for the Bluetooth environment. Discovery services are defined as the process whereby devices and services find each other and build on their individual capabilities to offer enhanced services. Because Bluetooth devices and services may move in and out of piconets, a mechanism that can locate what devices and negotiate what services are currently available is critical. Essentially, Bluetooth piconets self-configure Bluetooth resources.

SDP divides Bluetooth devices into two groups: service providers and service consumers (typically called clients). The service providers maintain a list of services available in a service registry. Each service is identified by universally unique identifiers (UUIDs). In practice, a client specifies what service is desired and a provider responds with any services that match the request, providing the appropriate UUIDs and messaging through the use of SDPPDUs.

Point-to-Point Protocol (PPP)

While not a Bluetooth-specific protocol, PPP deserves a brief mention for its membership in the middleware group. PPP runs over RFCOMM to establish point-to-point communications. Essentially, it stands between IP packets and Bluetooth BBPDUs. In this sense, PPP allows IP networks to use the Bluetooth transport group. The idea is that dial-up networking almost always uses a serial port. With RFCOMM and PPP, Bluetooth presents, respectively, a serial port and a mapping to a serial port through PPP that fits well into most LAN environments.

Telephony Control
Protocol and WAP

Bluetooth describes a telephony control protocol specification binary (TCSBIN) that emulates call control signaling. This protocol supports voice and data calls between Bluetooth devices. It is based on the widely used standard Q.931, published by the ITU.

TCSBIN supports the call control and management of either voice or data calls between Bluetooth devices. These calls can operate in a point-to-point topology or a point-to-multipoint topology. Since the calls go through the L2CAP component, the calls use the SCO BBPDUs, in a similar way that RFCOMM makes use of this capability, and are connection-oriented communications. There is one primary difference between how RFCOMM and TSCBIN operate. TSCBIN requires a separate instance of L2CAP for each call, while RFCOMM can multiplex many serial connections in a single L2CAP instance.

WAP (also shown in Figure 19-3) is not a Bluetooth-specific protocol. WAP is designed to work across a wide range of network types, and Bluetooth wants its transport group to be one of them. As discussed in Chapter 18, WAP may emerge as a standard application interface across many different wireless networks. Essentially, WAP provides another way for Bluetooth to build a wide base and interact with a wide range of device types and users. WAP is an optional part of the middleware group and its inclusion in end devices will be determined by the acceptance of WAP, not Bluetooth.

■ APPLICATION GROUP

In the Bluetooth sense, applications are called Bluetooth profiles. There are four main groups of profiles: generic, serial, telephony, and networking. A profile is designed to facilitate interoperability, independent of what choices are made by implementers in the middleware and transport groups. All profiles are driven by either existing or perceived user needs.

In one sense, the generic profile is a separate profile grouping, but it is also the "mother" of all other profiles. All other profiles are derived from the capabilities of the generic profile. Each specialized profile group is associated with certain aspects of the middleware group. For example, the serial profile group uses the RFCOMM component. Similarly, the telephony profile group uses the TCSBIN component. Finally, the networking profile group uses the PPP and RFCOMM components.

This section will examine briefly only the generic profile group because it is fundamental to Bluetooth and it is the basis for all other profiles. The generic profile group contains two members: the generic access profile (GAP) and the service discovery application profile (SDAP). GAP provides the model for using the transport group primitives, while SDAP is closely coupled with the middleware group component, SDP. Of course, SDAP also uses GAP functionality, as do all profiles. GAP and SDAP are designed to be implemented in almost all Bluetooth devices and, as such, they provide a baseline application set that should be common to any Bluetooth device.

Because GAP is in some sense the mother of all profiles, it contains the procedures that any profile uses to establish Bluetooth communications. GAP specifies and uses a lot of Bluetooth-specific terminology that is outside the scope of this overview. For this reason, we will not use those specific terms but will still try to achieve a general understanding of the main functional aspects of the profile.

Basically, GAP sorts Bluetooth communications into several classes. These classes of communications say a lot about what types or modes of communications are possible with each device. These classes include security considerations, and several security modes are also defined. Special modes of communication are defined for those Bluetooth devices in one of the power save modes. Depending on the GAP mode defined, other applications change the way they interact with the Bluetooth device.

SDAP provides a profile interface to the SDP. In many ways, it is similar to a set of application programming interface (API) calls. Because discovery of other Bluetooth nodes in range is a critical function, SDAP is considered part of the core or generic set of profiles. SDAP defines the way that SDP uses the transport group components. SDAP is intended to be used by other profiles to accomplish service discovery. Thus, all profiles have access to services that may exist in any Bluetooth device in the piconet.

■ SUMMARY

Bluetooth is an ambitious project. With its dual focus on low cost and ease of use, it could well be widely deployed. As discussed in the introduction, a certain critical number of interoperable products is vital to the success of Bluetooth. It is also true that the critical number be produced in a relatively short time for the unconscious networking vision to be realized. If this number is not reached, cable replacement will become the primary Bluetooth application domain, and much of the promise of Bluetooth will not be realized. Only time will tell if this specification will move forward.

There are no review questions for this chapter or any of the chapters in the fourth part of this book. The technology is not yet mature, so technological questions are inappropriate at the time of this writing. Business and adoption issues can best be addressed after the technology appears in the marketplace; again, review questions at this point are premature.

The best approach for learning more about the emerging technologies described in the fourth part of this book is to read trade press and journal publications as current sources of material and compare what has emerged into the marketplace with what was projected in this chapter. Analysis of how issues raised in this part were addressed by the manufacturers and the marketplace over time can provide a useful tool for additional exploration of the topics presented here.

Abbreviations, Acronyms, and Units Summary

ABS   alternative billing service
AC   authentication center
ACK   acknowledgment signal
ACL   asynchronous connectionless
A/D   analog to digital
ADDCP   advanced data communications control procedure
ADM   adaptive delta modulation
ADPCM   adaptive differential pulse code modulation
AGCH   access grant channel
AIN   advanced intelligent network
AM   amplitude modulation
AMPS   advanced mobile phone service
AP   access point
APDU   application protocol data unit
API   application programming interface
ARIB   Association of Radio Industries and Business
ARN   authentication random number
ARQ   automatic repeat request
ASK   amplitude shift keying
ASN   authentication sequence number
ATM   asynchronous transfer mechanism
AuC   authentication center

BBPDU   baseband packet data unit
BCCH   broadcast control channel
BCH   Bose Chaudhuri Hocquenghem; broadcast channel
BER   bit error rate
BFSK   binary frequency shift keying
BGP   border gateway protocol
BPSK   binary phase shift keying
BS   base station
BSC   base station controller
BSS   base station subsystem; basic service set
BSSGP   base station subsystem GPRS application protocol
BSSID   basic service set identification

BTS   base transceiver station
BTV   business TV
BW   bandwidth
BWA   broadband wireless access

CA   collision avoidance
CAM   constant access mode
CAMEL   customized applications for mobile network enhanced logic
CATV   community antenna TV
CB   citizens band radio
CC   call control
CCCH   common control channel
CCK   complementary code keying
CCSN   common channel signaling network
CD   carrier detect; compact disk
CDMA   code division multiple access
CDPD   cellular digital packet data
CE   connection endpoint
CID   channel identifier
CLASS   custom local area signaling service
CLEC   competitive local exchange carrier
CLNP   connectionless network protocol
CLS   connectionless service
CM   call management; connection management
CNI   core network interface
CO   central office
CODEC   coder-decoder
COS   connection-oriented service
CPCH   common packet channel specification
CRC   cyclic redundancy code
CS   code scheme
CSC   common signaling channel
CSMA   carrier sense multiple access
CSU   channel service unit
CTS   clear to send
CVSDM   continuously variable slope delta modulation
CW   continuous wave

CWTS   China wireless telecommunication standard
         group

DA   destination address
D/A   digital to analog
DAMA   demand assigned multiple access
DAMPS   digital advanced mobile phone service
dB   decibel
dBm   decibels relative to 1 mW
DBPSK   differential binary phase shift keying
DBS   direct broadcast satellite
DCCH   dedicated control channel
DCE   data communications equipment
DDN   defense data network
DDS   digital data service
DECT   digital enhanced cordless telecommunications
DES   data encryption standard
DHCP   dynamic host control protocol
DHDX   double half-duplex
DIFS   distributed interframe spacing
DLCI   data link connection identifier
DLH   data link header
DLPDU   data link protocol data unit
DLT   data link trailer
DM   delta modulation
DNS   domain name service
DOD   Department of Defense
DPCM   differential pulse code modulation
DQPSK   differential quadrature phase shift keying
DR   data rate; dynamic range
DSAP   destination service access point
DSB   double sideband
DSB-SC   double sideband suppressed carrier
DSL   digital subscriber line
DSMA   data sense multiple access
DSMA/CD   data sense multiple access/collision
         detection
DS-1   digital service—1
DSSS   direct sequence spread spectrum
DSU   digital service unit
DS0   digital signal level 0
DS-0   digital service—0
DS0A   digital signal level 0A
DS0B   digital signal level 0B
DTE   data terminal equipment
DTIM   delivery traffic identification map
DTMF   dual tone multifrequency

EAMPS   extended advanced mobile phone service
ECSD   enhanced circuit-switched data
EDGE   enhanced data rates for GSM evolution
EGC   enhanced group call
EGP   exterior gateway protocol
EGPRS   enhanced GPRS
EHF   extremely high frequency
EIN   electronic identification number
EIR   equipment identity register
ELF   extremely low frequency
EMF   electromotive force
ERP   effective radiated power

ESN   electronic serial number
ESS   extended service set
ESSID   extended service set identification
ETACS   European total access communication system
ETSI   European Telecommunications Standard
         Institute

FACCH   fast associated control channel
FCC   Federal Communications Commission
FCCH   frequency control channel
FCS   frame check sequence
FDD   frequency division duplex
FDM   frequency division multiplexing
FDMA   frequency division multiple access
FDX   full duplex
FEC   forward error correction
FES   fixed end station
FFSK   fast frequency shift keying
FFT   fast Fourier transform
FHSS   frequency hopping spread spectrum
FIFO   first in first out
FISU   fill in signal unit
FM   frequency modulation
FRF   frequency reuse factor
FSK   frequency shift keying
FTP   file transfer protocol

GAP   generic access profile
GEO   geosynchronous or geostationary earth orbit
GFSK   Gaussian frequency shift keying
GGSN   gateway GPRS support node
GHz   gigahertz
GMSC   gateway mobile switching center
GMSK   Gaussian minimum shift keying
GPRS   general packet radio service
GPS   global positioning satellite
GR   GPRS register
GSM   global system for mobile communications
GTP   GPRS tunnel protocol
GW   ground wave

HCI   host controller interface
HDLC   high level data link control
HDR   high data rate
HDX   half duplex
HF   high frequency
HLR   home location register
HPA   high power amplifier
HSCSD   high-speed circuit-switched data
HTML   hypertext markup language
Hz   hertz

IBM   International Business Machines
IBSS   independent service set
IDU   indoor unit; interface data unit
IEEE   Institute of Electrical and Electronic Engineers
IGRP   interior gateway routing protocol
IMEI   international mobile equipment identity
IMSI   international mobile subscriber identity
IMTS   international mobile telephone system

IN   intelligent network
IP   Internet protocol
IrDA   infrared data association
ISDN   integrated services digital network
IS-IS   intermediate system to intermediate system
ISM   industrial, scientific, and medical
ISO   International Organization for Standardization
ISP   Internet service provider
ITRO   initial retransmission timeout
ITU   International Telecommunications Union
ITU-D   International Telecommunications Union—Development
ITU-R   International Telecommunications Union—Radio
ITU-T   International Telecommunications Union—Telecommunications
IWF   interworking function
IXC   interexchange carrier

JTACS   Japanese total access communication system

kHz   kilohertz
km   kilometer

LA   location area
LAN   local area network
LAP   link access protocol
LAPB   link access protocol balanced
LAPD   link access protocol, D-channel
LAPDm   link access protocol, D-channel, mobile
LAPG   link access procedure on the G channel (GPRS channel)
LATA   local access transport area
LCS   location services
LEC   local exchange carrier
LEO   low earth orbit
LES   land earth station
LF   low frequency
LLC   logical link control
LMDS   local multipoint distribution system
LMP   link manager protocol
LMPPDU   link manager protocol packet data unit
LNA   low noise amplifier
LNP   local number portability
LOS   line of sight
LPF   low-pass filter
LSB   lower sideband
LSSU   link status signal unit
L2CAP   logical link control and adaptation protocol

m   meter
MAC   medium access control
MAP   mobile application part
M-ary   multiple-ary
Mbps   megabits per second
MCS   master control station; modulation coding scheme
MDBS   mobile data base station
MDIS   mobile data intermediate station
MDLP   mobile data link protocol

ME   mobile equipment
MES   mobile end station
MF   medium frequency
MHF   mobile home function
MHz   megahertz
MIB   management information base
MM   mobility management
MNLP   mobile network location protocol
MNP   microcom networking protocol
MNRP   mobile network registration protocol
MPLS   multiprotocol label switching
MS   mobile station
MSAT   mobile satellite
MSC   mobile switching center
MSF   mobile serving function
MSK   minimum shift keying
Msps   mega symbols per second
MSRN   mobile station roaming number
MSU   message signal unit
MTP   message transfer part
MTSO   mobile telephone switching office
MTU   maximum transferable unit
MU   mobile unit
mW   milliwatt

NACK   negative ACK
NAM   number assignment module
NAV   network allocation vector
NEI   network equipment identity
NID   network identification
NLPID   network layer protocol identifier
NNE   network node equipment
NOC   network operations center
NPDU   network protocol data unit
NSAP   network service access point

OBP   on-board processing
ODU   outdoor unit
OHG   operators harmonization group
OMAP   operations, maintenance, and administrative part
1G   first generation
OOK   on-off keying
OP   organizational partner
OQPSK   offset quadrature phase shift keying
OSA   open service architecture
OSI   open systems interconnection
OSPF   open shortest path first

PACCH   packet associated control channel
PAGCH   packet access grant channel
PAM   polled access mode; pulse amplitude modulation
PBCCH   packet broadcast control channel
PBX   private branch exchange
PCG   project coordination group
PCH   paging channel
PCM   pulse code modulation
PCMCIA   Personal Computer Memory Card International Association

PCN personal communications networks
PCS personal communications systems
PDA personal digital assistant
PDCH packet data channel
PDLP packet data link protocol
PDU protocol data unit
PIN personal identification number
PLCP physical layer convergence procedure
PLL phase locked loop
PMD physical medium dependent
PMP point to multipoint
PN pseudo-random
POTS plain old telephone service
PP plenipotentiary
PPCH packet paging channel
PPDU physical protocol data unit; presentation protocol data unit
PPM pulse position modulation
PPP point-to-point protocol
PRACH packet random access channel
PSK phase shift keying
PSTN public switched telephone network
PTCH packet traffic channel
PTP point to point
PTP-CNS point-to-point—connectionless network service
PTP-CONS point-to-point—connection-oriented network service
PUK personal unblocking key
PWM pulse width modulation

QAM quadrature amplitude modulation
QOS quality of service
QPSK quadrature phase shift keying

RA rate adoption; receiving station address
RACH random access channel
RAG Radio Advisory Group
RAN radio access network
RBHC regional Bell holding companies
RFCOMM RF communications
RIP routing information protocol
RLC radio link control
RLP radio link protocol
RMS root mean square
RNC radio network controller
RR radio resource management
RRMP radio resource management protocol
RS rate set
RSA Rivest Shamir Adelman algorithm
RTO retransmission timeout
RTS request to send
RTT round-trip time

SA source address
SACCH slow associated control channel
SAP service access point
SAPI service access point identifier
SAT supervisory audio tone
SC suppressed carrier

SCCP signaling connection control part
SCH synchronization channel
SCO synchronous connection oriented
SCORE signal communicating by orbital relay equipment
SCP signaling control part
SCPC single channel per carrier
SDAP service discovery application profile
SDCCH stand-alone dedicated control channel
SDLC synchronous data link control
SDP service discovery protocol
SDPPDU service discovery protocol packet data unit
SED single error detecting
SGSN serving GPRS support node
SHDX single half-duplex
SHF super-high frequency
SID system identification
SIFS short interframe spacing
SIG special interest group
SIM subscriber identity module
SIP session initiation protocol
SLA service level agreement
SMAP system management application process
SME security management entity
SMS short message service
SMSCB short message service cell broadcast
SMSMC short message service message center
SMTP simple mail transfer protocol
S/N signal to noise
SNA synchronous network architecture
SNDCP subnetwork-dependent convergence protocol
SNMP simple network management protocol
SOD slope overload distortion
SP signaling points
SPADE single channel per carrier PCM multiple access demand assignment equipment
SPC stored program control
SPDU session protocol data unit
sps symbols per second
SSAP source service access point
SSD shared secret data
SSL secure sockets layer
SSM supplementary services management
SSP signaling service part
SS7 signaling system seven
SS-TDMA satellite-switched TDMA
ST signaling tone
STP signal transfer point
SW sky wave
SWR standing wave ratio
SX simplex

TA transmitting station address
TACS total access communications system
TCAP transaction capabilities part
TCH traffic channel
TCM trellis coded modulation
TCP transport control protocol
TCQAM trellis coded quadrature amplitude modulation

TCSBIN telephony control protocol specification binary
TDAB Telecommunications Development Advisory Group
TDD time division duplex
TDM time division multiplexing
TDMA time division multiple access
TEI terminal equipment identifier
3G third generation
3GPP Third-Generation Partnership Project
3GPP2 Third-Generation Partnership Project 2
TIM traffic identification map
TOS type of service
TPDU transport protocol data unit
TSAG Telecommunications Standards Advisory Group
TSG technical study group
TSG-CN TSG—core network
TSG-RAN TSG—radio access network
TSG-SA TSG—services and architecture
TSG-T TSG—terminals
TTA Telecommunications Technology Association
TTC telecommunications technology committee
TV broadcast television
TVRO television receive only
2G second generation
TWT traveling wave tube

UCT universal coordinated time
UDP user datagram protocol
UHF ultra-high frequency
UI unacknowledged information

UMTS universal mobile telephone service
USB upper sideband
USF uplink state flag
UTRA UMTS terrestrial radio access
UTRAN UTRA network
UUID universally unique identifier

VF voice frequency
VHF very high frequency
VLF very low frequency
VLR visitor location register
VPN virtual private network
VSAT very small aperture terminal

W watt
WAE wireless application environment
WAN wide area network
WAP wireless application protocol
WARC World Radio Communications Conference
WCDMA wideband CDMA
WDP wireless datagram protocol
WEP wired equivalent privacy
WIN wireless intelligent network
WML wireless markup language
WSP wireless session protocol
WTLS wireless transport layer security
WTP wireless transaction protocol
WWW World Wide Web

XML extensible markup language

Index